Tanmoy Rana
Editor

Microbiomes in Animal Nutrition, Health, and Disease

nova
science publishers
www.novapublishers.com

DOI: https://doi.org/10.52305/DYLL0727

Library of Congress Cataloging-in-Publication Data

ISBN: 979-8-89530-425-9 (Hardcover)
ISBN: 979-8-89530-516-4 (eBook)

Published by Nova Science Publishers, Inc. † New York

Contents

Preface

This book provides useful information on animal nutrition, health, and disease. The book presents the basic structure and function of the gut microbiome with excellent technological intervention. The book chapters are written by leading experts in their field. The book also adds biotechnological interventions on gut microbiota and immune defense systems. The book will help graduate, postgraduate, and postdoctoral students, veterinarians, and researchers gain a better understanding of the subject. It specifically addresses the molecular and cellular mechanisms involved in gut health in animals. The editor welcomes any suggestions from the readers to improve this book.

Tanmoy Rana
Department Veterinary Clinical Complex
West Bengal University of Animal and Fishery Sciences
Kolkata, India

Acknowledgments

I would like to express my sincere gratitude and respect to the Hon'ble Vice Chancellor, West Bengal University of Animal & Fishery Sciences for editing the book. I also thank all the contributors who have written their respective chapters on time. I am really grateful to all my colleagues for their useful ideas and suggestions in this regard. I am also thankful to all the members of the publishing house for giving me a platform to edit the book. I am also grateful to my family members for sacrificing valuable time to complete the book.

Chapter 1

Introduction

Amitava Roy[1,*]
and Tanmoy Rana[2]
[1]Department of Livestock Farm Complex, W.B. University of Animal and Fishery Sciences, Kolkata, India
[2]Department of Veterinary Clinical Complex, W.B. University of Animal and Fishery Sciences, Kolkata, India

Abstract

Diet, the GIT barrier's effective structure and function (gut mucosal), efficient feed digestion and absorption, the host's immune status, and a normal, stable microbiota with bacteria, protozoa, and fungal zoospores coexisting to enable efficient fermentation, absorption, and immunity-boosting for better host performance—all of which require holistic considerations—are the main components of gut health. Methane is produced in the rumen, and microbial protein, and volatile fatty acids (VFA) are the fermentation products of ruminant digestion. The digestion, absorption, metabolism, general health, and growth performance of nutrients are all enhanced by the gut bacteria and the metabolic products they produce.

As natural growth promoters, microbial feed supplements may be very beneficial for improving health and productivity by preventing disease, promoting desirable microbial growth in the host environment, stabilizing gut and ruminal pH, changing ruminal fermentation patterns, improving nutrient retention, reducing stress through enhanced immune response, and increasing nutrient digestibility and flow to the small intestine. Livestock's gut microbiota is essential for feed conversion. The kind of microorganisms present, the type of feed, and other elements including the animal's age and genetic makeup all affect the fermentation results. The majority of livestock production expenditures are attributed to feed. For animal agriculture to become more profitable and sustainable, feed efficiency must be improved. Gaining knowledge of these roles and relationships will aid in the creation of fresh dietary and management approaches, which will ultimately boost animal development and feed efficiency. In order to meet the growing demand for livestock products, livestock farmers must prioritize high performance animals and feed efficiency.

Keywords: immune system, microbiota, health, nutrition, gastrointestinal tract

---

* Corresponding Author's Email: amitavanutri07@gmail.com

In: Microbiomes in Animal Nutrition, Health, and Disease
Editor: Tanmoy Rana
ISBN: 979-8-89530-425-9

Introduction

Having livestock is essential for feeding the growing population. Because of this expansion, resource consumption is always being constrained, which poses a barrier to the production of high-quality animal protein sources including meat, milk, and eggs. The cattle business is today faced with the challenge of meeting the demand for these goods while using fewer resources. The vast gut and/or rumen microbiome fulfills several metabolic needs for the growth of the animal. The gut microbiota has a big impact on metabolic requirements, which has led to a lot of research being done on it recently. The two most important objectives in host production are feed efficiency and good performance. To accomplish these objectives, it is also necessary to take into account the host's health, the environment, and the quality of the food (Belanche et al., 2021). The digestion, absorption, metabolism, and general health and growth performance of poultry are all enhanced by gut bacteria and the metabolic products they produce. Methane is produced in the rumen, microbial protein, and volatile fatty acids (VFA) are the end products of ruminant digestive fermentation. The effectiveness of rumen fermentation and immunity are directly correlated with the gastrointestinal tract (GIT) microbiota, which includes bacteria, protozoa, and fungal zoospores. Most of the time, people view animals as autonomous beings living in their habitats. But inside an organism, prokaryotes and eukaryotes interact dynamically to generate what is known as a holobiont, or meta-organism, where each partner plays a vital and varied role. The gastro-intestinal tract of host animals is home to dense and complex microbial populations that can include bacteria, protozoa, fungi, viruses, and archaea. Gastrointestinal tract microbial communities have several beneficial impacts, one of which is the digestion and fermentation of plant polymers, which is particularly important for herbivorous animals. In addition, the natural gut microbiome is responsible for vitamin production, the biotransformation of harmful substances into non-toxic byproducts, immune system stimulation, intestinal mucosal integrity preservation, gut peristalsis, and barrier function against pathogen colonization. In the past, the digestive tract was thought to be an organ that was only used for nutrient absorption and digestion. All of the host cells are outnumbered by the amount of immune cells and bacteria found in the GIT. Consequently, everyone agrees that a healthy gut results in a host that is in good health and functions at their best. Within an animal's GIT, the rumen may contain the most varied and intricate microbial environment. Ruminants benefit from this microbial community's evolutionary advantage, which enables them to use non-protein nitrogen and lignocellulosic materials to make high-quality food. Because of their capacity to digest a variety of forages, ruminants have an advantage over other animals in the hunt for foods fit for human consumption. However, there are a few disadvantages to rumen microbial fermentation: Low nitrogen efficiency can result from protozoa and some bacterial species' proteolysis. Excess ruminal ammonia that is not absorbed and expelled to the environment by the ruminal microbiota is not used by them for protein synthesis. In a similar vein, the methanogenic archaea's production of CH_4 in the rumen wastes feed energy and accelerates climate change. Furthermore, compared to monogastric animals, ruminants' milk and meat have a higher percentage of saturated fat due to the biohydrogenation of rumen fatty acid by rumen bacteria. Therefore, comprehending the interplay among the gut microbiota, nutrition, host genetics, and health is essential for formulating novel approaches to fulfill customer expectations for improved food quality, animal well-being, and more ecologically sustainable and productive animal husbandry. To increase the financial benefits of cattle farming, the study reviewed the roles played by gut microbial ecology in the poligastric and monogastric guts in terms of health, performance, and feeding practices (Bergamaschi et al., 2020).

Objectives

- To go over the characteristics of microbiome-immunity crosstalk and how they relate to both health and illness.
- To identify the function of the GIT microbiota in animal health and performance, evaluate variables influencing microbial function, and explore potential feeding approaches.

What Is Gut Microbiota

Yeoman and White define microbiome as the group of autochthonous bacteria that live in a specific environment, such as the gastrointestinal tract, and microbiota as the entire community of allochthonous and autochthonous microbes that live in a specific environment, such as the gastrointestinal tract. Many different microorganisms, collectively known as the microbiota, inhabit the oral cavity and gastrointestinal tract of vertebrates. These microorganisms include bacteria, fungi, archaea, and protozoa. Due to their adaptation to the intestinal environment, many of the resident bacteria form intricate relationships with other bacteria and occupy host niches to obtain nutrition. The gastrointestinal tract (GIT) is often referred to as "the gut," but in reality, it is composed of three main components: (1) an epithelium; (2) a robust and diverse immune arm that houses the majority of the body's immune cells; and (3) commensal bacteria, which have a higher cell count than the entire host organism. Comprehending the interdependencies of ALL of these gut-related elements is what ultimately establishes the gut as the foundation for animal health and the engine that propels their behavior. Hippocrates, who lived from 460 to 370 BC, said that "the microbiota in the gut determines health and all disease begins in the gut." "Let food be thy medicine and medicine be thy food," she added. It has been demonstrated that the gut microbiota, the greatest symbiotic ecosystem with the host, is crucial for preserving intestinal homeostasis. Animal gut microbiota have a complex relationship with the host, nutrition, and immune and mucous systems in the gut. The mammalian digestive tract is the body's largest immunological organ. It contains both hemopoietic (macrophages, dendritic cells, and T cells) and nonhemopoietic (epithelia, paneth cells, and goblet cells) cells. Trillions of bacteria live there together as the microbiota. To maximize the benefits of symbiotic partnerships and minimize the hazards associated with living in such proximity, the homeostasis of this enormous microbial biomass is a precondition for maintaining host health. It is possible to refer to the communication between the microbiota and the host immune system as a "love-hate relationship" in which both parties work to maintain equilibrium. Additionally, the immune system's host innate and adaptive arms work together and counterbalance one another to sustain a steady and strict balance in the very complex gut ecology. Dysbiosis and low-grade to robust gut inflammation can emerge from any imbalance caused by an innate or adaptive immune deficit or an abnormal immune response, which can ultimately contribute to metabolic disorders (Cholewinska et al., 2021).

Composition of Gut Microbiota

The gut microbiota is made up of roughly 1010–1014 cells, which include viruses, bacteria, fungi, archaea, protozoa, and bacteriophages. The gastrointestinal tract is home to these cells' genes and different metabolites. It has co-evolved with every species to support daily biological processes like immunity development and immunomodulation, xenobiotic metabolism, mucosal immunity, and defense against invasive pathogens. The range of factors that may affect gut microbiota diversity has been shown by the results of recent studies characterizing the gut microbiota of different species. These factors include animal strain, obesity, the types of enrichment used in bedding and housing, antimicrobial treatment, vendor source, specific animal housing, diet, and co-occurring diseases. Trillions of bacteria, collectively known as the gut microbiota, are among the many microorganisms that colonize the digestive tracts of mammals. These native microbes have a symbiotic co-evolutionary connection with the host. Apart from their metabolic advantages, symbiotic bacteria offer various other roles to the host that support immunological responses, immune homeostasis, and defense against pathogen invasion. Through a variety of strategies, such as direct eradication, rivalry for scarce nutrients, and bolstering of immune responses, symbiotic bacteria can prevent pathogen colonization. Pathogens have developed adaptations to facilitate their reproduction when the gut microbiota is present. Environmental and genetic perturbations that alter the structure of the gut microbiota raise the danger of infection by pathogens, encourage the proliferation of pathogenic microorganisms, and accelerate the onset of inflammatory diseases. Critical insight into the pathophysiology of disease and the creation of preventative and therapeutic measures for inflammatory conditions will come from an understanding of the interactions between the microbiota, pathogens, and the immune system. The nutritional needs of individual bacteria play a major role in determining the microbiota's composition, which varies greatly depending on where in the digestive tract it is found. While commensal bacteria are essential to the health and metabolism of the host, pathogenic bacteria can have detrimental direct or indirect consequences. The gut microbiota interacts with its host as well as with itself and its nutrition. Therefore, it is important to choose feed ingredients that support gut health and preserve the equilibrium of the host, environment, and microbiota. The rumen's bacteria transform fibrous, low-quality plant matter into energy that the host ruminant may use. The rumen is the most important microbial ecology in polygastric mammals, containing key groups such as Prevotella, Selenomonas, Streptococcus, Lactobacillus, and Megasphaera along with the bulk of fiber-degrading groups Fibrobacter, Ruminococcus, Butyrivibrio, and Bacteroides. To maintain the proper rumen microenvironmental condition, the rumen also harbors a significant number of methanogens, ciliate protozoa, and some anaerobic fungi. Each sequence is assigned to a microbial taxon (bacteria, archaea, or lower eukaryotes) at several taxonomic levels, ranging from phylum to species, using a range of methods.

Culture Dependent Approaches

Various selection and enrichment culture settings are used in culture-dependent approaches to mimic the natural environment of the microorganisms. Because they require a continuous culture system to reproduce the rumen environment, anaerobes are difficult to cultivate because of the necessity to exclude oxygen, the slow growth of microorganisms, and other complex growth requirements. Rumen microorganisms were counted and identified using this method as well as others of a similar kind. Conventional techniques for categorizing rumen bacteria relied on appearance, shape, and gram staining—standard approaches for bacterial identification. Other classification techniques included using nutritional needs and fermentation end products. Instead of using traditional agar plates, roll tubes were utilized to cultivate and isolate anaerobic species (Carrasco et al., 2019).

Culture-Independent Approaches

The analysis of microbial communities at the molecular level is made possible by culture-independent techniques, or more precisely, DNA-based techniques for microbe identification and detection. The gut microbial ecology of ruminants is currently being studied using both traditional (classical culture-based techniques, such as isolation, enumeration, and nutritional characterisation) and modern molecular techniques (the use of small sub-unit rDNA analysis, e.g. 16S rDNA (phylogenetic composition) using RT-PCR, oligonucleotide probes, denaturing Gradient Gel Electrophoresis, Northern Blot Analysis, etc.), which offer a more comprehensive understanding of the changes in the composition of the intestinal microbiota and their secreted metabolites, providing valuable information on the GIT. Large datasets are handled in culture-independent ways using bioinformatics tools. Among the molecular methods, the duodenum, jejunum, ileum, and ceca's microbiota were compared and contrasted using the Terminal Restriction Fragment Length Polymorphism (TRFLP). The field of microbial community analysis has changed as a result of advances in DNA/RNA, protein, and metabolite analytical platforms as well as increasing computational capabilities. Utilizing molecular techniques, the microbial census is determined by analyzing 16S rRNA genes, 18S rRNA marker genes, and genomic areas that have been amplified and sequenced from specific biological samples.

Role of Gut Microbiota

Consensual gut microbiota is essential for preserving the host animals' typical physiology. Some of the main functions include directing the normal development or formation of the gut's morphology and structure, enhancing immunological responses, providing protection against luminal pathogens, and actively participating in the digestion and absorption of nutrients. Both favorably and unfavorably, the microbiota affects host physiology and function in numerous crucial ways. Additionally, gut bacteria can negatively impact chicken growth and performance by increasing cell turnover, decreasing fat digestibility, producing hazardous metabolites from protein fermentation, and more. The primary function of the rumen microbiome, which is the first to come into touch with the forage's biomass, is to break down into smaller particles or chemicals. As our understanding of the microbiome grows, we have the opportunity to modify it in a way that maintains the animal's symbiotic relationship with it while decreasing greenhouse gas emissions into the atmosphere and boosting feed efficiency.

Feed Digestion and Physiology

A large number of animals from many orders have had a section of their digestive tract modified to make room for a microbial population that helps with digesting and offers several nutritional and physiological advantages. In all zoological classes, including herbivores, omnivores, and carnivores, microbial communities have been reported from their guts. In addition to forming a closely knit ecological unit with the host animal, this complex, mixed microbial culture—which includes bacteria, ciliate and flagellate protozoa, anaerobic phycomycete fungi, and bacteriophages—also plays a critical role in the host's physiological, immunological, and protective processes. Higher animals' digestive tracts begin to develop their microbial populations shortly after birth. This process includes numerous microbial host interactions as well as a complex microbial succession that ultimately results in dense, stable microbial populations occupying distinct gut regions.

The rumen, reticulum, and omasum are the three pregastric fermentation chambers that have developed from the ruminant foregut or stomach, with the rumen being by far the largest. After being hydrolyzed and fermented in the rumen, undigested plant material and microbial cells move into the abomasum, where gastric digestion starts. The most characteristic of ruminants is rumination, which is the process of regurgitating and swallowing foregut digest in a richweed-like manner up to 500 times a day. This allows for the reduction of particle size (combination) and maximum surface area exposure to microbial attack. The basis of mutualistic microbial fermentation is the hydrolysis of plant polymers (starch, cellulose, hemicellulose, pectin, and protein) by fermentative microbes, primarily bacteria, into short oligomers and monomers. Certain transport mechanisms carry these soluble substrates into the microorganism, where they are fermented to produce fermentation end products (acetate, propionate, butyrate, carbon dioxide and hydrogen, methane, ammonia, and occasionally lactic acid) and the synthesis of microbial cells.

Feed Intake and Feed Conversion Efficiency- The Role of Microbes in the Gut

The profitability and sustainability of animal production facilities are significantly impacted by feed efficiency, and solutions for improving traits are in high demand. Animals that have a high feed conversion need less feed to gain weight or to make milk or eggs. Enhancing feed conversion directly boosts farmers' finances because feed expenses account for a significant portion of their expenses. Improved feed conversion also results in less animal waste production and a smaller environmental "footprint." Livestock's gut microbiota is essential for feed conversion. Feed is transformed into nutrients that the animal can use by gut microbiological fermentation. The kind of microorganisms present, the kind of feed, and other elements including the animal's age and genetic makeup all affect the fermentation results. Improved immunity also has unintended consequences, such as helping healthy animals gain weight effectively. The majority of costs associated with livestock production are related to feed. Feed efficiency needs to be increased for animal husbandry to become more profitable and sustainable. The host phenotype can be influenced by both host genetics and the gut microbiota. Since some characteristics of the rumen microbiome are associated with host phenotypic diversity in feed efficiency, there is a chance that rumen microbes could serve as markers for feed efficiency. Methane output reduction may be the result of selection for higher feed efficiency.

Gut Microbiota and Health

Through a variety of processes, such as immune system modulation and ruminant health and productivity promotion, the GIT microbiome protects the host animal from environmental hazards and diseases. Disturbances in this commensal microbiota can lead to gastrointestinal tract problems such as diarrhea, bloat, rumen acidosis, and nutritional poisoning. The "one-pathogen, one-disease" theory has traditionally been used to address these health issues, which pose significant welfare and financial risks in the current intensive production systems. Nevertheless, there is mounting evidence that symbiotic microbiomes have a significant role in regulating and reducing the occurrence of these GIT disorders, as well as mastitis and respiratory illnesses. The interaction between the microbes and their host gives new insights into the ways the GIT microbiota can affect host stress, metabolic efficiency, or disease resistance. Animal health, development, and productivity are now strongly correlated with the Gastrointestinal Tract (GIT) microbiomes of production animals. Early gut colonization is especially crucial for the development of the GIT's morphology and immunology, the creation of a functioning fermentative environment, and the development of the newborn's resistance to pathogenic challenges. Changes in an animal's GIT microbiota can have a substantial impact at any age, but during early GIT development, these changes can be especially severe and have long-lasting effects. For ruminants to effectively ferment plant resources into consumable foods like milk and meat, the rumen must develop early.

Harmful Roles of Gut Microbiota

The condition of Dysbiosis in the gut microbiome is a result of an imbalance between the commensal and pathogenic bacteria within the microbiome. The pathogenic microbiome leads to immunological dysfunction and the development of disease, whereas the commensal microbiome controls the maturation of the mucosal immune system. An immune system that protects the integrity of the digestive tract is the gut mucosa, which is made up of lymph nodes, lamina propria, and epithelial cells. The typical mucosal immune system monitors the makeup of the gut microbiota. Intestinal disorders are the consequence of inflammation, which is brought on by aberrant immune responses and alters the balance of the gut microbiota. Immune system expenses, nutritional competition, the creation of toxic amino acids that catabolize reduced fat digestibility, and altered intestine architecture and function are among the detrimental impacts. The macrobiotic is essential for the growth of the host immune system. But when immunological activation is sustained at a chronic level—as seems to be the case in conventional versus germ-free animals—there is intrinsic inefficiency. It has been observed that more than 70% of an animal's total antibody production is attributed to the IgA produced across the gut mucosa. In high quantities, a variety of hazardous and poisonous chemicals produced by putrefaction may hurt the growth and performance of chickens. Ammonia, phenols, indoles, amines, and cresol are among the byproducts of protein fermentation that can be harmful to the health of the host or cell. Immune system expenses, nutritional competition, the creation of toxic amino acids that catabolize reduced fat digestibility, and altered intestine architecture and function are among the detrimental impacts. The macrobiotic is essential for the growth of the host immune system. But when immunological activation is sustained at a chronic level—as seems to be the case in conventional versus germ-free animals—there is intrinsic inefficiency. It has been observed that more than 70% of an animal's total antibody production is attributed to the IgA produced across the gut mucosa. In high quantities, a variety of hazardous and poisonous chemicals produced by putrefaction may harm the growth and performance of chickens. Ammonia, phenols, indoles, amines, and cresol are among the byproducts of protein fermentation that can be harmful to the health of the host or cell.

Factors Affecting Gut Microbiota

The body's largest group of organs is called the GIT. In addition to being the site of nutritional absorption and digestion, it also offers defense against poisons and pathogens. It also supports a sizable population of immune cells and bacteria. As a result, enhanced animal output and general good health depend greatly on a functioning digestive system. "A steady state where the microbiome and the intestinal tract exist in symbiotic equilibrium and where intestinal dysfunction does not limit the welfare and performance of the animal" is what is meant to be understood by gut health. This concept integrates the key elements of gut health, such as nutrition, the host's effective immune system, the normal and stable microbiota, the structure and function of the GIT barrier (gut mucosa), and efficient feed digestion and absorption. According to Conway, the mucosa, commensal microbiota, and nutrition all have a significant role in gut health. Subsequently, Montagne, et al. clarified that it comprises a diet that would supply enough nutrients, a microbial community that maintains a balanced, healthy environment, and mucosa that preserves the integrity of the gut. Pigs and poultry's gastrointestinal tracts (GITs) comprise over 70% of the body's immune cells, hence they should be considered when defining "intestinal health." Therefore, it is suggested that intestinal health should be viewed holistically, taking into account the immune system, microbiome, mucosa, and food (Figure 1).

One of the most stressful times in a pig's life is right after weaning, when they are most likely to be exposed to environmental, social, and psychological stressors that can affect their gut health and growth in general both directly and indirectly. Generally speaking, animal welfare concerns, upholding sanitary standards and disease control procedures, and nutrition and management techniques are all part of the husbandry procedures in the swine barn. Inadequate husbandry techniques can lead to decreased feed intake, stress, and illness, which in turn impacts the gut health and performance of weaned piglets. Increasing feed intake in recently weaned piglets is crucial because it reduces the risk of compromised gut structure and function, as mentioned. Hemopoietic cells (macrophages, dendritic cells, and T-cells) and non-hemopoietic cells (epithelia, Paneth cells, and goblet cells) make up the gastrointestinal tract (GIT). Additionally, the microbiome, is made up of viruses, bacteria, protists, fungi, and archaea, all of which support gut health. To preserve homeostasis, the innate and adaptive immune systems communicate with the microbiome continuously. Dysbiosis can be brought on by any disruption in the microbiome or immune system, which increases vulnerability to several illnesses. The digestive epithelium with its unique structure, the gut-associated lymphoid tissue (GALT), and the mucus covering the epithelium make up the intestinal mucosa. The interaction of the intestinal mucus layer, host epithelial cells, GALT, and microbiota forms a delicate yet dynamic equilibrium that is vital to the digestive system's effective operation and ability to absorb nutrients.

Animal age, the location of the gastrointestinal tract, and a range of dietary and environmental variables all affect the species composition and quantity of bacteria in the gastrointestinal microbiota. Diet, efficient gastrointestinal barrier structure and function, host-gastrointestinal microbiota interaction, efficient feed digestion and absorption, and immune system performance are the main determinants of gastrointestinal functionality. Even though there are many intricate relationships between these fields, developing nutritional strategies that will help farm animals become more resilient to the physiological and environmental stresses they will face throughout their productive lives will require a multidisciplinary approach.

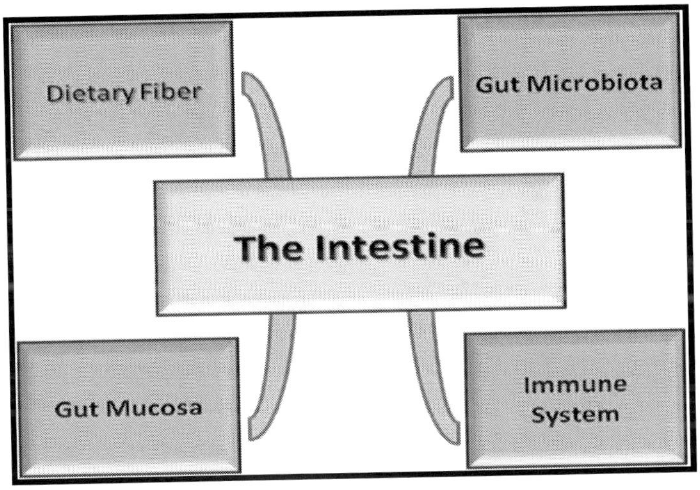

Figure 1. The components of gut health from a comprehensive standpoint.

The Gut Microbiota of Mongastric

The nutrition and general health of monstrous herbivorous animals, like as donkeys, are greatly influenced by the gut bacteria. Along the GIT, donkeys have a rich, varied, and multipurpose microbiota. Microbiota richness and diversity in the hindgut are generally far higher than in the foregut; at the phylum level, Bacteroides and Firmicutes are abundant in the hindgut, whereas at the genus level, Lactobacillus was more dominant in the foregut and Streptococcus was more dominant in the hindgut. Differentiating microbiota along the GIT is functionally compatible with the relevant physiological function of different GIT locations, according to the study. For instance, the microorganisms in the hindgut are more active in the metabolism of amino acids, whereas those in the foregut are more active in the metabolism of carbohydrates. The esophagus, crop, proventriculus, gizzard, duodenum, jejunum, ileum, cecum, colon, and cloaca make up the avian gastrointestinal tract (GIT). In comparison to other mammals, poultry have a substantially shorter GIT about their body length. To cling to the mucosal wall and multiply, microbiota that thrive in such a small GIT with a relatively short transit time need certain adaptations. The utilization of nutrients and general health of poultry are impacted by the ceca, which have a reduced passage rate and are favorable to various bacterial groups. Chickens have diverse intestinal microbiota, the makeup and metabolism of which differ between intestinal compartments with radically differing physicochemical microenvironments, like all homoeothermic animals.

The primary determinants of microbe fitness and colonization efficiency are the presence of appropriate growth media, the local pH and redox potential, and the host's antibacterial secretions in a particular intestinal segment. Bacteria find less easy-to-grow substrates as they proceed down the gastrointestinal tract. Because of this, bacteria found in the lower intestine are frequently experts at using feed ingredients such as non-starch polysaccharides, resistant starch, or resistant proteins that are incompatible with the host's natural digestion system. The proximal GIT (crop, proventriculus, and gizzard) is characterized by low pH, which strongly selects bacteria and limits the growth of many species. The microbiota composition of jejunum and caecum is highly different. Specialist microbiota that has been adapted to the physicochemical circumstances, host physiology, and available nutrients of the particular habitat inhabit the various regions of the digestive tract of broiler chickens. Lactic acid bacteria, which have intricate dietary needs similar to those of their chicken host, predominate in the small intestine. Because lactobacilli cannot synthesize amino acids for anabolism, the availability of amino acids in the growing environment is crucial to their survival. Consequently, the microbiota and the chicken host compete with one another for amino acids in the small intestine. Exogenous enzymes that facilitate the digestion of proteins are also likely to give the chicken a competitive edge by limiting the development potential of bacteria that are dependent on amino acids. The three types of protein that pass through the ileum are resistant protein derived from food, protein taken up by intestinal bacteria, and endogenous protein that is produced and released by the host. The latter is made in the host's tissues from amino acids found in food, making it authentic endogenous protein. The amount of the microbial protein fraction and the generation of endogenous proteins derived from mucin, epithelial cells, and antibodies are influenced by the activities of tiny intestine bacteria.

The Bovine Rumen Microbiome

The microbial community found in the rumen is rich and varied, with microorganisms collaborating to decompose the feedstuffs that ruminant animals consume. The animal's production efficiency is regulated by its microbiome, and some of its metabolic pathways—such as those linked to the creation of methane—cause it to lose energy. In addition to influencing the quality of final products like milk and meat, the microbiome also pollutes the environment. Ruminants may use materials like lignocellulose that the human body is unable to process thanks to the activity of their microbiota. The rumen plays a major part in releasing the planet's most abundant carbon polymer, lignocellulose, which is a huge energy storage. In the end, the rumen uses lignocellulose to produce goods (such as milk and meat) that people can eat to get a high-nutrient diet. In the rumen, the bacteria and the host work together to make protein and digest byproducts like volatile fatty acids (VFAs) that the host may consume. The host supplies food, moisture, and heat. The host uses the microorganisms to break down their protein, fat, and carbohydrate content in the final stage.

Conclusion

The gut microbiome plays a huge role in the digestion, fermentation, and absorption of feed as well as in assuring the host's survival by enhancing immunity. Because the GIT increases the host's productivity through increased feed intake and feed conversion efficiency, there will be less pollution in the environment from greenhouse gas emissions (GHGs) and the overall health and economic benefits of the world's rapidly expanding population, which requires livestock. Diet, the efficient structure and operation of the gastrointestinal barrier, the host's interaction with the gastrointestinal microbiota, efficient feed digestion and absorption, and a healthy immune system are the main factors that determine gastrointestinal functionality. However, since the GIT's health and effectiveness are regulated by several intricate mechanisms, it is critical to expand our understanding of these interactions to develop strategies for modulating these factors in the context of improved animal performance. The GIT's health and effectiveness are significant factors in determining animal performance. Even though there are many intricate relationships between these fields, developing nutritional strategies that will help farm animals become more resilient to the physiological and environmental stresses they will face throughout their productive lives will require a multidisciplinary approach.

References

Belanche A, Patra AK, Morgavi DP, Suen G, Newbold CJ, Yanez-Ruiz DR. Gut Microbiome Modulation in Ruminants: Enhancing Advantages and Minimizing Drawbacks. *Frontiers in microbial.* 2021;11:622002.

Bergamaschi M, Tiezzi F, Howard JYi, Huang J, Gray KA, Schillebeeckx C, et al. Gut microbiome composition differences among breeds impact feed efficiency in swine. *Microbiome*. 2020;8(1):110.

Carrasco JMD, Casanova NA, Miyakawa MEF. Microbiota, Gut Health and Chicken Productivity: What Is the Connection? *Microorganisms*. 2019;7(10):374.

Cholewinska P, Gorniak W, Wojnarowski K. Impact of selected environmental factors on microbiome of the digestive tract of ruminants. *BMC Vet Res*. 2021;17(1): 25.

Chapter 2

Biomarkers for Intestinal Health

Partha Sarathi Roy[1]
Amitava Roy[2,*]
and Tanmoy Rana[3]

[1]Murshidabad Krishi Vigyan Kendra, Milebasa, Murshidabad, W.B., India
[2]Department of Livestock Farm Complex, W.B. University of Animal and Fishery Sciences, Kolkata, India
[3]Department of Veterinary Clinical Complex, W.B. University of Animal and Fishery Sciences, Kolkata, India

Abstract

A major factor in influencing the health, welfare, and performance of animals is their effective gastrointestinal functions. Recently, a new definition of gastrointestinal functioning was introduced, outlining the essential elements that support both healthy and functional gastrointestinal systems. These elements include nutrition, the gastrointestinal barrier's efficient construction and function, the host's interaction with the microbiota in the gastrointestinal tract, efficient feed digestion and absorption, immune system performance, and the gut's neuroendocrine function. Although there are numerous intricate mechanisms and pathways connecting each of these elements, identifying some of the essential elements of gastrointestinal functionality provides a chance to assess possible biomarkers that might be used to gauge how well the gastrointestinal system functions in farm animals. An expanding array of methods, each with unique benefits and drawbacks, are yielding an ever-growing quantity and variety of biomarkers. Furthermore, even when studies appear to be quite similar, it can be challenging to extrapolate findings across species and to draw meaningful comparisons due to variations in models and methodology. The development of a panel of gastrointestinal function biomarkers that must be indicative of both the effective functionality and health of the gastrointestinal tract as well as animal performance, health, and welfare is necessary due to the complexity of the interactions between the key components of gastrointestinal functionality, making the use of a single biomarker impractical.

Keywords: gastrointestinal functionality, biomarkers, inflammation, microbiota, nutrition

---

* Corresponding Author's Email: amitavanutri07@gmail.com

In: Microbiomes in Animal Nutrition, Health, and Disease
Editor: Tanmoy Rana
ISBN: 979-8-89530-425-9

Introduction

Scientists, veterinarians, and nutritionists have all shown a great deal of interest in the idea of "gut health" in recent years (Kogut and Arsenault, 2016). The goal of this attention is to create dietary therapies that will modulate gastrointestinal functions towards optimal health in order to improve farm animal production performances, including growth, milk yield, meat quality, and egg quality. Even if the term "gut health" has been used frequently in the context of animal nutrition, a precise scientific definition is still missing (Kogut & Arsenault, 2016). To keep an eye on the wellbeing and health of animals, as well as to assess how any nutritional intervention affects their performance, it is imperative to have a precise understanding of gastrointestinal functions. Despite the difficulty in defining gut health, six main areas of gastrointestinal functionality have recently been identified. According to Celi et al. (2017), these categories include gut mucosa, effective immune system, normal and stable microbiota, effective digestion and absorption, and gut neuroendocrine and motor function. The proposed definition of gastrointestinal functionality integrates the three main elements of gastrointestinal functionality—diet, intestinal permeability, and gut microbiota—with feed digestion and absorption and effective immune status. It describes a steady state where the microbiome and the intestinal tract exist in symbiotic equilibrium and where intestinal dysfunction does not limit the welfare and performance of the animal.

According to Celi et al. (2017), every one of these domains is essential to gastrointestinal physiology, animal welfare, and performance. Finding biomarkers of gastrointestinal functionality in vivo is necessary, but developing and validating these biomarkers is difficult because the aforementioned domains can be influenced by a variety of circumstances. This is especially true if we attempt to assess the possible advantages of dietary interventions at the population level using the knowledge developed in human medicine at the individual level. The lack of reagents required to create and confirm the tests, as well as the absence of some markers in particular species, present further challenges (Niewold, 2015). To improve our comprehension of the processes influencing the intestinal barrier, its functioning, and the ecology of the gastrointestinal microbiota, biomarkers of gastrointestinal functionality must be developed. There is a significant gap in the literature regarding biomarkers of gastrointestinal permeability, gastrointestinal barrier function, biomarkers for the gut endocrine system, or biomarkers that are indicative of the functional presence of beneficial microbiota or their metabolites, despite our advanced understanding of the digestive and absorptive mechanisms of the major macro and micronutrients (Celi et al., 2017). Thus, that chapter is to provide veterinary scientists and animal nutritionists with a variety of gastrointestinal functionality markers, emphasizing non-invasive ones and emphasizing the potential of certain markers.

Nutrition

The growth and operation of the gastrointestinal tract, including its immune system and the microbiota it harbors, can be modulated by the composition of the diet (ingredients, nutrients, and additives) (Conway, 1994). Dietary variables can indeed affect inflammation by modifying pro- and anti-inflammatory pathways (Broom and Kogut, 2018). These include a variety of antinutrients that can damage the gut's structural and functional integrity, including some forms of dietary fiber, trypsin inhibitors, phytate, lectins, undigested protein in the distal gastrointestinal system, mycotoxins, and many more (Celi et al., 2017). Another important factor in immunonutrition is the makeup of the diet (Klasing, 2007). To improve animal health, welfare, and productivity, nutrients such as macronutrients, vitamins, minerals, and trace elements can be used to modulate the inflammatory response during nutritional and environmental challenges and diseases (Starkey, 2014). More and more research is demonstrating the advantages of dietary antioxidant supplementation in modifying the redox equilibrium in poultry and cattle (McGrath et al., 2018). All of these findings point to the potential for creating a scoring system that would assess the overall inflammatory characteristics of food while taking into account the contributions of both pro- and anti-inflammatory dietary components. The Dietary Inflammatory Index was developed by human nutrition researchers based on published data on food and inflammation, as the potential of the diet to influence the inflammatory response cannot be quantified with a piece of equipment in the lab (Shivappa et al., 2014). Using a continuum from maximally pro-inflammatory to maximally anti-inflammatory, the Dietary Inflammatory Index classifies people's meals based on how likely they are to cause inflammation. According to Shivappa et al. (2014), a diet that is more pro-inflammatory has a higher Dietary Inflammatory Index score than one that is more anti-inflammatory. No similar approaches have been developed in animal nutrition yet, however, it is worth mentioning that in vivo indices for predicting acidosis risk of grains in cattle have been compared to in vitro methods (Lean et al., 2013).

Ruminal acidosis is a nutritional condition that ruminants typically develop from eating large amounts of feeds high in carbohydrates that are easily fermented. This causes the rumen to produce large amounts of lactic acid and volatile fatty acids, which drastically lowers the pH of the rumen, causing rumenitis, metabolic acidosis, lameness, liver abscess, pneumonia, and even death (Lean et al., 2013). Lean et al., (2013) contrasted the animal assessments of ruminal acidity with an existing method of ranking grains for risk of acidosis using near-infrared reflectance spectroscopy (NIRS), which was established from in sacco and in vitro measures from cattle and grain chemical composition analysis. The approach yielded results that were compared with rankings derived from in vivo rumen fermentation measures of acidosis. These rankings were based on research conducted by Bramley et al., (2008), which found a significant correlation between cattle classified as "acidotic" and an increase in lameness, a decrease in the ratio of milk fat to protein, and diets that were higher in non-fiber carbohydrates and lower in NDF. According to Lean et al., (2013), grains that have very similar starch contents have different effects on ruminal fermentation and the risk of ruminal acidosis. They also suggested that a reliable non-invasive nutritional indicator (NIRS) method could be utilised to quickly and accurately determine the risk of acidosis related to feeding different grains.

Capacity for Efficient Digestion and Absorption

Nutrient digestion and absorption is the gastrointestinal system's primary job. From a physiological perspective, the process of digestion begins in the oral cavity and proceeds through various sections of the gastrointestinal system, aided by the action of endogenous and exogenous enzymes as well as stomach acid. Smaller molecules known as macronutrients (proteins, carbs, and fats) and micronutrients (vitamins and minerals) are formed during feeding and can be absorbed by a variety of physical and biological processes. Optimal digestion and absorption are typically associated with gastric functionality. On the other hand, gastrointestinal tract inflammation can lead to decreased absorption of macro- and micronutrients as well as decreased digestive efficiency (Celi et al., 2017). The macronutrients and micronutrients most affected by malabsorption include fat (Koutsos et al., 2006) and iron and zinc (Davin et al., 2013). Maldigestion and malabsorption are hence two possible outcomes when digestion and absorption are compromised. The incomplete breakdown of nutrients, or maldigestion, is frequently caused by insufficient or faulty stomach acid or enzyme secretion. Rather, malabsorption refers to a decrease in the ability of the final products of digestion to be absorbed. When nutrients are not properly broken down and absorbed, extra nutrients—especially protein and fat—are transferred to the distal portions of the gastrointestinal system, where the microbiota may ferment them in an unsuitable way. A summary of the different biomarkers of digestion and absorption is reported in Table 1. Because they cannot synthesize carotenoids, birds and mammals must rely on food sources to get them. Intestinal integrity is necessary for the absorption, digestion, and tissue insertion of carotenoids. A certain proportion of carotenoids might be transformed into vitamin A within the intestines; although, in cases where gastrointestinal function is compromised, the conversion of carotenoids into vitamin A may become exceedingly challenging for the gastrointestinal tract, leading to intestinal disorders (McGrath et al., 2018).

Table 1. Digestive and absorption biomarkers in the gastrointestinal tract

Biomarker	Biological sample	Test site	Method	Comments
Products of protein fermentation	Feces	Ileum and colon	high performance liquid chromatography; gas chromatography; nuclear magnetic resonance; electrophoresis in capillaries	Typically carried out in specialized labs; little information on farm animals
Fecal fat	Feces	Whole intestine	Sudan stain, colorimetry, spectrophotometry, microscopy, and nuclear magnetic resonance	Arduous and time-consuming, not always quantitative, and therefore not employed for serial analysis.
Total carotenoids	Blood	Duodenum and jejunum	Spectrophotometry; high-performance liquid chromatography	For usage on farms, quick, easy, and portable instruments are available. Restrictions brought on by blood sampling's invasiveness.
Fat-soluble vitamins	Blood	Duodenum and jejunum	High-performance liquid chromatography; spectrophotometry.	For usage on farms, quick, easy, and portable instruments are available. Restrictions brought on by blood sampling's invasiveness.

According to Vieira et al., (2008), carotenoids play a significant role in maintaining the integrity of the intestinal barrier and so lowering the transfer of toxins and germs. In fact, according to Conway (1994), blood levels of total carotenoids are thought to be a highly reliable indication of intestine physical integrity, and their levels are decreased in birds infected with coccidia (Windey et al., 2012). Given that total carotenoids and vitamin A are both readily measured using a quick, animal-side test that takes only a few minutes to complete (Raila et al., 2017), evaluating these two nutrients would make it easier to optimize their dietary supplementation as well as assess gastrointestinal functionality. Upon reaching the distal segments of the gastrointestinal tract, proteins or their breakdown products, such as oligopeptides and amino acids, undergo proteolytic fermentation by the gut microbiota. This process produces a variety of compounds, such as short-chain fatty acids (SCFAs), isovalerate, isobutyrate, and valerate (Windey et al., 2012). Physiologically, protein digestion and absorption is relatively complete in the stomach and small intestine and therefore only small amounts of protein-derived SCFAs are present in the distal segments of the gastrointestinal tract (Qaisrani et al., 2015).

Gastrointestinal Microbiota

The study of the gut microbiome is becoming a very interesting and useful field for controlling not only the health of the gastrointestinal tract but also the overall health of the body. The term "fifth organ" has been applied to the gut microbiome, and research indicates that the microbiome plays a critical function in absorbing environmental cues and signals and delivering them to the organism (Dietert and Silbergeld, 2015). According to some theories (Dietert and Silbergeld, 2015), one of the gastrointestinal microbiome's primary roles is to "teach" the immune system how to sustain a mutualistic interaction between the host immune system and the microbiome. The major mechanisms by which the gut microbiota modulates the operation of organs and systems, including the immune system and the brain, are the results of this intricate connection. Numerous physiological processes, including energy metabolism, immune system development, digestion and absorption, and infection prevention, are modulated by the gut microbiome (Marchesi et al., 2015). The characterisation of the gastrointestinal microbiome and its functions has been significantly impacted by the multiple intricate interactions that exist between the microbiota and the host. Our knowledge of the changes in the phylogenetic composition (16S sequencing), functional capability (metagenomics), intent (metatranscriptomics), and metabolic impact (metabolomics) of the intestinal microbiota is rapidly expanding due to the development and availability of high-throughput techniques (Ji and Nielsen, 2015). Although the application of "omics" technologies is advancing our understanding of intestinal ecology, these techniques do not allow us to identify the composition of the "normal" or "ideal" gastrointestinal microbiota. In light of this consideration, the "omics" approach might prove vital in shedding some light on the complex ecology of the gastric environment.

The gastrointestinal microbiome is characterized by a significant degree of functional redundancy (Moussavi et al., 2007); in other words, different bacteria can perform similar functions, such as metabolize the same substrates and producing similar metabolites. As a result, the analysis of the gastrointestinal microbiome's activity rather than its composition and structure may be more appropriate to evaluate the impact of eubiotic interventions. Specifically, enhanced comprehension of the microbiome's temporal variations throughout the various segments of the gastrointestinal system will facilitate a deeper grasp of the variables influencing the dynamic fluctuations in the microbiota's composition and activity. The fecal metabolome of horses is somewhat representative of the colonic metabolome, but not of the caecum, which is indicative of the various patterns of bacterial metabolism and product absorption in various sections of the gastrointestinal tract (Dougal et al., 2012). Additionally, research on pigs has shown that the microbiome's activity and composition vary depending on the site along the various gastrointestinal tract segments (Isaacson and Kim, 2012). Although the fecal microbiota and caeca microbiota are qualitatively similar, they differ numerically, according to studies on chicken. Nevertheless, feces samples can be utilized to efficiently detect various shifts and reactions of caeca microbiota (Stanley et al., 2015). A variety of metabolites are produced when proteins and carbohydrates are fermented by microorganisms. These include branched-chain fatty acids (BCFA), ammonia, amines, p-cresol, indole, hydrogen (H2), methane (CH4), and hydrogen sulfide (H2S), as well as SCFA (acetic, propionic, and butyric acid). Because lactate and succinate are often metabolized during physiological conditions to produce acetate, butyrate, and propionate, the gastrointestinal system has a notably low concentration of these substances (Macfarlane and Macfarlane, 2007). According to recent research, succinate is connected to several metabolic processes, such as the regulation of inflammation and redox balance (Tretter et al., 2016). Piglets with dyspepsia and those on antibiotic therapy have been found to have higher fecal concentrations of lactic and succinic acid (Tsukahara & Ushida, 2001). Furthermore, it has been documented that the colons of mice suffering from colitis have higher concentrations of succinate (Ariake et al., 2000). These colitis-affected mice's microbiota had higher concentrations of Bacteroidaceae cells, particularly B. caccae, which is a major producer of succinic acid.

Effective Immune Status

It is not unexpected that the gastrointestinal tract is regarded as the immune system's greatest organ given that it contains more than 70% of all immune system cells (Vighi et al., 2008). It is becoming more and more important to characterize the interaction between the immune system and the gastrointestinal tract in terms of gastrointestinal functionality as well as health, welfare, and illnesses (Celi et al., 2017). Since the gastrointestinal tract is the main interface between the host and the environment, it plays a crucial physiological role in acting as a barrier against antigens and infections. The fact that the gastrointestinal barrier is a living, breathing thing that not only divides the gut material from the host but also serves as a site of frequent communication and sampling between the two highlights how important it is. As a result, a variety of assays and indicators may be taken into account while evaluating the gastrointestinal barrier. The gastrointestinal tract's interactions with the environment and the degree to which the gastrointestinal barrier is functionally maintained can be inferred from biomarkers of gastrointestinal inflammation and immune function. The gastrointestinal barrier is made up of two parts: the functional immunological barrier, which is made up of digestive secretions, immune molecules, and cell products like cytokines, inflammatory mediators, and antimicrobial peptides. The functional immunological barrier is primarily produced by Paneth cells in the small intestine's crypts (Bischoff et al., 2014), and trefoil factor family proteins actively stimulate the process of mucosal epithelium repair, protecting the mucosal epithelium (Kim et al., 2009). Determining if intestinal inflammation is present is essential for evaluating the gastrointestinal barrier because intestinal inflammation can impair gastrointestinal function by causing disruptions to the structure and function of the intestinal mucosa

Gut Mucosa

According to Celi et al. (2017), maintaining good gastrointestinal functionality, animal health, and welfare requires effective intestinal barrier function. The mucus layer, which is made up of an inner layer with high concentrations of sIgA and mucin and an outside layer linked to the microbiota, makes up the first layer of the intestinal barrier. The gastrointestinal tract's innate immune system is made up of several components, all of which work together to maintain a delicate balance between the body's tolerance to commensal bacteria and its reaction to pathogens. Despite being constantly exposed to the gut microbiota, the gastrointestinal mucosa can selectively allow essential nutrients to enter while preserving the integrity of the intestinal barrier against various insults such as toxins, bacteria their cell debris, and anti-nutritional factors. Intestinal permeability and integrity can be evaluated using a variety of methods (Furness et al., 2013). They can measure a wide range of molecules (ions, carbohydrates of different sizes, macromolecules and antigens, bacterial products, and bacteria), in several biological matrices (peripheral blood, portal vein blood, urine), and they can be set up in vitro or in vivo. They have also been validated in various animal models. Oligosaccharides that the host cannot digest or metabolize, such as mannitol, L-rhamnose, and lactulose, are frequently used in permeability tests. In animals in good health, lactulose goes undigested through the small intestine and stomach. If there is damage to the intestinal barrier, then it can leak through the tight junctions and enter the blood stream, after which it is processed by the kidney and excreted from the body via urine or excreta in the case of mammals or chickens, respectively.

Neuroendocrine Function of the Gut

Information collecting is arguably the gastrointestinal system's most significant job; in fact, the gut should be viewed as a sensory organ (Furness et al., 2013). To match the volume and composition of food taken with the gastrointestinal tract's capacity for digestion and absorption, the brain receives signals from the sense of the contents of the gut during digestion (Furness and Cottrell, 2017). A key communication component of the so-called "gut-brain axis," which affects animal performance and welfare, is the neuroendocrine activity of the gastrointestinal tract (Celi et al., 2017). In response to many stimuli, including nutrition, neurological or hormonal stimulation, and metabolic products of the gut microbiota, the many enteroendocrine cells widely distributed across the intestinal mucosa release gastrointestinal peptide hormones. In addition to aiding in digestion and absorption, intestinal hormones also support healthy mucosal tissue, feed intake, metabolism, and intestinal barrier function. Even though humans and rats make up the majority of the research now in publication, several studies conducted on farm animals indicate that the mechanisms that have been identified thus far may apply to a variety of other species. The characterization of gastrointestinal hormones may also be used as valuable biomarkers of gastrointestinal functionality as the information gathered with this approach might prove to be a valuable approach to improve livestock health, welfare, and consequently productivity.

Conclusion

A review of the "what, where, and when" of biomarkers for gastrointestinal functionality has been conducted. Selecting which biomarker to test and where and when to measure it has inherent obstacles. Many of the methods now in use to track gastrointestinal function are intrusive since they necessitate the sacrifice of animals to gather sample data. Even though we don't think there is currently a single perfect biomarker, it would be great to have a biological marker that could be evaluated repeatedly from the same animal, enabling longitudinal evaluation of gastrointestinal performance. Numerous metabolites that are produced by the diverse microbial community that lives in farm animals' gastrointestinal tracts (GITs) have been linked to either better or worse gut health. A list of potential biomarkers for gut health in pigs and poultry has been compiled. Nevertheless, a single biomarker that can adequately capture the intricate physiology and pathology of the gastrointestinal system appears to be lacking, in addition to several technological constraints (such as lack of reagents and standardization of techniques and methodologies) that impede the advancement of knowledge in this field.

References

Ariake, K., Ohkusa, T., Sakurazawa, T., Kumagai, J., Eishi, Y., Hoshi, S., Yajima, T., 2000. Roles of mucosal bacteria and succinic acid in colitis caused by dextran sulfate sodium in mice. *J. Med. Dent. Sci.* 47, 233–241.

Broom, L. J., Kogut, M. H., 2018. Inflammation: friend or foe for animal production? *Poult. Sci.* 97 540–514.

Celi, P., Cowieson, A. J., Fru-Nji, F., Steinert, R. E., Kluenter, A. M., Verlhac, V., 2017. Gastrointestinal functionality in animal nutrition and health: new opportunities for sustainable animal production. *Anim. Feed Sci. Technol.* 234, 88–100.

Conway, P. L., 1994. Function and regulation of the gastrointestinal microbiota of the pig. In: Souffrant, W. B., Hagemeister, H. (Eds.), Proceedings of the VIth *International Symposium on Digestive Physiology in Pigs. EAAP Publication*, Dummerstof, pp. 231–240.

Davin, R., Manzanilla, E. G., Klasing, K. C., Perez, J. F., 2013. Effect of weaning and in-feed high doses of zinc oxide on zinc levels in different body compartments of *piglets. J. Anim. Physiol. Anim. Nutr. (Berl.)* 97 (Suppl. 1), 6–12.

Dietert, R. R., Silbergeld, E. K., 2015. Biomarkers for the 21st century: listening to the microbiome. *Toxicol. Sci.* 144, 208–216.

Dougal, K., Harris, P. A., Edwards, A., Pachebat, J. A., Blackmore, T. M., Worgan, H. J., Newbold, C. J., 2012. A comparison of the microbiome and the metabolome of different regions of the equine hindgut. *FEMS Microbiol. Ecol.* 82, 642–652.

Furness, J. B., Cottrell, J. J., 2017. Signalling from the gut lumen. *Anim. Prod. Sci.* 57, 2175–2187.

Furness, J. B., Rivera, L. R., Cho, H. J., Bravo, D. M., Callaghan, B., 2013. The gut as a sensory organ. *Nat. Rev. Gastroenterol. Hepatol.* 10, 729–740.

Isaacson, R., Kim, H. B., 2012. The intestinal microbiome of the pig. *Anim. Health Res. Rev.* 13, 100–109.

Ji, B., Nielsen, J., 2015. From next-generation sequencing to systematic modeling of the gut microbiome. *Front. Genet.* 6, 219.

Kim, C. H., Kim, D., Ha, Y., Cho, K. D., Lee, B. H., Seo, I. W., Kim, S. H., Chae, C., 2009. Expression of mucins and trefoil factor family protein 1 in the colon of pigs naturally infected with Salmonella typhimurium. *J. Comp. Pathol.* 140, 38–42.

Klasing, K. C., 2007. Nutrition and the immune system. *Br. Poult. Sci.* 48, 525–537.

Kogut, M. H., Arsenault, R. J., 2016. Editorial: gut health: the new paradigm in food animal production. *Front. Vet. Sci.* 3, 71.

Koutsos, E. A., Garcia Lopez, J. C., Klasing, K. C., 2006. Carotenoids from in ovo or dietary sources blunt systemic indices of the inflammatory response in growing chicks (Gallus gallus domesticus). *J. Nutr.* 136, 1027–1031.

Lean, I. J., Golder, H. M., Black, J. L., King, R., Rabiee, A. R., 2013. In vivo indices for predicting acidosis risk of grains in cattle: comparison with *in vitro* methods. *J. Anim. Sci.* 91.

Macfarlane, S., Macfarlane, G.T., 2007. Regulation of short-chain fatty acid production. *Proc. Nutr.* Soc. 62, 67–72.

Marchesi, J. R., Adams, D. H., Fava, F., Hermes, G. D. A., Hirschfield, G. M., Hold, G., Quraishi, M. N., Kinross, J., Smidt, H., Tuohy, K. M., Thomas, L. V., Zoetendal, E. G., Hart, A., 2015. The gut microbiota and host health: a new clinical frontier. *Gut* 65, 330–339.

McGrath, J., Duval, S. M., Tamassia, L. F. M., Kindermann, M., Stemmler, R. T., de Gouvea, V. N., Acedo, T. S., Immig, I., Williams, S. N., Celi, P., 2018. Nutritional strategies in ruminants: a lifetime approach. *Res. Vet. Sci.* 116, 28–39.

Moussavi, A. R. H., Gilbert, R. O., Overton, T. R., Bauman, D. E., Butler, W. R., 2007. Effects of feeding fish meal and n-3 fatty acids on milk yield and metabolic responses in early lactating dairy cows. *J. Dairy Sci.* 90, 136–144.

Niewold, T. A., 2015. Chapter 9: intestinal health biomarkers in vivo. *Intestinal Health.* pp. 219–228.

Qaisrani, S. N., Van Krimpen, M. M., Kwakkel, R. P., Verstegen, M. W. A., Hendriks, W. H., 2015. Dietary factors affecting hindgut protein fermentation in broilers: a review. *Worlds Poult. Sci.* J. 71, 139–160.

Raila, J., Kawashima, C., Sauerwein, H., Hülsmann, N., Knorr, C., Myamoto, A., Schweigert, F. J., 2017. Validation of blood vitamin A concentrations in cattle: comparison of a new cow-side test (iCheck™ FLUORO) with high-performance liquid chromatography (HPLC). *BMC Vet. Res.* 13, 126.

Shivappa, N., Steck, S. E., Hurley, T. G., Hussey, J. R., Hébert, J. R., 2014. Designing and developing a literature-derived, population-based dietary inflammatory index. *Public Health Nutr.* 17, 1689–1696.

Stanley, D., Geier, M. S., Chen, H., Hughes, R. J., Moore, R. J., 2015. Comparison of fecal and cecal microbiotas reveals qualitative similarities but quantitative *differences. BMC Microbiol.* 15, 51.

Starkey, J. D., 2014. Triennial growth symposium–a role for vitamin D in skeletal muscle development and growth. *J. Anim. Sci.* 92, 887–892.

Tretter, L., Patocs, A., Chinopoulos, C., 2016. Succinate, an intermediate in metabolism, signal transduction, ROS, hypoxia, and tumorigenesis. *Biochim. Biophys. Acta* 1857, 1086–1101.

Tsukahara, T., Ushida, K., 2001. Organic acid profiles in feces of pigs with pathogenic or non-pathogenic diarrhea. *J. Vet. Med. Sci.* 63, 1351–1354.

Vieira, M. M., Paik, J., Blaner, W. S., Soares, A. M., Mota, R. M. S., Guerrant, R. L., Lima, A. A. M., 2008. Carotenoids, Retinol, and Intestinal Barrier Function in Children From Northeastern Brazil. *J. Pediatr. Gastroenterol. Nutr.* 47, 652–659.

Vighi, G., Marcucci, F., Sensi, L., Di Cara, G., Frati, F., 2008. Allergy and the gastrointestinal system. *Clin. Exp. Immunol.* 153, 3–6.

Windey, K., De Preter, V., Verbeke, K., 2012. Relevance of protein fermentation to gut health. *Mol. Nutr. Food Res.* 56, 184–196.

Chapter 3

Gut Health in Commercial Feed Formulation

Amitava Roy[1,*]
and Tanmoy Rana[2]
[1]Department of Livestock Farm Complex, W.B. University of Animal and Fishery Sciences, Kolkata, India
[2]Department of Veterinary Clinical Complex, W.B. University of Animal and Fishery Sciences, Kolkata, India

Abstract

The need to see an animal's intestinal microbiome as its "last organ" has arisen as microbiome research has progressed. The gut microbiota may have a long-term or short-term impact on the host's eating habits through the synthesis of metabolites and/or the stimulation of hormone and neurotransmitter synthesis. According to available data, short-chain fatty acids (SCFA), peptide hormones including peptide YY (PYY) and glucagon-like peptide-1 (GLP-1), and the amino acid tryptophan, along with the related neurotransmitters serotonin, dopamine, and aminobutyrate (GABA), appear to be the main mediators. It appears to have an impact on taste receptor expression as well as central neural networks. The gut microbiota may potentially affect the host's senses, as there is a relationship between metabolic activities and taste sensory pathways. The present state of research on the microbiome in farm animals, specifically about the regulation of appetite in general and the impact of bacteria on feed intake, is summarized in this chapter. In order to determine the need for additional study, a brief comparison is made between this and the current results from studies using mouse models. Improving our knowledge of appetite regulation may help regulate feed intake, feed frustration, and anorexia in farm animals that are associated with unhygienic environments.

Keywords: appetite, eating behavior, farm animals, feed intake, intestine, microbiome, neurotransmitters, physiology, taste reception

---

* Corresponding Author's Email: amitavanutri07@gmail.com

In: Microbiomes in Animal Nutrition, Health, and Disease
Editor: Tanmoy Rana
ISBN: 979-8-89530-425-9

Introduction

In light of the recent rapid evolution of the area of microbiome research, a comprehensive understanding of microbe-host interaction grounded in the holobiont theory is now required (Berg, et al., 2020). The definitions of Berg et al. (2020), which state that the microbiota is the entirety of live microorganisms in a particular environment, are the foundation for this review. The phrase "microbiome" refers to the full range of compounds produced by microbes, including their structural components and metabolites, as well as their genes. A vast array of metabolites is produced by the microbiota, with the most well-known being biogenic amines, branched-chain fatty acids, amino acids (AA), and short-chain fatty acids (SCFA), among many others. According to Cryan et al. (2019), intestinal bacteria can start their host's production of hormones, neurotransmitters, and incretins. This allows the animal's microbiome to interact with it and support its metabolism, which in turn affects the animal's welfare, productivity, and feed intake. A small amount of research, mostly using rodent models, has been done on the impact of the microbiome on the nutritional intake of the host about obesity.

When it comes to environmental elements that could affect the microbiota, laboratory rodent models offer a higher degree of standardization than big animal models. Pigs' unique microbiome-influenced eating behavior has been linked to specific mechanisms (Fleming et al., 2019). Research has indicated that there may be differences in the microbial composition of the gastrointestinal tract (GIT) between pigs that feed at various rates. In this case, the variable feed efficiencies were found to be caused by variations in voluntary feed intake (Metzler-Zebeli et al., 2018). These excellent findings from numerous pig studies show that, even in an era of extremely efficient animal production, farm animals' microbiomes must be taken into account when determining what they should eat. Thus, the goal of this study is to present an overview of the current body of knowledge regarding the impact of the microbiome on farm animal feed intake.

Digestion Microbiota in Agricultural Animals

Animals have evolved over millions of years to be able to support diverse and dynamic microbial consortia throughout their life cycles (Wren, 2000). According to Lee et al. (2019), colonisation of the mammalian gut can start as early as embryogenesis and continue to develop into the diverse and dynamic microbial community that forms after birth (Kikuchi et al., 2009). On the other hand, in birds, the initial feeding triggers the concurrent colonization of the microbiota in the various GIT segments (Saxena et al., 2016). The species, breed, age, nutrition, environment, rearing styles, stocking density, stress, and particularly antibiotics all affect the GIT microbiota's composition (Shehata et al., 2022). Because of this, care should be taken while evaluating the microbial patterns of the different livestock species. The farm animal literature that is currently available is limited to what has been created through artificial raising and domestication. It has been established that domesticated animals cannot be fairly compared to their wild counterparts without restrictions. For example, the predominant bacterial families found in domesticated pigs are Lactobacillus and Enterobacteriaceae, whereas studies on wild Suidae have demonstrated a significant Bifidobacterium abundance (Ushida et al., 2016). 36% of GIT microbiota species were influenced by host genotype and sex in commercial chicken lines used for meat production (Zhao et al., 2013). Of them, the Lactobacillus genus included 15 impacted species. The microbiome of young chickens raised in a sterile environment away from older conspecifics after the industrial in-out process differs significantly from that of chicks raised for 24 hours with their adult hen (Kubasova et al., 2019). Furthermore, according to Vasaï et al. (2014), overfeeding affects the families Clostridiaceae, Lactobacillaceae, Streptococcaceae, and Enterococcaceae, whereas genotype influences the Lachnospiracecae, Betadacteroidaceae, and Desulfovibrionaceae in the ceca.

Over 90% of the known microbiota species in animal GITs are found in the groups Bacteroidetes, Firmicutes, and Actinobacteria. Fusobacteria, Proteobacteria, Verrucomicrobia, and Cyanobacteria make up the remaining percentage (Adak and Khan, 2019). In essence, the ratio of Firmicutes to Bacteroidetes is used as a gauge for the gut microbiome's makeup. As a result, in studies on human obesity, weight loss was directly correlated with a lower Firmicutes-to-Bacteroidetes ratio, whereas a higher ratio is linked to a greater ability to obtain energy from food (Ley et al., 2006). Through saccharolysis, the phyla Bacteroidetes and Firmicutes give the host useful compounds. While species of Bacteroidetes break down complex carbohydrates to synthesize primarily propionate, Firmicutes species break down polysaccharides to generate mostly butyrate (Turnbaugh et al., 2006). Proteolytic bacteria, in addition to fibre fermenters, also generate potentially advantageous but dangerous metabolites such ammonia and hydrogen sulphide (Ringseis et al., 2020). The availability of dietary carbohydrates, including fibre and protein, for animals, is a determining factor in the ratio of saccharolytic to proteolytic bacteria, based on their food base (Ellner et al., 2022). Ruminants' fore stomach mechanism makes them well-suited to diets high in crude fibre. The microbiota aids in the breakdown of complex carbohydrates in the anterior portion of the gastrointestinal tract (GIT), facilitating the animal's easy absorption of nutrients and bacterial metabolites in the intestine. Crude fibres are not broken down in the foregut of monogastric animals. The large intestine's posterior section serves as the primary organ for fermentation. Pigs' cecum and colon can only partially and exclusively break down complex carbs using bacteria (Roura et al., 2016). Because bacterial fermentation almost solely takes place in the ceca, hens' ability to digest crude fibre is even less developed (Shehata et al., 2022). Compared to monogastric animals, ruminants have a higher relative amount of saccharolytic phyla than proteobacteria, depending on their diet.

Regulation of Feed Consumption

Regulation at the Centre and Periphery

According to a thorough description by Camilleri, 2015, the mechanisms underlying hunger, satiation, and satiety are extremely complicated processes regulated by the interaction of humoral components and the central nervous system. In summary, the hypothalamus gets data from both central and peripheral functioning systems, such as hormones, circulating nutrients and metabolites, and stomach distension. Emotions, social variables, and sensory cues (including aroma and the appearance of a ration) can also influence eating behavior. The ARC, a group of neurons in the mediobasal hypothalamus, is crucial in identifying and incorporating peripheral feedback signals about nutritional and metabolic status. Peripheral impulses reach the ARC either directly after passing through the blood-brain barrier (BBB) or indirectly through the sympathetic and vagus afferent nerves. There are two functionally opposed populations of neurons in the ARC. Pro-opiomelanocortin (POMC) and cocaine-and amphetamine-regulated transcript (CART) are coexpressed in the anorexigenic neuronal population. By releasing different anorexic signals in the brain, including α-melanocyte-stimulating hormone (α-MSH), both might decrease appetite. The orexigenic neuronal population enhances appetite while blocking the anorexigenic-acting POMC-expressing neurons by coexpressing the potent orexigenic neuropeptide Y (NPY) and agouti-related peptide (AgRP) (Chaudhri et al., 2008).

Hormones released by endocrine cells in the GIT, including ghrelin, peptide YY (PYY), glucagon-like peptide-1 (GLP-1), and cholecystokinin (CCK), are responsible for peripheral signaling (Chaudhri et al., 2008). In the pre-absorptive stage, several receptors along the GIT detect nutrients like glucose and AA and release PYY, GLP-1, and CCK. These acute satiety-inducing signals help regulate the amount of food consumed by way of the brain's homeostatic and hedonic systems. In contrast, the hormone produced in the stomach, ghrelin, is a potent hunger stimulant whose release is reduced in response to meal consumption (Ringseis et al., 2020). Peptides that bear striking similarities to leptin, ghrelin, PYY, and NPY are synthesized by a variety of commensal and pathogenic bacteria (Alcock et al., 2014). These peptides may have an impact on the central regulation of appetite by activating the corresponding neurons. Moreover, one way that intestinal bacteria can intervene on the host is through the stimulation of the previously indicated peripheral signaling pathways by microbial metabolites. This approach has only been partially studied to far. It is established from rodent models that SCFA increases GLP-1 and PYY secretion (Tolhurst et al., 2012). Ileal infusions of acetate, propionate, and butyrate, which are SCFAs, were found to enhance PYY secretion (Cuche et al., 2000) and plasma CCK levels (Sileikiene et al., 2008) in pigs. Important peripheral feedback signals, like insulin, glucagon, and pancreatic peptide, are released by the pancreas during the post-absorptive phase. Insulin, an anabolic hormone, lowers hunger and food intake through a number of central mechanisms, such as the reduction of NPY and AgRP expression and the insulin receptor-dependent activation of CART and α-MSH. Leptin functions as a long-term feedback signal in the hypothalamus to decrease hunger and food intake, working in tandem with insulin, another peripheral feedback signal derived from white adipose tissues (Ringseis et al., 2020).

Microbial-derived SCFA increases the release of leptin in the gut (Zaibi et al., 2010). Furthermore, pigs' plasma leptin concentration increased after receiving a combination of SCFA orally (Jiao et al., 2018). Via a transport system connected to the leptin receptor, leptin traverses the blood-brain barrier. In the ARC, leptin activates POMC, which results in the production of the satiety signal α-MSH, while simultaneously inhibiting hunger signals like NPY and AgRP. Voigt and Fink reported synergistic effects between leptin and CCK in the regulation of food intake (Voigt, 2015).

Neurotransmission and Reward Mediation

The pleasant aspect of food as a stimulus to eat leads to food consumption over the coverage of nutrient requirements. The combination of seven signaling systems—the opioid, dopaminergic, cannabinoid, GABAergic, serotoninergic, noradrenergic, and neurotensin systems—has resulted in a complex reward circuitry, as Stanley et al., (2005) summarised. A preferential hunger for appealing items like sugar or fat is stimulated by certain modulators, like opioids and dopamine (deMacedo et al., 2016). On the other hand, even when there is appetising food present, stimulating the serotoninergic system in the ARC results in an anorectic response because serotonin inhibits the orexigenic peptides (NPY/AgRP) and stimulates the anorexigenic (POMC, α-MSH) in response to peripheral signals like ghrelin, leptin, and CCK that travel through vagal afferents to the hypothalamus (Voigt 2015). While non-homeostatic eating is commonly attributed to the neurotransmitter dopamine, serotonin is primarily considered a neurotransmitter within the homeostatic system. Homeostatic and hedonic mechanisms regulating food or feed intake are not independent. The gut microbiota is known to impact the host GIT serotonergic system, even though the GIT's enterochromaffin cells produce more than 95% of the serotonin in the body (Cryan et a., 2019). For example, the spore-forming bacterium Clostridium perfringens can increase colonic tryptophan hydroxylase 1 expression, which in turn increases the gut's ability to produce serotonin from tryptophan (Reigstad et al., 2015). Moreover, important bacterial metabolites like SCFA directly cause enterochromaffin cells to produce serotonin, stimulating its release into the gut lumen and the vasculature (Reigstad et al., 2015). Other neurotransmitters are produced by gut bacteria in addition to serotonin. It has been demonstrated that a variety of short-lived and long-lived gut flora, such as Bacillus cereus, Bacillus mycoides, Bacillus subtilis, Proteus vulgaris, Serratia marcescens, and Staphylococcus aureus, create dopamine (Alcock et al., 2014). It appears that B. subtilis secretes norepinephrine and dopamine into its surroundings, where it interacts with mammalian cells. It was shown that certain bacteria can synthesise gamma-aminobutyrate (GABA), including Bifidobacterium spp. (Barrett et al., 2012), Escherichia spp. (Richard and Foster, 2004), and Lactobacillus spp. (Siragusa et al., 2007). It is believed that GABA is an orexigenic neurotransmitter in pigs (Baldwin et al., 1990). Neurotransmitters generated by microbes are assumed to facilitate host-microbe communication.

Species-Specific Considerations

Monogastric Farm Animals

Poultry

It has previously been demonstrated that the gut microbiome of chickens affects the serotonergic system and brain functions like anxiety and memory (Yan et al., 2018). However, studies on how the gut microbiota affects eating habits give less attention to poultry. The colon is thought to be less significant for microbe–host interactions because of its quick transit time (Siegerstetter et al., 2017). The largest intestine fermentation occurs in the ceca of chickens and other animals (Wilkinson et al., 2016). In addition, the crop has a significant number of microorganisms in it. Firmicutes and Proteobacteria (~78% and ~16%, respectively) predominate in this thin-walled, expanded section of the fowl digestive tract that stores food before digestion (Saxena et al., 2016). Lactobacilli are also present in significant levels (de Cesare et al., 2020). According to Tremaroli and Bäckhed (2012), effective fermentation is linked to a greater host nutrient production. Consequently, information regarding the relationship between chicken feed efficiency and its microbiota in the cecum and faeces is accessible. While bacteria from the Enterobacteriaceae (Siegerstetter et al., 2017), but more recently, Lactobacillus (Stanley et al., 2016), were also reported to negatively affect feed efficiency in chickens, bacteria from the Bacteroides (Singh et al., 2014), Clostridium (Stanley et al., 2016), Ruminococcus (Stanley et al., 2013), Faecalibacterium (Stanley et al., 2016), and Lactobacillus (Stanley et al., 2013) families have been positively linked to feed efficiency in chickens. The results for microbial patterns linked to high feed efficiency are contradictory, as the two directions of Lactobacillus' impact suggest. This might be because the hygienic practices used in contemporary commercial hatcheries have the unintended consequence of enabling the intestines of chickens to get colonized with a very particular type of bacteria (Stanley et al., 2013).

Additional factors that may cause variation include the composition of the diet, the type of chicken line used, and the environmental microbes that are encountered (Huang et al., 2004). Additionally, Siegerstetter et al. (2017) found a correlation between the Lactobacillus genus and two Lactobacillus crispatus taxa and a high feed intake that is exclusive to female chickens. Generally, breeding has a significant impact on microbial colonization, with a focus on Lactobacillus species (Zhao et al., 2013). In chickens, quantitative trait loci for the presence of bacteria like Lactobacillus and *L. crispatus colocalize* with those for feeding behavior (Mignon-Grasteau et al., 2017), suggesting that these native bacteria may have an impact on eating behavior, though further research is necessary to confirm this relationship. Nonetheless, adding Lactobacillus strains to chickens' diets boosts their feed intake (Huang et al., 2004). The ability of Lactobacillus strains to synthesize GABA may be a potential mechanism of action (Siragusa et al., 2007), as GABA release has been shown to mediate the orexigenic effects of the hypothalamic NPY/AgRP signaling pathway (Sohn, 2015).

Pigs

There is likewise little data on how the microbiota-host interaction affects food habits in pigs. Variations in the GIT microbial community have been linked to varying feeding efficiency in poultry. Here, variations in voluntary feed intake were the root cause of the various feed efficiencies (Metzler-Zebeli et al., 2018). The colon (Ellner et al., 2022) and the cecum (Metzler-Zebeli et al., 2018) are thought to be the primary locations for microbial fermentation in pigs. Pigs with a low residual feed intake, a sign of improved feed efficiency, had enrichment of gut bacteria linked to a slimmer but healthier host (e.g., Christensenellaceae, Oscillibacter, and Cellulosilyticum), according to McCormack et al. (2017). In the Metzler-Zebeli et al. (2018) study, pigs assigned to the high feed intake group were linked to Escherichia, Shigella, Ruminobacter, and Veillonella, while the pig group with low residual feed intake had a higher abundance of Campylobacter in their cecal mucosa.

Fleming et al. (2019) found that peripheral amounts of butyrate and serotonin, as well as intestinal SCFA synthesis, are individual mechanisms by which the microbiome affects feeding behavior in pigs. Ileal infusions of SCFAs, such as acetate, butyrate, and propionate, raised plasma CCK levels or PYY secretion in pigs (Cuche et al., 2000; Sileikiene et al., 2008). Pigs that were given the peptide hormone CCK in a feed motivation test showed signs of satiety (Farmer et al., 2001). Jiao et al.'s research (Jiao et al., 2018) indicates that an additional mechanism for stopping feed intake is the rise in plasma leptin concentrations brought on by SCFA.

Ruminant Farm Animals

Since ruminants' primary fermentation organ is situated in the anterior region of the GIT, they can transform plant biomass—which is indigestible for monogastric animals—into edible food (Shabat et al., 2016). Because microbial fermentation in the rumen is thought to be responsible for energy production and the provision of nutrients to the host, research on ruminants in the past has concentrated on this organ (Myer et al., 2017). However, it should be highlighted that each segment of the GIT has distinct metabolic and functional characteristics that, according to the hoboing theory, work together to support the animal's overall health and nutritional status. The bacterial communities associated with each segment exhibit changes that could potentially affect the energy balance and nutrition of the host (Myer et al., 2017). According to a study by Myer et al (2017), the most of the lower GIT taxa were found in the Frimicutes genera Butyrivibrio and Ruminococcus, while the majority of the ruminal bacteria in beef cattle belonged to the Bacteroidetes genus Prevotella. Large amounts of fungi, protozoa, and archaea are also present in the rumen microbiota; these microorganisms are vital to the rumens and the host's functions (Elekwachi et al., 2017). It was previously believed that rumen filling served as the primary mechanical mechanism for controlling feed intake in cattle (Allen 2000). Nonetheless, current research suggests that microbes in the rumen may potentially play a role in controlling feed intake in ruminants. The bacterial communities associated with each segment exhibit changes that collectively could affect the energy balance and nutrition of the host (Myer et al., 2017). According to a study by Myer et al. (2017), the bulk of the lower GIT taxa were found in the Firmicutes genera Butyrivibrio and Ruminococcus, while the majority of the ruminal bacteria in beef cattle belonged to the Bacteroidetes genus Prevotella. Large amounts of fungi, protozoa, and archaea are also present in the rumen microbiota; these microorganisms are vital to the rumen's and the host's functions (Elekwachi et al., 2017). It was previously believed that rumen filling served as the primary mechanical mechanism for controlling feed intake in cattle (Allen 2000). Nonetheless, current research suggests that microbes in the rumen may potentially play a role in controlling feed intake in ruminants.

Conclusion

Feed intake and feeding behavior are critical to animal production. Therefore, more understanding of how to modify the GIT microbiota to condition feed intake regulation is needed. Future studies should concentrate on how and when an animal's microbiota might be altered in a way that is both practical and long-term beneficial, depending on its age or (patho-) physiological state. Even while we are aware that the pen community has a greater influence than maternal transfer or heredity, additional research is still needed to fully understand how early microbial colonization affects a later stage of animal life. At the moment, there is not enough cohesive research on the microbial regulation of farm animal behavior to develop practical plans for enhancing feed intake by gut microbiome modification. On the other hand, theoretical inferences can be made in specific situations. Weaned piglets, for example, are arguably the most susceptible category of farm animals when it comes to feeding intake, depression, and side effects like diarrhea. Introducing microorganisms linked to increased feed intake into the neonatal intestine on purpose could be one way to speed up the weaning process.

It is necessary to conduct a scientific verification process to determine whether or not extra feeding of the suckling piglets or maternal supplementation with probiotics or prebiotics can effectively modify the microbial profile of each individual or at least the group as a whole. Additionally, it might be possible to create unique antibiotic-associated feeds that, like those used in humans today, prevent the collateral loss of advantageous microorganisms. Finally, there is the question of whether farm animals should be fed in a way that promotes their microbiota or if the microbiota should be altered to encourage the animals to eat more, make better use of their feed, and grow more quickly. Regarding the most pressing social issues of the day, such as the fight against hunger in some parts of the world or the preservation of the environment and animal welfare in areas experiencing nutrient oversaturation, the answers to these questions will likely vary depending on where you live.

References

Adak, A., Khan, M. R. An insight into gut microbiota and its functionalities. *Cell. Mol. Life* Sci. **2019**, 76, 473–493.

Alcock, J., Maley, C. C., Aktipis, C.A. Is eating behavior manipulated by the gastrointestinal microbiota? Evolutionary pressures and potential mechanisms. *Bioessays* **2014**, 36, 940–949.

Allen, M. S. Effects of Diet on Short-Term Regulation of Feed Intake by Lactating Dairy Cattle. *J. Dairy Sci.* **2000**, 83, 1598–1624.

Baldwin, B. A., Ebenezer, I. S., de La Riva, C. Effects of intracerebroventricular injection of muscimol or GABA on operant feeding in pigs. *Physiol. Behav.* **1990**, 48, 417–421.

Barrett, E.; Ross, R. P., O'Toole, P. W., Fitzgerald, G. F., Stanton, C. -Aminobutyric acid production by culturable bacteria from the human intestine. *J. Appl. Microbiol.* **2012**, 113, 411–417.

Berg, G., Rybakova, D., Fischer, D., Cernava, T., Vergès, M. C. C., Charles, T., Chen, X., Cocolin, L., Eversole, K., Corral, G. H., et al. Microbiome definition re-visited: Old concepts and new challenges. *Microbiome* **2020**, 8, 103.

Camilleri, M. Peripheral mechanisms in appetite regulation. Gastroenterology **2015**, 148, 1219–1233.

Chaudhri, O. B., Salem, V., Murphy, K. G., Bloom, S. R. Gastrointestinal satiety signals. *Annu. Rev. Physiol.* **2008**, 70, 239–255.

Cryan, J. F., O'Riordan, K. J., Cowan, C. S. M., Sandhu, K. V., Bastiaanssen, T. F. S., Boehme, M., Codagnone, M. G., Cussotto, S., Fulling, C., Golubeva, A. V., et al. *The Microbiota-Gut-Brain Axis. Physiol. Rev.* **2019**, 99, 1877–2013.

Cuche, G.; Cuber, J. C., Malbert, C. H. Ileal short-chain fatty acids inhibit gastric motility by a humoral pathway. *Am. J. Physiol. Gastrointest. Liver Physiol.* **2000**, 279, G925-30.

de Cesare, A., Sala, C., Castellani, G.; Astolfi, A.; Indio, V.; Giardini, A.; Manfreda, G. Effect of Lactobacillus acidophilus D2/CSL (CECT 4529) supplementation in drinking water on chicken crop and caeca microbiome. *PLoS ONE* **2020**, 15, e0228338.

de Macedo, I. C., de Freitas, J. S., da Silva Torres, I. L. The Influence of Palatable Diets in Reward System Activation: A Mini Review. *Adv. Pharmacol. Sci.* **2016**, 2016, 7238679.

Elekwachi, C. O., Wang, Z.; Wu, X.; Rabee, A.; Forster, R. J. Total rRNA-Seq Analysis Gives Insight into Bacterial, Fungal, Protozoal and Archaeal Communities in the Rumen Using an Optimized RNA Isolation Method. *Front. Microbiol.* **2017**, 8, 1814.

Ellner, C.;Wessels, A. G., Zentek, J. Effects of Dietary Cereal and Protein Source on Fiber Digestibility, Composition, and Metabolic Activity of the Intestinal Microbiota in Weaner Piglets. *Animals* **2022**, 12, 109.

Farmer, C., Roy, N., Rushen, J., Robert, S. Feeding motivation in swine: Relation with insulin, glucose and free fatty acids in portal and jugular blood, and involvement of cholecystokinin. *Can. J. Anim. Sci.* **2001**, 81, 75–82.

Fleming, S. A., Monaikul, S.; Patsavas, A. J., Waworuntu, R. V., Berg, B. M., Dilger, R. N. Dietary polydextrose and galactooligosaccharide increase exploratory behavior, improve recognition memory, and alter neurochemistry in the young pig. *Nutr. Neurosci* **2019**, 22, 499–512.

Huang, M. K., Choi, Y. J., Houde, R., Lee, J. W., Lee, B., Zhao, X. Effects of Lactobacilli and an acidophilic fungus on the production performance and immune responses in broiler chickens. *Poult. Sci.* **2004**, 83, 788–795.

Jiao, A. R., Diao, H., Yu, B., He, J., Yu, J., Zheng, P., Huang, Z. Q., Luo, Y. H., Luo, J. Q., Mao, X. B., et al. Oral administration of short chain fatty acids could attenuate fat deposition of pigs. *PLoS ONE* **2018**, 13, e0196867.

Kikuchi, Y., Hosokawa, T., Nikoh, N., Meng, X. Y., Kamagata, Y., Fukatsu, T. Host-symbiont co-speciation and reductive genome evolution in gut symbiotic bacteria of acanthosomatid stinkbugs. *BMC Biol.* **2009**, 7, 2.

Kubasova, T., Kollarcikova, M., Crhanova, M., Karasova, D., Cejkova, D., Sebkova, A., Matiasovicova, J., Faldynova, M., Pokorna, A., Cizek, A., et al. Contact with adult hen affects development of caecal microbiota in newly hatched chicks. *PLoS ONE* **2019**, 14, e0212446.

Lee, S., La, T. M., Lee, H. J., Choi, I. S., Song, C. S., Park, S. Y., Lee, J. B., Lee, S. W. Characterization of microbial communities in the chicken oviduct and the origin of chicken embryo gut microbiota. *Sci. Rep.* **2019**, 9, 6838

Ley, R. E., Turnbaugh, P. J., Klein, S.; Gordon, J. I. Microbial ecology: Human gut microbes associated with obesity. *Nature* **2006**, 444,1022–1023.

McCormack, U. M., Curião, T.; Buzoianu, S. G., Prieto, M. L., Ryan, T., Varley, P., Crispie, F., Magowan, E., Metzler-Zebeli, B. U., Berry, D., et al. Exploring a Possible Link between the Intestinal Microbiota and Feed Efficiency in Pigs. Appl. Environ. *Microbiol.* **2017**, 83, e00380-17.

Metzler-Zebeli, B. U., Lawlor, P. G., Magowan, E., Zebeli, Q. Interactions between metabolically active bacteria and host gene expression at the cecal mucosa in pigs of diverging feed efficiency. *J. Anim. Sci.* **2018**, 96, 2249–2264.

Mignon-Grasteau, S., Chantry-Darmon, C.; Boscher, M. Y., Sellier, N., Le Bihan-Duval, E., Bertin, A. Genetic Determinism of Fearfulness, General Activity and Feeding Behavior in Chickens and Its Relationship with Digestive Efficiency. *Behav. Genet.* **2017**, 47, 114–124.

Myer, P. R., Freetly, H. C., Wells, J. E., Smith, T. P. L., Kuehn, L. A. Analysis of the gut bacterial communities in beef cattle and their association with feed intake, growth, and efficiency. *J. Anim. Sci.* **2017**, 95, 3215.

Reigstad, C. S., Salmonson, C. E., Rainey, J. F., Szurszewski, J. H., Linden, D. R., Sonnenburg, J. L., Farrugia, G., Kashyap, P. C. Gut microbes promote colonic serotonin production through an effect of short-chain fatty acids on enterochromaffin cells. *FASEB J.* **2015**, 29, 1395–1403.

Richard, H., Foster, J. W. Escherichia coli glutamate- and arginine-dependent acid resistance systems increase internal pH and reverse transmembrane potential. *J. Bacteriol.* **2004**, 186, 6032–6041.

Ringseis, R., Gessner, D. K., Eder, K. The Gut-Liver Axis in the Control of Energy Metabolism and Food Intake in Animals. *Annu. Rev. Anim. Biosci.* **2020**, 8, 295–319.

Roura, E., Koopmans, S. J., Lallès, J. P., Le Huerou-Luron, I., de Jager, N., Schuurman, T., Val-Laillet, D. Critical review evaluating the pig as a model for human nutritional physiology. *Nutr. Res. Rev.* **2016**, 29, 60–90.

Saxena, S., Saxena, V. K., Tomar, S., Sapcota, D., Gonmei, G. Characterisation of caecum and crop microbiota of Indian indigenous chicken targeting multiple hypervariable regions within 16S rRNA gene. *Br. Poult. Sci.* **2016**, 57, 381–389.

Shabat, S. K. B., Sasson, G., Doron-Faigenboim, A., Durman, T., Yaacoby, S., Berg Miller, M. E., White, B. A., Shterzer, N., Mizrahi, I. Specific microbiome-dependent mechanisms underlie the energy harvest efficiency of ruminants. *ISME J.* **2016**, 10, 2958–2972.

Shehata, A. A., Yalçın, S., Latorre, J. D., Basiouni, S., Attia, Y. A., Abd El-Wahab, A., Visscher, C.; El-Seedi, H. R., Huber, C., Hafez, H. M., et al. Probiotics, Prebiotics, and Phytogenic Substances for Optimizing Gut Health in Poultry. *Microorganisms* **2022**, 10, 395.

Siegerstetter, S. C., Schmitz-Esser, S., Magowan, E., Wetzels, S. U., Zebeli, Q.; Lawlor, P G., O'Connell, N. E., Metzler-Zebeli, B. U. Intestinal microbiota profiles associated with low and high residual feed intake in chickens across two geographical locations. *PLOS ONE* **2017**, 12, e0187766.

Sileikiene, V., Mosenthin, R., Bauer, E., Piepho, H. P., Tafaj, M., Kruszewska, D., Weström, B., Erlanson-Albertsson, C., Pierzynowski, S. G. Effect of ileal infusion of short-chain fatty acids on pancreatic prandial secretion and gastrointestinal hormones in pigs. *Pancreas* **2008**, 37, 196–202.

Singh, K. M., Shah, T. M., Reddy, B., Deshpande, S., Rank, D. N., Joshi, C. G. Taxonomic and gene-centric metagenomics of the fecal microbiome of low and high feed conversion ratio (FCR) broilers. *J. Appl. Genet.* **2014**, 55, 145–154.

Siragusa, S., de Angelis, M., Di Cagno, R., Rizzello, C. G., Coda, R., Gobbetti, M. Synthesis of gamma-aminobutyric acid by lactic acid bacteria isolated from a variety of Italian cheeses. *Appl. Environ. Microbiol.* **2007**, 73, 7283–7290.

Sohn, J. W. Network of hypothalamic neurons that control appetite. *BMB Rep.* **2015**, 48, 229–233.

Stanley, D., Geier, M. S., Denman, S. E., Haring, V. R., Crowley, T. M., Hughes, R. J., Moore, R. J. Identification of chicken intestinal microbiota correlated with the efficiency of energy extraction from feed. *Vet. Microbiol.* **2013**, 164, 85–92.

Stanley, D., Hughes, R. J., Geier, M. S., Moore, R. J. Bacteria within the Gastrointestinal Tract Microbiota Correlated with Improved Growth and Feed Conversion: Challenges Presented for the Identification of Performance Enhancing Probiotic Bacteria. *Front. Microbiol.* **2016**, 7, 187.

Stanley, S., Wynne, K., McGowan, B., Bloom, S. Hormonal regulation of food intake. *Physiol. Rev.* **2005**, 85, 1131–1158.

Tolhurst, G., Heffron, H., Lam, Y. S., Parker, H. E., Habib, A. M., Diakogiannaki, E.; Cameron, J.; Grosse, J.; Reimann, F.; Gribble, F. M. Short-chain fatty acids stimulate glucagon-like peptide-1 secretion via the G-protein-coupled receptor FFAR2. *Diabetes* **2012**, 61, 364–371.

Tremaroli, V., Bäckhed, F. Functional interactions between the gut microbiota and host metabolism. *Nature* **2012**, 489, 242–249.

Turnbaugh, P. J., Ley, R. E., Mahowald, M. A., Magrini, V., Mardis, E. R., Gordon, J. I. An obesity-associated gut microbiome with increased capacity for energy harvest. *Nature* **2006**, 444, 1027–1031.

Ushida, K., Tsuchida, S., Ogura, Y.; Toyoda, A.; Maruyama, F. Domestication and cereal feeding developed domestic pig-type intestinal microbiota in animals of suidae. *Anim. Sci. J.* **2016**, 87, 835–841.

Vasaï, F., Brugirard Ricaud, K., Bernadet, M. D., Cauquil, L., Bouchez, O., Combes, S., Davail, S. Overfeeding and genetics affect the composition of intestinal microbiota in Anas platyrhynchos (Pekin) and Cairina moschata (Muscovy) ducks. FEMS *Microbiol. Ecol.* **2014**, 87, 204–216.

Voigt, J. P., Fink, H. Serotonin controlling feeding and satiety. Behav. *Brain Res.* **2015**, 277, 14–31.

Wilkinson, N., Hughes, R. J., Aspden, W. J., Chapman, J., Moore, R. J., Stanley, D. The gastrointestinal tract microbiota of the Japanese quail, Coturnix japonica. *Appl. Microbiol. Biotechnol.* **2016**, 100, 4201–4209.

Wren, B. W. Microbial genome analysis: Insights into virulence, host adaptation and evolution. *Nat. Rev. Genet.* **2000**, 1, 30–39.

Yan, F. F., Wang, W. C., Cheng, H. W. Bacillus subtilis based probiotic improved bone mass and altered brain serotoninergic and dopaminergic systems in broiler chickens. *J. Funct. Foods* **2018**, 2018, 501–509.

Zaibi, M. S., Stocker, C. J., O'Dowd, J., Davies, A.; Bellahcene, M.; Cawthorne, M. A., Brown, A. J. H., Smith, D. M., Arch, J. R. S. Roles of GPR41 and GPR43 in leptin secretory responses of murine adipocytes to short chain fatty acids. *FEBS Lett.* **2010**, 584, 2381–2386.

Zhao, L., Wang, G., Siegel, P., He, C., Wang, H., Zhao, W., Zhai, Z., Tian, F., Zhao, J., Zhang, H., et al. Quantitative genetic background of the host influences gut microbiomes in chickens. *Sci. Rep.* **2013**, 3, 1163.

Chapter 4

The Composition and Dynamics of Microbiomes

Amitava Roy[1,*]
and Tanmoy Rana[2]
[1]Department of Livestock Farm Complex, W.B. University of Animal and Fishery Sciences, Kolkata, India
[2]Department of Veterinary Clinical Complex, W.B. University of Animal and Fishery Sciences, Kolkata, India

Abstract

Due to the rapidly growing worldwide population and increased awareness of their detrimental environmental impact, ruminant production methods are currently facing substantial problems. Recent research has highlighted the relationship between feed efficiency and methane emission—two economically significant traits—and the makeup and function of the rumen microbiome. While the past ten years have seen revolutionary changes in our understanding of host-associated microbial populations due to omics-based technical advancements, disagreements persist over the best method for analyzing vast amounts of omic data. To fully utilize the gastrointestinal microbiome for sustainable ruminant production, a comprehensive strategy that looks at host/microbiome interactions in the rumen and lower digestive tract is needed. To effectively apply omic data to animal health and production, this chapter outlines how the ruminant animal production community may recognize and take advantage of the causal links between the gut microbiota and host attributes of interest.

Keywords: gut, rumen, microbiota, microbiome, omics, statistics, cattle, sustainable agriculture

Introduction

According to Cammack et al., (2018), there are an estimated 3.9 billion ruminants worldwide. These animals play a significant role in sustainable agricultural practices because they can produce milk and meat from low-quality forages, use industrial byproducts like distillers grains as a food source, and make use of nonarable land through grazing. The complex microbial

* Corresponding Author's Email: amitavanutri07@gmail.com

In: Microbiomes in Animal Nutrition, Health, and Disease
Editor: Tanmoy Rana
ISBN: 979-8-89530-425-9

community known as the gut microbiome, which inhabits the ruminant gastrointestinal tract (GIT), is essential to ruminant production and health. It is widely acknowledged that the gut microbiome plays a critical role in maintaining intestinal homeostasis, developing mucosal and lymphoid structures, and stimulating the repertoire of host immune cells (Malmuthuge and Guan 2017). Furthermore, an essential component of both dairy and beef systems is the microbial fermentation of ingested plant biomass in the rumen, a specialized foregut fermentation chamber, which enables the animal to utilize the nutritional content of host-indigestible plant biomass (Mackie, 2002). Systems used for the production of livestock currently face many difficulties. According to Gerber et al., (2013), to feed the world's population, which is predicted to reach 9.15 billion people by 2050, there will need to be a 70% increase in food production from 2007 levels in affluent countries and possibly a doubling of output from underdeveloped countries. This is made worse by growing worries about how animal agriculture affects the environment. According to recent estimates based on a complete life cycle assessment, agriculture accounts for about 14.5% of worldwide anthropogenic greenhouse gas (GHG) emissions, with direct emissions from livestock accounting for less than 5% of the total (Intergov, 2014). The cattle themselves create nitrous oxide (N2O) from manure and methane (CH4) enterically, among other greenhouse gases (GHGs) across the beef and dairy production chains (Lynch and Pierrehumbert, 2019). Methanogenic archaea produce methane in the rumen and lower gut, which is a particularly notable greenhouse gas (GHG) linked to ruminant production. Methane has the potential to cause global warming that is around 28 times higher than that of carbon dioxide (Intergov, 2014). Enteric methanogenesis is a significant factor in decreased host feed efficiency (FE) due to its estimated 2-12% loss of gross dietary energy to the animal, in addition to its detrimental effects on the environment (Johnson and Johnson, 1995).

Studies of these microbial communities as a way to increase cow production efficiency while lowering or eliminating its environmental effect have been going on for many years due to the complex interactions that exist between the host animal and its resident gut microbiomes (Hungate, 1969). Recent advances in high-throughput sequencing technology have produced a wealth of information about the makeup and roles of the rumen microbiota in a variety of hosts and settings (Zhou et al., 2018). Although this has not been thoroughly investigated to date (O'Hara et al., 2018), there is growing evidence that the lower GIT and its resident bacteria also contribute significantly to cattle health and productivity (Malmuthuge and Guan, 2017).In this chapter, we discuss the contributions of both the rumen and hindgut bacteria on animal performance and highlight recent studies on the ruminant gut microbiome. Furthermore, we evaluate the consequences of current discoveries about the interactions between microbes and hosts in the rumen for the performance of the host animal. Finally, complementing the chapter on the application of omics technologies to study host-associated microbiomes (Li et al., 2018), we discuss the challenges associated with statistical analysis of data generated from such studies and provide guidelines for robust analysis of microbial data sets, to better understand the roles of the gut microbiome in cattle production.

The Gut Microbiome's Importance for Health and Cattle Production

The Function and Composition of the Rumen Microbiome

The anaerobic bacteria, fungus, methanogenic archaea, ciliate protozoa, and viruses that make up the rumen microbiome are varied in terms of phylogeny. This community's principal microbial constituents are listed in depth. This microbial cohort can efficiently break down host-indigestible plant fibre and includes cellulolytic, hemicellulolytic, amylolytic, proteolytic, and biohydrogenating (lipolytic) species. It also shows a high degree of functional redundancy (Firkins and Yu, 2015). The main byproduct of rumen microbial fermentation is volatile fatty acids (VFAs), which are taken by the host and used as energy sources. These VFAs include primarily acetate, propionate, and butyrate (Bergman, 1990). The generation of ruminally derived VFAs is crucial to animal performance since they can supply up to 70% of the host's energy needs (Bergman, 1990). The rumen microbiota's metabolism of nitrogen-containing substances, such as peptides, ammonia, and urea, is essential for supplying the host with microbial proteins for the synthesis of milk and muscle (Bach et al., 2005). First, main microbiota members hydrolyze ingested fibre, carbs, protein, and lipids to shorter chains (or oligomers) and monomers (e.g., glucose, amino acids), which are then utilized as substrates by different microbiota members (Millen, 2016). Research on the rumen microbiota's temporal colonization of ingested feed revealed distinct functional and taxonomic profiles among the primary and secondary colonizers, suggesting differences in their function(s) and/or substrate specialization (Wilkinson et al., 2018).

Li et al., (2019) identified several factors that influence the composition and function of rumen microbiota, including genetics, age, gender, and geography. Of these, the impact of nutrition has been examined the most thoroughly. This chapter does not include the makeup of the rumen microbiota under different production systems and life phases, as this has been thoroughly examined in recent years (Huws et al., 2018). There is interest in the rumen microbiome's role in animal production because of the significance of microbial metabolism in the rumen to the host's health. Variations in FE (Li and Guan, 2017), CH4 emission intensity (Kittelmann et al., 2014), health (Silberberg et al., 2013), and milk composition (Jami et al., 2014) are all correlated with the microbial composition of the rumen.

The Microbiome of Rumen and Feed Effectiveness

The efficiency of food production, both crop- and animal-derived, needs to be increased because it is anticipated that global food consumption will increase dramatically over the next several decades (Berry and Crowley, 2013). According to Berry and Crowley (2013), feed efficiency (FE) refers to the efficiency with which feed is converted into a product that can be used. It is a feature that is somewhat heritable in cattle. Improving FE is a way to increase output while lowering costs, as feed inputs can make up up to 75% of variable costs in beef operations and 40–60% of those in dairy systems (Bach, 2012). Numerous metrics have been employed to assess feed efficiency (FE) in cattle, such as feed conversion ratio (Sherman et al., 2008) and partial efficiency of growth (Sobrinho et al., 2011). However, the most widely used metric is now residual feed intake (RFI). RFI was first introduced in 1963 and is defined as the discrepancy between an animal's actual and expected feed intake for weight gain and maintenance (Richardson and Herd, 2004).

Animals can be categorised as Low-RFI (efficient) or High-RFI (inefficient) based only on their genetic makeup. The idea is to choose animals that will have lower input costs (feed) and the same or higher output values (meat yield/quality, for example). The rumen microbiome may be crucial in determining an animal's FE status, even though a variety of physiological processes can lead to divergence in FE within a population (Richardson and Herd, 2004). This is because the rumen microorganisms are responsible for converting ingested feedstuff to energy substrates (such as VFA). The rumen microbial ecology of efficient (low-RFI) cattle was shown to be different from that of inefficient (high-RFI) cattle in a seminal work by Guan and colleagues (Guan et al., 2008). Additionally, the microbial profiles of the efficient cattle were more comparable to each other. More recently, efficient cattle (dairy and beef) were shown to have decreased rumen microbial diversity and richness in terms of both microbial species and gene content (Shabat et al., 2016) and metabolic profile (Roehe et al., 2016) by the use of high-throughput sequencing. It's unclear if this is a cause or an effect of the efficiency phenotype, but it implies that the rumen of efficient animals has fewer non-essential bacteria. Variation in VFA concentration according to RFI classification has also been reported, but these differences appear to be diet-dependent (Shabat et al., 2016).

Untapped Potential to Enhance Animal Health and Performance: The Lower-Gut Microbiome

The essential role(s) of the lower-gut microbiota and its contribution to the health and productivity of ruminants are not as well understood as that of the rumen. The small intestine and hindgut areas are included in the lower gut, which is defined as the post-gastric intestinal tract for this chapter.

Feed Efficiency and the Lower Gut

Cattle's hindgut digest (cecum, colon, and rectum) contain 10^{12}–10^{14} cells/ml of bacteria (Krause et al., 2013). Up to 30% of the cellulose and hemicellulose breakdown in ruminants may be attributed to microbial fermentation in the hindgut (Hoover, 1978), while lower estimates have also been suggested (Gressley et al., 2011). The reduced retention time of digesta in the hindgut compartments compared to the rumen and the fact that substrates entering the colon and cecum have already been partially digested by enzymes in the small intestine (both host and microbial) and rumen (microbiological) are likely contributing factors to lower dietary energy production in the hindgut compartments. But hindgut fermentation may be especially significant for the calf in its early days and weeks of life, before the rumen fully develops, and hindgut energy from the diet is probably a significant factor in energy availability in cattle at all stages of production (Castro et al., 2016). According to the intestinal segment, the lower-gut microbiota differ in composition (Mao 2015), most likely due to variations in the physical, chemical, and biological circumstances within each compartment. The jejunum is a key location for the post-ruminal digestion and absorption of proteins and carbohydrates, with Firmicutes accounting for up to 90% of the prominent phyla found there (Myer at al., 2016). Firmicutes and Bacteroidetes dominate the microbial communities in the cecum and colon, two hindgut areas with comparable roles. Several taxa in the small and large intestines have been linked to feed efficiency status, supporting the theory that lower-gut microbes are important to animal performance. Abundances of Butyrivibrio, Pseudobutyrivibrio, Prevotella, Paludibacter, Faecalibacterium, and Succinivibrio in the hindgut and Butyrivibrio in the jejunum have been reported to differ across FE phenotypes (Myer at al., 2017).

Gastrointestinal Health of the Host Is Influenced by the Lower-Gut Microbiome

The lower-gut areas have a highly active immune system, in contrast to the rumen, where there is still disagreement regarding the existence of any strong host immune mechanisms that maintain gut health. The mucosal immune system consists of physical barriers (mucosal/epithelial layers), chemical barriers (antimicrobial peptides, secretory IgA), pattern-recognition receptors (such as toll-like receptors, TLRs), and a variety of immune cells that support host defense (Hooper et al., 2012). Consequently, there is growing evidence that the microbial communities of the lower gut contribute to immune system establishment and homeostasis in beef cattle (Malmuthuge and Guan, 2017), which has a direct impact on animal gut health in addition to their role(s) in feed digestion and energy production. The lower-gut regions are therefore known to be crucial to immune system development in monogastric animals (Mulder et al., 2011). According to Malmuthuge et al. (2013), starter feeding, which is a regular aspect of early-life calf care, has an impact on the expression of genes (TLR10 and TLR2) that are linked to the efficiency of the host mucosal immune response in the lower gut as well as the diversity of bacteria. A subsequent investigation revealed a strong correlation between the expression of genes encoding the host immune response and the total counts of mucosa-associated and luminal bacteria in the small intestine of pre-weaned dairy calves (Liang et al., 2016). Additionally, the same authors demonstrated that the interaction between commensal gut microbes and the expression of particular host microRNAs may aid in the development of the immune system in the gut of newborn calves (Liang et al., 2015). Genes implicated in "leukocyte and lymphocyte chemotaxis" and the "cytokine/chemokine-mediated signaling pathway" were found to have higher expression in a recent study of functional metagenomic profiles taken from the ileal tissue of *Lactobacillus*-dominant calves (Malmuthuge et al., 2019). Taken together, these observations suggest the importance of lower-gut microbiota in immune system development in dairy calves, which may lay the foundation for improving the health of neonatal calves through nutritional manipulation strategies.

The Gut Microbiome's Environmental Impact during the Production of Cattle

Rumen Methanogenesis

The production chains for food and agriculture release the most greenhouse gases (GHGs) into the atmosphere, primarily carbon dioxide (CO_2), methane (CH_4), and nitrous oxide (N_2O). Livestock industries are a major source of these gases. Since CH_4 has the strongest potential to cause global warming, it is the most researched greenhouse gas (GHG) when it comes to emissions from ruminants. Lowering intestinal methanogenesis rates is preferable for increased animal productivity and environmental stability. As previously mentioned, methanogenic archaea, which are thought to make up 0.3-2.3% of the rumen microbial community, create CH_4 in the rumen (Janssen and Kirs, 2008). This information is based on an examination of the

16S ribosomal RNA (rRNA) gene. Ruminal methanogenesis can be categorized into three main pathways: (a) hydrogenotrophic, which uses H_2 as an electron donor to reduce CO_2 to CH_4 (formate can also be used as an electron donor and may contribute to the production of up to 18% of ruminal CH_4 (Hungate, 1967)); (b) methylotrophic, which uses methanol or methylamines; and (c) acetoclastic, which uses acetate and H2 to produce CH4 (Leahy et al., 2013). The primary pathway in the rumen is hydrogenotrophic methanogenesis, which is primarily carried out by Methanobrevibacter species.

According to Hristov et al., (2012), these species typically account for over 90% of archaeal 16S rRNA gene reads. However, the rumen also contains a few other less common methanogen species (Tapio et al., 2017). The rate and intensity of rumen methanogenesis are determined by a variety of variables. The amount of detectable ruminal CH4 can be significantly influenced by dietary composition; high-forage diets encourage microbial acetate synthesis in the rumen, which increases H2 and subsequently CH4 production more than in concentrate-rich diets, where starch is primarily metabolized to propionate (Wolin, 1960). Reduced rumen pH under high-starch diets may contribute to microbial community imbalance, fermentation, and sub-acute ruminal acidosis, even though it may appear more cost-effective to just stop feeding forages to calves (Plaizier et al., 2008). Furthermore, several methods for lowering CH4 production across a range of diets are required, since the majority of livestock worldwide depend on fodder sources for growth.

Lower-Gut Microbiome's Harmful Effects on the Environment

The lower-gut microbiota contributes significantly to the rumen's generation of waste nitrate and CH4, as well as other greenhouse gases. According to earlier research, methanogens are present in the gastrointestinal tract (GIT) of newborn dairy calves, and their abundance varies between calves that are 0 and 3 days old (Guzman et al., 2015). It was also demonstrated by Zhou and associates (Zhou et al., 2014) that the predominant methanogenic taxon in the ileum of 3- to 4-week-old dairy calves was Methanobrevibacter. Given that methanogens are found in the newborn gut, these archaea and their metabolites may be crucial for intestinal development's early phases as well as for methane emissions in the hindgut. While the rumen is more powerful than the cecum, the latter can lose up to 10% of the energy from food when cattle undergo enteric methanogenesis. As a result, decreasing methane synthesis in the hindgut areas may also enhance production efficiency and lower the total amount of enteric GHG produced. As a result, the make-up and functional dynamics of the hindgut methanogenic community are of growing interest. Methanobacteriales are said to be the predominant group in the cecum, setting the hindgut archaea apart from those in the rumen from a compositional standpoint (Freetly 2015). Although there has been extensive research on the connection between dietary management techniques and the total amount of methane produced by the rumen, little is known about the connection between host production and methanogenesis in the lower gut.

The Actual Bovine Gut Microbiome: Features and Obstacles

The development of next-generation sequencing and other omics technologies in the last ten years has made it possible to examine host-associated microbial communities in ruminants at a depth never before imaginable, as we have recently reported (Bergman, 1990). To determine the metataxonomic, metagenomic, metatranscriptomic, metabolomic, and metaproteomic profiles of a microbial community and to spot trends or alterations linked to a biological state of interest, researchers can now employ a number of techniques. Nonetheless, there are several biases that affect high-throughput sequencing efforts, such as sample collection methods (Firkins and Yu, 2015), sample preservation techniques and duration before analysis (Wilkinson et al., 2018), nucleic acid extraction protocol selection (O'Hara et al., 2018), and sequencing technology (Malmuthuge and Guan, 2017). Moreover, a wide range of bioinformatic tools have been created recently for the interpretation of high-throughput sequencing data, although their consistency has not been extensively evaluated. Lastly, even if these technologies are effective means of producing sizable, high-quality data sets, there is ongoing discussion over the optimal technique to data analysis in order to derive significant and scientifically sound findings. This section summarises popular techniques for omic data analysis, and we leverage existing research to suggest more reliable procedures for best-practice statistical analysis of large omics data sets for ruminant gut microbiome studies.

Different Methods for Analysing Microbiome Data

Alternative statistical techniques have been developed to replace the conventional statistical procedures in the analysis of compositional omics data to get around the problems mentioned above. The identification of differentially abundant features and microbial signatures in this context should be done using the methods described below for MixMC (Krause et al., 2013) and Analysis of Composition of Microbiomes (ANCOM) (Hoover, 1978), or other similar methods. Since this section is not exhaustive, we refer the reader to a recent paper by Gloor and associates (Castro et al., 2016), which goes into more depth on these strategies than is appropriate for this study.

Analysis of Composition of Microbiomes

According to Janssen and Kirs (2008), ANCOM is a statistical method that contrasts Aitchison's log-ratio of each taxon's abundance with the abundance of each of the other taxa separately. Using this method, changes in the relative abundance of a taxon between two ecosystems are revealed by computing differential tests (e.g., Mann–Whitney U, ANOVA, ANOVA with Linear Mixed Effect Models, Friedman, Kruskal–Wallis, and Wilcoxon tests) on each log-ratio. The purpose of these differential tests is to determine if the null hypothesis of equality for taxonomic abundance across groups for the condition of interest (diet, for example) is accepted or rejected. ANCOM calculates the number of tests conducted for each taxon and produces a count random variable W, which indicates the number of null hypotheses that must be disproved.

MixMC

Currently implemented as the R package mixOmics (Wolin, 1960), MixMC (Freetly 2015) is a multivariate statistical framework that considers the intrinsic properties of microbiome data (sparsity and compositionality) to identify microbial signatures associated with the phenotype or condition being studied. To account for uneven sequencing depths across samples and the sparsity of the data set, preprocessing and normalization steps (such as total sum scaling) are carried out before data are centered log-ratio (CLR) transformed and analyzed via sparse partial-least-squares discriminant analysis (sPLS-DA models). To project the data from a simplex space to an Euclidian space, MixMC, sPLS-DA is used in conjunction with CLR transformations. This approach incorporates a multilevel decomposition approach for repeated measure designs, which are frequently found in microbiome investigations (Bergman, 1990).

Present Difficulties in Comparing Study Results

The lack of standardisation in the bioinformatic and statistical procedures used by the various research groups hinders the interpretation of the data generated across multiple studies, despite the fact that next-generation sequencing has led to an explosion of publications exploring the microbiological diversity in various ecosystems in the last few decades. When the rumen microbiome of efficient cattle was compared across research to identify consensus microbial taxa and/or genes that could act as worldwide biomarkers for forecasting ruminant FE and methane emissions, one example of this issue surfaced. According to Huws et al., (2018), the only way that the microbial gene correlations with RFI reported by Shabat et al., (2016) and Li & Guan (2017) overlapped was in respect to the rumen of feed-efficient animals having a lower number of genes involved in amino acid metabolism. These findings are consistent with the theory that feed-efficient cattle use rumen nitrogen more efficiently and excrete less ammonia in their urine than inefficient cattle (Janssen and Kirs, 2008). Notably, however, there were discrepancies between Li & Guan's (2017) and Shabat et al., (2016) findings. While variations in experimental design (different cattle, sample collection technique, nucleic acid selection (DNA versus RNA)) may be responsible for this, an additional factor that may contribute is the absence of a common method for data analysis, suggesting that it is currently impractical to make meaningful comparisons between studies.

Conclusion

A new era in the study of the livestock gut microbiome, whose makeup and function are closely linked to animal health and productive performance, has been made possible by significant advancements in high-throughput sequencing technology. The data has a significant social and economic influence. Cattle and chickens, or the microbial groups rather than their hosts, were the main subjects of earlier attempts to compile the body of knowledge regarding gut microbiome research in livestock. The composition of cattle's microbiomes has been extensively studied in relation to feeding-related alterations and their effects on production methods or environmental problems related to ruminal methane emissions. The first thing to think about is the subjects of the microbiome research that have been done thus far, where some species (like pigs and cattle) have been prioritized over others (like goats).

Furthermore, there is a blatant bias in the breeds that are studied: as might be expected, the majority of studies concentrate on a small number of highly selective, cosmopolitan breeds, while local breeds from rural areas are mainly ignored, despite the fact that livestock research is an essential part of improving development strategies and the socioeconomic status of related human communities. For example, describing the makeup of the microbiome and how it interacts with the host in non-intensive husbandry settings may offer helpful insights on how to maximize livestock productivity through nutrient supplementation. It should be mentioned that Africa and the tropical and subtropical regions in general have received far less attention than China or Europe, where a large portion of microbiome studies—also on indigenous breeds—have been conducted.

References

Bach A, Calsamiglia S, Stern MD. 2005. Nitrogen metabolism in the rumen. *J. Dairy Sci.* 88:E9–E21.

Bach A. 2012. Ruminant nutrition symposium: optimizing performance of the offspring: nourishing and managing the dam and postnatal calf for optimal lactation, reproduction, and immunity. *J. Anim. Sci.* 90(6):1835–45.

Bergman EN. 1990. Energy contributions of volatile fatty acids from the gastrointestinal tract in various species. *Physiol. Rev.* 70(2):567–90.

Berry DP, Crowley JJ. 2013. Cell biology symposium: genetics of feed efficiency in dairy and beef cattle. *J. Anim. Sci.* 91(4):1594–613.

Cammack KM, Austin KJ, Lamberson WR, Conant CG, Cunningham HC. 2018. Tiny but mighty: the role of the rumen microbes in livestock production. *J. Anim. Sci.* 96(2):752–70.

Castro JJ, Gomez A, White B, Loften JR, Drackley JK. 2016. Changes in the intestinal bacterial commu-nity, short-chain fatty acid profile, and intestinal development of preweaned Holstein calves. 2. Effects of gastrointestinal site and age. *J. Dairy Sci.* 99(12):9703–15.

Firkins JL, Yu Z. 2015. Ruminant nutrition symposium: how to use data on the rumen microbiome to improve our understanding of ruminant nutrition. *J. Anim. Sci.* 93(4):1450–70.

Freetly HC, Lindholm-Perry AK, Hales KE, Brown-Brandl TM, et al. 2015. Methane production and methanogen levels in steers that differ in residual gain. *J. Anim. Sci.* 93(5):2375–81.

Gerber PJ, Hristov AN, Henderson B, Makkar H, Oh J, et al. 2013. Technical options for the mitigation of direct methane and nitrous oxide emissions from livestock: a review. *Animal* 7(Suppl. 2):220–34.

Gressley TF, Hall MB, Armentano LE. 2011. Ruminant nutrition symposium: productivity, digestion, and health responses to hindgut acidosis in ruminants. *J. Anim. Sci.* 89(4):1120–30.

Guan LL, Nikrumah JD, Basarab JA, Moore SS. 2008. Linkage of microbial ecology to phenotype: correlation of rumen microbial ecology to cattle's feed efficiency. *FEMS Microbiol. Lett.* 288(1):85– 91.

Guzman CE, Bereza-Malcolm LT, De Groef B, Franks AE. 2015. Presence of selected methanogens, fibrolytic bacteria, and proteobacteria in the gastrointestinal tract of neonatal dairy calves from birth to 72 hours. *PLOS ONE* 10(7):e0133048.

Hooper LV, Littman DR, Macpherson AJ. 2012. Interactions between the microbiota and the immune system. *Science* 336(6086):1268–73.

Hoover WH. 1978. Digestion and absorption in the hindgut of ruminants. *J. Anim. Sci.* 46(6):1789–99.

Hristov AN, Callaway TR, Lee C, Dowd SE. 2012. Rumen bacterial, archaeal, and fungal diversity of dairy cows in response to ingestion of lauric or myristic acid. *J. Anim. Sci.* 90(12):4449–57.

Hungate RE. 1967. Hydrogen as an intermediate in the rumen fermentation. *Arch. Mikrobiol.* 59(1– 3):158–64.

Hungate RE. 1969. A roll tube method for cultivation of strict anaerobes. *Methods Microbiol.* 3(B):117– 32.

Huws SA, Creevey CJ, Oyama LB, Mizrahi I, Denman SE, et al. 2018. Addressing global ruminant agricultural challenges through understanding the rumen microbiome: past, present, and future. *Front. Microbiol.* 9:2161.

Intergov. Panel Clim. Change. 2014. *Climate Change 2014: Synthesis Report. Contribution of Working Groups I, II and III to the Fifth Assessment Report of the Intergovernmental Panel on Climate Change*, ed. RK Pachauri & LA Meyer. Geneva*: Intergov. Panel Clim. Change*. 151 pp.

Jami E, White BA, Mizrahi I. 2014. Potential role of the bovine rumen microbiome in modulating milk composition and feed efficiency. *PLOS ONE* 9(1):e85423.

Janssen PH, Kirs M. 2008. Structure of the archaeal community of the rumen. *Appl. Environ. Microbiol.* 74(12):3619–25.

Johnson KA, Johnson DE. 1995. Methane emissions from cattle. *J. Anim. Sci.* 73(8):2483–92.

Kittelmann S, Pinares-Patiño CS, Seedorf H, Kirk MR, Ganesh S, et al. 2014. Two different bacterial community types are linked with the low-methane emission trait in sheep. *PLOS ONE* 9(7):e103171.

Krause DO, Nagaraja TG, Wright AD, Callaway TR. 2013. Board-invited review: rumen microbiology: leading the way in microbial ecology. *J. Anim. Sci.* 91(1):331–41.

Li F, Guan LL. 2017. Metatranscriptomic profiling reveals linkages between the active rumen micro-biome and feed efficiency in beef cattle. *Appl. Environ. Microbiol.* 83(9):e00061-17.

Li F, Neves ALA, Ghoshal B, Guan LL. 2018. Symposium review: mining metagenomic and metatranscriptomic data for clues about microbial metabolic functions in ruminants. *J. Dairy Sci.* 101(6):5605–18.

Liang G, Malmathuge N, Bao H, Stothard P, Griebel PJ, Guan LL. 2016. Transcriptome analysis reveals regional and temporal differences in mucosal immune system development in the small intestine of neonatal calves. *BMC Genom.* 17(1):602.

Liang G, Malmuthuge N, Guan LL, Griebel P. 2015. Model systems to analyze the role of miRNAs and commensal microflora in bovine mucosal immune system development. *Mol. Immunol.* 66(1):57–67.

Lynch J, Pierrehumbert R. 2019. Climate impacts of cultured meat and beef cattle. *Front. Sustain. Food Syst.* 3(5).

Mackie RI. 2002. Mutualistic fermentative digestion in the gastrointestinal tract: diversity and evolution. *Integr. Comp. Biol.* 42(2):319–26.

Malmuthuge N, Guan LL. 2017. Understanding host-microbial interactions in rumen: searching the best opportunity for microbiota manipulation. *J. Anim. Sci.* Biotechnol. 8:8.

Malmuthuge N, Li M, Goonewardene LA, Oba M, Guan LL. 2013. Effect of calf starter feeding on gut microbial diversity and expression of genes involved in host immune responses and tight junctions in dairy calves during weaning transition. *J. Dairy Sci.* 96(5):3189–200.

Malmuthuge N, Liang G, Griebel PJ, Guan LL. 2019. Taxonomic and functional compositions of the small intestinal microbiome in neonatal calves provide a framework for understanding early life gut health. *Appl. Environ. Microbiol.* 85(6):e02534-18.

Mao S, Zhang M, Liu J, Zhu W. 2015. Characterising the bacterial microbiota across the gastrointestinal tracts of dairy cattle: membership and potential function. *Sci. Rep.* 5:16116.

Millen DD, De Beni Arrigoni M, Pacheco RDL, eds. 2016. *Rumenology*. Cham, Switz.: Springer Int.

Mulder IE, Schmidt B, Lewis M, Delday M, Stokes CR, et al. 2011. Restricting microbial exposure in early life negates the immune benefits associated with gut colonization in environments of high microbial diversity. *PLOS ONE* 6(12):e28279.

Myer PR, Freetly HC, Wells JE, Smith TPL, Kuehn LA. 2017. Analysis of the gut bacterial communities in beef cattle and their association with feed intake, growth, and efficiency. *J. Anim. Sci.* 95(7):3215–24.

Myer PR, Wells JE, Smith TPL, Kuehn LA, Freetly HC. 2016. Microbial community profiles of the jejunum from steers differing in feed efficiency. *J. Anim. Sci.* 94(1):327–38.

O'Hara E, Kelly A, McCabe MS, Kenny DA, Guan LL, Waters SM. 2018. Effect of a butyrate-fortified milk replacer on gastrointestinal microbiota and products of fermentation in artificially reared dairy calves at weaning. *Sci. Rep.* 8(1):14901.

Plaizier JC, Krauze DO, Gozho GN, McBride BW. 2008. Subacute ruminal acidosis in dairy cows: the physiological causes, incidence and consequences. *Vet. J.* 176(1):21–31.

Richardson EC, Herd RM. 2004. Biological basis for variation in residual feed intake in beef cattle. 2. Synthesis of results following divergent selection. *Aust. J. Exp. Agric.* 44(5):431–40.

Roehe R, Dewhurst RJ, Duthie C-A, Rooke JA, McKain N, et al. 2016. Bovine host genetic variation influences rumen microbial methane production with best selection criterion for low methane emitting and efficiently feed converting hosts based on metagenomic gene abundance. *PLOS Genet.* 12(2):e1005846.

Shabat SK, Sasson G, Doron-Faigenboim A, Durman T, Yaacoby S, et al. 2016. Specific microbiome-dependent mechanisms underlie the energy harvest efficiency of ruminants. *ISME J.* 10:2958–72.

Sherman EL, Nikrumah JD, Murdoch BM, Moore SS. 2008. Identification of polymorphisms influenc-ing feed intake and efficiency in beef cattle. *Anim. Genet.* 39(3):225–31.

Silberberg M, Chaucheyras-Durand F, Commun L, Mialon MM, Monteils V, et al. 2013. Repeated aci-dosis challenges and live yeast supplementation shape rumen microbiota and fermentations and modu-late inflammatory status in sheep. *Animal* 7(12):1910–20.

Sobrinho TL, Branco RH, Bonilha SFM, de Castilhos AM, de Figueiredo LA, et al. 2011. Residual feed intake and relationships with performance of Nellore cattle selected for post weaning weight. *Rev. Bras. Zootec.* 40:929–37.

Tapio I, Snelling TJ, Strozzi F, Wallace RJ. 2017. The ruminal microbiome associated with methane emissions from ruminant livestock. *J. Anim. Sci. Biotechnol.* 8:7.

Wilkinson TJ, Huws SA, Edwards JE, Kingston-Smith AH, Siu-Ting K, et al. 2018. CowPI: a rumen microbiome focused version of the PICRUSt functional inference software. *Front. Microbiol.* 9:1095.

Wolin MJ. 1960. A theoretical rumen fermentation balance. *J. Dairy Sci.* 43(10):1452–59.

Zhou M, Chen Y, Griebel PJ, Guan LL. 2014. Methanogen prevalence throughout the gastrointestinal tract of pre-weaned dairy calves. *Gut Microbes* 5(5):628–38.

Zhou M, Peng YJ, Chen Y, Klinger CM, Oba M, et al. 2018. Assessment of microbiome changes after rumen transfaunation: implications on improving feed efficiency in beef cattle. *Microbiome* 6:62.

Chapter 5

Microbiomes in Gut Functionality

Felix Uchenna Samuel[1,*]
Henry Olanrewaju Jegede[2]
John Fred Odhiambo[3]
and Ama Adadzewa Esun[3]

[1]Animal Science Program: Alabama Cooperative Extension, Alabama A&M University, USA
[2]Tilad Environmental Consultant, Ritadh, Saudi Arabia
[3]Department of Food and Animal Science, College of Agricultural, Live and Natural sciuence, Alabama A&M University, USA

Abstract

The gut microbiome plays a pivotal role in the overall health and disease states of animals, influencing a wide range of physiological functions. This complex microbial ecosystem, composed of bacteria, archaea, viruses, and fungi, begins to colonize the gut shortly after birth, shaped by factors such as genetics, diet, and environmental exposures. A diverse and balanced gut microbiome is essential for optimal digestive processes, nutrient absorption, and synthesis of essential vitamins and amino acids. Moreover, it is integral to the development and regulation of the host's immune system, offering protection against pathogenic invasions through competitive exclusion and the production of antimicrobial substances. Dysbiosis, or the imbalance of gut microbial communities, has been implicated in various animal diseases, including gastrointestinal disorders like inflammatory bowel disease (IBD), metabolic conditions such as obesity and diabetes, and even behavioral disorders through the gut-brain axis. The mechanisms by which the microbiome influences disease processes include modulation of immune responses, production of inflammatory mediators, and alteration of gut barrier integrity. Recent advances in metagenomics and high-throughput sequencing technologies have provided deeper insights into the composition and functional capabilities of the gut microbiome. These tools have facilitated the identification of specific microbial taxa and metabolic pathways associated with health and disease states, paving the way for potential therapeutic interventions. Probiotics, prebiotics, and dietary modifications are being explored to restore and maintain a healthy gut microbiome, offering promising avenues for disease prevention and management in animal populations.

---

* Corresponding Author's Email: felixsam75@yahoo.com

In: Microbiomes in Animal Nutrition, Health, and Disease
Editor: Tanmoy Rana
ISBN: 979-8-89530-425-9

Keywords: gut microbiome, animal health, dysbiosis, immune system, probiotics

Introduction

The gut microbiome, the vast and diverse community of microorganisms residing within the gastrointestinal tract, plays a crucial role in shaping animal gut functionality (Montalto et al. 2009). These symbiotic microbes have co-evolved with their hosts, forming intricate relationships that profoundly impact a wide range of physiological processes, from nutrient acquisition to immune modulation (Chow et al. 2010). The gut microbiome, a complex and dynamic community of microorganisms, plays an essential role in the health and disease of animals. This microbial ecosystem, composed of bacteria, archaea, viruses, and fungi, begins to establish itself shortly after birth, influenced by factors such as genetics, diet, and environmental exposures. The gut microbiome is integral to numerous physiological processes, including digestion, nutrient absorption, vitamin synthesis, and the development and regulation of the immune system. The symbiotic relationship between the host and its gut microbiota is crucial for maintaining homeostasis and overall well-being (Maritan et al., 2024). A healthy gut microbiome is characterized by a diverse and balanced microbial community, which contributes to the resilience and stability of the gut ecosystem. This diversity is essential for the efficient breakdown of complex dietary components, the production of short-chain fatty acids, and the synthesis of essential nutrients such as vitamins B and K. Moreover, the gut microbiome plays a protective role against pathogenic microorganisms through competitive exclusion, production of antimicrobial compounds, and modulation of the host immune responses. However, an imbalance in the gut microbial community, known as dysbiosis, can have detrimental effects on animal health (Ma et al., 2024). Dysbiosis has been linked to a wide range of diseases, including gastrointestinal disorders like inflammatory bowel disease (IBD), metabolic conditions such as obesity and diabetes, and even neurobehavioral disorders through the gut-brain axis. The mechanisms by which the gut microbiome influences these disease processes are complex and multifaceted, involving alterations in immune function, inflammatory responses, and gut barrier integrity. Advances in metagenomics and high-throughput sequencing technologies have revolutionized our understanding of the gut microbiome's composition and functional capabilities. These technologies allow for comprehensive analysis of microbial communities and their metabolic pathways, providing insights into the specific microbial taxa and functions associated with health and disease. Such knowledge is pivotal for the development of targeted therapeutic interventions, including the use of probiotics, prebiotics, and dietary modifications, to restore and maintain a healthy gut microbiome. The exploration of the gut microbiome's role in animal health and disease is a rapidly evolving field with significant implications for veterinary medicine and animal husbandry. Understanding the complex interplay between the gut microbiome and animal gut function is essential for gaining insights into both animal health and the mechanisms underlying various gastrointestinal disorders.

This chapter will delve into the multifaceted ways in which the gut microbiome influences gut functionality, highlighting the latest research in this rapidly evolving field. We will explore how the gut microbiome is established, the key metabolic functions it performs, its role in maintaining gut barrier integrity, the consequences of microbiome disruptions, and how these disruptions can be modulated to favor the animal's overall health and well-being. The gut microbiome's involvement in regulating nutrient absorption, immune system development, and even behavior will be discussed, underscoring its central importance in animal physiology and health (Figure 1).

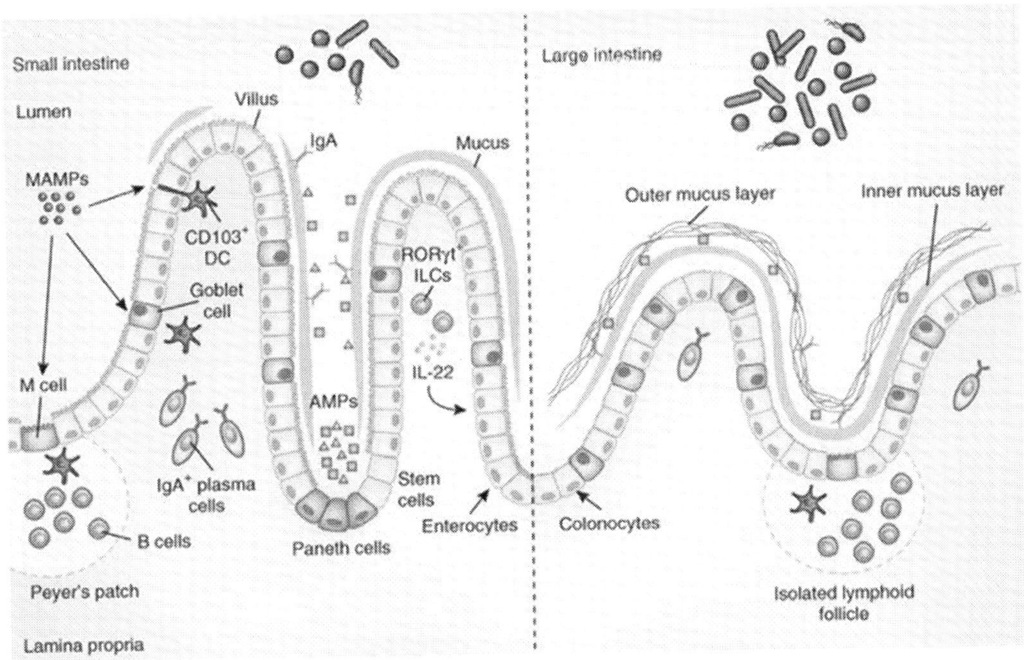

Figure 1. Diagram of the intestinal epithelial barrier, showcasing the various components and their interactions. Antigen sampling across the intestinal barrier is mediated by microfold (M) cells and dendritic cells (DC) extending their dendrites into the lumen. IgA, immunoglobulin A; ILC, innate lymphoid cell; IL-22, interleukin-22; MAMP, microbe-associated molecular pattern. Adapted from Van den Bossche, 2017.

Establishing the Gut Ecosystem

Development of the Gut Microbiome

The development of the gut microbiome is a critical factor in animal nutrition, health, and disease. This complex microbial community, which includes bacteria, archaea, viruses, and fungi, begins to establish itself immediately after birth and is influenced by various factors, such as the mode of delivery, early-life nutrition, and environmental exposures. The initial colonization of the gut microbiome is significantly affected by the mode of birth. Vaginally born animals typically receive their first microbial inoculum from their mother's vaginal and intestinal microbiota, whereas cesarean-born animals are colonized primarily by skin and environmental microbes (Maritan et al., 2024). Early nutrition, particularly whether the animal is breastfed or formula-fed, also plays a vital role in shaping the gut microbiota. Breast milk contains not only nutrients but also prebiotics and immunoglobulins that promote the growth of beneficial bacteria, such as Bifidobacterium and Lactobacillus species. The gut microbiome is integral to the digestive process and nutrient absorption. Microbes in the gut help break down complex carbohydrates, proteins, and lipids into simpler molecules that the host can absorb (Alshammary and Rusul, 2024). They also synthesize essential vitamins, such as B vitamins and vitamin K, and produce short-chain fatty acids (SCFAs) like acetate, propionate, and butyrate, which serve as energy sources for the host and play a role in regulating gut health and immune function. A well-balanced gut microbiome enhances feed efficiency and growth rates in livestock, thereby improving overall productivity in animal husbandry. The gut microbiome is crucial for the development and function of the host's immune system. It helps in the maturation of immune cells, production of antimicrobial peptides, and maintenance of gut barrier integrity. Microbial colonization during early life is essential for educating the immune system to distinguish between harmless commensals and pathogenic organisms. Dysbiosis, or an imbalance in the gut microbiota, can disrupt this process, leading to inappropriate immune responses and increased susceptibility to infections and inflammatory diseases (Di et al., 2024). Dysbiosis has been linked to various health issues in animals, including gastrointestinal disorders like inflammatory bowel disease (IBD), metabolic conditions such as obesity and diabetes, and even behavioral changes through the gut-brain axis. The mechanisms by which the microbiome influences these diseases include modulation of immune responses, alteration of gut permeability, and production of inflammatory mediators. For instance, a decrease in beneficial SCFA-producing bacteria can compromise gut barrier function, allowing pathogens and toxins to enter the bloodstream and trigger systemic inflammation. Recent advances in metagenomics and high-throughput sequencing technologies have revolutionized our understanding of the gut microbiome. These technologies allow for comprehensive profiling of microbial communities and their functional capabilities. Studies have identified specific microbial taxa and metabolic pathways associated with health and disease states, providing insights into the microbial contributions to host physiology. Probiotics, prebiotics, and dietary modifications are promising strategies for modulating the gut microbiome to enhance animal health. Probiotics are live beneficial microbes that, when administered in adequate amounts, confer health benefits to the host. Prebiotics are non-digestible food ingredients that selectively stimulate the growth and activity of beneficial bacteria. Together, these interventions can help restore microbial balance and improve gut health.

Factors Shaping Gut Microbiome Composition

Host Genetics

Host genetics significantly influence the composition of the gut microbiome by affecting various factors. Genetic variations can impact the availability of specific nutrients and energy sources that serve as substrates for microbial growth and metabolism (Dąbrowska and Witkiewicz 2016). Additionally, host genetics are involved in the production of antimicrobial compounds, such as defensins and immunoglobulins, which can selectively inhibit or promote the growth of certain bacterial taxa (Belizário, Faintuch, and Garay-Malpartida 2018). Furthermore, genetic factors contribute to the development and maturation of the gut-associated immune system, which regulates the gut microbiome's composition and diversity through immune-mediated interactions. These genetically determined factors can lead to significant variations in the abundance and relative proportions of different bacterial groups within the gut, ultimately influencing the gut microbiome's overall functionality and metabolic capabilities. Diet is another major determinant of gut microbiome composition. In ruminants, the microbiome plays a crucial role in digesting complex plant materials, such as cellulose, which the host cannot break down independently (Huws et al. 2018). The ruminant gastrointestinal tract contains a specialized forestomach called the rumen, providing an environment conducive to microbial fermentation of these fibrous plant materials. The diverse community of microbes in the rumen, including bacteria, archaea, protozoa, and fungi, produce enzymes that effectively break down cellulose and other complex carbohydrates, allowing the ruminant host to extract nutrients and energy from these otherwise indigestible components of their diet. In contrast, the microbiome of carnivores is better equipped to metabolize animal-derived proteins and fats, reflecting the distinct dietary needs and adaptations of these animal species (Dearing and Kohl 2017). The gut microbiome is highly responsive to dietary changes, with shifts in macronutrient intake, fiber content, and the presence of antinutritional factors all contributing to alterations in microbial community structure (Gentile and Weir 2018). These dietary influences on the gut microbiome composition are crucial for maintaining optimal gut functionality and host health across different animal species (Table 1).

Table 1. Dietary Components and Their Effects on the Gut Microbiome

Dietary Component	Examples	Impact on Gut Microbiome
Prebiotics	Inulin, Fructooligosaccharides	Promote the growth of beneficial bacteria
Probiotics	Lactobacillus, Bifidobacterium species	Introduce live beneficial bacteria
Fermented Foods	Yogurt, Sauerkraut, Kimchi	Sources of probiotics and prebiotics
Polyphenols	Berries, Tea, Cocoa	Promote the growth of specific beneficial bacteria

Environmental Factors

Environmental factors, such as stress, antibiotic exposure, and the presence of environmental toxins, can also significantly impact the gut microbiome (Youngblut et al. 2019). Stress, for example, can lead to changes in gut motility, secretion of antimicrobial compounds, and immune function, all of which can disrupt the delicate balance of the gut microbiome. Antibiotics can selectively target and kill certain bacterial taxa, leading to a loss of diversity and the overgrowth of opportunistic pathogens. Additionally, exposure to environmental toxins, such as heavy metals or pesticides, can have detrimental effects on the gut microbiome by altering microbial community composition and metabolic activities. These environmental stressors can impair the normal function of the gut microbiome, contributing to the development of various gastrointestinal disorders and other health issues in the host (Lobionda et al. 2019).

Metabolic Functions and Nutrient Acquisition

The gut microbiome plays a central role in facilitating the metabolism and acquisition of essential nutrients critical for host health. Nutrient acquisition and absorption are fundamental functions of the gut, and across various animal species, the microbiome's involvement in digestion, energy extraction, and nutrient synthesis is well-documented. How the gut microbiome supports these metabolic and nutrient acquisition functions varies depending on the specific nutrient or metabolic process involved. For example, the microbiome contributes uniquely to the fermentation of carbohydrates, the breakdown of dietary fiber, and the synthesis of vitamins and amino acids. These processes are tailored to the distinct digestive physiology of each host animal. In ruminants, the gut microbiome is indispensable for the digestion of complex plant materials, leveraging specialized microbial communities in the rumen to ferment fibrous plant material efficiently. In contrast, in monogastric animals, the microbiome's contributions to nutrient acquisition may emphasize the enzymatic breakdown of dietary components and the provision of essential nutrients that the host cannot synthesize independently. The specific mechanisms and microbial communities engaged in these nutrient-related functions can vary significantly across different animal species and gut environments, reflecting adaptations to diet and evolutionary history. Overall, the gut microbiome's intricate interactions with host physiology highlight its crucial role in sustaining optimal nutrient acquisition and metabolic processes across diverse animal species.

Carbohydrate Fermentation and SCFA Production

A key function of the gut microbiome is the fermentation of complex carbohydrates that the host cannot break down on its own especially in ruminants. This process results in the production of short-chain fatty acids (SCFAs), such as acetate, propionate, and butyrate, which serve as important energy sources for the host (Busnelli et al. 2018). These SCFAs have been shown to have numerous beneficial effects, including promoting the growth and differentiation of intestinal epithelial cells, regulating inflammatory pathways, and improving glucose and lipid metabolism. This process in monogastrics (like dogs and swine) unlike in ruminants is rather focused on active enzymatic breakdown of dietary carbohydrates using amylase although microbiome still plays a significant role in assisting the host in extracting maximum nutritional value from the diet (Flint 2004).

Digestion of Dietary Fibre

The gut microbiome plays a crucial role in the digestion of dietary fiber, which contains substances that serve as precursors to energy and other essential nutrients. However, these fibers can only be effectively fermented and broken down with the help of the gut microbiome. In ruminants, the gut microbiome is particularly essential for the digestion of cellulose and other plant polysaccharides, allowing the host to extract energy and nutrients from these otherwise indigestible materials.(Patel 2018). While fiber digestion in monogastric animals, such as humans and other mammals, is not as efficient as in ruminants, it still plays an important role in their gut functionality. In monogastrics, the gut microbiome assists the host in breaking down and extracting nutrients from dietary fiber, although the process is generally less extensive compared to the fermentation that occurs in the ruminant gut. The breakdown of fiber in monogastrics is still present and contributes to the host's overall nutrient acquisition and energy harvesting, albeit to a lesser degree than in ruminant species. One example is the colobine monkey, which has been observed to obtain energy from fiber during the winter, although the precise mechanisms enabling this remain largely unknown. It has been suggested that these species possess specialized adaptations to enhance fiber digestion in the hindgut. These colobines were found to have a significantly greater microbiome volume compared to other foregut-fermenting primates. While the microbial composition and abundance differed notably between the foregut and hindgut regions, both gut areas were dominated by taxa producing enzymes that facilitate the active digestion of complex carbohydrates. The data suggests that both the foregut and hindgut of these colobines contribute to fiber digestion, and an enlarged colon may be an evolutionary adaptation to this specialized function (Liu et al. 2022) (Liu et al. 2022).

Vitamin and Amino Acid Synthesis

In addition to carbohydrate fermentation and fiber digestion, the gut microbiome can synthesize a variety of vitamins, including vitamins K, B12, and folate, which are critical for various physiological processes in the host (Otaru et al. 2021) (Quigley 2013) (Montalto et al. 2009). The gut microbiome can also produce all essential amino acids, contributing to the host's amino acid pool and overall nitrogen balance. Examples of such amino acids include phenylalanine, tyrosine, and tryptophan, which serve as precursors for neurotransmitters and other bioactive molecules (Montalto et al. 2009) (Otaru et al. 2021) (Quigley 2013). Examples of major microbiomes responsible for this include Bacteroides, Bifidobacterium, and Clostridium species which are involved in amino acid and vitamin biosynthesis (Quigley 2013)(Montalto et al. 2009) (Bull and Plummer 2014). In monogastrics like humans and other mammals, these microbes are essential for providing essential nutrients that the host cannot synthesize on their own. Pathways for monogastrics differ from ruminants, but both are integral to host physiology and gut function.

In summary, the gut microbiome plays a critical role in supporting the digestive and metabolic functions of its animal host through various mechanisms, including carbohydrate fermentation, fiber digestion, and the synthesis of essential vitamins and amino acids. These microbial processes are often tailored to the unique digestive physiology of the host species, with ruminants relying more heavily on microbial fermentation of complex plant materials, while monogastric animals utilize their microbiomes in more targeted ways to supplement their host's nutrient acquisition.(Karasov and Douglas 2013) (Quigley 2013).

The specific mechanisms and microbial communities involved can vary substantially across different animal species and gut environments, highlighting the importance of understanding the interplay between the host and its resident microbiome (Quigley 2013) (Otaru et al. 2021) (Karasov and Douglas 2013). Continued research in this field promises to yield valuable insights into the role of the gut microbiome in maintaining animal health and productivity.

The Gut Microbiome and Gut Barrier Function

The gut microbiome also plays a crucial role in maintaining the integrity and function of the intestinal barrier, which acts as the primary interface between the host and the external environment. The gut epithelium, along with the mucus layer and tight junctions between epithelial cells, forms a selectively permeable barrier that regulates the absorption of nutrients, electrolytes, and water, while preventing the entry of harmful pathogens, toxins, and antigens (Parker, Fonseca, and Carding 2019).

The gut microbiome contributes to barrier function in several ways: (Parker, Fonseca, and Carding 2019)

- *Production of antimicrobial compounds:* Certain commensal bacteria, such as Lactobacillus and Bifidobacterium species, can produce antimicrobial peptides and bacteriocins that inhibit the growth of potentially pathogenic microbes, thereby preventing their translocation across the intestinal epithelium (Parker, Fonseca, and Carding 2019). Pathogenic microbes have been known to disrupt the integrity of the gut barrier by, and the production of antimicrobials by commensal bacteria helps maintain a healthy barrier (Montalto et al. 2009) (Bull and Plummer 2014). Pathogenic microbes have been known to disrupt the integrity of the gut barrier by damaging epithelial cells, disrupting tight junctions, and adversely altering the mucus layer (Awad et al. 2017). The production of antimicrobial compounds by commensal bacteria, such as Lactobacillus and Bifidobacterium species, helps counteract these disruptive effects and maintain a healthy intestinal barrier.
- *Modulation of host immune responses:* The gut microbiome can influence the development and regulation of the host's innate and adaptive immune responses, which are critical for maintaining intestinal homeostasis and barrier integrity. For example, certain commensal bacteria like can enhance the production of antimicrobial proteins, mucin, and secretory IgA, all of which contribute to gut barrier function (Pope, Tomkovich, and Jobin 2020). Additionally, the gut microbiome can induce regulatory T cells and anti-inflammatory cytokines, which help prevent excessive inflammation that can compromise barrier integrity (Montalto et al. 2009). .
- *Maintenance of the Mucus Layer:* Beyond the epithelial cell layer lies a protective mucus layer, primarily composed of mucins secreted by goblet cells. This viscous layer acts as a physical barrier, trapping microbes and preventing their direct contact with the epithelium. The gut microbiome influences both the production and composition of this mucus layer (Hansson 2019). For instance, certain bacteria can stimulate goblet cell differentiation and mucin production, thereby enhancing the barrier function (Yang and Yu 2021). Furthermore, the gut microbiome plays a crucial role in stimulating the host's production of antimicrobial peptides (AMPs). AMPs are small proteins with broad-spectrum antimicrobial activity, effectively targeting and killing bacteria, fungi, and even viruses. The presence of commensal bacteria in the gut triggers the expression of genes responsible for AMP production by the host's epithelial cells (Kazemzadeh-Narbat et al. 2020). This continuous stimulation ensures a basal level of AMP production, contributing to the constant defense against pathogens. The mucus layer covering the intestinal epithelium is a key component of the gut barrier, providing a physical and biochemical barrier against pathogen invasion. Additionally, certain commensal bacteria may degrade and remodel the mucus layer, maintaining its optimal thickness and permeability to support nutrient absorption while preventing pathogen translocation.
- *Gut-Brain Axis Communication:* The gut microbiome's influence on gut barrier function extends beyond the local intestinal environment, as it can also impact the brain-gut axis. Gut microbes can produce neuroactive compounds, such as neurotransmitters and metabolites, that can signal to the central nervous system and modulate brain function (Kim and Ho 2010). This gut-brain axis communication is critical for maintaining intestinal homeostasis, as the brain can send signals back to the

gut that regulate epithelial barrier integrity, motility, secretion, and immune responses (Parodi and Rosbo 2021).

Overall, the gut microbiome plays a multifaceted role in supporting the integrity and function of the intestinal barrier, which is essential for maintaining gut homeostasis and preventing the entry of harmful substances or pathogens (Faderl et al. 2015) (Kim and Ho 2010).

Disruptions to Gut Functionality

Dysbiosis

Disruptions to the gut microbiome, known as dysbiosis, have been associated with various gastrointestinal and systemic diseases characterized by impaired gut barrier function, such as inflammatory bowel disease, irritable bowel syndrome, obesity, and metabolic disorders (Kelly et al. 2017) (Parker, Fonseca, and Carding 2019). The loss of microbial diversity, the overgrowth of opportunistic pathogens, and the imbalance of commensal communities can all contribute to the breakdown of the intestinal barrier (Parodi and Rosbo 2021). Dysbiosis can lead to increased intestinal permeability, allowing the passage of harmful molecules, antigens, and pathogens across the epithelium, triggering local and systemic inflammatory responses (Kim and Ho 2010) (Vancamelbeke and Vermeire 2017) (Table 2).

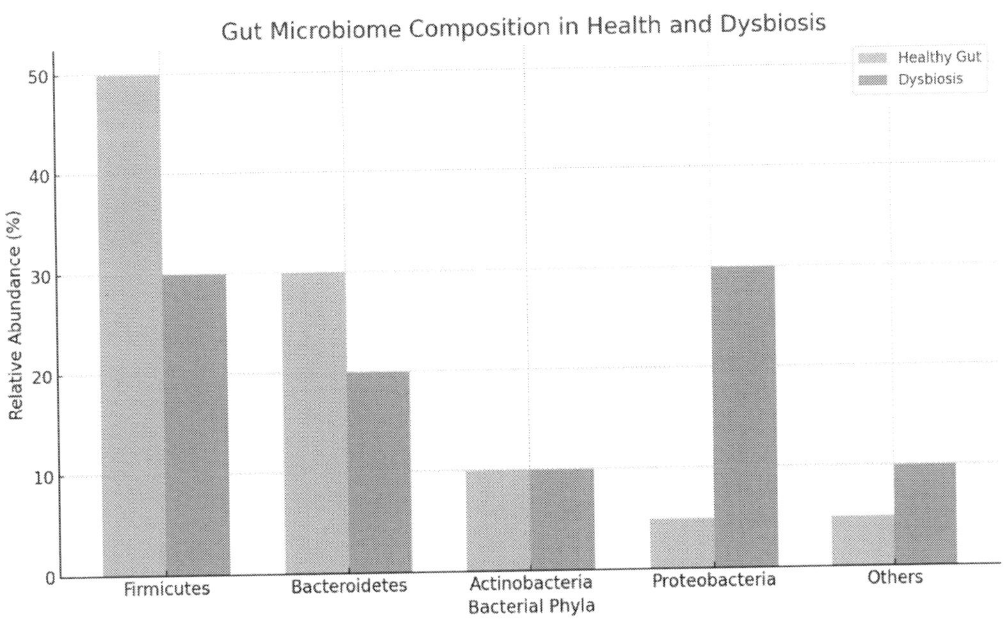

Figure 2. Bar graph showing Gut Microbiome Composition in Health and Dysbiosis.

Table 2. Factors contributing to dysbiosis in animals

Category	Factor	Description	Examples
Behavioral Factors	Diet	The type and quality of food consumed can significantly impact gut microbiota composition and function.	High-fat diet, low-fiber diet, processed foods, specific nutrient deficiencies, high-sugar diet
Behavioral Factors	Stress	Chronic stress can alter gut microbiota through the gut-brain axis, impacting immune response and gut permeability.	Psychological stress, social stress, environmental stress
Behavioral Factors	Exercise	Physical activity influences gut microbiota diversity and metabolism.	Sedentary lifestyle, moderate exercise, excessive exercise
Behavioral Factors	Sleep	Sleep patterns affect circadian rhythms and gut microbiota balance.	Sleep deprivation, irregular sleep patterns, sleep quality
Medical Interventions	Antibiotic Use	Antibiotics can disrupt the gut microbiome by killing both pathogenic and beneficial bacteria.	Broad-spectrum antibiotics, repeated antibiotic courses
Medical Interventions	Other Medications	Non-antibiotic medications can also affect the gut microbiome, altering its composition and function.	Nonsteroidal anti-inflammatory drugs (NSAIDs), proton pump inhibitors (PPIs), immunosuppressants
Environmental Factors	Exposure to Toxins	Environmental toxins can impact gut microbiota and gut health.	Pesticides, heavy metals, pollutants
Environmental Factors	Hygiene Hypothesis	Reduced microbial exposure in early life due to high hygiene standards can lead to a less diverse microbiome.	Overuse of disinfectants, lack of exposure to natural environments
Host Genetics and Health Status	Genetics	Genetic makeup influences the composition and function of the gut microbiome.	Genetic predispositions, hereditary conditions affecting the gut
Host Genetics and Health Status	Health Status	The overall health and existing medical conditions of the host can affect gut microbiota.	Chronic diseases (e.g., diabetes, IBD), immune status, nutritional status

Several factors can contribute to dysbiosis, including:

- Dietary changes, such as a high-fat, high-sugar diet, which can alter the composition and metabolic activity of the gut microbiome (Yoo et al. 2020)
- Antibiotic use, which can disrupt the delicate balance of the gut microbial community and select for the growth of resistant strains
- Stress, which can modulate the gut-brain axis and lead to changes in the gut microbiome composition and function (Sudo 2019)
- Aging, as the gut microbiome tends to become less diverse and less resilient with advanced age, is another factor that can contribute to dysbiosis (Seidel and Valenzano 2018). The aging process is associated with gradual changes in the composition and diversity of the gut microbial community, often leading to a decrease in beneficial commensal bacteria and an increase in opportunistic pathogens. This age-related dysbiosis can impair the gut barrier function and increase susceptibility to various age-related diseases.

These disruptions to the gut microbiome can have far-reaching consequences, as they are linked to the development of various neurological, metabolic, and inflammatory disorders (Shabbir et al. 2021) (Parodi and Rosbo 2021) (Yoo et al. 2020) (Figure 2).

Consequences of Dysbiosis

Dysbiosis can lead to various gastrointestinal disorders, such as inflammatory bowel disease, irritable bowel syndrome, and infectious diarrhea by disrupting the delicate balance of the gut microbiome, increasing intestinal permeability, and enabling the passage of harmful substances and pathogens across the epithelial barrier (Arrieta 2006). This can trigger inflammatory responses that further compromise gut functionality and contribute to the development of these conditions.

Potential systemic effects of dysbiosis, such as metabolic disorders, cardiovascular diseases, and neurological conditions. Since the gut microbiome is involved in the regulation of various physiological processes, including energy metabolism, immune function, and neurotransmitter production, its disruption can have far-reaching consequences beyond the gastrointestinal tract (Birt et al. 2013) (Ullah et al. 2023).

The role of the gut microbiome in the development of obesity and associated metabolic disorders has been particularly well-studied. Dysbiosis has been linked to increased energy harvesting from the diet, altered regulation of appetite and satiety, and low-grade inflammation - all of which can contribute to the development of obesity and related metabolic disorders like type 2 diabetes (Agustí et al. 2018)(Kim et al. 2022).

Furthermore, dysbiosis has been associated with neurological conditions, such as depression, anxiety, and neurodegenerative diseases. The gut-brain axis, a bidirectional communication system between the gastrointestinal tract and the central nervous system, is a key mechanism by which the gut microbiome can influence brain function and behavior (Choi, Hur, and Hong 2018) (Yoo et al. 2020).

Table 3. Therapeutic strategy and approach of microbiome

Therapeutic Strategy	Approach	Description	Potential Benefits	Limitations
Dietary Interventions	Prebiotics	Non-digestible food ingredients that promote the growth of beneficial bacteria in the gut.	Enhance gut health, improve digestion, boost immune function.	May not be effective for all individuals, but specific types and doses needed for effectiveness.
Dietary Interventions	Probiotics	Live beneficial bacteria that can colonize the gut and confer health benefits.	Improve gut health, enhance immune response, potentially treat or prevent certain diseases.	Survival and colonization of bacteria can be affected by various factors, not all strains are equally beneficial.
Dietary Interventions	Fermented Foods	Foods that have undergone fermentation and contain beneficial bacteria and/or prebiotics.	Provide a natural source of probiotics and prebiotics, improve gut health.	Variability in bacterial content, potential for contamination, not all fermented foods have the same benefits.
Fecal Microbiota Transplantation	Fecal Microbiota Transplantation	Transplanting fecal bacteria from a healthy donor to a recipient to restore gut microbiota balance.	Restore gut microbiota balance, potentially effective in treating Clostridium difficile infections and other conditions.	Potential for transfer of undesirable microbes, requires screening and matching of donors.
Pharmabiotics	Pharmabiotics	Drugs designed to target and modulate the gut microbiome to achieve therapeutic effects.	Precisely target and modulate gut microbiome, potentially treat a variety of conditions.	Still an emerging field, potential side effects, and long-term impacts not fully understood.
Postbiotics	Postbiotics	Beneficial metabolites produced by gut microbes that can have health-promoting effects.	Directly provide health benefits without the need for live bacteria, potential for targeted therapy.	Research is still in early stages, identification and production of beneficial metabolites can be challenging.

Modulating the Gut Microbiome for Improved Gut Functionality

Given the crucial role of the gut microbiome in maintaining intestinal barrier function and overall health, strategies to modulate the gut microbial community have become an area of intense research and clinical interest.

One common approach in animal production and care is the use of probiotics, which are live microorganisms that, when administered in adequate amounts, can provide health benefits to the host. Probiotics have been shown to improve gut barrier function, reduce inflammation, and enhance nutrient absorption by modulating the gut microbiome in a vast variety of animal species including.

- *Dietary interventions:* In animal nutrition, dietary adjustments, such as the inclusion of prebiotics, fibers, and other fermentable substrates, can selectively stimulate the growth and activity of beneficial gut microbes, thereby strengthening the intestinal barrier. Probiotics have been used widely in animal nutrition from pets to animal production and these products based on gut microbiome modulation are gaining traction as alternatives to conventional antimicrobial growth promoters (Groisman, Han, and Krypotou 2023).(Aoun, Darwish, and Hamod 2020)
- *Fecal microbiota transplantation (FMT):* This emerging therapeutic approach involves the transfer of fecal material from a healthy donor into the gastrointestinal tract of an individual with dysbiosis, to restore a healthy gut microbial community. While the use of fecal microbiota transplantation in animal health and production is still in its infancy, it has shown promising results in the treatment of recurrent *Clostridioides difficile* infections in companion animals and livestock.
- *Phage therapy:* The use of bacteriophages, viruses that specifically target and kill bacterial pathogens, is another avenue being explored to modulate the gut microbiome and improve gut functionality. Phage therapy has been studied as a potential alternative to antibiotics in animal production, with the potential to selectively target and eliminate harmful bacteria while preserving beneficial commensals (Table 3).

Conclusion

The gut microbiome plays a crucial role in maintaining intestinal barrier function and overall host health. In animal and veterinary science, the study of microbiomes has led to a better understanding of the complex interactions between gut commensals and the host, and has paved the way for the development of targeted interventions to modulate the gut microbial community for improved gut functionality and animal health.

Disruptions to the gut microbiome, known as dysbiosis, can have far-reaching consequences, including the development of gastrointestinal disorders, metabolic diseases, and neurological conditions.

To address the challenges posed by dysbiosis, various strategies have been explored to modulate the gut microbiome, such as the use of probiotics, dietary interventions, fecal microbiota transplantation, and phage therapy.

As our understanding of the gut microbiome and its intricate relationship with host physiology continues to evolve, these approaches hold great promise for improving gut functionality and overall animal health and well-being (Groisman, Han, and Krypotou 2023) (Zhang et al. 2023) (Tungland 2018).

In the future, a deeper understanding of the gut microbiome and its interactions with the host will likely lead to more targeted and personalized interventions to maintain a healthy gut ecosystem and promote animal health and productivity. Also can give a better understanding and insight in gut-brain axis and help develop more effective treatments for neurodegenerative diseases in animals (Parker 2016) (Pérez-Sánchez, Mora-Sánchez, and Balcázar 2018) (Groisman, Han, and Krypotou 2023) (Rabetafika et al. 2023).

References

Agustí, Ana, María Pilar García-Pardo, Inmaculada López-Almela, Isabel Campillo, Michaël Maes, Marina Romaní-Pérez, and Yolanda Sanz. 2018. "Interplay Between the Gut-Brain Axis, Obesity and Cognitive Function." *Frontiers Media* 12. https://doi.org/10.3389/fnins.2018.00155.

Alshammary, Assal Ghazi, and Rusul AA Alshammary. 2024. "The Role of Microbiota in Animal Physiology: An Integrative Review." *Al-Salam Journal for Medical Science* 3, no. 2: 53-61.

Aoun, A., Fatima Darwish, and Natacha Hamod. 2020. "The Influence of the Gut Microbiome on Obesity in Adults and the Role of Probiotics, Prebiotics, and Synbiotics for Weight Loss." *Korean Society of Food Science and Nutrition* 25 (2): 113-123. https://doi.org/10.3746/pnf.2020.25.2.113.

Arrieta, Marie-Claire. 2006. "Alterations in intestinal permeability." https://doi.org/10.1136/gut.2005.085373.

Belizário, Ernesto, José, Joel Faintuch, and Miguel Garay-Malpartida. 2018. "Gut Microbiome Dysbiosis and Immunometabolism: New Frontiers for Treatment of Metabolic Diseases." https://doi.org/10.1155/2018/2037838.

Birt, F., Diane, Terri D. Boylston, Suzanne Hendrich, Jay-lin Jane, James Hollis, Li Li, John F. McClelland, Samuel A. Moore, Gregory J. Phillips, Matthew J. Rowling, Kevin L. Schalinske, M. Paul Scott, and Elizabeth M. Whitley. 2013. "Resistant Starch: Promise for Improving Human Health." https://doi.org/10.3945/an.113.004325.

Bosco, Nabil, and Mario Noti. 2021. "The aging gut microbiome and its impact on host immunity." https://doi.org/10.1038/s41435-021-00126-8.

Bull, Matthew, and Nigel T Plummer. 2014. "Part 1: The Human Gut Microbiome in Health and Disease." *National Institutes of Health* 13 (6): 17-22.

Busnelli, M., Stefano Manzini, Cesare R. Sirtori, Giulia Chiesa, and Cinzia Parolini. 2018. "Effects of Vegetable Proteins on Hypercholesterolemia and Gut Microbiota Modulation." *Multidisciplinary Digital Publishing Institute* 10 (9): 1249-1249. https://doi.org/10.3390/nu10091249.

Choi, Jeong-Hyun, Tai-Young Hur, and Yonggeun Hong. 2018. "Influence of Altered Gut Microbiota Composition on Aging and Aging-Related Diseases." 8 (1): 1-7. https://doi.org/10.15280/jlm.2018.8.1.1.

Chow, Janet, S. Melanie Lee, Yue Shen, Arya Khosravi, and Sarkis K. Mazmanian. 2010. "Host–Bacterial Symbiosis in Health and Disease." https://doi.org/10.1016/b978-0-12-381300-8.00008-3.

Dąbrowska, Krystyna, and Wojciech Witkiewicz. 2016. "Correlations of Host Genetics and Gut Microbiome Composition." https://doi.org/10.3389/fmicb.2016.01357.

Dearing, Denise, M., and Kevin D. Kohl. 2017. "Beyond Fermentation: Other Important Services Provided to Endothermic Herbivores by their Gut Microbiota." https://doi.org/10.1093/icb/icx020.

Di Vincenzo, Federica, Angelo Del Gaudio, Valentina Petito, Loris Riccardo Lopetuso, and Franco Scaldaferri. 2024. "Gut microbiota, intestinal permeability, and systemic inflammation: A narrative review." *Internal and Emergency Medicine* 19, no. 2: 275-293.

Faderl, Martin, Mario Noti, Nadia Corazza, and Michael Mueller. 2015. "Keeping bugs in check: The mucus layer as a critical component in maintaining intestinal homeostasis." *Wiley-Blackwell* 67 (4): 275-285. https://doi.org/10.1002/iub.1374.

Flint, J., Harry. 2004. "Polysaccharide Breakdown by Anaerobic Microorganisms Inhabiting the Mammalian Gut." https://doi.org/10.1016/s0065-2164(04)56003-3.

Gentile, L., Christopher, and Tiffany L. Weir. 2018. "The gut microbiota at the intersection of diet and human health." https://doi.org/10.1126/science.aau5812.

Groisman, A., Eduardo, Weiwei Han, and Emilia Krypotou. 2023. "Advancing the fitness of gut commensal bacteria." *American Association for the Advancement of Science* 382 (6672): 766-768. https://doi.org/10.1126/science.adh9165.

Hansson, C., Gunnar. 2019. "Mucus and mucins in diseases of the intestinal and respiratory tracts." https://doi.org/10.1111/joim.12910.

Huws, Sharon, Christopher J. Creevey, Linda Oyama, Itzhak Mizrahi, Stuart E. Denman, Milka Popova, Rafael Muñoz-Tamayo, Evelyne Forano, Sinéad M. Waters, Matthias Hess, Ilma Tapio, Hauke Smidt, S. J. Krizsan, D. R. Yáñez-Ruíz, Alejandro Belanche, Leluo Guan, Robert J. Gruninger, Tim A. McAllister, Charles J. Newbold, R. Roehe, R. J. Dewhurst, Timothy J. Snelling, Mick Watson, Garret Suen, Elizabeth H. Hart, Alison H. Kingston-Smith, Nigel Scollan, Rodolpho Martin do Prado, Eduardo Jorge Pilau, Hilário Cuquetto Mantovani, Graeme T. Attwood, Joan E. Edwards, Neil R. McEwan, Steven Morrisson, Olga Mayorga, Christopher L. Elliott, and Diego Morgavi. 2018. "Addressing Global Ruminant Agricultural Challenges Through Understanding the Rumen Microbiome: Past, Present, and Future." https://doi.org/10.3389/fmicb.2018.02161.

Karasov, H., William, and Angela E. Douglas. 2013. "Comparative Digestive Physiology.": 741-783. https://doi.org/10.1002/cphy.c110054.

Kazemzadeh-Narbat, Mehdi, Hao Cheng, Rosa Chabok, Mario Moisés Álvarez, César de la Fuente-Núñez, K. Scott Phillips, and Ali Khademhosseini. 2020. "Strategies for antimicrobial peptide coatings on medical devices: A review and regulatory science perspective." https://doi.org/10.1080/07388551.2020.1828810.

Kelly, R., John, Chiara Minuto, John F. Cryan, Gerard Clarke, and Timothy G. Dinan. 2017. "Cross Talk: The Microbiota and Neurodevelopmental Disorders." *Frontiers Media* 11. https://doi.org/10.3389/fnins.2017.00490.

Kim, Donghyun, Ye Jin Jeong, Yerin Lee, Jihoon Choi, Young Min Park, Oh Chan Kwon, Yong Ji, Sung Jun Ahn, Hyung Keun Lee, Min-Chan Park, and Jae-Yol Lim. 2022. "Correlation Between Salivary Microbiome of Parotid Glands and Clinical Features in Primary Sjögren's Syndrome and Non-Sjögren's Sicca Subjects." *Frontiers Media* 13. https://doi.org/10.3389/fimmu.2022.874285.

Kim, S., Young, and Samuel B. Ho. 2010. "Intestinal Goblet Cells and Mucins in Health and Disease: Recent Insights and Progress." *Springer Science+Business Media* 12 (5): 319-330. https://doi.org/10.1007/s11894-010-0131-2.

Liu, Rui, Katherine R. Amato, Rong Hou, Andrés Gómez, Derek W. Dunn, Jun Zhang, Paul A. Garber, Colin A. Chapman, Nicoletta Righini, Gang He, Gu Fang, Yuhang Li, Baoguo Li, and Songtao Guo. 2022. "Specialized digestive adaptations within the hindgut of a colobine monkey." *Elsevier BV* 3 (2): 100207-100207. https://doi.org/10.1016/j.xinn.2022.100207.

Lobionda, Stefani, Panida Sittipo, Hyog Young Kwon, and Yun Kyung Lee. 2019. "The Role of Gut Microbiota in Intestinal Inflammation with Respect to Diet and Extrinsic Stressors." https://doi.org/10.3390/microorganisms7080271.

Ma, Zhanshan, and Peng Shi. 2024. "Critical complex network structures in animal gastrointestinal tract microbiomes." *Animal Microbiome* 6, no. 1: 23.

Maritan, Elisa, Andrea Quagliariello, Enric Frago, Tomaso Patarnello, and Maria Elena Martino. 2024. "The role of animal hosts in shaping gut microbiome variation." *Philosophical Transactions of the Royal Society B* 379, no. 1901: 20230071.

Montalto, Massimo, Dario Gasbarrini, and Antonio Gasbarrini. 2016. "Small Intestinal Bacterial Overgrowth." *John Wiley & Sons, Inc* 23 (3): 101-102. https://doi.org/10.1002/9781118981754.ch7.

Pascale, Alessia, Serena Marchesi, Valeria Marelli, Maurizio Coppola, and Enzo Bianchi. 2020. "The role of gut microbiota in health and disease: In vitro studies on specific dietary compounds and future perspectives." https://doi.org/10.1016/j.fitote.2020.104626.

Rinninella, Emanuele, Loris Riccardo Lopetuso, Silvia Grazia De Gaetano, Pauline Mele, Stefano Vincenzo Isoldi, Alessandra Cintoni, Maria Cristina Miggiano, and Antonio Gasbarrini. 2019. "Food Components and Dietary Habits: Keys for a Healthy Gut Microbiota Composition." https://doi.org/10.1016/j.fshw.2019.07.001.

Rittiphairoj, Thanet, Alessandra Nardin, Joerg Graf, and Angela E. Douglas. 2019. "Anatomical Features That Promote the Coexistence of Symbiotic Bacteria in the Aphid Gut." *John Wiley & Sons, Inc* 51 (1): 1-7. https://doi.org/10.1111/eth.12849.

Scheiman, Jonathan, Katherine Luber, Artem Shalapour, Niv Zitomersky, Caitlin A. Seekatz, Daniel A. Friedman, Reiko T. Lung, Yunwei Wang, Sanjeev Marietta, Joanne R. Glickman, Inbal Winer, Martin Kaldunski, Paul K. Intlekofer, Lawrence Garrett, Christopher S. Brown, Jeffrey Strober, Jonathan Lightfoot, Christina M. C. Lewis, Sean T. Krongold, Roderick Nair, Surya Tripathi, Thaddeus S. Stappenbeck, Mark A. McGovern, Kenneth S. Simpson, Bruce E. Sands, Ramnik J. Xavier, Lloyd Mayer, and Eric J. Alm. 2019. "Functional characterization of the gut microbiome in colorectal cancer." https://doi.org/10.1073/pnas.1904243116.

Sekirov, Inna, Shuang Jin, Eran G. Pichersky, Karen R. M. Cheung, John S. Sullivan, David A. Higginbottom, Dinesh T. Valji, Kim A. Garrity, Roland J. Piasecki, and B. Brett Finlay. 2023. "Lactobacillus reuteri maintains a functional mucosal barrier during DSS treatment despite moderate alteration of the gut microbiome." *American Society for Clinical Investigation* 5 (4): e117687. https://doi.org/10.1172/jci.insight.117687.

Sengupta, R., S., and Alexander Khoruts. 2019. "Fecal Microbiota Transplantation." https://doi.org/10.1016/b978-0-12-811495-5.00024-3.

Sonnenburg, Erica, and Justin L. Sonnenburg. 2018. "The ancestral and industrialized gut microbiota and implications for human health." https://doi.org/10.1111/joim.12853.

Walter, Johann. 2019. "Gut microbiota and intestinal health: A regulatory perspective." https://doi.org/10.1128/microbiolspec.ARBA-0025-2019.

Chapter 6

Microbiomes and the Immunological Paradigm

Felix Uchenna Samuel[1,*]
Anthony Ananga[2]
Joy Iyojo Itodo[4]
and Rachael Alao[3]

[1]Animal Science Program, Alabama Cooperative Extension System, Alabama A&M University, Normal, USA
[2]College of Agriculture and Food Sciences, Florida A&M University Tallahassee, FL, USA
[3]Swine and Rabbit Research Program, National Animal Production Research Institute, Shika-Zaria, Nigeria
[4]Department of Animal Science, Faculty of Agriculture, Federal University Lafia, Nigeria

Abstract

The microbiome is considered the community of microorganisms inhabiting animals internally and externally, influencing development and functioning of the immune system. This abstract hence explores the relationship between microbiomes and the immunological paradigm in health and disease in animals. New findings from research have demonstrated that the role of the microbiome in influencing the immune system is through its ability to alter development and activity of various immune cells such as macrophages, T cells, and B cells. This early colonization, for instance, within the gut has been shown to have a prevailing influence on macrophage maturation and subsequent immune responses that affect susceptibility to disease and health. Thus, perturbations of the microbiome are at least associated with generalized immune dysbiosis, predisposing the host to infection and autoimmunity. This influence of the microbiome is also exerted through its role in metabolizing dietary components and producing bioactive molecules that aid in modulating immune function. The deeper these complex interactions are understood, the more potential therapeutic strategies for the management and prevention of diseases in animals are brought to the fore by targeting the microbiome. In this respect, such an integrative approach positions microbiome health as an important consideration for the maintenance of optimal immune function and in enhancing animal welfare.

Keywords: microbiome, immune system, microbiome, immune dysbiosis, bioactive molecules

---

* Corresponding Author's Email: felix.samuel@aamu.edu

In: Microbiomes in Animal Nutrition, Health, and Disease
Editor: Tanmoy Rana
ISBN: 979-8-89530-425-9

Introduction

Microbiomes are communities of microorganisms, including bacteria, fungi, viruses, and protozoa, that live in and on various environments, including animals. In the context of animals, microbiomes are the diverse assemblages of these microorganisms that inhabit different body sites, such as the gut, skin, mouth, and respiratory tract (Sharma & Yumnam, 2024; Muhammad et al., 2024).

- *Microbiome Composition:* The microbiome is composed of trillions of microbial cells, each contributing to a complex and dynamic ecosystem. These microorganisms can vary widely between different body sites and between individuals.
- *Microbial Diversity:* The diversity of the microbiome is influenced by numerous factors, including diet, environment, and genetics. A healthy microbiome typically has a high level of diversity, which is crucial for maintaining balanced microbial communities and overall health.
- *Microbiome vs. Microbiota:* While the term "microbiota" refers specifically to the microorganisms themselves, "microbiome" encompasses both the microorganisms and their genetic material, including the interactions among them and with their host.

Importance of Microbiomes

Microbiomes play a critical role in animal health, disease resistance, and overall well-being (Maritan et al., 2024) in the following ways:

- *Digestive Health*: In the gut, microbiomes are essential for the digestion and absorption of nutrients. They help break down complex carbohydrates, synthesize essential vitamins (e.g., vitamin K, B vitamins), and aid in the fermentation of dietary fibers. For example, gut microbiota in ruminants (like cows and goats) are crucial for breaking down plant material in the rumen, a process essential for their nutrition (Figure 1).
- *Barrier Function*: Microbiomes on the skin and mucosal surfaces act as a physical and biochemical barrier against pathogenic microbes. For instance, the skin microbiome can produce antimicrobial peptides that inhibit the growth of harmful bacteria.
- *Immune System Modulation*: Microbiomes interact closely with the immune system, helping to regulate immune responses. They influence the development of immune cells and the production of immune-related molecules. For instance, gut microbiota can shape the immune system's ability to respond to infections and prevent excessive inflammation (Figure 2).
- *Pathogen Competition*: Microbiomes compete with pathogenic microorganisms for resources and space, which helps prevent infections. For example, beneficial bacteria in the gut can outcompete pathogenic bacteria, reducing the likelihood of gastrointestinal diseases.
- *Toxin Neutralization*: Certain microbiota can detoxify harmful substances, such as toxins produced by pathogens. This can help prevent disease by neutralizing potential threats before they cause harm.

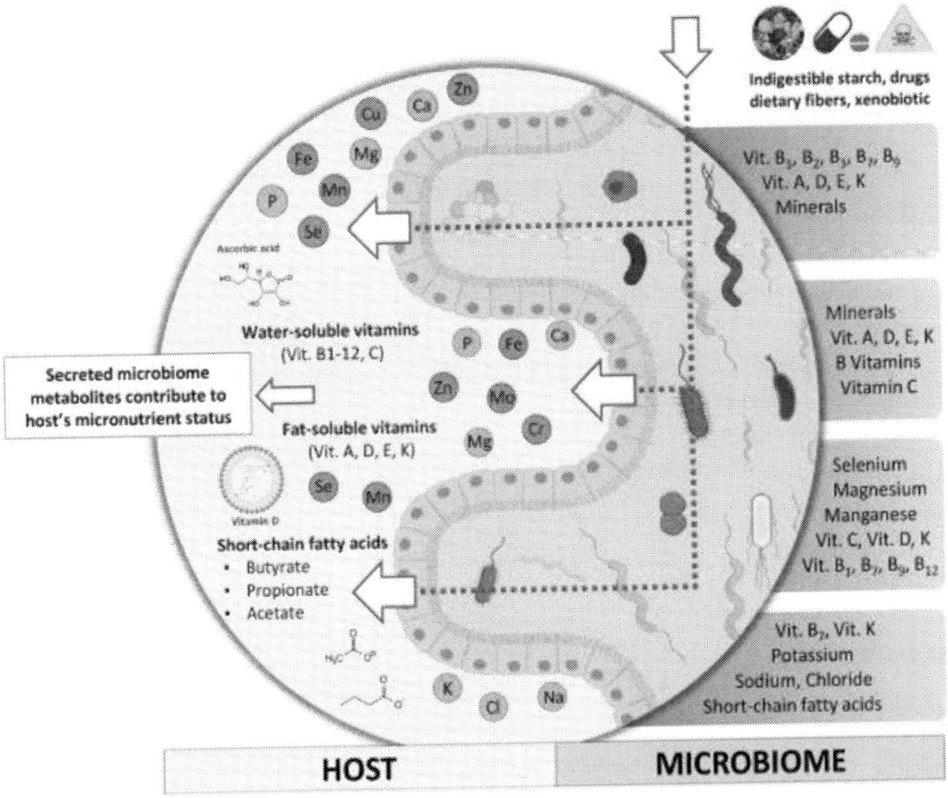

Figure 1. Diagram showing the role of the gut microbiome in nutrient digestion and absorption, and the production of vitamins.

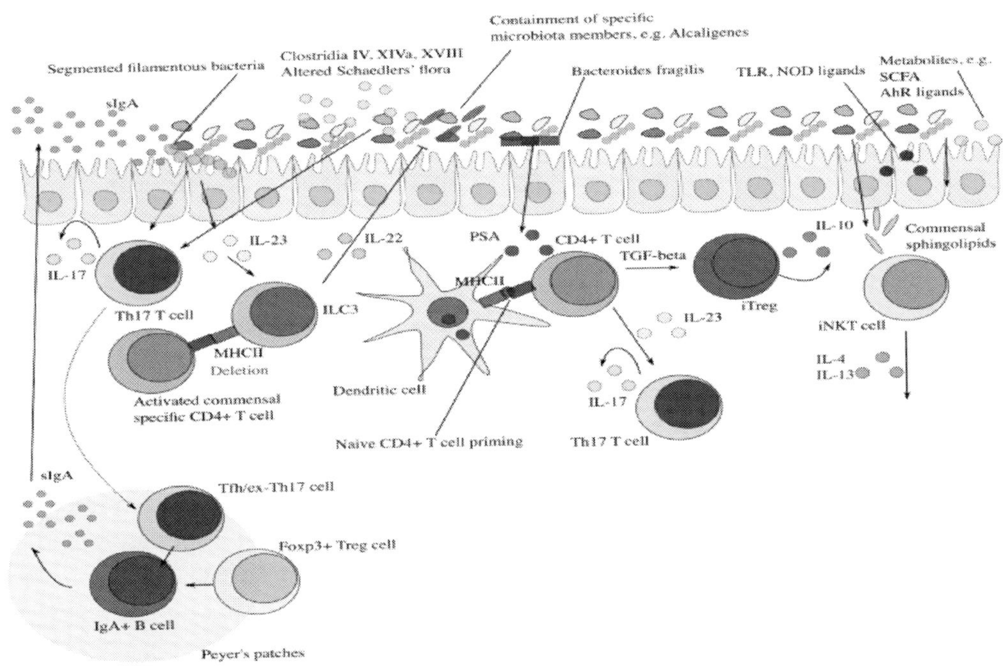

Figure 2. Graph illustrating the relationship between microbiome composition and immune system regulation, showing how diverse microbiomes can enhance immune responses and disease resistance.

- *Behavioral and Psychological Health*: Emerging research suggests that microbiomes can influence behavioral and psychological well-being. For instance, gut microbiota has been linked to the production of neurotransmitters and mood-regulating hormones, potentially affecting stress levels and overall mental health in animals.
- *Growth and Productivity*: In agricultural animals, such as livestock, a healthy microbiome is associated with improved growth rates, feed efficiency, and reproductive performance. This is due to the microbiome's role in optimizing nutrient utilization and maintaining overall health.
- *Adaptation to Environmental Changes*: Microbiomes help animals adapt to environmental changes by modulating metabolic pathways and stress responses. For instance, changes in diet or environmental conditions can lead to shifts in the microbiome composition, which can affect the animal's ability to cope with these changes.

Immunological Paradigms

Innate Immunity

Innate immunity, often described as the first line of defense against infectious agents, is therefore an essential systemic response to prevent infection and maintain homeostasis (Zheng, 2024). It also has an important role in priming and shaping adaptive immune responses, which may remain in the background but have the capacity for rapid induced expression when the host becomes infected by pathogens and can result in all inflammatory conditions. This is a specific response since cells bear on the surface pattern recognition receptors that can recognize complex molecules—including polysaccharides, glycolipids, lipoproteins, and nucleic acids.

These innate immune effectors represent conserved elements of immunity in all jawed vertebrates, although there are specific variations. Most efforts have been primarily directed to studying innate immunity in mammals because of its importance in human and veterinary sciences. However, the study of innate immune mechanisms in nonmammalian species is important and crucially needed for clear understanding of innate immunity and for the discovery of new classes of therapeutic molecules. Recent research has revealed new insights on the mechanisms of the innate immune system. What was formerly regarded as a non-inducible, a pre-existing mechanism of immunity, has now turned out to be evolving into a more specific immune response through molecular recognition and even developing memory, giving increased responses against secondary infections. This innate memory is mainly due to restrained activation of NK cells and macrophages in a way that provides protection against reinfection independently of T/B cells. It involves the development of that immunological memory comprising certain epigenetic changes, including methylation and acetylation, in a manner reprogramming cells to be expressed again upon restimulation. Not only vertebrates but also invertebrates, such as plants and insects, have been shown to have this form of memory (Lewis et al., 2024).

Components and Mechanisms

Physical Barriers: Skin and Mucosal Membranes Act as Primary Barriers

The primary mechanisms of innate immunity are physical and chemical barriers that present the first line of defense to a pathogen by stopping it from propagating and spreading. These barriers morphologically differ among species anatomically but undertake the basic work of shielding the internal environment from dangerous exogenous materials and pathogens. It is the main armor of vertebrates against the environment. In other words, the skin is composed of stratified epithelial layers joined by tight junctions and structural proteins, serving to provide mechanical protection and prevent loss of water, with limited infiltration of pathogens. The skin of vertebrates basically has two major layers: an outer layer called the epidermis that forms the outer portion of the skin interfacing with the outer environment, and under this layer lies another major region called the dermis. This general structure is shared among classes of vertebrates, although adaptations to various habitats gave rise to functional and structural variations. For example, fish have an epidermis armored by ossified scales and actin-rich microridges modulated for aquatic habitats. In the case of tetrapods, those which moved to land required their epidermis to be strengthened and a keratinized outer layer. In amphibians, the corneous layer is thin, with few keratinized cells, while in reptiles, it is thick, more robust, with many layers of keratin and overlapping scales. Feathers replace the scales on birds, which are very resistant to pathogens and offer lift when flying. The epidermis is more flexible in mammals: the stratum corneum of their epidermis is soft, moist, with only alpha-keratin that makes it less rigid but suitable for both locomotion and sensory functions. In addition to structural barriers, vertebrate skin is also constantly being lubricated by extrinsic secretions that not only function to maintain its integrity and water content but also exert antibacterial activity. The skin of aquatic/semiaquatic vertebrates—fish and amphibians—secretes mucous cuticles rich in mucopolysaccharides and glycoproteins, whereas terrestrial amniotes, including reptiles, birds, and mammals, form sebaceous fluids made up of glycolipids and lipids. Even though these secretions differ in their chemical composition, they are rich in microbicidal molecules like lysozyme and antimicrobial peptides that play a significant role in innate defense. Cellular responses to pathogen invasion are also found in vertebrate skin. Epithelial cells thicken and proliferate, and immature dendritic cells with dendritic projections that extend into the epidermis to capture antigens can migrate into the epidermis, preventing local infections. Very well characterized in mammals, they are found in other vertebrates to have the same functions. Furthermore, intraepithelial lymphocytes and macrophages could destroy pathogens and infected cells. The second important vertebrate epithelial barrier is mucosal membranes, sealing off body cavities from pathogen invasion. In fish, the mucosal epithelium lining gills and the alimentary tract prevents the entry of microbes from aquatic environments. Specialized mucosa in respiratory and gastrointestinal tracts in terrestrial vertebrates prevents the entry of foreign particles and pathogens. All mucosal surfaces, in general, exhibit biochemical defenses such as low pH, hydrolytic enzymes, mucus, and antimicrobial molecules like lysozyme, AMPs, and pro-inflammatory cytokines. Furthermore, gut-associated lymphoid tissue containing high levels of intraepithelial lymphocytes, which help destroy pathogen elements and maintain epithelial integrity, can be found in all jawed vertebrates. In many vertebrates, symbiotic microbiomes prevent pathogenic colonization by competing for attachment sites on mucosal surfaces and skin, and their products are often antimicrobial (Figure 3).

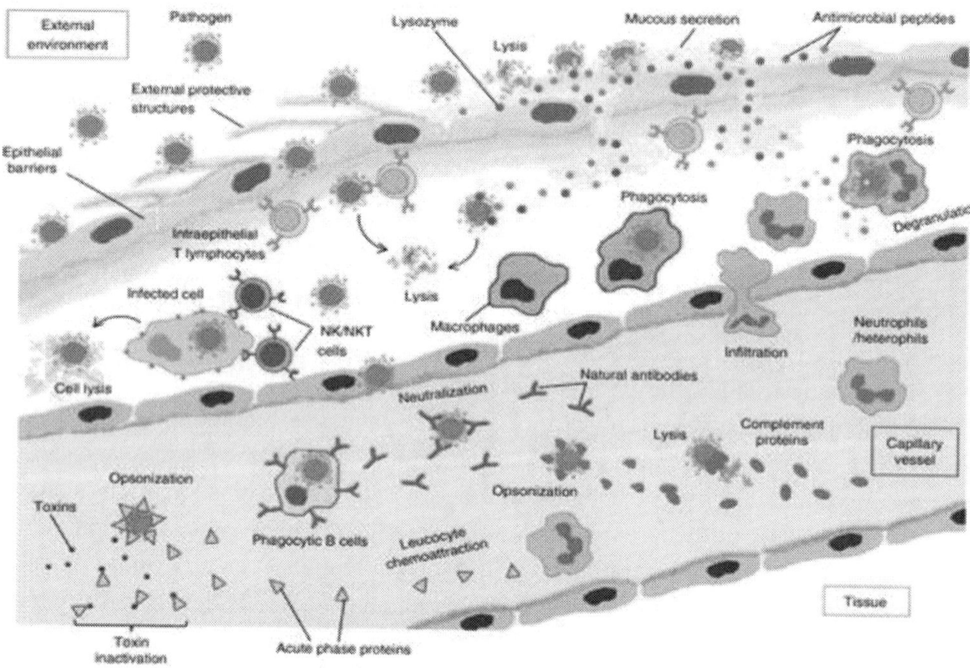

Figure 3. Innate immune sytem of vertebrate (Riera Romo et al., 2016).

Humoral Innate Immunity

Lysozyme

Lysozyme, also known as muramidase or N-acetyl muramide glycanohydrolase, is a potent antimicrobial protein found in the mucous secretions and plasma of all vertebrates. This enzyme is crucial for lysing bacterial cells. In fish, lysozyme is commonly associated with the mucous cuticle. In amphibians, such as Rana pipiens, eight different lysozyme isoforms have been identified through chromatographic and electrophoretic techniques in various tissues including skin, serum, liver, kidney, ovary, and spleen, all showing lytic activity against Micrococcus lysodeikticus. Similar findings have been reported for anurans like Xenopus laevis and Xenopus tropicalis (Riera Romo et al., 206).

Reptiles, such as the Chinese soft-shelled turtle (Pelodiscus sinensis), Asian soft-shelled turtle (Amyda cartilagenea), and green sea turtle (Chelonia mydas), also produce lysozyme homologues with lytic activity against both Gram-positive and Gram-negative bacteria. In birds, lysozyme has been detected and quantified in the plasma of various species including chickens (Gallus gallus), American kestrels (Falco sparverius), and several other species, showing bactericidal activity particularly against Escherichia coli.

In mammals, lysozyme is expressed in various tissues and secreted by glandular serous cells, surface epithelial cells, and macrophages into the airway lumen. It has been well-characterized at both the gene and protein levels, with different isoforms identified in rodents. The bacteriolytic activity of rat lysozyme has been demonstrated against Gram-positive bacteria such as Pseudomonas aeruginosa and Staphylococcus aureus. In humans, high concentrations of lysozyme are found in airway surface liquid, contributing to the microbicidal properties of nasal secretions. It is also produced by neutrophils, macrophages, and Paneth cells in the intestine and can be secreted by gastrointestinal epithelial cells during inflammatory conditions like ulcerative colitis and Crohn's disease. Phylogenetic studies reveal that lysozyme is widespread across the animal kingdom, with amino acid sequences identified in a range of vertebrates including humans, mice, chickens, turtles, frogs, and fish, as well as in cephalochordates like amphioxus and various invertebrates and plants. This broad conservation underscores its essential role in innate immunity.

Antimicrobial Peptides

Antimicrobial peptides (AMPs) are critical components of the innate humoral defense system, highly conserved across vertebrates. These peptides exhibit a broad spectrum of antimicrobial activities, including antibacterial, antifungal, and antiparasitic effects. They achieve this through various mechanisms such as disrupting microbial membranes, forming pores, aggregating proteins, inhibiting intracellular targets, interfering with DNA transcription, and blocking protein synthesis and folding. Key examples of conserved AMPs include defensins, cathelicidins, and hepcidins. Defensins and cathelicidins are present in the mucous secretions of fish, as well as in reptiles like the European pond turtle (Emys orbicularis) and the banded krait (Bungarus fasciatus). In reptiles, beta-defensin and cathelicidin-BF show bactericidal activity against Salmonella typhimurium, Listeria monocytogenes, Escherichia coli, and methicillin-resistant Staphylococcus aureus, as well as antifungal activity against Candida albicans. Fish-derived defensin-like peptides and cathelicidins also exhibit strong antimicrobial activity against various pathogens, including bacteria, fungi, viruses, and parasites. In birds and mammals, AMPs like alpha- and beta-defensins are widespread. For example, humans produce specific beta-defensins, such as human beta-defensins 1, 2, and 3, in epithelial cells and neutrophils. Neutrophils also produce azurocidins and alpha-defensins like human neutrophil peptides 1, 2, and 3. Mammals have additional cathelicidins, such as PR-39 in pigs and LL-37 in humans. Unique defensins, such as h-defensins, have been identified in the rhesus macaque (Macaca mulatta). These mammalian AMPs are effective against bacterial infections caused by Salmonella typhimurium, group A streptococcus, some Staphylococcus species, and Pseudomonas aeruginosa. They also perform immunomodulatory functions, including stimulating chemokine production, inhibiting pro-inflammatory cytokines, directing leukocyte chemotaxis, and modulating host cell gene expression. Hepcidin is considered one of the most conserved AMPs across vertebrates. This small peptide, rich in cysteine residues, is found in over 51 vertebrate species and is synthesized by the liver as an acute phase reactant. However, the presence of hepcidin in birds remains uncertain, as recent searches in mRNA expression and various genomic databases have not confirmed its presence in avian species. Additionally, other cationic peptides with linear or helical structures have been identified in different vertebrates. Piscidin and pleurocidin are found in fish, while pelovaterins are present in the eggs of the Chinese soft-shelled turtle (Pelodiscus sinensis). Amphibians produce various AMPs, including ascaphine-1 and ascaphine-8 from Ascaphus truei and Ascaphus montanus, bombinin H6 from Bombina species, and alyteserin-1 and alyteserin-2 from Alytes obstetricans. Mammals also produce cationic peptides such as profenins in pigs and histatins in human saliva. AMPs derived from protein domains, including hipposin (a histone-derived peptide from Atlantic halibut), lactoferricin (from mammalian lactoferrin I), casodicin I (from human casein), buforin II (from histone 2A), and antimicrobial domains from bovine lactoalbumin, human hemoglobin, lysozyme, and ovalbumin, have also been described in vertebrates.

Acute Phase Protein

Acute phase proteins (APPs) are a group of proteins found in vertebrate plasma and epithelial secretions, synthesized in response to infection and inflammation. These proteins play a pivotal role in innate immunity and host defense (Safi, et al., 2012). Key APPs in mammals include C-reactive protein (CRP), serum amyloid A and P, haptoglobin, α1-acid glycoprotein, α2-macroglobulin, ceruloplasmin, fibrinogen, and transferrin. CRP and serum amyloid P are pentraxins involved in pathogen recognition, agglutination, opsonization, and complement activation. CRP is a major acute phase protein in various mammals, including sheep, cattle, rabbits, dogs, mice, rats, humans, and other primates, and has homologues in teleost fish, although it is absent in reptiles and birds. Conversely, serum amyloid P serves as a CRP counterpart in some species, notably in mice, and has recently been identified in the serum of the iridescent shark (Pangasianodon hypophthalmus), where it shows agglutination activity against rabbit erythrocytes and pathogenic bacteria like Edwardsiella ictaluri and Aeromonas hydrophila. Similar CRP-like pentraxins have been reported in amphibians of the Xenopus genus. Serum amyloid A is a broadly conserved APP present in fish, reptiles, birds, and mammals, but not in amphibians. Its gene expression is upregulated during bacterial infections in species such as zebrafish (Danio rerio), rainbow trout (Oncorhynchus mykiss), and the Chinese soft-shelled turtle (Pelodiscus sinensis). It is a major APP in mammals like sheep, goats, cattle, horses, cats, dogs, pigs, mice, and humans, and a moderate APP in chickens and non-human primates. Serum amyloid A functions in leukocyte chemotaxis and inflammatory response regulation. A trypsin inhibitor unique to pigs also serves as a major acute phase protein, with levels increasing more than tenfold during infections. Haptoglobin and α1-acid glycoprotein are significant acute phase reactants found in fish and mammals, but haptoglobin has recently been identified in birds as well. Haptoglobin binds free hemoglobin to prevent oxidative damage and has antimicrobial and immunomodulatory effects. α1-Acid glycoprotein, present in birds and mammals, primarily inactivates harmful molecules like lipopolysaccharides and various drugs. Other plasma proteins such as α2-macroglobulin, ceruloplasmin, fibrinogen, and transferrin can also function as APPs, with elevated levels observed during inflammation and infection. α2-Macroglobulin, a protease inhibitor that neutralizes toxins and proteolytic enzymes, has been identified in teleost fish and mammals. Ceruloplasmin, a copper-binding protein that scavenges free radicals, is a positive APP in fish, birds, and mammals. Fibrinogen, involved in tissue repair, is upregulated in response to acute infections across fish, reptiles, birds, and mammals. Transferrin acts as a negative APP in mammals but serves as a positive acute phase reactant in fish and birds. Conversely, albumin levels decrease during acute infections or inflammation, making it a negative APP in domestic animals and mammals, with reduced hepatic synthesis and renal or gastrointestinal losses during infection, preserving amino acids for the production of positive APPs (Hamar et al., 2024).

Complement System

The complement system is composed of serum proteins which, upon contact with the pathogen, trigger a molecular cascade that leads to opsonization, enhancement of inflammatory responses, and the formation of the membrane attack complex, which disrupts microbial membranes and leads to lysis. In its best-characterized form in mammals, this system is activated through three main routes: the classical, alternative, and lectin pathways.

The classical pathway is triggered in the presence of antigen-antibody complexes, usually including IgG or IgM. These activate a proteolytic cascade. The mannose-binding lectin recognizes microbial surface carbohydrates and triggers the lectin pathway. It contains the activation of mannose-binding-lectin-associated serine proteases, which in turn fiddles the complement cascade. In comparison, the alternative pathway is self-activated by spontaneous hydrolysis of the complement protein C3 at microbial membranes. A new function of properdin identified in research has activated the pathway and interactions between complement and coagulation-associated proteases, such as kallikrein and thrombin. The importance of the complement system in mammalian immunity is underlined by studies in immunodeficient patients or knockout mouse models that demonstrate complement proteins or their fragments to be the main and indispensable activators of B-cell and T-cell functions, not only for innate host defense against each of the following pathogenic microorganisms—Streptococcus pneumoniae, Pseudomonas aeruginosa, Neisseria meningitidis, Haemophilus influenzae—but also against herpes simplex virus and West Nile virus. Different vertebrates have had their complement systems studied, showing diversified reactivities and antimicrobial properties. For example, in teleost fish, such as the common carp and catfish, Ictalurus punctatus and salmonids respectively, complement-mediated opsonization and activation through antibodies have been featured. In amphibians, serum hemolytic activity in Rana pipiens and Lithobates berlandieri shows antibody-stimulated lysis that may equal mammalian levels and be dependent on such factors as heat, EDTA, hydrazine, carrageenan, and precipitates of antigen-antibody complexes. Reptilian serum also exhibits antimicrobial properties. Comparative studies demonstrated that the complement system of the American alligator has an extended action spectrum against Gram-negative bacteria and showed antiviral activity against human immunodeficiency virus type 1, West Nile virus, and herpes simplex virus. Alligator sera also demonstrated amoebicidal activity against the human complement-resistant Naegleria and Acanthamoeba. Cell lysis in birds appears to be complement-mediated. In one study, Matson et al. (2005) found that sera from three bird species—the American kestrel, Falco sparverius; Japanese quail, Coturnix coturnix japonica; and zebra finch, Taeniopygia guttata—induced complement-mediated hemolysis of rabbit erythrocytes. The complement-mediated nature of this activity was noted by its heat-decaying attribute and differed between age classes as well as species. In the mammalian complement system, nine major proteins have been well characterised: C1, C2, C3, C4, C5, C6, C7, C8, and C9. Several inhibitors of complement act on the complement cascade to ensure its high specificity and prevent host-cell damage. These include Factor I, decay-accelerating factor, complement receptor 1, Factor H, C4-binding protein, CD59, vitronectin, and S protein. Complement proteins have also been identified in other vertebrates. For instance, important components like C3 and C5 have been isolated from the anurans of genus Xenopus and recognized as down-regulated genes in teleosts like Oncorhynchus mykiss during a bacterial infection. Complement proteins C1, C3, C4, C5, C7, C8, and C9 have been isolated in osteichthyes like catfish (Ictalurus punctatus). Factor B, required for the alternative pathway, has been found in several fish and in poultry. However, very little information is available on the molecular details of the reptilian complement cascade, with only one complement protein, C3, being described to date.

Cytokines

Cytokines are proteins that facilitate cell-to-cell communication and play a crucial role in the immune response. They are among the most conserved components of vertebrate innate immunity. Cytokines are categorized based on their genomic encoding regions, target receptors, signaling pathways, and biological functions into several classes, including interferons (IFNs), interleukins (ILs), tumor necrosis factors (TNFs), and transforming growth factors (TGFs). Interferons are primarily involved in antiviral defense, although they also have additional immunomodulatory roles. In mammals, three types of interferons have been identified: type I IFNs (including IFN-α, IFN-β, IFN-ω, IFN-γ, and IFN-ε), type II IFN (IFN-γ), and type III IFNs (IFN-λ). Type I IFNs are conserved across vertebrates but show different genomic organizations between lower and higher vertebrates. For instance, fish and amphibian type I IFNs are encoded by five-exon/four-intron transcripts, whereas reptilian, avian, and mammalian type I IFNs are encoded by intronless transcripts. In lower vertebrates, type I IFNs are divided into groups I and II based on cysteine residue patterns. These interferons are crucial for antiviral immunity in various vertebrates, such as teleost fish, amphibians, and birds. In fish like the crucian carp, they offer significant antiviral protection, while in amphibians, they inhibit viral replication and enhance survival during infections. Type I IFNs have been well-characterized in birds, including chickens, where they are encoded by intronless transcripts and exhibit strong antiviral activity. Type II IFNs, including IFN-γ, are also conserved and perform functions such as antiviral defense, modulation of MHC expression, phagocytosis stimulation, and regulation of cell growth and apoptosis. In mammals, IFN-γ is the only type II IFN identified, but multiple type II IFNs have been found in lower vertebrates such as teleost fish and amphibians. These IFNs share antiviral properties with their mammalian counterparts and have been detected in various species, including lizards and birds. Type III IFNs (IFN-λ) have been identified in mammals and recently in amphibians, where they participate in antiviral defense and share functions with type I IFNs. The presence of IFN-λ in other vertebrates, including reptiles and birds, remains underexplored, suggesting the need for further genomic screening. Interleukins are another class of cytokines that are conserved across vertebrates but vary in representation. IL-1, IL-2, and IL-8 are among the most conserved interleukins, found in amphibians, reptiles, birds, and mammals. IL-1 has similar functions in teleost fish and other vertebrates. The mammalian IL-1 family includes IL-1α, IL-1β, and IL-1 receptor antagonist (IL-1Ra). Only IL-1β has been identified in other vertebrates, retaining its pro-inflammatory functions. IL-2, important for lymphoid and myeloid cell differentiation in mammals, is found in reptiles and chickens, where it influences cell proliferation and differentiation. IL-8, a chemokine involved in leukocyte attraction, is present in various vertebrates but has not been identified in amphibians. Other interleukins, such as IL-15, IL-16, IL-17, and IL-18, are found mainly in birds and mammals, with functions related to cell differentiation, T lymphocyte chemoattraction, cytokine expression induction, and IFN-γ stimulation. Mammalian interleukins like IL-4, IL-5, IL-10, and IL-13 are involved in T-helper type 2 responses, and cytokines such as IL-12, IL-20, IL-21, IL-22, IL-26, IL-25, and IL-33 have been characterized for their roles in inflammation, leukocyte proliferation, and lymphocyte differentiation. Some of these interleukins have also been discovered in teleost fish, indicating their potential presence in other vertebrates. Among pro-inflammatory cytokines, TNF-α is highly conserved and plays a role in leukocyte attraction and macrophage activation. It has been identified in teleosts, amphibians, and mammals, with evidence of similar activity in Xenopus laevis. Although TNF-α-like activity has been reported in chickens, its exact gene sequence has not been described, and its presence in reptiles remains unreported. Transforming growth factors (TGFs), particularly TGF-β, are well-preserved across vertebrate species and are involved in various functions including cell growth, differentiation, and immune regulation. The mammalian TGF-β family includes three isoforms that regulate development, wound repair, and immune responses. TGF-β has been identified in various vertebrates, including fish, birds, and reptiles, with roles extending beyond immunity to reproduction and embryonic development. For example, the TGF-β4 isoform is induced during coccidial infections in chickens (Rathinam et al., 2024).

Natural Antibodies

In vertebrate plasma, there are circulating antibodies known as natural antibodies (NAbs), primarily of the IgM isotype. These antibodies, which have a restricted genetic repertoire, are produced by B cells even in the absence of pathogens. NAbs are crucial components of humoral innate immunity, reacting to foreign antigens and microbial substances, and can activate the complement system through the classical pathway. NAbs have been observed across various vertebrate groups. In fish, for instance, natural IgM antibodies recognize microbial surface proteins and offer protection against pathogens like Aeromonas salmonicida. This IgM isotype is found in chondrichthyes (such as sturgeons and sharks) and some osteichthyes like Oncorhynchus mykiss and Pagrus auratus. Amphibians also produce these antibodies, where they help neutralize modified bacteriophages and other viral particles, as well as cause hemagglutination of foreign red blood cells from species such as chickens, sheep, rats, and humans. High levels of NAbs are present in anurans like Bufo regularis, Bufo arenarum, and Xenopus laevis. Reptiles also produce NAbs, which are important for their humoral immune response. In some species, such as the water python (Liasis fuscus), NAb levels increase with age, potentially compensating for immunosenescence and decreased effectiveness of the adaptive immune response. In birds, NAbs are produced by gut-associated lymphoid tissue, stimulated by the intestinal microbiota. These antibodies help eliminate bacteria and spirochetes in vivo and aid in clearing bacterial substances like lipopolysaccharides. They play a significant role in defense against pathogens, such as Amblyceran lice in species like the Galápagos hawk (Buteo galapagoensis). NAbs against parasites, such as Plasmodium lophurae, have also been identified in White Leghorn chickens (Gallus domesticus). In mammals, NAbs represent a conserved element of humoral innate immunity, neutralizing pathogens and helminth parasites, and providing resistance against viruses like influenza. Additionally, these antibodies have novel functions in complex vertebrates. They bind to oxidized proteins, such as oxidized low-density lipoprotein, reducing macrophage uptake and mitigating the atherogenic process. Some reports also indicate that natural IgMs are involved in immune surveillance, recognizing tumor-associated or tumor-specific antigens, such as CFR-1 in epithelial cancers, and may induce apoptosis in malignant cells.

Cell-Mediated Innate Immunity

Intraepithelial T lymphocytes

Intraepithelial T lymphocytes (ITLs) are specialized immune cells found within the mucosal and epithelial barriers of vertebrates, where they play a role in recognizing and eliminating pathogens. These lymphocytes are particularly well-characterized in mammals and are categorized as γδ T lymphocytes. Unlike typical lymphocytes involved in adaptive immunity, ITLs exhibit characteristics more aligned with innate immune cells, including a restricted antigenic repertoire and less variability in T-cell receptors.

ITLs have been isolated from the intestinal epithelium of fish, such as Oncorhynchus mykiss, and shown to possess cytotoxic activity against EL4 mouse thymoma cells. Similarly, in amphibians, ITLs are present in specialized epithelial tissues like the gut-associated lymphoid tissue, where they share many phenotypic traits with mammalian ITLs.

In reptiles, ITLs are a component of protective immunity within the intestinal epithelium and gut-associated lymphoid tissue, a characteristic inherited from fish and amphibians. For instance, in the platypus (Ornithorhynchus anatinus), ITLs found in the lamina propria of the intestinal mucosa resemble those in mammals. Birds also have ITLs in their gut-associated lymphoid tissue, which include various subsets such as CD8+ T cells (CD8αα) and γδ T CD8αβ cells. These lymphocytes serve as a primary defense against pathogen spread and are crucial for immunity against avian coccidiosis caused by Eimeria. Research indicates that the number of ITLs increases following infection with intestinal parasites like Eimeria acervulina and Eimeria tenella.

In humans, two subsets of γδ T lymphocytes, Vδ1 and Vδ2 T cells, have been identified. Vδ1 T cells are predominantly located in the epithelial mucosa, while Vδ2 T cells are mainly found in the lymphatic system and peripheral blood. These cells are capable of recognizing non-peptidic antigens, soluble proteins, and MHC molecules, and perform various functions including cytokine secretion, dendritic cell maturation, macrophage recruitment, cytotoxic activity against transformed cells, maintenance of epidermal integrity, and support of B- and T-cell interactions.

Myeloid Phagocytic Cells

Myeloid phagocytic cells are essential components of vertebrate innate immunity, often referred to as 'professional' phagocytes. This category includes monocyte-macrophages, neutrophils/heterophils, basophils, and eosinophils. These cells are involved in pathogen phagocytosis and destruction, cytokine secretion, and the release of soluble mediators such as histamine, reactive oxygen species, reactive nitrogen species, lysozyme, and antimicrobial peptides (AMPs). They identify and engulf senescent cells and microbes through pattern recognition receptors and present antigens to B and T cells, thus acting as the initial defense against infection and aiding in the activation of the adaptive immune response. Pattern recognition receptors on phagocytes include Toll-like receptors, C-type lectin receptors, NOD-like receptors, and Rig I-helicases. These receptors recognize a range of microbial structures such as lipopolysaccharides, lipoproteins, complex glycans, formylmethionine, and nucleic acids. They can distinguish between different chemical isomers and polysaccharides, lipids, or nucleotides, working together to trigger specific signaling pathways that initiate pro-inflammatory and antimicrobial responses. In mammals, myeloid phagocytes are critical for combating infections caused by both Gram-negative (e.g., Eschcrichia coli, Salmonella typhimurium, Yersinia enterocolitica) and Gram-positive bacteria (e.g., Listeria monocytogenes, Staphylococcus aureus). Research using mouse models has highlighted their roles in phagocytosis, antigen presentation, histamine release, and the production of pro-inflammatory cytokines, AMPs, and granule-derived proteins that destroy microorganisms. Additionally, these cells contribute to wound healing and tissue homeostasis. Neutrophils are the first responders in human innate immunity, engaging in microbial phagocytosis, collaboration with other immune cells, and the release of AMPs, proteases, hydrolytic enzymes, and phospholipases. In other vertebrates, macrophages, heterophils (which are similar to mammalian neutrophils), basophils, and eosinophils are essential for effective innate immunity. These cells have been identified in fish, amphibians, and reptiles, maintaining their roles in phagocytosis, inflammation, and the release of soluble mediators. In ectothermic vertebrates, phagocytosis plays a significant role in the immune response due to its relative independence from temperature compared to adaptive immune mechanisms. Recent advancements in imaging techniques for fish models, such as Danio rerio, have provided new insights into the phagocytic process and immune cell recruitment during inflammation. These techniques, combined with genetic manipulation, have led to the discovery of novel inflammatory pathways, danger signals, and myeloid development-related genes, contributing to a deeper understanding of the mammalian immune system. In birds, myeloid phagocytes, including macrophages, heterophils, eosinophils, and basophils, have preserved their functions from early vertebrate ancestors. These cells are involved in pathogen recognition and phagocytosis, with their pattern recognition receptor-mediated recognition triggering pathogen destruction, antigen presentation, adaptive immune response stimulation, and pro-inflammatory cytokine production. Additionally, recent research has focused on the lymphoid lineage of vertebrate leukocytes to understand developmental relationships and functionality. A new type of leukocyte, nuocytes, has been identified in mammals, playing a role in type 2 immunity against helminth parasites. These cells, characterized in IL13-eGFP reporter mice, exhibit IL-25/IL-33-dependent proliferation and IL-13 secretion during helminth infection. Furthermore, lineage-marker-negative innate lymphoid cell subsets have been discovered, which do not express traditional immune cell surface markers but are involved in immune responses and tissue homeostasis regulation (Riera Romo et al., 2016).

Non-Specific Cytotoxic Cells

Non-specific cytotoxicity is a crucial component of innate immunity in vertebrates, serving a key role in antiviral defense and immune surveillance. In mammals, this function is primarily carried out by natural killer (NK) cells, which target and destroy cells infected by viruses or intracellular pathogens. NK cells also produce antiviral cytokines and contribute to liver defense, tissue regeneration, and remodeling. Similar non-specific cytotoxic cells have been observed in lower vertebrates. For example, in fish such as Ictalurus punctatus, these cells are characterized by their small, non-granular appearance and are found abundantly in lymphoid tissues like the kidney and spleen. Additionally, zebrafish (Danio rerio) have been shown to possess NK-like cells, supported by the discovery of a new gene family called novel immune-type receptors. These receptors share structural similarities with mammalian NK receptors, are expressed in lymphoid cells, and provide lytic capabilities. In amphibians, NK-like lymphocytes play a vital role in combating virus-infected cells. In Xenopus laevis, multiple NK cell subsets have been identified using specific monoclonal antibodies, demonstrating their strong cytotoxic activity against tumor cells. These findings have been corroborated by further research, showing that Xenopus NK and NKT-like cells contribute to antitumor immune responses and are valuable for studying tumorigenesis and immune responses due to their genetic manipulability. Birds also exhibit non-specific cytotoxic activity through NK-like lymphocytes known as TRC0 cells. These cells, which are found in the intestinal mucosa of chickens, have surface antigens similar to mammalian NK cells and are divided into two populations: 28-4+ CD3− NK cells and 28-4+ CD3+ NK cells, the latter resembling mammalian NKT cells. These cells are conserved across vertebrates and play significant roles in anti-parasitic immune responses and mucosal immunity. Although non-specific cytotoxic cells have not been thoroughly studied in reptiles, their presence is suggested by observations of NK-mediated cytotoxicity in the turtle Mauremys caspica. This activity, found in thymic cells, has not been attributed to a specific subset. Additionally, a specialized NK cell type with T lymphocyte characteristics, known as NKT lymphocytes, has been identified in mammals. These cells recognize glycolipid antigens via the CD1 molecule and release cytokines involved in the innate immune response. Functional analogues of NKT cells have been found in amphibians and birds, indicating that this cell type is a conserved component across vertebrates and links innate and adaptive immunity. NKT cells contribute to cytotoxic responses, antitumor activity, and the secretion of antiviral cytokines. Recent studies have revealed new functional properties of NK cells and other innate immune cells, such as monocyte-macrophages. These findings highlight the concept of "trained immunity," where innate cells exhibit enhanced responses to secondary infections through epigenetic and biochemical changes, providing a novel perspective on innate immune memory (Fischer and Takizawa et al., 2022).

Phagocytic B Cells

Phagocytic B cells are a specialized type of B lymphocyte involved in the innate immune responses of vertebrates. In lower vertebrates, these cells exhibit phagocytic activity and engage in degradative processes leading to the fusion of phagosomes and lysosomes, resulting in the destruction of internalized bacteria. While B cells in these organisms function similarly to leukocytes, they are also capable of producing antibodies of the IgG and IgA isotypes, which are crucial for classical adaptive immunity, as well as natural antibodies (NAbs) of the IgM isotype. Phagocytic B cells have been identified in teleost fish, such as Oncorhynchus mykiss and Ictalurus punctatus, as well as in amphibians like Xenopus laevis and reptiles such as the red-eared slider (Trachemys scripta). In mammals, however, B cells generally do not exhibit phagocytic activity and are primarily involved in adaptive immunity. Nevertheless, a specific subpopulation of B lymphocytes known as B-1 cells retains the phagocytic capabilities and antimicrobial properties found in lower vertebrate B cells. B-1 cells are predominantly located in the peritoneal and pleural cavities, where they produce NAbs and present phagocytosed antigens to CD4+ T cells, thereby playing a role in both innate and adaptive immune responses. The preservation of phagocytic function in B-1 cells suggests an evolutionary link to the phagocytic B cells of lower vertebrates. Furthermore, the presence of a bipotent B macrophage precursor in the fetal liver and adult bone marrow of mammals supports the notion of a common evolutionary origin for B cells and macrophages, derived from a phagocytic ancestor. Conversely, the phagocytic abilities of B cells in birds remain unexplored, making it challenging to trace the evolution of this function in higher amniotes. However, similar to mammalian B-1 cells, some B cell subsets in birds, particularly within gut-associated lymphoid tissue, produce NAbs.

Pattern Recognition Receptors (PRRs)

Pattern recognition receptors form a critical part of the innate immune response and function primarily to identify and respond to a variety of pathogens. By recognizing specific molecular patterns associated with various pathogens, PRRs trigger immune responses that the body can use to defend itself against infection. There are three main families that constitute PRRs: TLRs, NOD-like receptors, and RIG-I-like receptors. Each family recognizes different types of PAMPs, and the mechanisms of action vary.

Toll-like Receptors (TLRs)

TLRs comprise one of the best-characterized families of PRRs centrally involved in the innate immune response. They are named after the Toll protein in Drosophila melanogaster, first identified as a central component in the fly's immune system. The TLRs are mainly expressed by macrophages, dendritic cells, and epithelial cells but are positioned at the vanguard, functionally representing sentinels for the detection of pathogenic invasion.

Structure and Function

TLRs are transmembrane receptors characterized by an extracellular domain that recognizes Pathogen-associated molecular patterns (PAMPs) and an intracellular Toll/IL-1 receptor (TIR) domain, which initiates downstream signaling pathways. The ectodomain is composed of repetitive structures known as leucine-rich repeats that bind specific PAMP molecules, including bacterial LPS, flagellin, and viral nucleic acids.

The TIR domain interacts with downstream adaptor proteins such as MyD88, TIRAP, TRIF, and TRAM to activate various signaling pathways. These pathways activate transcription factors like NF-κB, AP-1, and IRFs, which finally lead to the induction of genes involved in the production of inflammatory cytokines and type I interferons.

Recognition by Pathogen-Associated Molecular Patterns

TLRs are primarily classified based on their ligand specificity:

- TLR1 and TLR2: Bacterial lipoproteins and peptidoglycans. TLR2 forms heterodimers with TLR1 or TLR6 to detect triacylated and diacylated lipoproteins, respectively.

- TLR3: Complexes the double-stranded RNA produced by viruses. Moreover, this receptor plays a critical role in identifying RNA viruses like the influenza virus.
- TLR4: LPS of Gram-negative bacteria. To complex well with LPS, the accessory protein MD-2 is needed for TLR4.
- TLR5: Recognizes bacterial flagellin, which is the protein component of bacterial flagellum.
- TLR7 and TLR8 recognize single-stranded RNA (ssRNA) derived from viruses. They have been implicated in the recognition of ssRNA viruses, such as HIV and influenza.
- TLR9 recognizes unmethylated CpG DNA motifs in bacterial and viral DNA.

Signaling Pathways and Immune Response

The recognition of PAMPs by the TLRs then induces a variety of signaling pathways intracellularly, leading to the production of a number of pro-inflammatory cytokines, chemokines, and antimicrobial peptides. This response will recruit and activate other immune cells, such as neutrophils and macrophages, to the site of infection. Moreover, TLRs also play a critical role in controlling adaptive immune responses. By inducing dendritic cell maturation and cytokine production, TLRs initiate and regulate T-cell responses to generate effective antigen-specific immunity.

Nucleotide-Binding Oligomerization Domain-Containing Protein (NOD)-Like Receptors

NOD-like receptors are cytoplasmic PRRs that detect pathogens in the cytosol. As opposed to TLRs, which are cell membrane-associated, Nucleotide-binding oligomerization domain-like receptors (NLRs) are cytosolic and continuously monitor intracellular compartments for the presence of infection.

Structure and Function

An NLR comprises a central nucleotide-binding oligomerization domain with an associated C-terminal leucine-rich repeat domain and other variable C-terminal domains in different NLRs. The NOD domain engages in oligomerization and signaling, while the LRR domain is responsible for the recognition of ligands. NLRs are usually divided into several subfamilies based on their domain structure:

- NOD1 and NOD2 identify bacterial peptidoglycans. Specifically, NOD1 recognizes certain dipeptides from Gram-negative bacteria, while NOD2 is more broadly reactive to muramyl dipeptides from both Gram-positive and Gram-negative bacteria.
- NLRP: It is a family of NLR family, pyrin domain containing. There are a number of members that include NLRP1, NLRP3, and NLRP6. They are receptors that detect several danger signals, including microbial products and host-derived signals of cellular stress or damage.

Recognition of Pathogen-Associated Molecular Patterns (PAMPs) and Damage-Associated Molecular Patterns (DAMPs)

NLRs mainly recognize PAMPs and DAMPs. PAMPs are pathogen origin molecules, whereas DAMPs are host molecules that get released during cell damage or stress.

- NOD1 and NOD2: The peptidoglycan fragments are detected, which are the part of bacterial cell wall.
- NLRP3: This receptor is responsive to the entire series of microbial and non-microbial stimuli, ranging from bacterial toxins to crystalline substances to host-derived danger signals, which include ATP and uric acid crystals. Inflammation causes inflammasome formation—a multiprotein complex responsible for activating caspase 1, leading to the secretion of pro-inflammatory cytokines involving IL-1β and IL-18.

Signaling Pathways and Immune Response

Many of these NLRs trigger signaling that leads to the activation of inflammatory responses. Members of the family of NOD-like receptors, including NOD1 and NOD2, trigger NF-κB and MAPK pathways, leading to the induction of cytokines and antimicrobial peptides. Activation of NLRP3 results in the formation of inflammasomes, which triggers the processing and release of IL-1β and IL-18, leading to an acute inflammatory response. NLRs also regulate immune homeostasis to prevent excessive inflammation. Their dysregulation in signaling has been linked to inflammatory disorders such as Crohn's disease and rheumatoid arthritis.

RIG-I-like Receptors (RLRs)

RIG-I-like receptors are a family of cytoplasmic PRRs that specialize in the recognition of viral RNA. They are named as such because they were first characterized with the RIG-I protein, which was originally discovered for its functions in identifying RNA viruses.

Structure and Function

They are characterized by a domain of RNA helicase, which is critical for the recognition and binding of viral RNA. They also include a C-terminal domain that interacts with downstream signaling molecules. The main members of the RLR family are RIG-I, MDA5, and LGP2.

- RIG-I: It recognizes short, 5'-triphosphate RNA molecules commonly occurring in RNA viruses. RIG-I is involved in the recognition of a diversity of RNA viruses, particularly the influenza viruses and other negative-sense RNA viruses.
- MDA5: It recognises long double-stranded RNA molecules. It plays a crucial role in the recognition of picornaviruses and other RNA viruses having long dsRNA.
- LGP2: doesn't have CARD domains necessary for the signalling and acts to modulate RIGI and MDA5 activity. The modulation by LGP2 of RLR dependent signalling can be positive or negative, depending on the context.

Viral Recognition Properties

RLRs recognize different signatures of viral RNA, which involve the 5'-triphosphate group and double-stranded structures. Their recognition is absolutely necessary for detecting and responding to viral infections.

- RIG-I: This molecule recognizes short RNA molecules with a 5'-triphosphate modification, which is a characteristic of replicating viruses.
- MDA5: This molecule recognizes long dsRNA structures which are part of the genome of most RNA viruses.

Signaling Pathways and Immune Response

After binding to viral RNA, RLRs become conformationally changed, which allows them to interact with the adaptor protein MAVS. MAVS triggers the nuclear factor κB (NF-κB) and IRF3/7 pathways, resulting in the induction of the type I IFN genes and other cytokines. Type I IFNs are required for the establishment of an antiviral state in infected cells and also in cells neighboring an infected cell by increasing the adaptive immune response. RLRs are involved in the early host response to viral infections and the initiation of antiviral responses. Over-activation of RLR signaling thus predisposes to continuous virus persistence and chronic inflammation.

Inflammatory Response

The inflammatory response is a fundamental aspect of the innate immune system, serving as the body's first line of defense against infections, injuries, and other harmful stimuli. This complex biological process involves the activation of innate immune cells and the release of signaling molecules, leading to a series of physiological changes designed to eliminate pathogens and repair damaged tissues. The hallmark signs of inflammation—redness, heat, swelling, and pain—reflect the underlying biochemical and cellular events that occur during this response. While inflammation is essential for pathogen clearance and tissue repair, it can also contribute to tissue damage and chronic diseases if not properly regulated.

Mechanisms of Inflammation

Initiation of Inflammation

The inflammatory response begins when innate immune cells, such as macrophages, dendritic cells, and mast cells, detect pathogen-associated molecular patterns (PAMPs) or damage-associated molecular patterns (DAMPs) through pattern recognition receptors (PRRs). These receptors, including Toll-like receptors (TLRs), NOD-like receptors (NLRs), and RIG-I-like receptors (RLRs), recognize specific molecular signatures associated with pathogens or tissue damage. Upon recognition of these signals, innate immune cells become activated and release a variety of signaling molecules, such as cytokines, chemokines, and prostaglandins. These mediators act to recruit additional immune cells to the site of infection or injury, enhance vascular permeability, and induce local and systemic changes aimed at containing and resolving the inflammatory response.

Cellular and Molecular Events

1. *Recruitment of Immune Cells*
   One of the key features of inflammation is the recruitment of immune cells to the affected area. This process is mediated by chemokines, which are small signaling proteins that attract immune cells to the site of inflammation. For example, cytokines like tumor necrosis factor-alpha (TNF-α) and interleukin-1β (IL-1β) play a role in inducing the expression of adhesion molecules on the endothelial cells of blood vessels, such as selectins and integrins. These molecules facilitate the binding of circulating leukocytes (e.g., neutrophils and monocytes) to the endothelium, allowing them to migrate into the tissue.
2. *Vascular Changes*
   Inflammation leads to significant changes in the blood vessels in the affected area. The release of mediators like histamine, bradykinin, and prostaglandins causes vasodilation, resulting in increased blood flow to the site of inflammation. This increased blood flow contributes to the redness (rubor) and heat (calor) observed during inflammation. Additionally, the increased permeability of blood vessels allows plasma proteins and immune cells to extravasate into the tissue, leading to the accumulation of fluid and proteins in the interstitial space. This fluid accumulation causes swelling (tumor).
3. *Activation of Acute-Phase Response*
   The acute-phase response is a systemic reaction to inflammation that involves the liver's production of acute-phase proteins, such as C-reactive protein (CRP) and fibrinogen. These proteins help to enhance the immune response by promoting pathogen clearance and modulating inflammation. For example, CRP can bind to pathogens and facilitate their recognition and phagocytosis by immune cells. The acute-phase response also includes systemic changes such as fever, which is induced by pyrogens (fever-inducing substances) that act on the hypothalamus to raise the body's temperature.

Clinical Manifestations of Inflammation

1. *Redness (Rubor)*
   Redness is caused by increased blood flow to the inflamed area. The dilation of blood vessels, driven by mediators like histamine and prostaglandins, leads to an influx of red blood cells and an accumulation of blood in the affected tissue. This increased blood flow results in the characteristic redness observed in inflamed tissues.

2. *Heat (Calor)*
   Heat is a consequence of increased blood flow and metabolic activity in the inflamed area. The higher volume of blood reaching the site of inflammation raises the local temperature, which contributes to the sensation of warmth.
3. *Swelling (Tumor)*
   Swelling is caused by the accumulation of fluid in the interstitial space. Increased vascular permeability allows plasma proteins and fluids to leak out of the blood vessels and into the surrounding tissue. This accumulation of fluid leads to edema, which manifests as swelling.
4. *Pain (Dolor)*
   Pain is a result of the release of inflammatory mediators, such as prostaglandins and bradykinin, that sensitize nerve endings and increase their sensitivity to mechanical and chemical stimuli. Additionally, the pressure exerted by the swollen tissue can compress nerve endings and contribute to the sensation of pain.

Resolution of Inflammation

The resolution of inflammation is an active process that involves the cessation of pro-inflammatory signals and the promotion of tissue repair. Several mechanisms contribute to the resolution of inflammation:

1. *Removal of Inflammatory Mediators*
   The degradation and removal of inflammatory mediators, such as cytokines and chemokines, are essential for resolving inflammation. Enzymatic degradation and uptake by macrophages help to reduce the levels of these mediators and terminate the inflammatory response.
2. *Phagocytosis of Dead Cells and Debris*
   Macrophages and other phagocytic cells play a crucial role in clearing apoptotic cells, cellular debris, and pathogens from the inflamed tissue. This process helps to restore tissue homeostasis and prevents the accumulation of harmful substances.
3. *Production of Anti-inflammatory Mediators*
   The release of anti-inflammatory mediators, such as IL-10 and transforming growth factor-beta (TGF-β), promotes the resolution of inflammation and tissue repair. These mediators inhibit the production of pro-inflammatory cytokines and promote the repair and remodeling of damaged tissues.
4. *Restoration of Tissue Architecture*
   The repair and regeneration of damaged tissues involve the proliferation of fibroblasts, the synthesis of extracellular matrix components, and the formation of new blood vessels (angiogenesis). This process helps to restore tissue function and integrity.

Dysregulation and Chronic Inflammation

While inflammation is a protective response, its dysregulation can lead to chronic inflammation and tissue damage. Chronic inflammation is characterized by the persistence of inflammatory mediators and the continuous recruitment of immune cells, leading to ongoing tissue damage and fibrosis.

1. *Chronic Inflammatory Diseases*
   Chronic inflammation is associated with various diseases, including rheumatoid arthritis, inflammatory bowel disease, and chronic obstructive pulmonary disease (COPD). In these conditions, the inflammatory response becomes maladaptive and contributes to tissue destruction and functional impairment.
2. *Autoimmune Diseases*
   In autoimmune diseases, the immune system mistakenly targets and attacks the body's own tissues, leading to chronic inflammation and damage. Conditions such as systemic lupus erythematosus (SLE) and multiple sclerosis (MS) are examples of autoimmune diseases characterized by persistent inflammation.
3. *Metabolic Disorders*
   Chronic low-grade inflammation is also linked to metabolic disorders such as obesity, type 2 diabetes, and atherosclerosis. In these conditions, inflammation contributes to insulin resistance, endothelial dysfunction, and cardiovascular disease.

Adaptive Immunity

Adaptive immunity, also known as acquired immunity, is a specialized defense mechanism that provides a highly specific and long-lasting response to pathogens. Unlike innate immunity, which offers a general defense against a broad range of pathogens, adaptive immunity is tailored to recognize and combat specific antigens, the unique molecular structures found on pathogens. This targeted approach allows the immune system to mount a more effective response and remember previous encounters with pathogens, leading to improved protection upon subsequent exposures.

Components and Mechanisms

Lymphoid Organs in Vertebrates

Lymphoid organs are essential components of the vertebrate immune system, present across all vertebrate species. They are typically classified into two categories based on their functions: primary and secondary lymphoid organs. Each category plays a distinct role in the immune system, contributing to the development, organization, and activation of immune cells.

Primary Lymphoid Organs

Primary lymphoid organs are specialized sites where lymphocytes, the key players in the adaptive immune response, develop and mature. These organs provide the necessary microenvironments for the generation and initial selection of lymphocytes.

1. *Bone Marrow*
   The bone marrow is a primary lymphoid organ found in all vertebrates. It serves as the site for the generation of hematopoietic stem cells (HSCs), which give rise to all blood cells, including lymphocytes. In mammals, the bone marrow is the principal site for the development of B cells, which undergo maturation and differentiation before entering the peripheral circulation.
2. *Thymus*
   The thymus is another crucial primary lymphoid organ, where T cells mature. It is located in the pharyngeal region and is thought to have evolved from the pharyngeal endoderm. The thymus provides a specialized environment for the maturation of T cells and their selection for self-tolerance, ensuring that T cells do not attack the body's own tissues. The development of T cells involves interaction with stromal cells and the expression of key transcription factors, such as FOXN1, which regulate the development of thymic epithelial cells. The thymus is found in various forms across different vertebrate species. In lampreys, thymic tissues, known as thymoids, are located at the tips of gill filaments. In sharks, thymic tissues are associated with some pharyngeal arches, while in teleosts, most thymic tissue originates from the third pharyngeal pouch. Birds and mammals have more complex thymic structures, with birds exhibiting multiple thymic lobes along the neck and mammals having a primarily thoracic thymus, with additional tissues sometimes found in the neck.

Secondary Lymphoid Organs

Secondary lymphoid organs are involved in the coordination of immune responses and the organization of immune cell interactions. They act as sites where immune cells can encounter antigens, become activated, and mount a targeted immune response.

1. *Spleen*
   The spleen is one of the most ancient secondary lymphoid organs. It plays a crucial role in filtering the blood and coordinating immune responses. The spleen's white pulp contains lymphoid tissues that interact with various immune effector cells. In jawed vertebrates, the spleen's structure becomes increasingly complex over evolutionary time, paralleling the diversification of immune cells and their functions. In lampreys, small foci of lymphoid cells in the gut-associated areas are considered to be early forms of the spleen.
2. *Lymph Nodes*
   Lymph nodes represent a significant innovation in the vertebrate immune system, evolving with the advent of reptiles and birds. They are distributed throughout the body and are essential for T cell activation, antibody affinity maturation, and memory responses. Lymph nodes are thought to have evolved from regional specializations of lymph vessels, which facilitated the formation of lymphoid aggregations at these sites. The development of lymph nodes is closely linked to the evolution of certain gene families, such as those encoding tumor necrosis factor (TNF) ligands and receptors, as well as chemokine receptors. Lymphoid tissue inducer (LTi) cells, which are crucial

for the development of secondary lymphoid organs, are essential for lymph node formation but not for primordial lymphoid organs like the spleen.

Evolutionary Considerations

The evolutionary development of lymphoid organs reveals a pattern of increasing complexity and specialization. Primary lymphoid organs, such as the bone marrow and thymus, have ancient origins and are fundamental to lymphocyte development. Secondary lymphoid organs, including the spleen and lymph nodes, exhibit greater morphological diversity and complexity, reflecting the evolution of more sophisticated immune responses.

1. *Emergence of Primary Lymphoid Organs*
   The primary lymphoid organs have emerged early in vertebrate evolution, providing essential functions for the generation and maturation of lymphocytes. These organs have remained relatively conserved in their basic roles across different vertebrate species.
2. *Development of Secondary Lymphoid Tissues*
   Secondary lymphoid organs have undergone significant evolutionary changes, with new structures and functions emerging over time. For example, lymph nodes represent a major evolutionary advancement, providing specialized microenvironments for immune cell activation and coordination. The spleen, while ancient, has also evolved to exhibit increasing morphological complexity, reflecting its role in the regulation of immune responses. The presence of lymphoid tissue organizer (LTo) cells, which play a critical role in the formation of secondary lymphoid organs, further illustrates the diversity and specialization of lymphoid tissues. These cells have different origins and functions depending on the type of lymphoid organ they support. For example, the LTo cells of the spleen originate from dorsal pancreatic mesenchyme, while those of peripheral lymph nodes are derived from adipocytes reprogrammed by lymphotoxin β receptor (LTβR) signaling.

Lymphocytes

Lymphocytes are the primary cellular effectors in adaptive immune responses and represent a relatively recent development in hematopoietic cell types. They are found universally across vertebrates, but similar cells in chordates are less well-documented. While cells analogous to vertebrate lymphocytes have been observed in amphioxus, there is no evidence that these cells undergo proliferation in response to antigens, a key characteristic of vertebrate lymphocytes. The immune systems of vertebrates are characterized by two major lymphocyte lineages, which are distinguished by their distinct genetic regulation, developmental sites, and antigen receptor functions. In jawed vertebrates, B cells develop in the bone marrow or its functional equivalents. Key transcription factors such as PAX5 and E2A, crucial for B cell development in mice, are also present and active during B cell development in lower vertebrates. This suggests that the genetic networks responsible for B cell development are evolutionarily conserved. Indeed, homologs of these genes are found in the lamprey genome and are expressed in its B-like lineage, indicating that the genetic program for antibody-producing lymphocytes predates the divergence of vertebrate lineages. B cell development in mice relies on the interaction of interleukin 7 (IL7) and FLT3 ligand, which are critical components of the B cell niche, originating from the mesenchymal germ layer. However, it remains unclear whether homologs of IL7 and FLT3 ligand exist in basal vertebrates or if alternative factors regulate B cell development in these species. The Notch signaling pathway, involving the NOTCH1 receptor and its ligand DLL4, is essential for T cell lineage specification. Lymphoid progenitor cells are attracted to the thymus—the central site of T cell differentiation—by chemokine gradients from the thymic microenvironment. In the thymus, the Notch ligand initiates the T cell differentiation program. Experimental evidence shows that CXCL12, a chemokine and growth factor for early thymocytes, plays a crucial role in the initial stages of T cell development. The presence of CXCL12 and DLL4 in the thymic microenvironment can support T cell development up to the CD4+/CD8+ double-positive stage. However, this system, which uses transgenic expression of individual molecules in Foxn1-deficient thymic epithelia, does not fully address the molecular requirements for later stages of differentiation. More refined experimental models are needed to study the expression of self-peptide/MHC complexes essential for positive selection of immature T cells. While these interactions are vital for developing T cells with αβ T cell receptors (TCRs), they are not required for γδ T cell development. When only CXCL12 and DLL4 are present in the thymic environment, progenitor cells do not follow the γδ T cell developmental pathway, resulting in the absence of γδ T cells. IL7, which is crucial for γδ T cell lineage as well as B and αβ T cell development, might be a key factor to investigate. However, neither IL7 nor its receptor (IL7R) genes have been identified in cartilaginous fishes, raising questions about whether IL7/IL7R signaling is less conserved compared to CXCL12/CXCR4 and DLL4/NOTCH1 pathways. Detailed genomic and transcriptional studies of these early vertebrates could clarify the evolutionary conservation of these signaling pathways. These experimental approaches could help identify the minimal set of factors necessary for developing mature T cell subsets in the thymus and provide insights into the essential requirements for B cell lineage development.

Microbiome Influence on Immune System Development

Early Microbial Colonization and Immune Priming

The development of the immune system in domestic animals is a complex and finely tuned process that begins at birth and is significantly influenced by early microbial colonization. Neonatal animals, much like human infants, enter the world with a relatively sterile internal environment. However, this changes rapidly as they are exposed to a variety of microbes from their immediate surroundings. This initial microbial exposure plays a crucial role in immune priming, setting the foundation for a robust and responsive immune system that will protect the animal throughout its life (Ignacio et al., 2024).

The Role of Maternal Microbiota in Initial Colonization

The process of microbial colonization begins at birth, where neonates are exposed to the maternal microbiota. The mode of delivery, whether vaginal or via cesarean section, significantly influences the composition of the initial microbial community that colonizes the neonate's gut. During vaginal birth, neonates are exposed to the maternal vaginal and fecal microbiota, which typically includes beneficial bacteria such as Lactobacillus and Bifidobacterium species. These bacteria are essential for establishing a healthy gut microbiome and play a pivotal role in the early development of the immune system. In contrast, neonates delivered by cesarean section are often colonized by skin-associated bacteria, such as Staphylococcus and Corynebacterium species. This difference in early microbial exposure can have long-term implications for immune system development, with cesarean-born animals potentially having a delayed or altered immune response compared to those born via vaginal delivery. The transfer of maternal microbiota is not limited to the birth process. Maternal milk also serves as a vital source of beneficial microbes and immunomodulatory factors, such as immunoglobulins and cytokines. Colostrum, the first milk produced after birth, is particularly rich in these components and is crucial for providing passive immunity to the neonate. It also contains prebiotics, which selectively promote the growth of beneficial gut bacteria, further supporting immune system development (Koren et al., 2024).

Microbial Succession in the Neonatal Gut

Following the initial colonization, the neonatal gut undergoes a dynamic process of microbial succession, where the composition of the gut microbiome changes over time in response to various factors, including diet, environment, and host genetics. During the first few weeks of life, the gut microbiome is relatively simple, dominated by a few key bacterial taxa. However, as the animal matures and its diet transitions from milk to solid food, the gut microbiome becomes more complex and diverse. This microbial succession is crucial for the development of the gut-associated lymphoid tissue (GALT), which is a key component of the immune system. The GALT includes structures such as Peyer's patches, isolated lymphoid follicles, and mesenteric lymph nodes, all of which are essential for initiating and regulating immune responses in the gut. The interaction between the developing immune system and the gut microbiome helps to shape the immune repertoire, promoting the maturation of immune cells and the establishment of immune tolerance (Chong et al., 2018).

Impact of Germ-Free Conditions on Immune Development

Studies using germ-free animals, which are raised in a sterile environment without exposure to microbes, have provided valuable insights into the critical role of the microbiome in immune development. Germ-free animals exhibit several immune deficiencies, highlighting the importance of microbial colonization in immune system maturation. For instance, germ-free animals have reduced lymphoid tissue, including smaller Peyer's patches and fewer isolated lymphoid follicles, which are critical for mounting effective immune responses against pathogens. Moreover, germ-free animals have fewer and less diverse T cells and B cells, which are essential components of the adaptive immune system. These animals also exhibit impaired production of secretory immunoglobulin A (IgA), an antibody that plays a crucial role in mucosal immunity by neutralizing pathogens and toxins at mucosal surfaces. The lack of microbial stimulation in germ-free animals leads to a failure to fully develop the immune system, leaving them more susceptible to infections and less capable of mounting effective immune responses. Interestingly, when germ-free animals are colonized with a diverse microbial community, many of these immune deficiencies can be reversed. This re-colonization process, known as microbial reconstitution, leads to the rapid expansion of lymphoid tissue, the development of a diverse and functional repertoire of immune cells, and the production of IgA. This demonstrates the remarkable plasticity of the immune system and its ability to respond to microbial signals even after a delayed start in microbial colonization.

Microbial Influence on Mucosal Immunity

Mucosal surfaces, such as the lining of the gut, are the primary interfaces where the host and the microbiome interact. The development of mucosal immunity is a key aspect of immune system maturation, and it is heavily influenced by early microbial colonization. The gut microbiome stimulates the production of IgA, which is secreted into the gut lumen and plays a vital role in neutralizing pathogens and preventing their attachment to the gut epithelium.In addition to IgA production, the microbiome influences the development of other components of mucosal immunity, such as the induction of regulatory T cells (Tregs). Tregs are essential for maintaining immune tolerance to harmless antigens, such as food proteins and commensal bacteria, while preventing excessive inflammatory responses that could damage the gut tissue. The presence of certain bacterial species, such as Bacteroides fragilis, has been shown to promote the development of Tregs, highlighting the importance of specific microbial taxa in shaping mucosal immune responses.The interaction between the microbiome and the mucosal immune system is bidirectional, with the immune system also influencing the composition of the microbiome. For example, the production of antimicrobial peptides by gut epithelial cells helps to regulate microbial populations and prevent the overgrowth of pathogenic bacteria. This intricate balance between the microbiome and the immune system is essential for maintaining gut homeostasis and preventing the development of inflammatory diseases (Figure 4).

Long-Term Implications of Early Microbial Colonization

The effects of early microbial colonization on immune development extend beyond the neonatal period, with long-term implications for the health and disease resistance of domestic animals. The composition of the gut microbiome during early life has been linked to the risk of developing various immune-related conditions later in life, including allergies, autoimmune diseases, and chronic inflammatory conditions. For example, early colonization with a diverse and balanced microbiome is associated with a lower risk of developing allergies, as it promotes the development of immune tolerance. In contrast, disruptions in early microbial colonization, such as those caused by antibiotic use or cesarean delivery, have been linked to an increased risk of allergic diseases and other immune-related disorders.

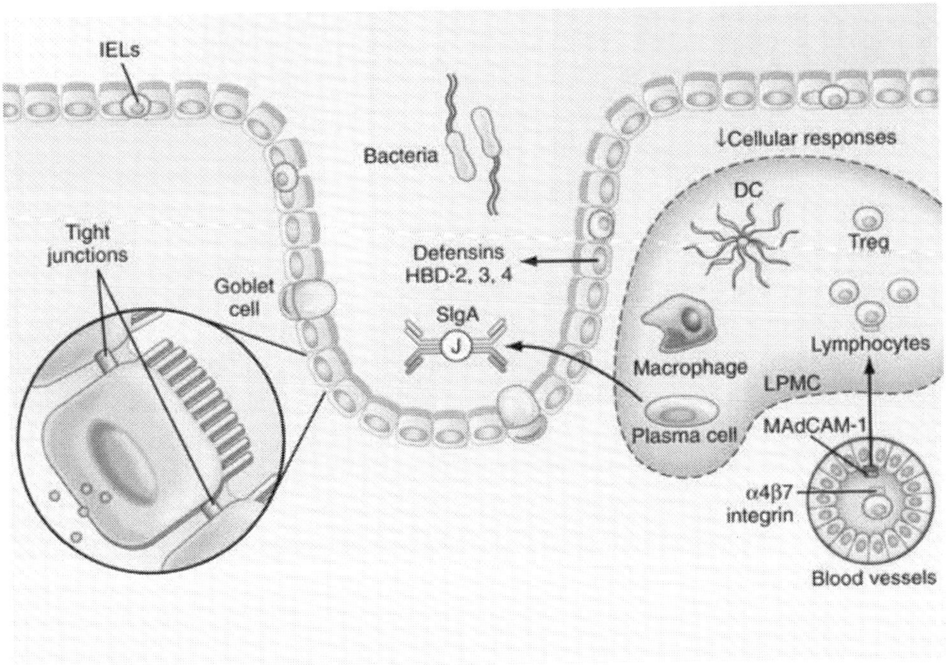

Figure 4. Mucosal immunity.

Role of Specific Microbial Taxa in Immune Development

The gut microbiome is composed of a vast array of microbial taxa, each contributing uniquely to the development and function of the host's immune system. Understanding the role of specific microbial species or groups in immune development is essential for deciphering the complex interactions that shape immune responses and maintain homeostasis. These microbes do not merely coexist with the host but actively participate in training and modulating the immune system, influencing everything from immune cell maturation to the establishment of immune tolerance (Pickard et al., 2017).

Bacteroides Species and Regulatory T Cells (Tregs)

One of the most studied microbial taxa in the context of immune development is the Bacteroides genus, which includes species such as Bacteroides fragilis. These bacteria play a pivotal role in promoting the maturation and function of regulatory T cells (Tregs), a subset of T cells that are essential for maintaining immune tolerance. Tregs prevent excessive immune responses that could lead to autoimmunity by suppressing the activity of potentially harmful T cells that recognize self-antigens.

Bacteroides fragilis has been shown to produce a specific polysaccharide, polysaccharide A (PSA), which is critical for the induction of Tregs. PSA interacts with receptors on the surface of dendritic cells (DCs), leading to the presentation of antigens in a way that promotes Treg differentiation. This interaction is crucial for establishing immune tolerance, particularly in the gut, where the immune system must distinguish between harmful pathogens and beneficial commensals or dietary antigens. The presence of Bacteroides species during critical windows of immune development is therefore essential for preventing the development of inflammatory and autoimmune conditions later in life. In addition to promoting Treg development, Bacteroides species contribute to the overall stability and resilience of the gut microbiome. Their ability to break down complex carbohydrates and produce short-chain fatty acids (SCFAs) such as butyrate also supports gut health by providing energy to colonocytes and promoting anti-inflammatory responses.

Lactobacilli and Mucosal Immunity

Lactobacilli are another group of beneficial bacteria that are well-known for their role in promoting mucosal immunity, particularly in the gut. These bacteria are often found in the gastrointestinal tract and other mucosal surfaces, where they contribute to the development and maintenance of a healthy immune environment.One of the key ways in which Lactobacilli contribute to immune development is by stimulating the production of secretory immunoglobulin A (IgA). IgA is the most abundant antibody at mucosal surfaces and plays a crucial role in neutralizing pathogens and preventing their attachment to and invasion of epithelial cells. Lactobacilli promote IgA production by interacting with gut-associated lymphoid tissue (GALT), where they stimulate the differentiation of B cells into IgA-secreting plasma cells. The role of Lactobacilli in promoting mucosal immunity extends beyond IgA production. These bacteria also produce a variety of metabolites, such as lactic acid, which create an acidic environment that is inhospitable to many pathogens. Additionally, Lactobacilli can modulate the activity of dendritic cells and other immune cells, further enhancing the host's ability to mount effective immune responses at mucosal surfaces. The presence of Lactobacilli in the gut during early life is particularly important for establishing a robust mucosal immune system. Animals that harbor these bacteria from an early age tend to have better-developed GALT and a more effective mucosal barrier, which helps protect against a wide range of gastrointestinal infections.

Segmented Filamentous Bacteria (SFB) and Th17 Cell Differentiation

Segmented filamentous bacteria (SFB) are a unique group of commensal bacteria that have been shown to play a critical role in the differentiation of T helper 17 (Th17) cells, a subset of CD4+ T cells that are important for the defense against extracellular pathogens, particularly at mucosal surfaces. Th17 cells produce the pro-inflammatory cytokine interleukin-17 (IL-17), which recruits neutrophils and other immune cells to the site of infection, enhancing the host's ability to clear bacterial and fungal pathogens. SFB are particularly interesting because of their intimate association with the gut epithelium. They form tight attachments to the epithelial cells lining the small intestine, where they induce the production of Th17 cells. This process involves the activation of specific pattern recognition receptors (PRRs) on epithelial cells and dendritic cells, leading to the release of cytokines such as IL-6 and IL-23, which are essential for Th17 cell differentiation. The induction of Th17 cells by SFB is a double-edged sword. While these cells are important for protecting against infections, an overactive Th17 response can contribute to inflammatory diseases such as inflammatory bowel disease (IBD) and autoimmune conditions. Therefore, the presence and activity of SFB must be tightly regulated to maintain immune balance. In the context of immune development, the colonization of the gut by SFB during early life is associated with the establishment of a functional Th17 cell population. This is crucial for developing a balanced immune response that can effectively respond to pathogens without causing excessive inflammation. The absence of SFB or disruptions in their colonization can lead to impaired Th17 cell development, which may increase susceptibility to infections and dysregulate immune responses.

Clostridia and Regulatory T Cells (Tregs)

Beyond Bacteroides species, certain members of the Clostridia class, particularly clusters IV and XIVa, have also been identified as important in the promotion of Treg differentiation. These bacteria produce butyrate, a short-chain fatty acid (SCFA) that plays a significant role in enhancing the Treg population in the colon. Butyrate has been shown to influence the expression of genes involved in Treg differentiation and function, contributing to immune tolerance and anti-inflammatory responses in the gut. The colonization of the gut by Clostridia during early life is associated with an increase in colonic Tregs, which help maintain gut homeostasis and protect against inflammatory diseases. The presence of these bacteria in the gut microbiome also supports the production of other SCFAs, such as acetate and propionate, which further contribute to the regulation of immune responses and the maintenance of a healthy gut environment.

Faecalibacterium Prausnitzii and Anti-Inflammatory Effects

Faecalibacterium prausnitzii is another key member of the gut microbiome that has been recognized for its potent anti-inflammatory properties. This bacterium is a major producer of butyrate, which not only supports Treg differentiation but also has direct anti-inflammatory effects on the gut epithelium. Butyrate enhances the integrity of the gut barrier by promoting the production of tight junction proteins, reducing the translocation of harmful bacteria and antigens into the bloodstream. In addition to its role in maintaining gut barrier function, Faecalibacterium prausnitzii has been shown to suppress the production of pro-inflammatory cytokines, such as IL-12 and interferon-gamma (IFN-γ), while promoting the secretion of anti-inflammatory cytokines like IL-10. This balance of cytokines is crucial for preventing chronic inflammation and maintaining immune homeostasis. The presence of Faecalibacterium prausnitzii in the gut microbiome is therefore associated with a lower risk of developing inflammatory conditions such as IBD. Ensuring that animals harbor a healthy population of this bacterium during early life can contribute to the long-term stability of the immune system and reduce the risk of immune-mediated diseases.

Helicobacter Species and Immune Modulation

While many commensal bacteria contribute positively to immune development, some microbial taxa, such as certain Helicobacter species, have a more complex relationship with the immune system. Helicobacter species are known to interact closely with the host's immune system, often leading to chronic immune activation and inflammation. However, in some contexts, these bacteria can also modulate the immune response in ways that may protect against other forms of immune-mediated diseases, such as allergies and autoimmunity. Helicobacter species have been shown to influence the balance between different subsets of T cells, including Tregs and Th17 cells. The presence of Helicobacter can promote the expansion of Tregs, which may help to limit excessive immune responses and maintain tolerance. However, these bacteria can also drive the differentiation of Th17 cells, particularly in the context of chronic infection, leading to persistent inflammation. The impact of Helicobacter species on immune development is therefore highly context-dependent. In some cases, these bacteria may contribute to the development of immune tolerance and protection against autoimmunity, while in other situations, they may promote chronic inflammation and contribute to the pathogenesis of diseases such as gastritis and gastric cancer.

Long-Term Implications of Microbial Taxa on Immune Competency

The presence or absence of specific microbial taxa during critical periods of immune development can have profound and lasting effects on the host's immune competency. Early colonization with beneficial microbes such as Bacteroides, Lactobacilli, SFB, and Clostridia can promote the development of a balanced and responsive immune system, characterized by effective pathogen defense, immune tolerance, and the prevention of chronic inflammation. Conversely, disruptions in the colonization process, whether due to antibiotic use, diet, or other environmental factors, can lead to the loss of key microbial taxa and the subsequent dysregulation of immune responses. This can increase the risk of developing a wide range of immune-related conditions, including allergies, autoimmune diseases, and chronic inflammatory disorders.

Microbial Metabolites and Immune Modulation

The interplay between the gut microbiome and the immune system is profoundly influenced by microbial metabolites, which serve as critical mediators of immune modulation. Among these metabolites, short-chain fatty acids (SCFAs) such as butyrate, acetate, and propionate are of particular importance due to their significant impact on immune development and function. These SCFAs are primarily produced through the fermentation of dietary fibers by gut bacteria, and their effects on the host's immune system are wide-ranging (Mann et al., 2024), (Figure 5).

Short-Chain Fatty Acids (SCFAs) and Immune Cell Differentiation

SCFAs, especially butyrate, play a crucial role in the differentiation and function of various immune cells. Butyrate, in particular, has been shown to promote the differentiation of regulatory T cells (Tregs), which are essential for maintaining immune tolerance and preventing excessive inflammatory responses. Tregs help to suppress immune reactions against self-antigens and commensal microbes, thereby preventing autoimmune diseases and maintaining gut homeostasis. Butyrate influences Treg differentiation through several mechanisms. One key pathway involves the inhibition of histone deacetylases (HDACs), which leads to increased acetylation of histones and the upregulation of genes associated with Treg development. This epigenetic regulation by butyrate enhances the expression of Foxp3, a transcription factor that is critical for the development and function of Tregs. As a result, butyrate not only increases the number of Tregs but also enhances their suppressive function, contributing to the establishment of immune tolerance in the gut. In addition to promoting Treg differentiation, SCFAs also impact other immune cells. For instance, propionate and acetate have been shown to influence the activity of neutrophils, enhancing their chemotaxis and phagocytic abilities, which are vital for the rapid clearance of pathogens. SCFAs also modulate the differentiation of B cells into plasma cells that produce immunoglobulin A (IgA), an antibody crucial for mucosal immunity. This IgA production helps to neutralize pathogens at mucosal surfaces and prevent their invasion into the host tissues.

Microbial Metabolites and Dendritic Cell Maturation

Dendritic cells (DCs) are antigen-presenting cells that play a pivotal role in bridging the innate and adaptive immune systems. They capture, process, and present antigens to T cells, initiating and shaping the adaptive immune response. The maturation and function of DCs are heavily influenced by microbial metabolites, particularly SCFAs. Butyrate and other SCFAs modulate dendritic cell maturation in several ways. Butyrate, for example, has been shown to inhibit the production of pro-inflammatory cytokines by DCs, such as interleukin-12 (IL-12) and tumor necrosis factor-alpha (TNF-α). This inhibition occurs through the suppression of the nuclear factor-kappa B (NF-κB) signaling pathway, which is a key regulator of inflammation. By dampening the production of these cytokines, butyrate promotes a more anti-inflammatory environment, which is conducive to the development of immune tolerance rather than a pro-inflammatory response. Moreover, SCFAs can influence the surface expression of co-stimulatory molecules on dendritic cells, such as CD80 and CD86, which are essential for T cell activation. By modulating the expression of these molecules, microbial metabolites can influence whether the immune response is skewed towards tolerance (characterized by Treg induction) or towards activation and inflammation (characterized by effector T cell induction). The impact of SCFAs on dendritic cell maturation is also linked to the metabolic programming of these cells. For instance, butyrate has been shown to enhance oxidative phosphorylation and reduce glycolysis in DCs, leading to a metabolic state that favors tolerance induction. This metabolic reprogramming is crucial because the energy demands and metabolic pathways of DCs influence their function and the type of immune response they promote.

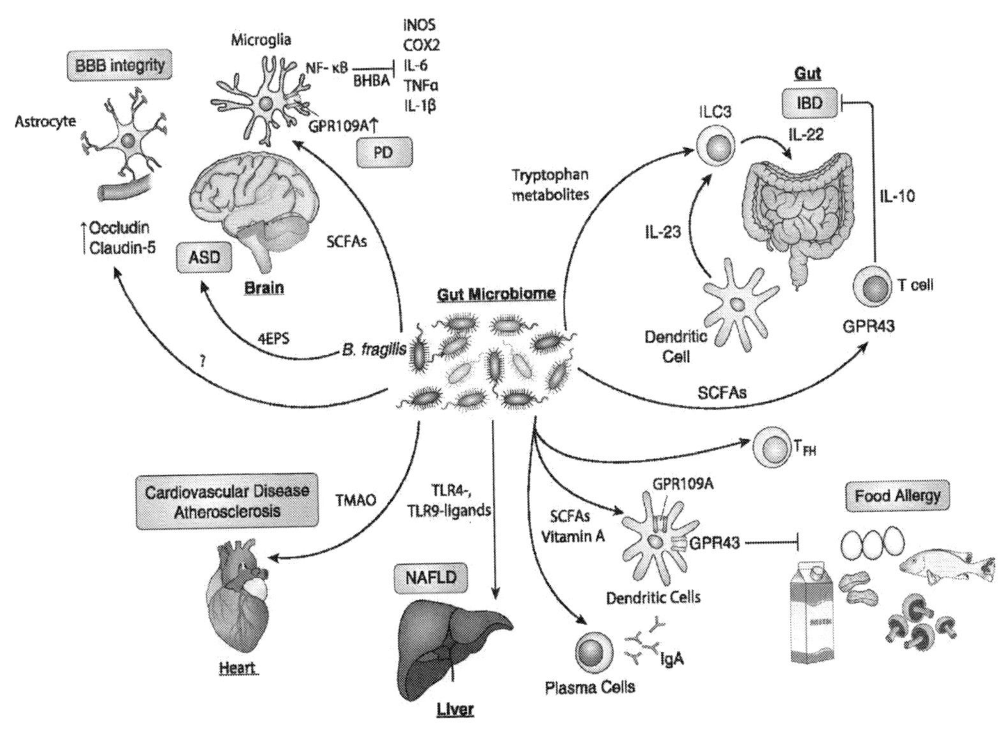

Figure 5. Microbial metabolite and immune system.

Systemic Effects of Microbial Metabolites on Immune Regulation

While the effects of microbial metabolites like SCFAs are most pronounced in the gut, their influence extends beyond the gastrointestinal tract, impacting systemic immune regulation. SCFAs can enter the bloodstream and exert effects on distant organs, including the liver, lungs, and brain. In the liver, for instance, SCFAs have been shown to modulate the production of acute-phase proteins and influence systemic inflammation. In the lungs, SCFAs can affect the function of alveolar macrophages, which are crucial for maintaining respiratory immune homeostasis and defending against inhaled pathogens. One of the key systemic effects of SCFAs is their role in modulating the immune system's response to infections and inflammatory stimuli. For example, studies have shown that butyrate can protect against systemic inflammation and sepsis by enhancing the barrier function of the gut epithelium, thereby preventing the translocation of bacteria and endotoxins into the bloodstream. This barrier function is crucial for preventing systemic inflammation and the development of conditions such as sepsis. Furthermore, SCFAs have been implicated in the regulation of the blood-brain barrier (BBB) and neuroinflammation. Butyrate, in particular, has been shown to enhance the integrity of the BBB and reduce neuroinflammation, which has implications for neuroimmune interactions and the development of neurodegenerative diseases. The systemic anti-inflammatory effects of SCFAs thus highlight their importance not only in gut immunity but also in overall immune homeostasis.

Microbial Metabolites as Therapeutic Targets

Given the significant role of microbial metabolites in immune modulation, there is growing interest in targeting these metabolites for therapeutic purposes. Modulating the production of SCFAs through diet, probiotics, or prebiotics is one potential strategy for enhancing immune health and preventing or treating immune-related diseases. For instance, increasing the intake of dietary fibers that are fermented by gut bacteria to produce SCFAs could enhance the production of butyrate and other beneficial metabolites. This approach could be particularly useful in conditions characterized by chronic inflammation, such as inflammatory bowel disease (IBD), where boosting SCFA levels could help restore immune balance and promote mucosal healing. Probiotic supplementation with butyrate-producing bacteria, such as certain strains of Clostridia or Faecalibacterium prausnitzii, is another potential strategy for enhancing SCFA production and improving immune function. These probiotics could help re-establish a healthy microbiome composition in individuals with dysbiosis, thereby enhancing the production of SCFAs and other beneficial metabolites. Additionally, the development of SCFA analogs or butyrate-releasing drugs is being explored as a direct means of modulating immune responses. Such therapies could be tailored to target specific immune cells or tissues, offering a precise approach to immune modulation in various diseases.

Modulation of Immune Responses by Microbiomes

Microbiome-Mediated Enhancement of Immune Responses

The gut microbiome, a complex community of microorganisms residing in the gastrointestinal tract, exerts profound effects on the immune system beyond its initial development. Throughout the life of domestic animals, the microbiome continues to modulate immune responses, enhancing the body's ability to detect and respond to pathogens. This modulation involves several mechanisms that lead to a more robust and efficient immune response, contributing to the animal's overall health and resistance to diseases (Tewari et al., 2024).

Trained Immunity: Priming of Innate Immune Cells

One of the key ways the microbiome enhances immune responses is through the phenomenon of "trained immunity." Traditionally, immunological memory was thought to be a feature exclusive to the adaptive immune system, particularly through the action of memory T and B cells. However, recent studies have shown that components of the innate immune system, such as macrophages and neutrophils, can also develop a form of memory that allows them to respond more effectively to subsequent infections. Trained immunity is characterized by an epigenetic reprogramming of innate immune cells, which leads to enhanced antimicrobial functions upon re-exposure to pathogens. This reprogramming can be triggered by microbial signals from the gut microbiome, including microbial metabolites, cell wall components like lipopolysaccharides (LPS), and other microbial antigens. These signals stimulate innate immune cells, leading to increased production of cytokines, enhanced phagocytic activity, and improved pathogen recognition. For example, certain components of the microbiome, such as β-glucans from fungal cell walls, can induce trained immunity in macrophages. Upon re-exposure to a pathogen, these macrophages exhibit heightened responsiveness, including increased production of reactive oxygen species (ROS) and enhanced phagocytic activity. This heightened state of readiness allows for a more rapid and effective clearance of infections, reducing the severity and duration of disease. Moreover, trained immunity is not limited to the site of initial microbial exposure. The systemic circulation of trained immune cells, including macrophages and monocytes, allows these cells to confer enhanced immunity at distant sites, such as the lungs, liver, and skin. This systemic effect highlights the far-reaching influence of the gut microbiome on the overall immune competence of domestic animals.

Microbiome-Enhanced Adaptive Immune Responses

In addition to modulating innate immunity, the microbiome also enhances adaptive immune responses, which are critical for long-term immunity and protection against reinfection. The gut microbiome influences the development and function of various adaptive immune cells, including T cells and B cells, through the production of microbial antigens and metabolites. One of the key ways the microbiome enhances adaptive immunity is by promoting the maturation and differentiation of T cells, particularly effector T cells that are crucial for responding to pathogens. The gut microbiome stimulates the production of cytokines such as interleukin-17 (IL-17) and interleukin-22 (IL-22), which are important for the differentiation of Th17 cells. Th17 cells play a vital role in defending against extracellular bacteria and fungi, particularly at mucosal surfaces. The presence of segmented filamentous bacteria (SFB) in the gut, for instance, has been shown to drive the differentiation of Th17 cells, leading to enhanced mucosal immunity. Furthermore, the microbiome influences the generation of memory T cells, which are essential for long-term immune protection. The interaction between microbial antigens and dendritic cells in the gut-associated lymphoid tissue (GALT) leads to the activation and differentiation of naïve T cells into memory T cells. These memory T cells can rapidly respond to previously encountered pathogens, providing effective immunity against reinfections. The gut microbiome also enhances B cell function and antibody production, particularly the production of immunoglobulin A (IgA). IgA is the primary antibody isotype found in mucosal secretions and plays a crucial role in neutralizing pathogens and toxins at mucosal surfaces. The microbiome stimulates the differentiation of B cells into IgA-producing plasma cells, ensuring a robust humoral response against enteric pathogens. This microbiome-driven enhancement of IgA production is particularly important for protecting the host from gastrointestinal infections and maintaining gut homeostasis.

Microbiome-Mediated Modulation of Inflammatory Responses

While the microbiome enhances immune responses to pathogens, it also plays a critical role in modulating inflammatory responses, ensuring that the immune system does not overreact to harmless antigens or self-tissues. This balance between immune activation and regulation is essential for preventing chronic inflammation and autoimmune diseases. The gut microbiome influences the production of anti-inflammatory cytokines, such as interleukin-10 (IL-10) and transforming growth factor-beta (TGF-β), which help to maintain immune tolerance. Regulatory T cells (Tregs), which are critical for suppressing excessive immune responses, are particularly influenced by the gut microbiome. As mentioned earlier, SCFAs like butyrate promote the differentiation and function of Tregs, enhancing their ability to produce IL-10 and suppress inflammation. In addition to promoting Treg function, the microbiome can also influence the balance between pro-inflammatory and anti-inflammatory immune responses through its effect on dendritic cells. By modulating the maturation and cytokine production of dendritic cells, the microbiome can skew immune responses towards tolerance rather than inflammation. This modulation is crucial for maintaining gut homeostasis and preventing inflammatory diseases such as colitis. Moreover, the gut microbiome plays a role in shaping the systemic inflammatory response. The integrity of the gut barrier, which is influenced by the microbiome, prevents the translocation of bacteria and their products into the bloodstream. A compromised gut barrier, often referred to as "leaky gut," can lead to systemic inflammation and the development of diseases such as sepsis, metabolic syndrome, and chronic inflammatory conditions. By maintaining gut barrier integrity, the microbiome helps to prevent the systemic dissemination of inflammatory signals, thereby protecting the host from chronic inflammation.

Impact of Microbiome Diversity on Immune Responses

The diversity of the gut microbiome is a key factor in determining the effectiveness of immune modulation. A diverse microbiome is associated with a more robust and balanced immune response, while a loss of microbial diversity, often seen in dysbiosis, can lead to impaired immune function and increased susceptibility to infections and inflammatory diseases. Microbiome diversity ensures that a wide range of microbial signals and metabolites are available to influence immune cell function. For example, different microbial species produce different SCFAs and other metabolites, each with unique effects on immune cells. A diverse microbiome can therefore provide a broader range of immune-modulating signals, leading to a more balanced and effective immune response. Conversely, a loss of microbiome diversity, which can result from factors such as antibiotic use, poor diet, or environmental stressors, can disrupt immune modulation and lead to immune dysregulation. This dysregulation can manifest as increased susceptibility to infections, chronic inflammation, and autoimmune diseases. Restoring microbiome diversity through dietary interventions, probiotics, or fecal microbiota transplantation (FMT) has been shown to improve immune function and reduce the risk of immune-related diseases in domestic animals.

Immune Modulation via Microbial-Derived Signals

The interaction between the host immune system and the gut microbiome is heavily influenced by microbial-derived signals, which play a crucial role in shaping immune responses. These signals, derived from various microbial components, engage with pattern recognition receptors (PRRs) on immune cells, leading to the activation of immune pathways that can either stimulate or regulate immune responses. The outcome of these interactions depends on the context, including the nature of the microbial signal, the type of immune cell involved, and the overall microbial environment (Oldereid et al., 2024).

Pattern Recognition Receptors and Microbial-Derived Signals

Pattern recognition receptors (PRRs) are essential components of the innate immune system, designed to detect conserved microbial structures known as pathogen-associated molecular patterns (PAMPs). These receptors include toll-like receptors (TLRs), NOD-like receptors (NLRs), and RIG-I-like receptors (RLRs), each of which recognizes specific PAMPs. The engagement of PRRs by microbial-derived signals triggers downstream signaling pathways that lead to the production of cytokines, chemokines, and other immune mediators. One of the most well-studied microbial-derived signals is lipopolysaccharide (LPS), a component of the outer membrane of Gram-negative bacteria. LPS is recognized by TLR4, a PRR expressed on various immune cells, including macrophages, dendritic cells, and epithelial cells. The binding of LPS to TLR4 initiates a signaling cascade that leads to the activation of nuclear factor-kappa B (NF-κB) and the production of pro-inflammatory cytokines such as interleukin-1 (IL-1), tumor necrosis factor-alpha (TNF-α), and interleukin-6 (IL-6). These cytokines play a critical role in mounting an effective immune response against invading pathogens. Peptidoglycans, another class of microbial-derived signals, are found in the cell walls of Gram-positive bacteria and are recognized by PRRs such as TLR2 and NOD1/NOD2. The engagement of these receptors by peptidoglycans leads to the activation of similar inflammatory pathways, promoting the recruitment of immune cells to the site of infection and enhancing pathogen clearance.

Immune Tolerance and Chronic Exposure to Microbial Signals

While the acute activation of immune pathways by microbial-derived signals is essential for defending against infections, chronic exposure to these signals can lead to immune tolerance. Immune tolerance is a state in which immune cells become less responsive to subsequent stimulation by the same signals, thereby preventing excessive inflammation and tissue damage. For example, chronic exposure to low levels of LPS, which occurs in the context of a stable gut microbiome, can induce a state of tolerance in immune cells such as macrophages and dendritic cells. This tolerance is characterized by a reduced production of pro-inflammatory cytokines in response to subsequent LPS stimulation. The underlying mechanisms involve epigenetic changes, such as alterations in histone modifications and DNA methylation, which lead to the downregulation of genes associated with inflammation.

Immune tolerance induced by chronic microbial exposure is a crucial mechanism for maintaining immune homeostasis in the gut. The gut is constantly exposed to a vast array of microbial-derived signals, and without tolerance mechanisms, the immune system would be perpetually activated, leading to chronic inflammation and tissue damage. By inducing tolerance, the immune system can coexist with commensal microbes, allowing for a balanced immune response that protects against pathogens while avoiding unnecessary inflammation.

Balancing Pro-Inflammatory and Anti-Inflammatory Pathways

The outcome of immune modulation by microbial-derived signals is not solely determined by the nature of the signal but also by the context in which the signal is received. The balance between pro-inflammatory and anti-inflammatory pathways is influenced by the overall composition of the microbiome, the presence of other microbial signals, and the state of the host immune system. For instance, while LPS and peptidoglycans typically promote inflammation, their effects can be modulated by other microbial-derived signals that induce anti-inflammatory pathways. Short-chain fatty acids (SCFAs), produced by the fermentation of dietary fibers by gut bacteria, are known to enhance the function of regulatory T cells (Tregs) and promote the production of anti-inflammatory cytokines such as interleukin-10 (IL-10). The presence of SCFAs in the gut can dampen the inflammatory response to LPS and peptidoglycans, contributing to the maintenance of immune homeostasis. Similarly, certain microbial-derived signals can activate PRRs that are associated with anti-inflammatory responses. For example, TLR9, which recognizes unmethylated CpG DNA from bacteria, has been shown to promote the production of IL-10 and TGF-β, cytokines that are critical for immune regulation and tolerance. The engagement of TLR9 by microbial DNA can thus counterbalance the pro-inflammatory effects of LPS and other PAMPs, preventing excessive inflammation and promoting tissue repair. The ability of the microbiome to modulate the balance between pro-inflammatory and anti-inflammatory pathways is essential for the prevention of chronic inflammatory diseases. Dysbiosis, a state of imbalance in the microbiome, is often associated with a shift towards a more pro-inflammatory environment, leading to conditions such as inflammatory bowel disease (IBD), metabolic syndrome, and autoimmune diseases. Restoring the balance of microbial-derived signals through dietary interventions, probiotics, or prebiotics can help to re-establish immune homeostasis and prevent the progression of these diseases.

Microbial Signals and Immune System Development

In addition to modulating immune responses in mature animals, microbial-derived signals play a crucial role in the development of the immune system during early life. The initial colonization of the gut microbiome in neonates provides the first exposure to microbial signals, which are essential for the maturation of the immune system. As discussed earlier, germ-free animals, which lack exposure to microbial signals, have underdeveloped immune systems characterized by reduced lymphoid tissue, impaired T and B cell function, and a lack of fully mature dendritic cells. The introduction of microbial-derived signals, such as LPS and peptidoglycans, into these animals can partially restore immune function by promoting the maturation of immune cells and the development of gut-associated lymphoid tissue (GALT). Microbial-derived signals are also crucial for the establishment of immune tolerance during early life. The exposure of neonatal immune cells to low levels of microbial signals helps to "train" the immune system to distinguish between harmful pathogens and harmless commensal microbes. This training is essential for the prevention of autoimmune diseases and the development of a balanced immune response. The timing and nature of microbial exposure during early life are critical for proper immune development. Disruptions in the normal colonization process, such as those caused by antibiotic use or cesarean section delivery, can lead to long-term alterations in immune function and an increased risk of immune-related diseases. Understanding the role of microbial-derived signals in immune development can inform strategies to promote healthy immune maturation in domestic animals, including the use of probiotics or microbiome transplants in early life.

Microbiome and Vaccine Responses

The interaction between the microbiome and the immune system extends to the realm of vaccine responses, where the composition and diversity of the gut microbiome can significantly influence the efficacy and durability of vaccines in domestic animals. The gut microbiome acts as an adjuvant, modulating both the innate and adaptive immune responses elicited by vaccines, which is essential for generating effective and long-lasting immunity against infectious diseases.

Impact of Microbiome Composition on Vaccine Efficacy

Research has demonstrated that the composition of the gut microbiome can profoundly affect the immune system's ability to respond to vaccines. Animals with a rich and diverse microbiome tend to exhibit stronger and more sustained immune responses to vaccination. This includes heightened production of antibodies, which are crucial for neutralizing pathogens, and enhanced T-cell responses, which are vital for long-term immune memory. For instance, studies in livestock such as pigs and cattle have shown that animals with a more diverse gut microbiome exhibit a better response to vaccines, as evidenced by higher levels of vaccine-specific antibodies and a more robust cellular immune response. Conversely, animals with a less diverse or dysbiotic microbiome may have a weaker vaccine response, which can result in reduced protection against diseases. The mechanisms underlying these effects are complex and involve multiple interactions between the microbiome, gut-associated lymphoid tissue (GALT), and systemic immune responses. The microbiome contributes to the maturation and function of dendritic cells and other antigen-presenting cells, which are critical for the initiation of vaccine-induced immune responses. Moreover, microbial metabolites, such as short-chain fatty acids (SCFAs), play a role in shaping the immune environment, influencing the effectiveness of the immune response to vaccines.

Microbial Adjuvants and Vaccine Response

The microbiome can act as a natural adjuvant, enhancing the immune response to vaccines. Adjuvants are substances that enhance the body's immune response to an antigen, and microbial components can fulfill this role by stimulating the immune system. For example, microbial-derived signals like lipopolysaccharides (LPS) and peptidoglycans can activate pattern recognition receptors (PRRs) on immune cells, leading to the production of cytokines and chemokines that boost the immune response. In addition to naturally occurring microbial adjuvants, probiotics and prebiotics have been explored as strategies to modulate the microbiome in a way that enhances vaccine efficacy. Probiotics, which are live beneficial bacteria, can improve the diversity and composition of the gut microbiome, creating a more favorable environment for vaccine responses. Prebiotics, which are non-digestible fibers that promote the growth of beneficial bacteria, can also support a healthy microbiome, potentially enhancing the immune response to vaccination. Studies have shown that administering probiotics before vaccination can lead to improved antibody responses in animals. For example, in poultry, the use of probiotic supplements has been associated with enhanced immune responses to vaccines against pathogens like Newcastle disease virus and avian influenza. Similarly, in cattle, probiotics have been shown to enhance the response to vaccines against respiratory pathogens.

Timing and Microbiome Manipulation for Optimal Vaccine Responses

The timing of microbiome manipulation, such as the administration of probiotics or prebiotics, is critical for optimizing vaccine responses. Administering these interventions at specific time points, such as before or during vaccination, can maximize their beneficial effects. For instance, giving probiotics during the neonatal period, when the immune system is still developing, can have long-lasting effects on vaccine responses later in life. In addition to probiotics and prebiotics, other strategies for manipulating the microbiome to enhance vaccine responses are being explored. These include fecal microbiota transplantation (FMT) and the use of synbiotics (a combination of probiotics and prebiotics). FMT, which involves the transfer of fecal material from a healthy donor to a recipient, has been shown to restore microbiome diversity and improve immune function in animals with dysbiosis, potentially enhancing vaccine efficacy. Furthermore, understanding the specific microbial taxa that are associated with improved vaccine responses can inform targeted interventions. For example, certain bacteria within the Firmicutes and Bacteroidetes phyla have been linked to better immune responses to vaccines, suggesting that promoting the growth of these taxa could enhance vaccine efficacy.

Challenges and Considerations in Microbiome-Based Vaccine Modulation

While the potential for microbiome-based modulation of vaccine responses is promising, there are challenges and considerations that need to be addressed. One of the main challenges is the variability in microbiome composition among individual animals, which can lead to differences in vaccine responses. This variability is influenced by factors such as diet, environment, genetics, and previous exposure to microbes and antibiotics. Another consideration is the potential for unintended consequences of microbiome manipulation. While probiotics and prebiotics are generally considered safe, their effects on the microbiome and immune system can be complex and context-dependent. For example, introducing specific probiotic strains may promote the growth of certain beneficial microbes but could also inhibit others, potentially leading to unintended shifts in the microbiome that could affect vaccine responses. Additionally, the optimal composition of the microbiome for vaccine responses may differ depending on the type of vaccine and the pathogen being targeted. Live attenuated vaccines, for example, may interact differently with the microbiome compared to inactivated or subunit vaccines. Understanding these interactions is crucial for developing effective microbiome-based strategies for enhancing vaccine efficacy.

Future Directions in Microbiome and Vaccine Research

Future research on the microbiome's role in vaccine responses in domestic animals is likely to focus on several key areas. These include identifying specific microbial taxa and metabolites that enhance vaccine responses, optimizing the timing and dosage of microbiome interventions, and developing personalized approaches to microbiome modulation based on individual animal microbiome profiles. Additionally, there is growing interest in the use of next-generation sequencing and other advanced technologies to characterize the microbiome and its interactions with the immune system in more detail. These technologies can provide insights into the specific mechanisms by which the microbiome influences vaccine responses, leading to more targeted and effective interventions. The development of microbiome-based adjuvants, which could be included in vaccines to enhance their efficacy, is another promising area of research. These adjuvants could be designed to mimic the beneficial effects of certain microbial-derived signals, potentially improving vaccine responses in animals with suboptimal microbiomes.

Gut-Immune Axis: Central Role of the Gut Microbiome

The Gut as an Immune Organ

The gut is often referred to as the largest immune organ in the body, due to the extensive and diverse population of immune cells and tissues present within it. This characterization highlights the gut's central role in both local and systemic immune responses, with the gut-associated lymphoid tissue (GALT) being a key component of this immune landscape. The interplay between the gut microbiome and the immune system is crucial for maintaining health and preventing disease, underscoring the significance of the gut as an immunological hub.

Gut-Associated Lymphoid Tissue (GALT)

The GALT comprises various structures and tissues that contribute to the immune surveillance and response in the gut. Major components of the GALT include:

- *Peyer's Patches*: These are organized lymphoid follicles located in the mucosa of the small intestine. They contain specialized epithelial cells known as M cells, which are responsible for sampling antigens from the gut lumen and delivering them to underlying immune cells.
- *Mesenteric Lymph Nodes*: Located in the mesentery, these lymph nodes filter lymphatic fluid from the intestines and serve as sites where immune responses can be initiated and regulated.
- *Lamina Propria*: This layer of connective tissue lies beneath the epithelial lining of the gut and is rich in immune cells, including macrophages, dendritic cells, and lymphocytes. The lamina propria is a crucial site for antigen presentation and immune cell activation.
- *Intestinal Epithelium*: The epithelial layer of the gut, while primarily serving as a physical barrier, also plays an active role in immune defense through the secretion of antimicrobial peptides and cytokines.

Role of the Gut Microbiome in Immune Regulation

The gut microbiome significantly influences the function of the GALT and overall immune responses. The interaction between microbiota and the gut mucosal immune system is a dynamic process that involves several mechanisms:

- *Microbial Interactions with Epithelial Cells*: The gut epithelium serves as the primary interface between the microbiome and the immune system. Microbial components, such as lipopolysaccharides (LPS) and peptidoglycans, interact with epithelial cells through pattern recognition receptors (PRRs), such as Toll-like receptors (TLRs). These interactions trigger the production of antimicrobial peptides and cytokines that help modulate local immune responses and maintain mucosal integrity.
- *Antigen Sampling and Immune Activation*: Specialized epithelial cells, including M cells in Peyer's patches, facilitate the uptake of antigens and microorganisms from the gut lumen. This antigen sampling is critical for initiating immune responses and promoting the development of mucosal immunity. Antigens are delivered to underlying immune cells, including dendritic cells and T cells, which then process and present these antigens to other immune cells.
- *Influence on Immune Cell Differentiation and Function*: The gut microbiome impacts the differentiation and function of various immune cells. For example, microbial-derived signals can promote the development of regulatory T cells (Tregs) and influence the balance between pro-inflammatory and anti-inflammatory responses. The production of short-chain fatty acids (SCFAs) by microbial fermentation supports the growth of Tregs and enhances the barrier function of the gut epithelium.

Barrier Function and Immune Defense

The gut epithelium is not only a physical barrier but also an active participant in immune defense. It provides a selective permeability that regulates the passage of nutrients and immune cells while preventing the entry of pathogens and toxins. Several mechanisms contribute to the barrier function of the gut epithelium:

- *Tight Junctions*: Epithelial cells are connected by tight junctions, which seal the space between cells and prevent the leakage of luminal contents into the underlying tissue. Disruption of tight junctions can lead to increased intestinal permeability, also known as "leaky gut," which is associated with various inflammatory and autoimmune conditions.
- *Antimicrobial Peptides*: Epithelial cells produce antimicrobial peptides, such as defensins and cathelicidins, which inhibit the growth of pathogenic microorganisms and support the maintenance of a balanced microbiome. These peptides play a role in preventing infections and modulating the composition of the gut microbiome.
- *Cytokine Production*: In response to microbial signals, epithelial cells secrete cytokines and chemokines that regulate local immune responses. For example, the production of interleukin-6 (IL-6) and tumor necrosis factor-alpha (TNF-α) can recruit

and activate immune cells, while anti-inflammatory cytokines like interleukin-10 (IL-10) help to maintain immune homeostasis.

Systemic Immune Responses and Gut Microbiome

The influence of the gut microbiome extends beyond the local mucosal immune responses to impact systemic immunity. Microbial-derived signals and metabolites can enter the bloodstream and affect immune responses throughout the body:

- *Systemic Immunity and Microbial Metabolites*: Short-chain fatty acids (SCFAs), such as butyrate, propionate, and acetate, are produced by the fermentation of dietary fibers by gut microbiota. These metabolites have systemic effects on immune function, including modulation of T cell activity and regulation of inflammation. SCFAs can influence the differentiation of T cells and promote the development of Tregs, which are important for maintaining immune tolerance and preventing autoimmune diseases.
- *Immune System Training*: The gut microbiome contributes to the concept of "trained immunity," where the innate immune system becomes more responsive to subsequent infections following initial microbial exposure. This training effect can enhance the ability of the immune system to respond to future microbial challenges, providing a form of immune protection that extends beyond the gut.
- *Systemic Inflammation and Dysbiosis*: Imbalances in the gut microbiome, known as dysbiosis, can lead to systemic inflammation and contribute to the development of various chronic diseases. Dysbiosis can result in increased intestinal permeability and the translocation of microbial components into the bloodstream, leading to systemic inflammatory responses and contributing to conditions such as metabolic syndrome, cardiovascular disease, and autoimmune disorders.

Microbiome and Intestinal Barrier Function

The gut microbiome plays a pivotal role in maintaining the integrity and functionality of the intestinal barrier, which is crucial for preventing the translocation of pathogens, toxins, and other harmful substances from the gut lumen into the systemic circulation. A well-maintained intestinal barrier supports overall health and immune function, while disruptions to this barrier, often referred to as "leaky gut," can have significant pathological consequences.

The Intestinal Barrier and Its Components

The intestinal barrier is a complex and dynamic structure composed of several key elements:

- *Epithelial Cells:* The epithelial layer lines the intestinal tract and serves as the primary physical barrier between the gut lumen and the underlying tissues. Epithelial cells are connected by tight junctions, which regulate the permeability of the intestinal barrier.
- *Tight Junctions*: Tight junctions are protein complexes that seal the spaces between adjacent epithelial cells, maintaining the selective permeability of the intestinal barrier.

These junctions prevent the leakage of luminal contents and help to control the passage of ions, nutrients, and water.

- *Mucus Layer*: The mucus layer, produced by goblet cells, acts as an additional physical barrier that traps pathogens and facilitates their removal from the gut. It also provides a habitat for beneficial microbiota.
- *Antimicrobial Peptides*: Epithelial cells produce antimicrobial peptides, such as defensins and cathelicidins, that help to control microbial populations and prevent infections.

Role of the Microbiome in Maintaining Barrier Integrity

The gut microbiome influences the intestinal barrier through various mechanisms:

- *Microbial Metabolites*: One of the most significant ways the microbiome supports barrier function is through the production of microbial metabolites, particularly short-chain fatty acids (SCFAs). SCFAs, such as butyrate, propionate, and acetate, are generated by the fermentation of dietary fibers by gut bacteria. These metabolites have several beneficial effects on the intestinal barrier:
- *Butyrate*: Butyrate is a primary energy source for colonocytes (colon epithelial cells) and plays a crucial role in maintaining barrier integrity. It enhances the production of tight junction proteins, such as occludin and claudins, which strengthen the junctions between epithelial cells and reduce intestinal permeability. Additionally, butyrate has anti-inflammatory properties that help to maintain a balanced immune response and prevent excessive inflammation.
- *Other SCFAs:* Propionate and acetate also contribute to barrier function by influencing epithelial cell proliferation and differentiation. These SCFAs can promote the production of mucus and antimicrobial peptides, further supporting the gut's defensive mechanisms.
- *Microbial Diversity*: A diverse microbiome helps to maintain the balance of beneficial and pathogenic microbes, which is essential for a healthy intestinal barrier. Dysbiosis, characterized by a reduction in microbial diversity and an imbalance in microbial populations, can disrupt barrier function and lead to increased intestinal permeability.
- *Immune Modulation*: The gut microbiome modulates the local immune environment, influencing the production of cytokines and other factors that affect barrier function. For example, beneficial microbes can promote the development of regulatory T cells (Tregs) and the production of anti-inflammatory cytokines, which help to maintain barrier integrity and prevent inflammation.

Dysbiosis and Increased Intestinal Permeability

Dysbiosis, an imbalance in the gut microbiome, is often associated with increased intestinal permeability and a range of health issues:

- *Leaky Gut:* Dysbiosis can disrupt the balance of tight junction proteins and lead to the loosening of tight junctions between epithelial cells. This increased permeability, often referred to as "leaky gut," allows the passage of microbial products, such as lipopolysaccharides (LPS), into the bloodstream.

- *Systemic Inflammation:* The entry of LPS and other microbial products into the systemic circulation triggers an inflammatory response, known as endotoxemia. This systemic inflammation can contribute to the development of various chronic diseases, including metabolic syndrome, cardiovascular disease, and autoimmune disorders.
- *Immune Activation*: Increased intestinal permeability can lead to the activation of the immune system and the development of immune-related diseases. For example, the translocation of microbial antigens can result in the activation of pro-inflammatory pathways and the production of cytokines that contribute to chronic inflammation.

Strategies to Support Gut Barrier Function

Maintaining a healthy gut microbiome and supporting intestinal barrier function can help to prevent and manage conditions associated with dysbiosis and increased permeability. Some strategies include:

- *Dietary Interventions*: Consuming a diet rich in dietary fibers, prebiotics, and fermented foods can promote the growth of beneficial microbes and the production of SCFAs. This dietary approach supports barrier integrity and overall gut health.
- *Probiotics and Prebiotics*: Probiotic supplements containing beneficial bacteria can help restore microbial balance and enhance barrier function. Prebiotics, which promote the growth of beneficial microbes, can also support a healthy microbiome and improve barrier integrity.
- *Avoiding Antibiotic Overuse*: Excessive use of antibiotics can disrupt the microbiome and contribute to dysbiosis. Using antibiotics judiciously and only when necessary can help preserve microbial diversity and support gut health.
- *Monitoring and Managing Stress*: Chronic stress can negatively impact the gut microbiome and barrier function. Implementing stress management techniques, such as mindfulness and relaxation exercises, can support overall gut health.

Gut Microbiome and Systemic Immunity

The gut microbiome, often considered a local regulator of immune function, has profound effects on systemic immunity. The interactions between the gut microbiome and systemic immune responses involve complex mechanisms that impact various immune organs and overall health. Here, we explore how microbial signals from the gut influence systemic immunity, including the roles of short-chain fatty acids (SCFAs), cytokine production, and the regulation of immune responses.

Microbial Signals and Systemic Immune Modulation

Microbial signals originating from the gut can profoundly impact systemic immune responses by influencing immune cells and tissues beyond the gut. Key mechanisms include:

- *Short-Chain Fatty Acids (SCFAs):* SCFAs, such as butyrate, propionate, and acetate, are produced through the fermentation of dietary fibers by gut microbiota. These metabolites have systemic effects on immune function:

- *T Cell Differentiation:* SCFAs can influence the differentiation of T cells in secondary lymphoid organs, such as the spleen and lymph nodes. For example, butyrate promotes the development of regulatory T cells (Tregs), which are essential for maintaining immune tolerance and preventing autoimmune responses. Propionate and acetate can also impact T cell differentiation and function.
- *Immune Cell Function*: SCFAs can modulate the function of various immune cells, including macrophages and dendritic cells. These effects can enhance the ability of immune cells to respond to infections and maintain immune homeostasis.
- *Microbial Metabolites and Immune Signaling*: In addition to SCFAs, other microbial metabolites, such as bile acids and polyamines, can influence systemic immune responses. These metabolites can affect immune cell function and contribute to the regulation of inflammation.

Systemic Inflammation and Cytokine Production

The gut microbiome plays a critical role in modulating systemic inflammation and cytokine production:

- *Anti-Inflammatory Cytokines*: A balanced gut microbiome promotes the production of anti-inflammatory cytokines, such as interleukin-10 (IL-10). IL-10 helps to suppress excessive inflammation and prevent immune-mediated damage. It also plays a role in maintaining mucosal and systemic immune homeostasis.
- *Pro-Inflammatory Cytokines*: Dysbiosis, or an imbalance in the gut microbiome, can lead to increased production of pro-inflammatory cytokines, such as tumor necrosis factor-alpha (TNF-α) and interleukin-6 (IL-6). This imbalance can contribute to chronic systemic inflammation and the development of inflammatory diseases.
- *Chemokines and Immune Cell Recruitment*: The gut microbiome can influence the production of chemokines, which are signaling molecules that attract immune cells to sites of infection or inflammation. This modulation can affect the distribution and activation of immune cells throughout the body.

Influence on Immune Organs and Systems

The effects of the gut microbiome extend to various immune organs and systems:

- *Liver:* The liver is a major immune organ that interacts with gut-derived microbial signals. Gut-derived LPS and other microbial components can enter the liver via the portal circulation, influencing liver immune responses and contributing to conditions such as non-alcoholic fatty liver disease (NAFLD) and hepatitis.
- *Spleen*: The spleen is involved in filtering blood and supporting immune responses. SCFAs can impact the differentiation and function of T cells in the spleen, influencing systemic immune responses and tolerance.
- *Lymph Nodes*: Gut-derived signals can influence the function of lymph nodes, which are critical for immune surveillance and response. Microbial metabolites can affect the activation and differentiation of immune cells within lymph nodes.

Clinical Implications and Therapeutic Approaches

The relationship between the gut microbiome and systemic immunity has important clinical implications and potential therapeutic applications:

- *Probiotics and Prebiotics*: Probiotics and prebiotics can modulate the gut microbiome and influence systemic immune responses. By promoting the growth of beneficial microbes and the production of SCFAs, these interventions may enhance immune function and reduce inflammation.
- *Dietary Modifications*: Dietary changes, such as increasing fiber intake and consuming fermented foods, can support a healthy microbiome and improve systemic immune function. These dietary strategies may help manage chronic inflammatory conditions and support overall health.
- *Microbiome-Based Therapies*: Emerging research is exploring microbiome-based therapies, such as fecal microbiota transplantation (FMT), to restore microbial balance and improve systemic immune responses. These therapies hold promise for treating conditions associated with dysbiosis and systemic inflammation

Immune Tolerance and Regulation by Microbiomes

Mechanisms of Immune Tolerance

Immune tolerance is a critical mechanism by which the immune system avoids attacking the body's own tissues (self-tolerance) and harmless antigens, such as food proteins and commensal microbes (peripheral tolerance). The microbiome plays a fundamental role in shaping and maintaining immune tolerance, which is essential for preventing autoimmune diseases and excessive inflammatory responses.

Concepts of Immune Tolerance

- *Self-Tolerance*: This refers to the immune system's ability to distinguish between self and non-self, preventing autoimmunity. Self-tolerance is primarily established during the development of immune cells in the thymus, where T cells that react strongly to self-antigens are typically eliminated or inactivated.
- *Peripheral Tolerance*: This type of tolerance occurs outside the thymus and involves mechanisms that prevent the activation of self-reactive T cells that have escaped thymic selection. Peripheral tolerance also involves the regulation of responses to non-self antigens that are normally harmless, such as food proteins and commensal bacteria.

Role of Regulatory T Cells (Tregs) in Immune Tolerance

Regulatory T cells (Tregs) are key players in maintaining immune tolerance. They are a subset of T cells characterized by the expression of the transcription factor Foxp3, which is crucial for their development and function. Tregs contribute to tolerance through several mechanisms:

- *Production of Anti-Inflammatory Cytokines*: Tregs produce cytokines such as interleukin-10 (IL-10) and transforming growth factor-beta (TGF-β). These cytokines help to suppress inflammatory responses and inhibit the activation of other immune cells, thereby promoting tolerance to self-antigens and non-harmful external antigens.
- *Suppression of Effector T Cells:* Tregs can directly suppress the proliferation and function of effector T cells through cell-cell contact and the secretion of inhibitory molecules. This interaction helps to prevent inappropriate immune responses and inflammation.
- *Induction of Tolerance in Other Cells:* Tregs also play a role in inducing tolerance in other immune cells and non-immune cells by modulating their activation and responses.

Microbiome Influence on Treg Development and Function

The gut microbiome significantly influences the development and function of Tregs, impacting overall immune tolerance:

- *Bacterial Species and Treg Induction*: Certain bacterial species are particularly effective at promoting Treg development. For instance, Bacteroides fragilis has been shown to enhance the differentiation of Tregs. This bacterium produces polysaccharide A (PSA), which interacts with TLR2 (Toll-like receptor 2) on immune cells, leading to an increase in Tregs and the production of anti-inflammatory cytokines like IL-10.
- *Microbial Metabolites and Treg Function*: Microbial metabolites, particularly short-chain fatty acids (SCFAs) like butyrate, also play a crucial role in Treg function. Butyrate enhances Treg differentiation and stability by promoting histone acetylation and the expression of genes associated with Treg function.
- *Microbiome Diversity and Tolerance*: A diverse microbiome contributes to the development of a balanced immune system by promoting a variety of microbial interactions that support Treg function and immune tolerance. Dysbiosis, or an imbalance in the microbiome, can impair Treg development and function, leading to a compromised tolerance to self and harmless antigens.

Clinical Implications of Microbiome-Induced Tolerance

The interaction between the microbiome and immune tolerance has several important clinical implications:

- *Autoimmune Diseases:* Disruptions in microbiome-induced tolerance can contribute to the development of autoimmune diseases. For example, dysbiosis has been linked to conditions such as rheumatoid arthritis and multiple sclerosis, where impaired tolerance leads to self-reactive immune responses.
- *Allergic Diseases:* An imbalance in microbial communities can also affect tolerance to environmental antigens, contributing to allergic diseases. The "hygiene hypothesis" suggests that reduced microbial exposure early in life can impair the development of immune tolerance and increase the risk of allergies.
- *Therapeutic Interventions*: Strategies that modulate the microbiome, such as probiotics, prebiotics, and fecal microbiota transplantation (FMT), have potential

therapeutic benefits for restoring immune tolerance and treating autoimmune and allergic conditions. By promoting beneficial microbial communities, these interventions aim to support Treg function and overall immune balance.

Microbiome-Driven Regulation of Inflammatory Responses

The gut microbiome is a key regulator of the host's inflammatory responses, influencing both local and systemic inflammation. The interaction between microbiota and the immune system involves complex mechanisms that modulate inflammation, including the production of anti-inflammatory metabolites, the induction of regulatory immune cells, and the prevention of dysregulated immune responses. Here, we delve into how the microbiome drives the regulation of inflammation and the implications of dysbiosis for chronic inflammatory conditions.

Microbial Metabolites and Inflammatory Regulation

Short-Chain Fatty Acids (SCFAs): SCFAs, primarily butyrate, propionate, and acetate, are produced by the fermentation of dietary fibers by gut bacteria. These metabolites have significant anti-inflammatory effects:

- *Butyrate:* Butyrate is a key SCFA with potent anti-inflammatory properties. It enhances the differentiation and function of regulatory T cells (Tregs), which are crucial for maintaining immune tolerance and suppressing inflammation. Butyrate also upregulates the expression of tight junction proteins, which strengthens the intestinal barrier and prevents the leakage of pro-inflammatory agents into the systemic circulation. Furthermore, butyrate inhibits the production of pro-inflammatory cytokines such as TNF-α and IL-6 by immune cells.
- *Propionate and Acetate*: Propionate and acetate also contribute to anti-inflammatory effects, though their mechanisms are less well-understood compared to butyrate. These SCFAs can modulate immune cell function and influence systemic inflammation. For example, propionate has been shown to affect the activation and function of macrophages and other immune cells.
- *Other Microbial Metabolites*: In addition to SCFAs, other microbial metabolites such as bile acids and polyamines also play roles in regulating inflammation. Bile acids, for example, can influence immune responses through their effects on the gut barrier and immune cell function.

Induction of Regulatory Immune Cells

The microbiome influences the production and function of regulatory immune cells, which play a crucial role in controlling inflammation:

- *Regulatory T Cells (Tregs):* As mentioned previously, Tregs are essential for maintaining immune tolerance and suppressing inflammatory responses. The microbiome, through specific bacterial species and their metabolites, promotes Treg differentiation and function. For instance, Bacteroides fragilis and its polysaccharide A (PSA) have been shown to enhance Treg development and function, thereby contributing to immune homeostasis and reducing inflammation.

- *Other Regulatory Cells:* Besides Tregs, the microbiome also affects other regulatory cells, such as macrophages and dendritic cells. These cells can adopt anti-inflammatory phenotypes in response to microbial signals, which helps in modulating inflammation and promoting tolerance.

Impact of Dysbiosis on Inflammatory Responses

Dysbiosis, an imbalance in the gut microbiome, can disrupt the regulatory mechanisms that control inflammation:

- *Increased Intestinal Permeability:* Dysbiosis often leads to increased intestinal permeability, or "leaky gut," which allows the translocation of microbial products like lipopolysaccharides (LPS) into the bloodstream. This systemic exposure to LPS can trigger chronic inflammation and contribute to various inflammatory conditions.
- *Impaired Immune Regulation*: Dysbiosis can impair the development and function of regulatory immune cells, such as Tregs. The imbalance in microbial communities can reduce the production of anti-inflammatory metabolites and cytokines, leading to unchecked inflammation and immune dysregulation.
- *Chronic Inflammatory Diseases*: Conditions such as inflammatory bowel disease (IBD), including Crohn's disease and ulcerative colitis, are associated with dysbiosis. In these diseases, the altered microbiome composition contributes to chronic inflammation in the gut. Similarly, dysbiosis has been linked to other inflammatory and autoimmune conditions, including rheumatoid arthritis and systemic lupus erythematosus.

Restoring Balance and Modulating Inflammation

Restoring a balanced microbiome can help manage inflammation and support immune tolerance:

- *Dietary Interventions*: Consuming a diet rich in dietary fibers, prebiotics, and fermented foods can promote the growth of beneficial microbes and enhance the production of SCFAs. These dietary strategies support a healthy microbiome and help regulate inflammation.
- *Probiotics and Prebiotics*: Probiotic supplements containing beneficial bacteria can help restore microbial balance and reduce inflammation. Prebiotics, which feed beneficial microbes, can also support a healthy microbiome and improve inflammatory responses.
- *Fecal Microbiota Transplantation (FMT):* FMT involves transferring fecal material from a healthy donor to a patient with dysbiosis. This approach aims to restore a balanced microbiome and has shown promise in treating conditions like recurrent Clostridium difficile infection and, potentially, other inflammatory diseases.

Role of Microbiome in Autoimmune Diseases

Autoimmune diseases are characterized by an inappropriate immune response in which the immune system mistakenly attacks the body's own tissues. The microbiome, particularly the gut microbiome, plays a significant role in the development and progression of autoimmune diseases by influencing immune tolerance and immune system homeostasis. This relationship underscores the importance of maintaining a balanced microbiome to prevent or manage autoimmune conditions.

Mechanisms Linking the Microbiome to Autoimmune Diseases

1. *Dysregulation of Immune Tolerance*: The microbiome helps establish and maintain immune tolerance, a process crucial for distinguishing between self and non-self. Dysbiosis, or an imbalance in the microbiome, can disrupt this tolerance by Activating Autoreactive T Cells (Alterations in the gut microbiome can lead to the activation of autoreactive T cells, which target the body's own tissues) and Promoting Autoantibody Production (Dysbiosis can stimulate B cells to produce autoantibodies that target self-antigens, contributing to autoimmune pathology).
2. *Molecular Mimicry*: Molecular mimicry occurs when microbial antigens closely resemble self-antigens, leading to cross-reactive immune responses. In dysbiosis, certain microbial components may mimic self-antigens, triggering autoimmune reactions.
3. *Increased Intestinal Permeability*: Dysbiosis often results in increased intestinal permeability or "leaky gut," allowing microbial products, such as lipopolysaccharides (LPS), to enter systemic circulation. This can lead to systemic inflammation and contribute to the development of autoimmune diseases by activating Systemic Inflammation (LPS and other microbial products can activate systemic inflammatory pathways, exacerbating autoimmune conditions) and Disrupting Immune Regulation (The increased entry of microbial products can overwhelm the immune system, leading to a breakdown in tolerance mechanisms).

Autoimmune Diseases Linked to Microbiome Alterations in Domestic Animals

1. *Autoimmune Hemolytic Anemia (AIHA):* AIHA is a condition where the immune system destroys red blood cells, leading to anemia. In domestic animals, alterations in the gut microbiome have been associated with AIHA, suggesting that dysbiosis may contribute to the development or exacerbation of this condition.
2. *Immune-Mediated Polyarthritis*: Immune-mediated polyarthritis is characterized by inflammation of multiple joints due to an autoimmune response. Changes in the gut microbiome have been linked to immune-mediated polyarthritis in domestic animals, highlighting the role of microbial imbalances in the pathogenesis of this disease.

Therapeutic Strategies Targeting the Microbiome

1. *Dietary Interventions:*
   Diet plays a crucial role in shaping the gut microbiome. Dietary changes, such as increasing fiber intake or consuming anti-inflammatory foods, can help restore microbial balance and potentially reduce the risk or severity of autoimmune diseases. Specific dietary components that support a healthy microbiome include:
   - Prebiotics: Foods high in prebiotics, such as inulin and fructooligosaccharides, support the growth of beneficial microbes.
   - Anti-inflammatory Foods: Omega-3 fatty acids, found in fish oil and flaxseeds, have anti-inflammatory properties that may benefit autoimmune patients.
2. *Probiotics:*
   Probiotic supplements containing beneficial bacteria can help restore a balanced microbiome and improve immune function. Probiotics have been investigated for their potential to:
   - *Modulate Immune Responses*: Probiotics can influence immune cell activity and cytokine production, which may help reduce autoimmune symptoms.
   - *Improve Gut Barrier Function*: By enhancing the integrity of the intestinal barrier, probiotics may prevent the translocation of harmful microbial products and reduce systemic inflammation.
3. *Fecal Microbiota Transplantation (FMT):*
   FMT involves transferring fecal material from a healthy donor to an individual with dysbiosis. This procedure aims to restore a balanced microbiome and has shown promise in treating various conditions, including autoimmune diseases. FMT may benefit autoimmune patients by Restoring Microbial Diversity (Increasing microbial diversity can help reestablish a healthy immune balance) and Improving Immune Tolerance (FMT may help restore tolerance mechanisms and reduce autoimmune responses).
4. *Prebiotics and Synbiotics:*
   Prebiotics and synbiotics (a combination of probiotics and prebiotics) offer additional therapeutic options for managing autoimmune diseases by supporting the growth of beneficial microbes and enhancing the effects of probiotics.

Implications for Animal Health and Disease Management

Microbiome-Targeted Therapies

The growing understanding of the interplay between the microbiome and the immune system has led to innovative approaches in developing therapies aimed at modulating the microbiome to enhance animal health. Microbiome-targeted therapies, including the use of probiotics, prebiotics, and synbiotics, offer promising strategies for improving immune function, preventing diseases, and supporting overall well-being in domestic animals.

Probiotics

Probiotics are live microorganisms that, when administered in adequate amounts, confer health benefits to the host. They can positively influence the gut microbiome and support immune function through several mechanisms:

1. *Enhancement of Vaccine Responses:*
   - Immune System Activation: Probiotics can stimulate the immune system, leading to a more robust response to vaccinations. For instance, specific probiotic strains like Lactobacillus and Bifidobacterium have been shown to enhance antibody production and improve the efficacy of vaccines in animals.
   - Gut-Mucosal Immunity: By strengthening the mucosal immune system, probiotics can improve the overall immune response to pathogens and vaccines.
2. *Reduction of Infections:*
   - Competitive Exclusion: Probiotics can outcompete pathogenic microorganisms for resources and adhesion sites in the gut, reducing the likelihood of infections.
   - Production of Antimicrobial Substances: Probiotics produce substances such as lactic acid and bacteriocins that inhibit the growth of harmful bacteria.
3. *Overall Immune Health Improvement:*
   - Regulation of Inflammation: Probiotics can help modulate systemic inflammation by influencing the production of inflammatory cytokines and enhancing the function of regulatory T cells.
   - Restoration of Microbial Balance: By restoring a balanced microbiome, probiotics can help prevent or manage dysbiosis-related conditions.

Prebiotics

Prebiotics are non-digestible food ingredients that selectively stimulate the growth and activity of beneficial microorganisms in the gut. They support immune function through several key mechanisms:

1. *Promotion of Beneficial Microbes:*
   - Selective Growth: Prebiotics such as inulin, fructooligosaccharides (FOS), and galactooligosaccharides (GOS) foster the growth of beneficial bacteria like Bifidobacterium and Lactobacillus, which are associated with improved immune health.
   - Microbial Diversity: By promoting the growth of a diverse range of beneficial microbes, prebiotics contribute to a more stable and resilient microbiome.
2. *Production of Short-Chain Fatty Acids (SCFAs):*
   Enhanced SCFA Production: Prebiotics are fermented by gut bacteria to produce SCFAs such as butyrate, propionate, and acetate. SCFAs have been shown to support immune function by promoting Treg differentiation, enhancing gut barrier integrity, and modulating systemic inflammation.

3. *Support for Immune Function:*
   - Modulation of Immune Responses: SCFAs and other metabolites produced from prebiotic fermentation can influence the production of cytokines and the activity of immune cells, contributing to a balanced immune response.
   - Prevention of Infections: By enhancing gut health and immune function, prebiotics can help reduce the risk of infections.

Synbiotics

Synbiotics combine probiotics and prebiotics to synergistically enhance the health benefits provided by each component. They offer a comprehensive approach to microbiome modulation:

1. *Enhanced Microbial Colonization:*
   - Support for Probiotic Growth: Prebiotics in synbiotics provide a food source for the probiotics, supporting their growth and activity in the gut.
   - Sustained Benefits: The combination of probiotics and prebiotics can lead to more sustained improvements in gut microbiome composition and function.
2. *Improved Immune Responses:*
   - Synergistic Effects: Synbiotics can enhance the effects of probiotics on immune function and microbial balance, leading to improved resistance to infections and betteroverall immune health.
   - Balanced Inflammation: By combining the anti-inflammatory effects of probiotics and the SCFA-producing effects of prebiotics, synbiotics can help regulate systemic inflammation.

Fecal Microbiota Transplantation (FMT)

FMT involves transferring fecal material from a healthy donor to an individual with dysbiosis, aiming to restore a balanced microbiome. This therapy has shown potential in:

1. *Restoring Microbial Diversity:*
   - Recolonization: FMT can reintroduce a diverse range of beneficial microbes, helping to restore a balanced microbiome and improve gut health.
   - Correction of Dysbiosis: By addressing the underlying microbial imbalance, FMT can help alleviate symptoms of dysbiosis-related conditions.
2. *Modulating Immune Function:*
   - Restoration of Immune Tolerance: FMT may help restore immune tolerance and improve immune responses by reestablishing a healthy microbiome and supporting regulatory mechanisms.
   - Potential Benefits for Autoimmune Diseases: FMT has shown promise in managing autoimmune conditions by modulating the gut microbiome and reducing systemic inflammation.

Nutritional Strategies to Support the Microbiome

Diet is a crucial determinant of gut microbiome composition and function, which in turn impacts immune system health in domestic animals. Nutritional strategies that support a diverse and balanced microbiome can enhance immune function, promote tolerance, and help manage conditions related to dysbiosis and immune dysfunction.

Fiber-Rich Diets

1. *Role of Dietary Fiber:*
   - Prebiotic Effects: Dietary fibers, including soluble and insoluble fibers, act as prebiotics by providing a substrate for beneficial gut bacteria. These fibers are fermented by gut microbiota to produce short-chain fatty acids (SCFAs) such as butyrate, propionate, and acetate.
   - SCFA Production: SCFAs play a crucial role in maintaining gut health by supporting the integrity of the intestinal barrier, modulating inflammation, and promoting the differentiation of regulatory T cells (Tregs). Butyrate, in particular, is known for its anti-inflammatory properties and its role in enhancing mucosal immunity.
2. *Sources of Fiber:*
   - Forage and Hay: High-fiber forages and hays are excellent sources of dietary fiber for ruminants. They support the growth of beneficial microbes and the production of SCFAs.
   - Whole Grains and Vegetables: For non-ruminants, whole grains, vegetables, and fruits provide dietary fiber that supports a healthy microbiome.

Functional Foods

1. *Fermented Feeds:*
   - Probiotic Effects: Fermented feeds, such as silage and certain types of grains, contain live beneficial microorganisms that can positively influence the gut microbiome. These probiotics help maintain microbial diversity and balance, enhance nutrient absorption, and support immune function.
   - Microbial Inoculants: Adding microbial inoculants to animal feeds can introduce beneficial bacteria and yeasts that support gut health and improve overall immune responses.
2. *Bioactive Compounds:*
   - Polyphenols: Found in fruits, vegetables, and herbs, polyphenols possess anti-inflammatory and antioxidant properties. They can modulate the gut microbiome by promoting the growth of beneficial bacteria and inhibiting pathogenic strains.
   - Phytochemicals: Compounds such as flavonoids, carotenoids, and saponins have been shown to support immune health and improve gut microbiota composition.

Balanced Nutrient Intake

1. *Protein and Amino Acids:*
   - Nutrient Balance: Ensuring adequate and balanced protein and amino acid intake is essential for maintaining a healthy gut microbiome. Proteins and amino acids support the growth of beneficial microbes and contribute to overall immune health.
   - Impact on Microbiota: Both the quantity and quality of protein in the diet can influence the composition and activity of gut microbiota.
2. *Vitamins and Minerals:*
   - Essential Nutrients: Vitamins and minerals play roles in immune function and microbiome health. For example, vitamins A, D, and E have been shown to influence immune responses and gut health.
   - Mineral Balance: Adequate intake of minerals such as zinc, copper, and selenium supports immune function and microbiome stability.

Managing Dysbiosis

1. *Dietary Adjustments:*
   - Identifying Imbalances: Dietary strategies to manage dysbiosis include identifying and correcting imbalances in the microbiome caused by poor diet, such as high levels of sugars or processed foods.
   - Tailored Diets: Implementing tailored diets that address specific microbial imbalances can help restore a healthy gut microbiome and improve immune function.
2. *Integrating Probiotics and Prebiotics:*
   - Supplementation: Combining probiotics and prebiotics with a high-fiber diet can enhance microbial diversity and support a balanced microbiome. This approach helps to restore and maintain gut health, particularly after disruptions such as antibiotic treatments or illness.

Hydration and Feeding Practices

1. *Adequate Water Intake*:
   - Hydration: Ensuring that animals have access to clean, fresh water is crucial for maintaining gut health. Proper hydration supports the mucosal lining of the gut and aids in the digestion and absorption of nutrients.
2. *Feeding Frequency and Management:*
   - Feeding Practices: Regular feeding schedules and avoiding abrupt changes in diet can help maintain a stable gut microbiome. Consistent feeding practices reduce stress on the gut and support healthy microbial populations.

Future Directions and Research Opportunities

As our understanding of the microbiome-immune system interactions in domestic animals expands, several key areas of future research hold promise for advancing animal health and well-being. These include identifying specific microbial species and metabolites that influence immune function, exploring the role of the microbiome in disease resistance, and developing microbiome-based diagnostics. Longitudinal studies and innovative research methodologies will be crucial for uncovering the complexities of these interactions and translating findings into practical applications.

Identifying Specific Microbial Species and Metabolites

1. *Microbial Taxa and Immune Function:*
   - Species-Specific Effects: Research should focus on identifying specific microbial species that have a direct impact on immune function. Understanding the roles of particular bacteria, fungi, and viruses in modulating immune responses can lead to targeted probiotic and prebiotic interventions.
   - Metabolite Profiling: Investigating the metabolic products of gut microbiota, such as short-chain fatty acids (SCFAs) and other bioactive compounds, can provide insights into how these metabolites influence immune system development and function.
2. *Microbiome Composition and Disease Outcomes:*
   - Disease Correlations: Exploring correlations between microbiome composition and disease outcomes can help identify microbial signatures associated with health and disease states. This information could lead to the development of microbiome-based biomarkers for disease risk and progression.

Microbiome-Based Diagnostics

1. *Diagnostic Tools:*
   - Microbiome Profiling: Advances in metagenomic and metabolomic technologies could facilitate the development of diagnostic tools that profile the microbiome to assess health status, predict disease risk, and guide treatment decisions.
   - Personalized Medicine: Tailoring interventions based on individual microbiome profiles could enhance treatment efficacy and reduce the risk of adverse effects. This personalized approach could be applied to nutrition, probiotics, and other therapeutic strategies.
2. *Early Detection and Monitoring:*
   - Preventive Measures: Using microbiome data for early detection of dysbiosis or potential disease can enable preventive measures and interventions before clinical symptoms develop.

   - Monitoring Treatment Outcomes: Tracking changes in microbiome composition and function during and after treatment can provide insights into the effectiveness of therapies and guide adjustments.
3. *Long-Term Health Outcomes:*
   - Chronic Disease Risk: Long-term studies can help determine how early-life microbial exposures influence susceptibility to chronic diseases and conditions later in life. This knowledge can guide interventions aimed at reducing disease risk and improving overall health.

Conclusion

The vital connection between the microbiome and the immune system highlights its critical role in maintaining animal health and preventing disease. Microbial communities contribute significantly to processes such as digestion, nutrient absorption, immune regulation, and disease resistance. Disruptions to the microbiome, often associated with immune dysregulation, can increase susceptibility to infections and chronic diseases.

Leveraging the therapeutic potential of microbiome-focused strategies offers opportunities to tackle key challenges in veterinary medicine, including enhancing disease resistance, boosting productivity, and promoting animal welfare. As our knowledge of microbiome-immune system interactions expands, holistic approaches that prioritize microbiome health will become essential for advancing sustainable and efficient animal management practices.

References

Chong, Clara Yieh Lin, Frank H. Bloomfield, and Justin M. O'Sullivan. "Factors affecting gastrointestinal microbiome development in neonates." *Nutrients* 10, no. 3 (2018): 274.

Fischer, Uwe, and Fumio Takizawa. "Cellular immune responses." In *Principles of Fish Immunology: From Cells and Molecules to Host Protection*, pp. 141-176. Cham: Springer International Publishing, 2022.

Hamar, Péter. "Local Production of Acute Phase Proteins: A Defense Reaction of Cancer Cells to Injury with Focus on Fibrinogen." *International Journal of Molecular Sciences* 25, no. 6 (2024): 3435.

Ignacio, Aline, Sonia Czyz, and Kathy D. McCoy. "Early life microbiome influences on development of the mucosal innate immune system." In *Seminars in Immunology*, vol. 73, p. 101885. Academic Press, 2024.

Ikeda-Ohtsubo, Wakako, Sylvia Brugman, Craig H. Warden, Johanna MJ Rebel, Gert Folkerts, and Corné MJ Pieterse. "How can we define "optimal microbiota?": a comparative review of structure and functions of microbiota of animals, fish, and plants in agriculture." *Frontiers in nutrition* 5 (2018): 90.

Koren, Omry, Liza Konnikova, Petter Brodin, Indira U. Mysorekar, and Maria Carmen Collado. "The maternal gut microbiome in pregnancy: implications for the developing immune system." *Nature Reviews Gastroenterology & Hepatology* 21, no. 1 (2024): 35-45.

Lewis, David B., Joern-Hendrik Weitkamp, and Ofer Levy. "Developmental immunology and role of host defenses in fetal and neonatal susceptibility to infection." In *Remington and Klein's Infectious Diseases of the Fetus and Newborn Infant*, pp. 73-159. Elsevier, 2025.

Mann, Elizabeth R., Ying Ka Lam, and Holm H. Uhlig. "Short-chain fatty acids: linking diet, the microbiome and immunity." *Nature Reviews Immunology* (2024): 1-19.

Maritan, Elisa, Andrea Quagliariello, Enric Frago, Tomaso Patarnello, and Maria Elena Martino. "The role of animal hosts in shaping gut microbiome variation." *Philosophical Transactions of the Royal Society B* 379, no. 1901 (2024): 20230071.

Muhammad, Murad, Wen-Jun Li, Li Li, Yong-Hong Liu, Kashif Ali, and Iftikhar Ahmed. "Introduction of Microbiomes, Viromes and Biofilms." In *Microbial Ecology*, pp. 1-30. CRC Press.

Oldereid, Tine Simensen, Xiaojun Jiang, Kathrine Sivertsen Nordhus, Andrea Ponzetta, Jørgen Vildershøj Bjørnholt, Niklas K. Björkström, Espen Melum, and Henrik Rasmussen. "Role of bacteria and microbial metabolites in immune modulation during early life." *Scandinavian Journal of Immunology* 99, no. 2 (2024): e13336.

Pickard, Joseph M., Melody Y. Zeng, Roberta Caruso, and Gabriel Núñez. "Gut microbiota: Role in pathogen colonization, immune responses, and inflammatory disease." *Immunological reviews* 279, no. 1 (2017): 70-89.

Rathinam, R. Bharathi, Arpit Acharya, A. Jancy Robina, Husne Banu, and Gayatri Tripathi. "The immune system of marine invertebrates: Earliest adaptation of animals." *Comparative Immunology Reports* (2024): 200163.

Riera Romo, Mario, Dayana Pérez-Martínez, and Camila Castillo Ferrer. "Innate immunity in vertebrates: an overview." *Immunology* 148, no. 2 (2016): 125-139.

Riera Romo, Mario, Dayana Pérez-Martínez, and Camila Castillo Ferrer. "Innate immunity in vertebrates: an overview." *Immunology* 148, no. 2 (2016): 125-139.

Safi, Shahabeddin. "Acute phase proteins-analysis, clinical applications and potentials." *Inflammatory Diseases-Immunopathology, Clinical and Pharmacological Bases* (2012): 351-381.

Sharma, I., & Yumnam, H. (2024). Introduction to Microorganisms and Pathogenesis. In *Antimicrobial Photodynamic Therapy* (pp. 1-17). CRC Press.

Tewari, Nisha, and Priyankar Dey. "Navigating Commensal Dysbiosis: Gastrointestinal Host-Pathogen Interplay in Orchestrating Opportunistic Infections." *Microbiological Research* (2024): 127832.

Zheng, Shijun J. "Innate Immunity." In *Veterinary Molecular Immunology*, pp. 37-80. Singapore: Springer Nature Singapore, 2024.

Chapter 7

Microbiomes in Stress Regulation

Amitava Roy[1,*]
Tanmoy Rana[2]
and Madhuchhanda Khan[3]
[1]Department of Livestock Farm Complex, W.B. University of Animal and Fishery Sciences, Kolkata, India
[2]Department of Veterinary Clinical Complex, W.B. University of Animal and Fishery Sciences, Kolkata, India
[3]Rathindra Krishi Vigyan Kendra, Visva-Bharati, Birbhum, W. B., India

Abstract

An increasing body of research is examining the connection between animal gastrointestinal microbiota (the complexity and diversity of microorganisms that live in guts) and many health conditions, including illnesses of the central nervous system and the brain. The brain, glands, gut, immune system, and gastrointestinal microbiota are all part of the microbiota-gut-brain axis, a dynamic matrix of tissues and organs that interact intricately in multiple directions to preserve homeostasis. A wide range of physiological and behavioral impacts, including as the activation of the hypothalamic-pituitary-adrenal (HPA) axis, changes in the functioning of neurotransmitter systems, and immune system dysfunction, can result from changes in this environment. A malfunctioning physiological response can be harmful to the host and contribute to the development of several central nervous system illnesses, even if a proper, coordinated physiological response—such as an immunological or stress response—is required for survival. This chapter highlights the role of the gut microbiota in immune-mediated and stress-mediated regulation of the neuroendocrine, immunological, and neurotransmitter systems, as well as the resulting behavior. Additionally, the ways in which commensal gut microbiota can control neuroinflammation, and the knowledge of function in illnesses linked to stress that arise from neuroinflammatory processes.

Keywords: microbiome, stress, gastrointestine, immune system, environment

---

* Corresponding Author's Email: amitavanutri07@gmail.com

In: Microbiomes in Animal Nutrition, Health, and Disease
Editor: Tanmoy Rana
ISBN: 979-8-89530-425-9

Introduction

The Microbiome

The animal gut is home to a diverse range of microorganisms known as the microbiota, which includes bacteria, yeasts, archaea, single-celled eukaryotes, helminth parasites, viruses, and bacteriophages (Williamson et al., 2015). Together with the enteric nervous system, the sympathetic and parasympathetic divisions of the autonomic nervous system, and the neuroendocrine and neuroimmune components of the central nervous system, the gut microbiota is part of a complex network known as the microbiota-gut-brain axis (Cryan and Dinan, 2015b). The composition of the core bacteria that harbor throughout our adult life is established early in the first few years of life and is shaped by several factors including mode of delivery (vaginal or C-section), whether breastfed or bottle-fed, diet, medication (in particular antibiotic medication), and exposure to viral or bacterial infections and stress (Borre et al., 2014b). It is noteworthy that the growth, maturation, and sprouting of neurones in the developing brain occur concurrently with the critical developmental period in which our gut's bacterial community and core microbiota are seeded (Borre et al., 2014b). In old age, a similar profile is evident, with a decline in microbiota complexity and diversity occurring concurrently with a decline in neuronal complexity (Biagi et al., 2012). These bacteria, which have an estimated mass of 1e2 kg, are thought to be in their trillions (Frank and Pace, 2008), and collectively have 100 times more genes than the animal genome (Kurokawa et al., 2007). As a result of their interactions with their host, these bacteria can affect almost every aspect of animal physiology and biology. The beneficial relationship between the bacteria and host is based on their shared roles in energy regulation, immune system function, gut barrier function, and pathogen defence, among other things (Burokas et al., 2015b). The host gives the bacteria the nutrients and conditions they need to grow and thrive. Stress and neuroinflammation are two examples of centrally mediated phenomena that the gut microbiota has recently been shown to be a critical regulator of (Sampson and Mazmanian, 2015). According to a recent theory, central microglia's maturation and activation are controlled by the microbiota, which in turn influences how neurones develop. The gut microbiota plays a critical role in regulating the immune system and/or stress response, helping the body become accustomed to certain circumstances and preventing the emergence of diseases and dysfunction linked to inflammation and/or stress.

The Stress Response

According to McEwen et al. (2015), stress is characterized as a complicated dynamic state in which an organism's homeostasis, or steady state internal milieu, is threatened or disturbed. All living things experience stress responses, whether they are physiological, psychological, or immunological, at some point in their lives. These responses arise from factors that exceed a homeostatic threshold. Most species have evolved to have the biological mechanisms required to mount a defence against an acute stressful input and to return homeostasis to normal when the stress or perceived threat has passed. The hypothalamic-pituitary-adrenal (HPA) axis, which connects the hypothalamus, pituitary, and adrenal glands, regulates the coordination of this defensive response. It ultimately causes the release of chemicals that change behavior, such as catecholamines, glucocorticoids, and mineralocorticoids. Numerous afferent sympathetic, parasympathetic, and limbic circuits (such as the hippocampus, amygdala, and medial prefrontal cortex) that innervate the paraventricular nucleus (PVN) of the hypothalamus either directly or indirectly control the activity of the HPA axis (Smith and Vale, 2006). According to Dickmeis et al. (2013), the HPA axis normally displays a continuous oscillatory activity that is coordinated with both circadian and ultradian rhythms. The primary neurotransmitter and neuroendocrine components of the stress response are the sympathetic nervous system (SNS) and HPA axis activation, respectively (Mayer, 2000).

Stress causes the nervous system (SNS) to recruit higher levels of catecholamine's in the tissues and systemic circulation. Simultaneously, the hypothalamus's periventricular neurons release corticosterone releasing factor (CRF), which in turn stimulates the anterior pituitary's release of adrenocorticotrophic hormone (ACTH). This hormone then travels throughout the systemic circulation to induce the synthesis and release of glucocorticoids from the adrenal cortex—cortisol in animals and corticosterone in animals—from the adrenal cortex. The main purpose of activating the SNS and HPA axis is to prepare the body for a "fight or flight" reaction by enhancing the metabolism of fat and protein, suppressing the immune system through cytokine suppression, and raising blood sugar levels through gluconeogenesis (Mayer, 2000). Glucocorticoids enter the brain more easily in response to a stressful stimulus, and they exert their effects via high-affinity mineralo-corticoid receptors and lower-affinity glucocorticoid receptors, the latter of which become activated at higher stress levels. Since the same brain regions and neuronal circuits involved in mounting a stress-mediated neuroendocrine response also express receptors sensitive to glucocorticoids throughout the central nervous system (CNS), an autoregulatory feedback loop is in place to prevent this response from being over activated once the stressor is removed. According to Kleiman and Tuckermann (2007), long-term stress exposure and the glucocorticoid hormone it produces can cause a number of negative side effects, including osteoporosis, diabetes, hypertension, dyslipidemia, and even neurodegeneration. One of the biggest risk factors for irritable bowel syndrome is stress as well. On the other hand, a variety of autoimmune and inflammatory disorders frequently present in the clinic with a muted or weak HPA axis. Additionally, glucocorticoids have a significant impact on the survival, phenotype, and functions of glial cells, macrophages, and monocytes. Moreover, almost all types of immune cells express their receptors (Bellavance and Rivest, 2014).

Stress and Neuroimmune Function

Microglia E-Key Regulators of Stress and Neuroinflammation

Microglias are the innate sentinel immune cells in the central nervous system (CNS) that are ultimately in charge of neuroinflammatory processes (Ransohoff et al., 2015). They are able to detect minute changes in molecules in their immediate environment (Tremblay et al., 2011). These resident cells make up 5–12% of all brain cells. They are highly ramified and plastic, and when they become activated, they can express a wide range of antigenic markers, release a variety of cytokines and chemokines, control neurotransmitters, and change their morphology dramatically (Pocock and Kettenmann, 2007). The maintenance of neuro-immune-endocrine system homeostasis is significantly aided by cytokines, neurotransmitters, and hormones (Schafer and Stevens, 2015). The cytokine system, which is made up of cytokines, their receptors, and regulators of their activity, is expressed all over the brain and is controlled from brain development until old age. Nearly every group of cells in the brain can be influenced by the cytokine system. By converting environmental cues into molecular signals, cytokines normally contribute to brain growth and plasticity under physiologically normal settings. They are also essential for maintaining normal brain function and plasticity (Bilbo and Schwarz, 2012). In addition, cytokines and chemokines are essential for immune cell signaling, the healing of injured tissue, and the maintenance of homeostasis (Charo and Ransohoff, 2006). In general, cytokines are categorized as either pro- or anti-inflammatory, meaning they either promote or suppress inflammatory processes. While IL-4 and IL-10 are well-known anti-inflammatory cytokines, interleukin (IL)-1b, IL-6, and tumor necrosis factor a (TNFa) are some of the proinflammatory cytokines that have been the subject of extensive research. Transcripts for all other cytokines were shown to be constitutively present in the brain, except IFN-gamma mRNA. The host benefits from the mobilization of these pro- and anti-inflammatory cytokines and chemical messengers because they enable the body to fight against infection or tissue injury. However, a careful balance between changes that are pro- or anti-inflammatory needs to be maintained because an unchecked and prolonged reaction can cause serious inflammatory collateral damage as well as negatively impact mood and behavior.

Mutual Interaction between the Immune System and Stress

Glucocorticoid receptors are highly expressed on microglia throughout the brain (Sierra et al., 2008), which is not surprising given the intricate nature of the stress and neuro-inflammatory responses. Additionally, microglia-mediated neuro-inflammatory signaling facilitates the integration of stress-induced signaling in the brain (Wohleb et al., 2014b). Neuroendocrine pathways change peripheral and central immune responses in stressful environments. This results in changes in microglia phenotype, monocyte priming, and trafficking, which in turn leads to neuroinflammation (Delpech et al., 2015). Neuroinflammation can then cause a variety of psychiatric disorders, such as stress, anxiety, and depression (Hodes et al., 2015). According to reports, the release of cytokines after microglia activation is responsible for the behavioral phenotype in stress models (Kreisel et al., 2014). It's interesting to note that distinct brain areas and certain neuronal circuits that play a major role in stress-mediated neuroendocrine activation also experience stress-induced microglia activation. As a result, much data explicitly connects cytokines to HPA axis activation (Bellavance and Rivest, 2014). Immunomodulators across the BBB and chorus plexus can also be used by an immunogenic stimulus to initiate HPA axis activation (Kunis et al., 2013). The choroid plexus is often regarded as a doorway for leukocyte trafficking to the central nervous system (CNS). Its activity is controlled by resident CD4þ IFNg-producing helper T-cells. This gate opens in response to an increase in circulating immunomodulators, which promotes neuroinflammation and the development of a stress response. Increases in systemic immunomodulators may also indirectly lead to neuroinflammation, depending on the type and severity of the insult. This can happen through cytokines from capillary endothelial cells at the blood-brain barrier (Verma et al., 2006) or through peripheral inflammatory mediators stimulating the vagal nerve (Eskandari et al., 2003).

Immune and Stress Response Are Regulated by the Microbiota-Gut-Brain Axis

Behaviour in Reaction to Stress Is Modulated by the Gut Microbiome

Preclinical research has shed light on the role of the microbiota in behavioral responses; however, for this chapter, will concentrate on behavioral responses to neuroinflammation. It has been demonstrated that various forms of psychological stress, such as chronic social defeat, crowding, heat stress, acoustic stress, and maternal separation, can modify the composition of the gastrointestinal microbiota (Bharwani et al., 2015), as can stress-related disorders like irritable bowel syndrome (IBS) (Shankar et al., 2015). Research has demonstrated that when exposed to a mild novel arena stressor (Clarke et al., 2013) or the more stressful restraint test (Sudo, 2012), animals raised in a sterile environment from birth and thus lacking gastrointestinal bacteria (germ-free) exhibit an exaggerated hypothalamic-pituitary-adrenal axis activity with elevated ACTH and corticosterone levels. These effects worsened after colonization with commensal bacteria from control mice. Additionally, in slightly stressful novel and unpleasant situations such as the elevated plus maze, light/dark box, and open field test, germ-free rats exhibit an anxiolytic phenotype (Clarke et al., 2013). Only in males was it possible to undo some of the behavioral and neurochemical effects that had been noted after recolonizing germ-free mice with commensal bacteria during their early years of life (Clarke et al., 2013). Conversely, a recent study employing a different breed of mice showed that, in comparison to the germ-free controls, short-term colonization with intragastric inoculation of commensal bacteria (*Bifidobacterium infantis* and *Blautia coccoides*) decreased anxiety-like behavior (Nishino et al., 2013). A higher corticosterone response was observed in germ-free animals after exposure to an open-field test in a different study using stress-sensitive germ-free rats (Crumeyrolle-Arias et al., 2014). When combined, these findings imply that species, sex, strain, and temporal dependence may play a role in the association between gut microbiota and stress- and/or anxiety-related phenotypes.

Control of the Neuroinflammatory Response by the Microbiota in the Gut

A number of recent studies have demonstrated how the gut's microbial population can modulate innate and adaptive immune responses at mucosal surfaces during infection, inflammation, and autoimmunity (El Aidy et al., 2014). The gut microbiota also affects the relative populations, migration, and function of different subsets of immune cells (Dorrestein et al., 2014). In recent times, it has been acknowledged that alterations in the communication of the intestinal epithelium, the intestinal immune system, and gut bacteria can influence systemic immunity (Belkaid and Naik, 2013). However, research on the relationship between CNS-restricted immune cells and the gut microbiota had not been done until relatively recently. Constitutively active microglia have a well-established role in host defense, but they also play a crucial part in neuronal processes during development and adulthood, such as synaptic remodeling to enhance neural network communication (Schafer and Stevens, 2015). Even though the BBB protects the microglia, a recent comprehensive study (Erny et al., 2015) has shown that the gastrointestinal microbiota plays a crucial role in the development, morphology, and immunological function of the microglia. A varied gut microbiome is required under normal homeostatic conditions to keep microglia in a functionally healthy state.

Possible Ways That the Gut Microbiota Affects the Immune System and Stress Response

The epithelial layer of the gut has many functions, including regulating the absorption of nutrients and fluid from the lumen of the intestines and serving as a physical barrier to invasive pathogens or harmful substances (Farhadi et al., 2003). This monolayer is interspersed with complex protein structures known as tight junction proteins (Madara, 1998), establishing a barrier to prevent diffusion of fluids and solvents between adjacent epithelial cells (Ivanov and Naydenov, 2013). Gastrointestinal microorganisms are limited to the gut by definition and interact with the host tissue along the alimentary canal as well as with their cohabiting microbial consortiums. Limiting the amount of contact between the intestinal microbiota and the visceral tissue is one of the intestines' other vital functions. It accomplishes this by secreting a viscous mucous layer that protects the goblet cells of the epithelium. By helping the immune system recognize both self and non-self antigens, the mucous layer and epithelium exchange chemicals, facilitating communication between the gut and the immune system and priming it to recognize potentially hazardous pathogens. Although it is clear that the gastrointestinal microbiota is important for immune response and HPA axis activity, the question of how the gut microbiota directly influences physiological response still needs to be answered. Several mechanisms have been suggested, however, the precise mechanism or mechanisms of action are, as yet, unresolved. To mediate a complex coordinated response on multiple physiological systems, crosstalk in the bi-directional microbiome-gut-brain axis, in homeostatic conditions, and response to environmental changes likely involves dynamic molecules with multiple effector mechanisms that can traverse different anatomical environments and communicate within their local environment. Numerous molecules, including neurotransmitters, neuropeptides, endocrine hormones, and immunomodulators, have been suggested as potential possibilities. The probable role of the microbiota in mediating or modifying stress and neuro-inflammation, as well as the behavioral reactions linked to these conditions, will be discussed in the final section. Because few studies definitively link changes in the gut microbiota and the peripheral to central effects, this section is necessarily conjectural.

The Gastrointestinal Microbiota's Environmental Niche

One must first comprehend the habitat in which these commensal microorganisms dwell to comprehend the mechanisms by which they affect our daily life. Through vagal and pelvic connections between the central nervous system (CNS), enteric nervous system (ENS), sympathetic prevertebral ganglia, and gut motility, permeability, release of gut hormones, and immune function, the gastrointestinal tract can regulate microbial density. The digestive tract's luminal-mucosal interface, which is heavily reliant on the immune system, is the first significant site of host-microorganism interaction. According to Pott and Hornef (2012), the mucosal surface covers the gut's epithelial lining, which is mostly made up of enterocytes, secretory cells, and gut-associated lymphoid tissue (GALT). The GALT uses lymphocytes to build a more focused immune response when necessary, whereas the enterocytes express innate immune receptors and can produce cytokines and chemokines.

Physical Interaction between the Epithelium and Bacteria

In addition to shielding the gastrointestinal bacteria from destruction, their polysaccharide covering helps the host recognize the many different types of bacteria present. This enables the immune surveillance system that protects the host to keep an eye on and control commensal bacteria as well as recognize and classify invasive pathogens. Research indicates that an imbalance in the diversity of homeostatic bacteria may lead to an overabundance of specific cell wall components, like peptidoglycans, which in turn may cause an increase in immune mediators within the gut. When activated, epithelial pattern recognition receptors (PRRs) can recruit inflammatory mediators, cytokine production, and chemokine-mediated recruitment of acute inflammatory cells. PRRs recognize molecular patterns specific to bacteria and other microorganisms (known as pathogen-associated molecular patterns, or PAMPs) (Duerkop et al., 2009). The Toll-like receptor family is the most studied of these receptors. Thus, the quantity and variety of bacteria in the stomach are regulated.

Biochemical Exchange between the Epithelium and the Microbiota

The average animal gut is thought to contain more than 100 species (Qin et al., 2010) and more than 7000 strains (Ley et al., 2008) of bacteria, with Bacter-iodates and Firmicute phylotypes predominating and Proteo-, Actino-, Fuso-, and Verrucomicrobia being less common (Eckburg et al., 2005). The native microbiota diversity varies throughout the gut's various locations (Lozupone et al., 2012). These beneficial bacteria can produce a wide range of neurotransmitters and neuromodulators on their own, as well as to stimulate enteroendocrine cells to produce neuropeptides. For instance, certain species of Bifidobacterium and Lactobacillus can generate GABA, or gamma-aminobutyric acid: Norepinephrine can be produced by Escheridia, Bacillus, and Saccharomyces species; 5-HT can be produced by Candida, Streptococcus, Escheridia, and Enterococcus species. Dopamine is produced by Bacillus, while acetylcholine is produced by Lactobacillus (Lyte, 2014). These neurotransmitters that are produced by microbes can pass through the intestinal mucosa and potentially mediate actions locally in the gut, or they can enter the bloodstream and affect physiological processes in the brain.

Effects of Microbiological Interaction

In a state of homeostasis, the microbiota stimulates the release of cytokines and chemokines, which in turn controls the local concentrations of bacteria in the gut. This results in a healthy, resting inflammatory tone. Microbiota can synthesise and release neurotransmitters and SCFAs, as well as affect the release of neuropeptides and hormones from enteroendocrine cells of the intestines, in addition to regulating local immune responses at the epithelium. Gut peptides including neuropeptide Y, cholecystokinin, orexin, galanin, ghrelin, gastrin, and leptin are believed to affect both central and peripheral neuronal transmission. Current theories propose that these circulating chemokines, cytokines, endocrine messengers, and microbial byproducts may enter the bloodstream and lymphatic systems, or they may influence neural messages transmitted by the vagal and spinal afferent neurons, which may have an impact on events mediated centrally, such as the regulation of the HPA axis and neuroinflammation.

Microbiota's Peripheral Influence on HPA Axis and Neuroinflammation Regulation

Recent data suggests a unique neuroimmune circuit that recruits inflammatory monocytes to the brain and reinforces stress-related behaviors utilizing microglia activation and altered sympathetic neuronal tone to the peripheral immune system (Wohleb et al., 2014b). According to D'Mello et al. (2009), TNF-mediated microglia activation appeared to control the trafficking of these monocytes to the brain. This trafficking may be reversed by probiotic intervention with VSL#3 (D'Mello et al., 2009). The trafficking of monocytes from the spleen—a region with a high density of FFAR2 receptors—plays a role in mediating specific behavioural abnormalities that result from prolonged HPA activation (Wohleb et al., 2014b). (Erny et al., 2015). These Gi/Go G-protein coupled FFAR2 receptors have been identified as being located on lymphocytes (Nilsson et al., 2003). In the absence of a complex microbiota, it has been demonstrated that SCFAs, the natural ligands for FFARs, can counteract the negative effects on microglia density, morphology, and maturity (Erny et al., 2015). When combined, these data imply that the gut microbiota may indirectly control centrally mediated processes by controlling the trafficking of monocytes to the brain, which in turn controls the activation of microglia and the activity of the HPA axis, potentially through SCFA-driven pathways.

Conclusion

It has recently come to light that the gut microbiota plays a major role in the body's ability to adapt to stress and mount an immunological response. In terms of behavior, gastrointestinal complexity, and content might influence fear and anxiety reactions as well as lessen behavioral anhedonia or despair. The data provided here implies that the richness and diversity of the gut microbiota may be associated with the immune system and stress-related illness resistance as well as the immune system and stress-related system dysfunction. The majority of the research on the impact of gut microbiota has been preclinical in character, and more studies using pro- and pre-biotic interventions are necessary. The results in clinical cohorts have been diagnostic in nature. The ultimate objective is to improve our comprehension of the function of gut microbiota in basic physiological and pathological processes to treat and comprehend a variety of immunological and stress-related illnesses. This chapter demonstrates the very intricate connections between the microbiota of the gastrointestinal tract and the immune system and stress reactions. It is still unclear, nevertheless, whether the physiological and behavioral reactions to stress and neuroinflammation are caused by commensal bacteria.

References

Belkaid, Y., Naik, S., 2013. Compartmentalized and systemic control of tissue im-*munity by commensals. Nat. Immunol.* 14 (7), 646e653.

Bellavance, M. A., Rivest, S., 2014. The HPA - immune Axis and the immunomodu-latory actions of glucocorticoids in the brain. *Front. Immunol.* 5, 136.

Bharwani, A., Mian, M. F., Foster, J. A., Surette, M. G., Bienenstock, J., Forsythe, P., 2015. Structural & functional consequences of chronic psychosocial stress on the microbiome & *host. Psychoneuroendocrinology* 63, 217e227.

Biagi, E., Candela, M., Fairweather-Tait, S., Franceschi, C., Brigidi, P., 2012. Aging of the animal metaorganism: the microbial counterpart. *Age* (Dordr) 34 (1), 247e267.

Bilbo, S. D., Schwarz, J. M., 2012. The immune system and developmental program-ming of brain and behavior. *Front. Neuroendocrinol.* 33 (3), 267e286.

Borre, Y. E., O'Keeffe, G. W., Clarke, G., Stanton, C., Dinan, T. G., Cryan, J. F., 2014b. Microbiota and neurodevelopmental windows: implications for brain disorders. *Trends Mol. Med.* 20 (9), 509e518.

Burokas, A., Moloney, R. D., Dinan, T. G., Cryan, J. F., 2015b. Microbiota regulation of the Mammalian gut-brain axis. *Adv. Appl. Microbiol.* 91, 1e62.

Charo, I. F., Ransohoff, R. M., 2006. The many roles of chemokines and chemokine receptors in inflammation. *N. Engl. J. Med.* 354 (6), 610e621.

Clarke, G., Grenham, S., Scully, P., Fitzgerald, P., Moloney, R. D., Shanahan, F., Dinan, T. G., Cryan, J. F., 2013. The microbiome-gut-brain axis during early life regulates the hippocampal serotonergic system in a sex-dependent manner. *Mol. Psychiatry* 18 (6), 666e673.

Crumeyrolle-Arias, M., Jaglin, M., Bruneau, A., Vancassel, S., Cardona, A., Dauge, V., Naudon, L., Rabot, S., 2014. Absence of the gut microbiota enhances anxiety-like behavior and neuroendocrine response to acute stress in rats. *Psychoneur-oendocrinology 42, 207*e217.

Cryan, J. F., Dinan, T. G., 2015b. More than a gut feeling: the microbiota regulates neurodevelopment and behavior. *Neuropsychopharmacology* 40 (1), 241e242.

Delpech, J. C., Madore, C., Nadjar, A., Joffre, C., Wohleb, E. S., Laye, S., 2015. Microglia in neuronal plasticity: influence of stress. *Neuropharmacology* 96 (Pt A), 19e28.

Dickmeis, T., Weger, B. D., Weger, M., 2013. The circadian clock and glucocorti-coidseinteractions across many time scales. *Mol. Cell Endocrinol.* 380 (1e2), 2e15.

D'Mello, C., Le, T., Swain, M. G., 2009. Cerebral microglia recruit monocytes into the brain in response to tumor necrosis factoralpha signaling during peripheral organ inflammation. *J. Neurosci.* 29 (7), 2089e2102.

Dorrestein, P. C., Mazmanian, S. K., Knight, R., 2014. Finding the missing links among metabolites, microbes, and the host. *Immunity* 40 (6), 824e832.

Duerkop, B. A., Vaishnava, S., Hooper, L. V., 2009. Immune responses to the micro-biota at the intestinal mucosal surface. *Immunity* 31 (3), 368e376.

Eckburg, P. B., Bik, E. M., Bernstein, C. N., Purdom, E., Dethlefsen, L., Sargent, M., Gill, S. R., Nelson, K. E., Relman, D. A., 2005. Diversity of the animal intestinal microbial flora. *Science* 308 (5728), 1635e1638.

El Aidy, S., Dinan, T. G., Cryan, J. F., 2014. Immune modulation of the brain-gut-microbe axis. *Front. Microbiol.* 5, 146.

Erny, D., Hrabe de Angelis, A. L., Jaitin, D., Wieghofer, P., Staszewski, O., David, E., Keren-Shaul, H., Mahlakoiv, T., Jakobshagen, K., Buch, T., Schwierzeck, V., Utermohlen, O., Chun, E., Garrett, W. S., McCoy, K. D., Diefenbach, A., Staeheli, P., Stecher, B., Amit, I., Prinz, M., 2015. Host microbiota constantly control matu-ration and function of microglia in the CNS. *Nat. Neurosci.* 18 (7), 965e977.

Eskandari, F., Webster, J. I., Sternberg, E. M., 2003. Neural immune pathways and their connection to inflammatory diseases. *Arthritis Res. Ther. 5 (6), 251*e265.

Farhadi, A., Banan, A., Fields, J., Keshavarzian, A., 2003. Intestinal barrier: an inter-face between health and disease. *J. Gastroenterol. Hepatol. 18 (5), 479*e497.

Frank, D. N., Pace, N. R., 2008. Gastrointestinal microbiology enters the meta-genomics era. *Curr. Opin. Gastroenterol.* 24 (1), 4e10.

Hodes, G. E., Kana, V., Menard, C., Merad, M., Russo, S. J., 2015. Neuroimmune mechanisms of depression. *Nat. Neurosci.* 18 (10), 1386e1393.

Ivanov, A. I., Naydenov, N. G., 2013. Dynamics and regulation of epithelial adherens junctions: recent discoveries and controversies. *Int. Rev. Cell Mol. Biol.* 303, 27e99.

Kleiman, A., Tuckermann, J. P., 2007. Glucocorticoid receptor action in beneficial and side effects of steroid therapy: lessons from conditional knockout mice. *Mol. Cell Endocrinol.* 275 (1e2), 98e108.

Kreisel, T., Frank, M. G., Licht, T., Reshef, R., Ben-Menachem-Zidon, O., Baratta, M. V., Maier, S. F., Yirmiya, R., 2014. Dynamic microglial alterations underlie stress-induced depressive-like behavior and suppressed neurogenesis. *Mol. Psychiatry* 19 (6), 699e709.

Kunis, G., Baruch, K., Rosenzweig, N., Kertser, A., Miller, O., Berkutzki, T., Schwartz, M., 2013. IFN-gamma-dependent activation of the brain's choroid plexus for CNS immune surveillance and repair. *Brain* 136 (Pt 11), 3427e3440.

Kurokawa, K., Itoh, T., Kuwahara, T., Oshima, K., Toh, H., Toyoda, A., Takami, H., Morita, H., Sharma, V. K., Srivastava, T. P., Taylor, T. D., Noguchi, H., Mori, H., Ogura, Y., Ehrlich, D. S., Itoh, K., Takagi, T., Sakaki, Y., Hayashi, T., Hattori, M., 2007. Comparative metagenomics revealed commonly enriched gene sets in animal gut microbiomes. *DNA Res. 14 (4), 169*e181.

Ley, R. E., Hamady, M., Lozupone, C., Turnbaugh, P. J., Ramey, R. R., Bircher, J. S., Schlegel, M. L., Tucker, T. A., Schrenzel, M. D., Knight, R., Gordon, J. I., 2008. Evo-lution of mammals and their gut microbes. *Science* 320 (5883), 1647e1651.

Lozupone, C. A., Stombaugh, J. I., Gordon, J. I., Jansson, J. K., Knight, R., 2012. Diversity, stability and resilience of the animal gut microbiota. *Nature* 489 (7415), 220e230.

Lyte, M., 2014. Microbial endocrinology and the microbiota-gut-brain axis. *Adv. Exp. Med. Biol.* 817, 3e24.

Madara, J. L., 1998. Regulation of the movement of solutes across tight junctions. *Annu. Rev. Physiol.* 60, 143e159.

Mayer, E. A., 2000. The neurobiology of stress and gastrointestinal disease. *Gut* 47 (6), 861e869.

McEwen, B. S., Gray, J. D., Nasca, C., 2015. 60 Years of Neuroendocrinology: redefining neuroendocrinology: stress, sex and cognitive and emotional regu-lation. *J. Endocrinol.* 226 (2), T67eT83.

Nilsson, N. E., Kotarsky, K., Owman, C., Olde, B., 2003. Identification of a free fatty acid receptor, FFA2R, expressed on leukocytes and activated by short-chain fatty acids. *Biochem. Biophys. Res. Commun.* 303 (4), 1047e1052.

Nishino, R., Mikami, K., Takahashi, H., Tomonaga, S., Furuse, M., Hiramoto, T., Aiba, Y., Koga, Y., Sudo, N., 2013. Commensal microbiota modulate murine be-haviors in a strictly contamination-free environment confirmed by culture-based methods. *Neurogastroenterol. Motil.* 25 (6), 521e528.

Pocock, J. M., Kettenmann, H., 2007. Neurotransmitter receptors on microglia. *Trends Neurosci.* 30 (10), 527e535.

Pott, J., Hornef, M., 2012. Innate immune signalling at the intestinal epithelium in homeostasis and disease. *EMBO Rep.* 13 (8), 684e698.

Qin, J., Li, R., Raes, J., Arumugam, M., Burgdorf, K. S., Manichanh, C., Nielsen, T., Pons, N., Levenez, F., Yamada, T., Mende, D. R., Li, J., Xu, J., Li, S., Li, D., Cao, J., Wang, B., Liang, H., Zheng, H., Xie, Y., Tap, J., Lepage, P., Bertalan, M., Batto, J. M., Hansen, T., Le Paslier, D., Linneberg, A., Nielsen, H. B., Pelletier, E., Renault, P., Sicheritz-Ponten, T., Turner, K., Zhu, H., Yu, C., Li, S., Jian, M., Zhou, Y., Li, Y., Zhang, X., Li, S., Qin, N., Yang, H., Wang, J., Brunak, S., Dore, J., Guarner, F., Kristiansen, K., Pedersen, O., Parkhill, J., Weissenbach, J., Meta, H. I. T. C., Bork, P., Ehrlich, S. D., Wang, J., 2010. A human gut microbial gene catalogue established by metagenomic sequencing. *Nature* 464 (7285), 59e65

Ransohoff, R. M., Schafer, D., Vincent, A., Blachere, N.E., Bar-Or, A., 2015. Neuro-inflammation: ways in which the immune system affects the brain. *Neurotherapeutics.*

Sampson, T. R., Mazmanian, S. K., 2015. Control of brain development, function, and behavior by the microbiome. *Cell Host Microbe* 17 (5), 565e576.

Schafer, D. P., Stevens, B., 2015. Microglia function in Central nervous system development and plasticity. Cold Spring Harb. *Perspect. Biol.*

Shankar, V., Homer, D., Rigsbee, L., Khamis, H. J., Michail, S., Raymer, M., Reo, N. V., Paliy, O., 2015. The networks of animal gut microbe-metabolite associations are different between health and irritable bowel syndrome. *ISME J.* 9 (8), 1899e1903.

Sierra, A., Gottfried-Blackmore, A., Milner, T. A., McEwen, B. S., Bulloch, K., 2008. Steroid hormone receptor expression and function in microglia. *Glia* 56 (6), 659e674.

Smith, S. M., Vale, W. W., 2006. The role of the hypothalamic-pituitary-adrenal axisin neuroendocrine responses to stress. Dialogues Clin. Neurosci. 8 (4), 383e395. Souza, D. G., Vieira, A. T., Soares, A. C., Pinho, V., Nicoli, J. R., Vieira, L. Q., Teixeira, M. M., 2004. The essential role of the intestinal microbiota in facilitating acute in- flammatory responses. *J. Immunol.* 173 (6), 4137e4146.

Sudo, N., 2012. Role of microbiome in regulating the HPA axis and its relevance to allergy. *Chem. Immunol. Allergy* 98, 163e175.

Tremblay, M. E., Stevens, B., Sierra, A., Wake, H., Bessis, A., Nimmerjahn, A., 2011. The role of microglia in the healthy brain. *J. Neurosci.* 31 (45), 16064e16069.

Verma, S., Nakaoke, R., Dohgu, S., Banks, W. A., 2006. Release of cytokines by brain endothelial cells: a polarized response to lipopolysaccharide. *Brain Behav. Immun.* 20 (5), 449e455.

Williamson, L. L., McKenney, E. A., Holzknecht, Z. E., Belliveau, C., Rawls, J. F., Poulton, S., Parker, W., Bilbo, S.D., 2015. Got worms? Perinatal exposure to helminths prevents persistent immune sensitization and cognitive dysfunction induced by early-life infection. *Brain Behav. Immun.*

Wohleb, E. S., McKim, D. B., Sheridan, J. F., Godbout, J. P., 2014b. Monocyte trafficking to the brain with stress and inflammation: a novel axis of immune-to-brain communication that influences mood and behavior. *Front. Neurosci.* 8, 447.

Chapter 8

Gut Microbiota and the Gut-Brain Axis

S. Yogeshpriya[1,*]
S. Sivaraman[2]
A. Arivuchelvan[3]
M. Saravanan[1]
and A. Latchumikanthan[4]
[1]Department of Veterinary Medicine, Veterinary College and Research Institute, Orathanad, India
[2]Department of Veterinary Clinical Medicine, Veterinary College and Research Institute, Salem, India
[3]Department of Veterinary Pharmacology and Toxicology, Veterinary College and Research Institute, Namakkal, India
[4]Department of Veterinary Parasitology, Veterinary College and Research Institute, Orathanadu, India

Abstract

The gut microbiota, a complex community of microorganisms residing in the gastrointestinal tract, plays a pivotal role in maintaining host health and influencing various physiological processes. One of its critical functions is its interaction with the central nervous system (CNS) through the gut-brain axis (GBA). This bidirectional communication pathway significantly impacts brain function, behavior, and the development of neurological and psychiatric disorders. This review aims to provide a comprehensive overview of the gut microbiota's role in the gut-brain axis, focusing on its impact on animal health, nutrition, and disease. It explores the mechanisms of interaction between gut microbiota and the CNS, the effects on various health conditions, and potential therapeutic strategies involving modulation of the microbiota. Despite advancements, challenges remain in understanding the precise mechanisms of the GBA and individual variability in responses to microbiota modulation. Further research is needed to elucidate the long-term effects of interventions and to optimize therapeutic strategies. Personalized approaches based on individual microbiota profiles may enhance the efficacy of treatments. The gut microbiota and the gut-brain axis play critical roles in maintaining health and managing diseases. By exploring and harnessing these interactions, new therapeutic avenues can be developed to address a wide range of health issues, potentially transforming approaches to disease prevention and treatment. The microbiota-gut-brain axis (MGBA)

* Corresponding Author's Email: dryogeshpriya@gmail.com

In: Microbiomes in Animal Nutrition, Health, and Disease
Editor: Tanmoy Rana
ISBN: 979-8-89530-425-9

has been increasing in popularity as a tool to modulate brain function and, ultimately, host health. This chapter enlightens the recent literature regarding the gut-brain axis.

Keywords: gut microbiota, complex mechanism, host, gut-brain- axis

Introduction

The gut microbiome comprises a vast ecology of commensal bacteria, archaea, fungi, and viruses in and on the body. However, understanding both the nature and function of the microbiota is still very much in its infancy as the exact composition varies across each individual and there is no definitive understanding of what truly constitutes a healthy adult gut microbial profile beyond exhibiting both diversity and stability as per the project data available from Human Microbiome Project C (2012). Due to this wide variety in microbiota composition, functional capacity, defined by metabolic pathways, may be a better metric for the health of one's microbiome provided the conservation of basic metabolic pathway categories across individuals (Vernocchi et al., 2016).

The signaling mechanism behind the microbiota-gut-brain axis communication is of particular interest when considering therapeutic design. The brain modulates gut function through the hypothalamic-pituitary-adrenal axis and the autonomic nervous system; norepinephrine (Martyanov et al., 2021). Conversely, the gut modulates CNS functions by involving a variety of microbiota-derived metabolites and products, neuroactive substances, and gut hormones that traverse through the enteric nervous system, vagus nerve, circulatory system, or immune system to reach the brain (Long-Smith et al., 2020). Collectively, these pathways are referred to as the microbiota-gut-brain axis (MGBA). The gut-brain axis describes the two-way biochemical communication between the gastrointestinal tract (GI tract) and the central nervous system (CNS). The term "microbiota-gut-brain axis" highlights the role of gut microbiota in this communication. Among various pathways, the vagus nerve offers the most direct connection, as its receptors on the vagal afferents detect and transmit signals from the gut to the brain. Recent research has even linked the vagus nerve to changes in CNS mood and behavior by affecting CNS reward neurons (Fulling et al., 2019). This communication can occur in two scenarios: when an infection in the gastrointestinal tract (GIT) microbiota adversely affects the brain and increases sickness behavior, or when a healthy GIT microbiota supports brain function. Studies with germ-free animals have demonstrated that probiotics and prebiotics can lead to behavioral changes (Kraimi et al., 2019). A recent review has examined the effects of gut microbiota on brain development and behavior in both laboratory and human studies, offering a deeper understanding of its implications for farm animals. In this review, we aim to provide new insights into the microbiota-gut-brain axis to systematically assess animal welfare.

Microflora of Rumnanat Stomach

Ruminants consume plants that contain complex polysaccharides such as hemicellulose, cellulose, and lignin. Since ruminants lack the enzymes needed to break down these polysaccharides, they rely on microflora in their gut to hydrolyze these compounds and generate energy. These commensal microbes use the ingested feed for their own survival, establishing a beneficial symbiotic relationship with the host (Flint and Bayer, 2008). The microbial community resides primarily in the reticulorumen compartment, which is an anaerobic environment. This compartment hosts a dense and diverse array of microorganisms, including eukaryotes (fungi and protozoa), archaea (methanogens), and bacteria, with their respective concentrations being 10^4, 10^6, and 10^10. The bacterial populations in the rumen are highly sensitive and can be affected by even minor changes in the rumen environment. Fermentation by the rumen microflora is a complex process where microorganisms work together to break down complex compounds into simpler forms that can be easily utilized by the host. Polysaccharides are converted into two main products: one is the proteins needed for bacterial cell wall synthesis, and the other is volatile fatty acids (VFAs), which are the end products of fermentation. VFAs are vital for ruminant metabolism, serving as a primary energy source for the host. They are involved in essential metabolic pathways, including fatty acid synthesis and gluconeogenesis. Recent research suggests that VFAs account for approximately 70% of the energy content derived from plant polysaccharides.

Gut Microbiota - Gut Brain Axis

The body hosts trillions of microorganisms, approximately 1.3 times the number of human cells. In adults, these microorganisms can weigh around 1 kg, with the majority residing in the intestines and collectively referred to as the gut microbiota (Krautkramer et al., 2021). The gut microbiome includes a diverse array of organisms such as commensal bacteria, fungi, phages, yeasts, archaeal parasites, and viruses, with over 1,000 species identified (Nicholson et al., 2019). This microbiota plays a crucial role in regulating innate immunity and inflammation (Campbell et al., 2023). Although recent research has focused on the composition of the gut microbiota, further investigation is needed to fully understand its characteristics and functions (Balmus et al., 2020). The gut microbiota influences the development of inflammatory bowel disease (IBD), innate immune disorders, and neuroinflammatory conditions through immune signaling (Gilbert et al., 2018). Additional studies have demonstrated that nerve development and microglial activation are dependent on the gut microbiota (Li et al., 2023). Maintaining gut microbiota homeostasis positively affects the regulation of both the immune and nervous systems, with potential impacts on pathways related to mucus homeostasis in neurological diseases (Li et al., 2023).

Metabolites and Neurotransmitters

Metabolites and neurotransmitters are crucial for regulating the bidirectional communication of the microbiota-gut-brain axis (MGBA). However, many neurotransmitters produced by microorganisms have a short half-life and limited ability to cross the blood-brain barrier (BBB) (Taj and Jamil, 2018). Consequently, it is still uncertain whether these neurotransmitters can reach specific targets within the central nervous system (CNS) and if they are present in sufficient concentrations to influence it (Channer et al., 2023). Conversely, short-chain fatty acids (SCFAs) and neurotransmitters produced by certain bacteria can activate the innate immune system, impacting the CNS and regulating brain function by modulating immune and inflammatory cell activity (Foster et al., 2017). Serotonin, SCFAs, and other tryptophan metabolites have been shown to significantly affect the stability of both the enteric nervous system and the CNS (Brummelte et al., 2017). These metabolites and products can convey information to the CNS via the nervous, circulatory, and immune systems (Sonner et al., 2019). Research indicates that the diversity and relative abundance of gut microbiota influence the types and concentrations of microbial metabolites and products, and are linked to various inflammatory diseases in humans (Bastiaanssen et al., 2019). In summary, the regulation of inflammatory diseases by metabolites and neurotransmitters is complex, and the clear relationship between these substances and the gut microbiota offers valuable insights for managing MGBA communication.

Tryptophan

Bacteria transform tryptophan from dietary sources into tryptamine by expressing tryptophan decarboxylase. Recent research highlights tryptamine as a crucial signaling molecule in host-microbe interactions, influencing gastrointestinal motility through 5-HT4 receptors (Bhattarai et al., 2018). However, it remains uncertain whether tryptamine can reach the central nervous system (CNS) and affect its functions. The kynurenine pathway is the primary route for tryptophan metabolism, and studies have shown that reduced activation of this pathway in the periphery can enhance tryptophan utilization in germ-free (GF) mice (Wei et al., 2021). Conversion of tryptophan to kynurenine may lead to neuroinflammation and negatively impact brain health (Kennedy et al., 2017). Conversely, the gut microbiota can convert tryptophan into indole, which supports intestinal homeostasis (Gheorghe et al., 2019). Understanding how tryptophan, produced by gut microbiota, regulates the nervous system could offer new approaches for mitigating inflammatory responses.

Short Chain Fatty Acids (SCFA)

Short-chain fatty acids (SCFAs) are linked to various inflammatory diseases and can impact intestinal epithelial integrity, glycolipid metabolism, and immune system balance. SCFAs, primarily acetate, propionate, and butyrate, account for about 95% of the gut microbiota's metabolic products (Zaky et al., 2021). The levels of SCFAs in the intestine are influenced by the abundance of bacteria such as Clostridia, Bifidobacteria, Bacteroides, and Lactobacillus (Xiao et al., 2022). SCFAs can readily enter the circulatory system from the intestine and affect the central nervous system (CNS) by crossing the blood-brain barrier (BBB) through monocarboxylate transporters (Wenzel et al., 2020). Additionally, SCFAs interact with enzymes involved in their own biosynthesis, influencing neurotransmitter production in the brain (Zhong et al., 2023). Treatment with acetate, propionate, and butyrate has been shown to correct morphological defects in microglia in germ-free (GF) mice (Erny et al., 2015). However, given that SCFAs have a short half-life of 25 minutes to 3 hours, it remains unclear whether these metabolites can effectively regulate neurotransmission in the body (Mandaliya and Seshadri, 2019). Therefore, determining the optimal concentration of SCFAs needed to impact the brain is crucial.

Histamine

Most bacteria possess the histidine decarboxylase gene, which is present in many species such as Lactobacillus, Streptococcus, and Enterococcus, and is essential for histamine production (De Palma et al., 2022). Histamine can induce anti-inflammatory responses through H4 receptors, which are particularly significant in the central nervous system (CNS). However, depending on the receptors it interacts with, histamine may also exhibit pro-inflammatory effects by promoting the production of various chemokines (Kim et al., 2022). Histamine has been linked to neuroinflammatory diseases, potentially worsening the severity of these conditions (Flores-Clemente et al., 2021). The concentration and distribution of histamine receptors in central or systemic areas significantly influence neuroinflammation. Consequently, modifying the gut microbiota to alter histamine levels has emerged as a potential strategy for managing inflammation (Carthy and Ellender, 2021).

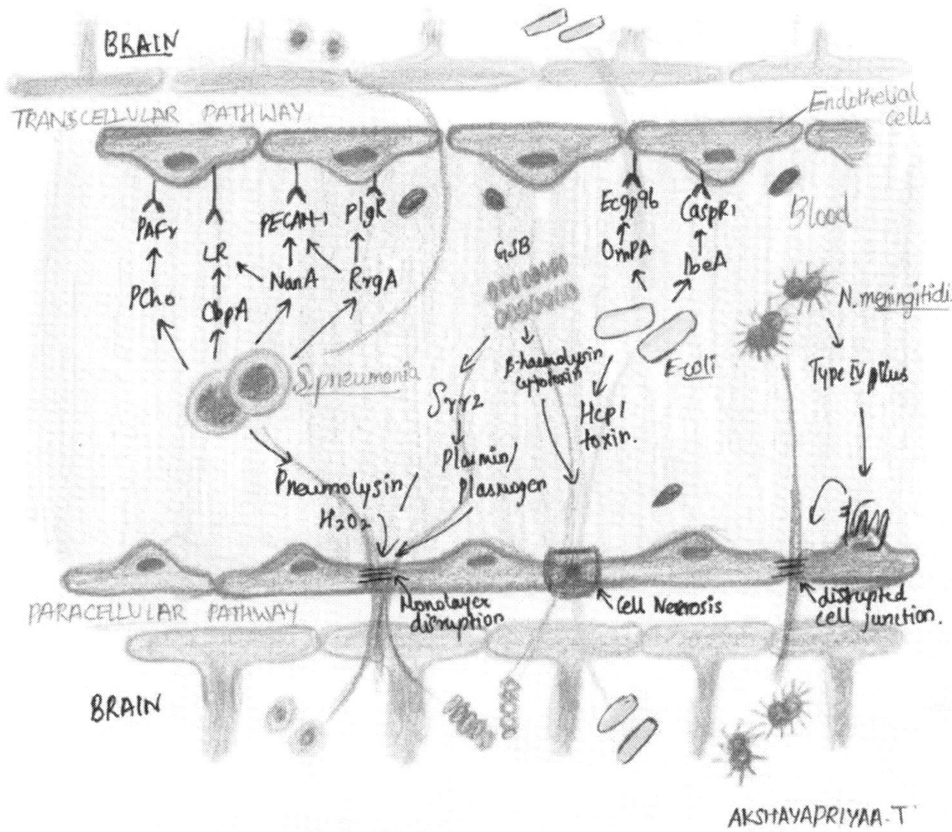

Figure 1. Mechanisms for bacteria crossing the blood brain barrier. The mechanisms for four common meningitis-causing bacteria have been elucidated in recent years through a mixture of largely in vitro and in vivo data. While a pre-requisite for these bacteria is the survival in the bloodstream long enough to interact with brain endothelial cells, the mechanisms and pathways differ.

Serotonin

Serotonin (5-HT), a key neurotransmitter, influences the development and function of both the enteric nervous system (ENS) and the central nervous system (CNS) through endocrine and paracrine signaling processes (Xiao et al., 2021). It has been observed that 5-HT can enhance cytokine secretion by lymphocytes and monocytes, and communicate with the CNS by stimulating the vagus nerve (McVey et al., 2019). Notably, the gut microbiota plays a significant regulatory role in 5-HT production. Metabolomics data show that approximately 20% of microorganisms can synthesize 5-HT (Valles-Colomer et al., 2019). Conversely, enterochromaffin cells (ECCs) contain over 90% of the body's 5-HT, and the autonomic nervous system (ANS) can release 5-HT into the intestine by activating ECCs, thereby affecting the characteristics and functions of the gut microbiota (Sgritta et al., 2019). Studies have shown that 5-HT levels are significantly lower in germ-free (GF) mice compared to normal mice (Xiao et al., 2021). Specific spore-forming bacteria have been found to increase 5-HT levels in the colon and serum of GF mice, with elevated short-chain fatty acids (SCFAs) enhancing 5-HT production and alleviating GF-related intestinal dysfunction by upregulating Tph1 expression in ECCs (Beyder, 2018). Additionally, irritable bowel syndrome (IBS) has been associated with increased blood 5-HT levels, which correlates with the severity of visceral pain in patients (Luo et al., 2021). Similarly, 5-HT released by endothelial cells may contribute to IBS abdominal pain by influencing vagal and inflammatory responses (Gao et al., 2022). Therefore, understanding the relationship between 5-HT and the gut microbiota could advance the treatment of inflammatory diseases.

Ghrelin

Ghrelin is present in a wide range of animals, including fish, amphibians, birds, and mammals. It acts as a natural ligand for the growth hormone secretagogue receptor in the pituitary gland, where it plays a key role in regulating growth, feeding, and metabolism. Ghrelin secretion increases when the stomach is empty, and its release can be influenced by gut microbiota dynamics. For instance, supplementation with Bifidobacterium has been shown to decrease ghrelin secretion (Chen C. et al., 2022). Additionally, gastrin, a neuropeptide in the central nervous system (CNS), is a critical regulator of nerve function and inflammation (Yuan MJ. et al., 2021). Ghrelin's protective effects against neuroinflammatory diseases such as Alzheimer's disease (AD) and Parkinson's disease (PD) have been well-documented. In AD, reduced ghrelin levels in the brain are associated with decreased synaptic plasticity and memory deficits, while increased ghrelin can enhance synaptic plasticity, mitigate memory loss, and suppress excessive inflammatory responses (Russo et al., 2022). In PD, ghrelin protects against neurotoxicity by safeguarding dopaminergic cells and mitochondrial function (Wang et al., 2021). Ghrelin plays a crucial role in CNS regulation and is closely linked to the gut microbiota. It is hypothesized that ghrelin may influence neuroinflammatory diseases through the microbiota-gut-brain axis.

Signaling Mechanisms of Microbiota Gut-Brain Axis Communication

Communication Pathways

The MGBA operates through several pathways:

- *Neural Pathways*: Vagus nerve transmission, connecting the gut and brain, facilitates communication between the microbiota and CNS.
- *Endocrine Pathways*: Microbial metabolites influence hormone levels, impacting brain function.
- *Immune Pathways*: Gut microbiota modulate immune responses, affecting neuroinflammation and brain health

Vagus Nerve

The vagus nerve, comprising 80% afferent fibers and 20% efferent fibers, is a crucial component of the peripheral nervous system. It serves as a primary communication channel between the gut and the brain, playing a significant role in the microbiota-gut-brain axis (MGBA) signaling (Fulling et al., 2019). While the afferent fibers of the vagus nerve are extensively distributed throughout the intestine, they do not directly interact with gut microbiota (Bonaz et al., 2018). Instead, the vagus nerve detects microbial signals via bacterial metabolites and relays this information to the central nervous system (CNS) (Bonaz et al., 2018). Intestinal endocrine cells release glutamate and serotonin, which can activate the vagus nerve and send feedback to the CNS (Kaelberer et al., 2018).

Among the various communication pathways, the vagus nerve is the most direct route for bidirectional signaling in the MGBA, making it a focal point in research on inflammatory diseases. Historically, vagotomy, a procedure that involved cutting the vagus nerve, was commonly used to treat peptic ulcer disease before the advent of widespread drug treatments for Helicobacter pylori infection (Wu et al., 2020). Inhibiting the vagus nerve can disrupt gut microbiota homeostasis and contribute to inflammatory conditions such as irritable bowel syndrome (IBS) and inflammatory bowel disease (IBD) (Oswiecimska et al., 2017). Preclinical studies have shown that the beneficial effects of Lactobacillus rhamnosus in mice were lost following vagotomy (Liu et al., 2021). Recent research has also highlighted the role of the vagus nerve in regulating emotions and behavior by modulating CNS responses (Fulling et al., 2019). Disruption of the vagus nerve can impair central signaling in species of Lactobacillus and Bifidobacterium, leading to altered emotional regulation (Zhao et al., 2023). In summary, the vagus nerve is essential for transmitting information between the gut microbiota and the CNS, and its integrity is crucial for effective bidirectional communication within the MGBA.

Immune Mechanism

In addition to the vagus nerve, the immune system plays a crucial role in facilitating bidirectional communication within the microbiota-gut-brain axis (MGB axis) (Muller et al., 2020). The integrity of the immune system is vital for maintaining dynamic immunity and defending the body against pathogenic microorganisms (Agusti et al., 2018). Approximately 70%-80% of immune cells are located in the mesenteric lymph nodes, which are populated by dendritic cells, macrophages, neutrophils, natural killer cells, and mast cells (Zhou et al., 2021). Among these, macrophages are extensively distributed throughout the intestine and, in conjunction with gut microbiota, play a significant role in regulating communication between the gut and the nervous system (Jarret et al., 2020).

Activated microglia in the CNS can secrete various antigen markers that influence neurotransmitter regulation and trigger neuroinflammatory responses (Erny et al., 2015). Gut microbiota are thought to modulate innate and adaptive immune responses and inflammatory processes, potentially affecting the activation of gut-extrinsic sympathetic nerves through the gut-brain circuit. Furthermore, maintaining gut microbiota homeostasis is crucial for the maturation, health, and normal function of microglia (Mossad and Blank, 2021). Gut microbiota are also considered key factors in microglia maturation and astrocyte activation (Sun et al., 2018). Research has shown that germ-free (GF) mice exhibit immune abnormalities, including reduced T and B cell populations and cytokine levels (Fujita et al., 2020). However, recolonization with Bacteroides fragilis can support the immune maturation of gut-associated lymphoid tissue (Gomaa, 2020).

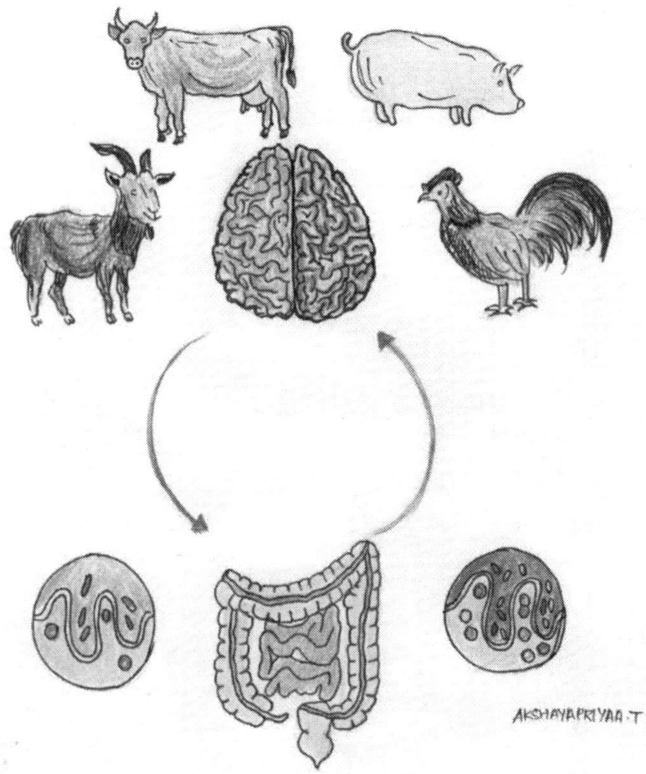

Figure 2. Animal welfare in the implication and perspective of the gut microbiome (hand drawn image courtesy: Ms. Akshayapriyaa t).

Pattern recognition receptors (PRRs), such as toll-like receptors (TLRs), peptidoglycans (PGN), and nucleotide-binding oligomerization domain-like receptors (NLRs), are crucial for innate immune responses. These receptors mediate the immune response to microorganisms (Yuan M. et al., 2021). TLRs, as sensors of gut microbiota presence, transmit information to the enteric nervous system (ENS), affecting its development and function. For instance, the absence of TLR2 signaling in mice results in abnormal neurochemical coding in the ENS, which can be reversed by TLR2 agonist supplementation (Yarandi et al., 2020). Additionally, reduced expression of PGN-sensitive receptors in the striatum of GF and antibiotic-treated mice has been linked to increased social behavior and behavioral changes (Arentsen et al., 2017). This indicates a close relationship between the immune system and gut microbiota homeostasis, suggesting that the immune system might be a key mediator in bidirectional communication between the gut microbiota and the nervous system, offering an additional avenue for exploring MGB axis communication.

Therapeutically Targeting the Microbiota Gut Brain Axis

Nutritional Interventions

In addition to investigating the relationship between innate immune signaling pathways and the microbiota-gut-brain (MGB) axis, there is growing interest in exploring therapeutic approaches that target the MGB axis. Intestinal microbiota, as a central component of the MGB axis, represents a promising breakthrough for managing inflammatory and infectious diseases. Various methods for modulating gut microbiota—such as probiotics, prebiotics, symbiotic, and dietary interventions—have been extensively studied for their potential effects on these conditions (Figure 2). Despite numerous reports highlighting the positive outcomes of gut microbiota regulation, each approach has its limitations, and their overall effectiveness remains a subject of ongoing debate.

Fecal Microbiota Transplantation

Fecal microbiota transplantation (FMT) involves the transfer of a small amount of liquefied or filtered feces to a recipient and has emerged as a potential treatment method for various conditions, including inflammation (Johnsen et al., 2018). Research by Sun et al., demonstrated that FMT can reduce the activity of microglia and astrocytes by modulating the TLR4/TNF-α signaling pathway (Sun et al., 2018). Additionally, FMT has been shown to temporarily alleviate leg tremors and neuroinflammation in models of Parkinson's disease (PD) (Huang et al., 2019). Moreover, transplanting fecal microbiota from PD patients into mice that overexpress α-synuclein exacerbates motor function issues, highlighting the connection between gut microbiota and PD (Sampson et al., 2016).

FMT has also been found to improve symptoms in children with autism spectrum disorder (ASD), including gastrointestinal issues such as constipation, diarrhea, and indigestion, potentially by enhancing gut microbiota diversity (Kang et al., 2017). Furthermore, clinical studies suggest that reduced gut microbiota diversity and abundance are linked to depression (Liu et al., 2020b). FMT has been shown to mitigate depressive symptoms by restoring gut microbiota diversity (Kurokawa et al., 2018). In animal studies, FMT from healthy animals has alleviated depressive symptoms in animals with spinal cord injuries, while FMT from depressed individuals has induced depressive behavior in germ-free mice (Pearson-Leary et al., 2020; Schmidt et al., 2020). These findings suggest that FMT could potentially target the gut-brain axis to relieve depressive symptoms.

However, the FMT process may introduce potential pathogenic bacteria or viruses that could disrupt the beneficial microbiota, and uncertainties remain regarding its safety and efficacy (Metta et al., 2022). Therefore, extensive research is necessary to evaluate the effectiveness and safety of FMT before it can be widely applied in clinical settings.

Antibiotics

Antibiotics can influence the immune response by altering gut microbiota, and their effects can vary depending on the specific antibiotic used. For instance, antibiotics such as amoxicillin, azithromycin, clarithromycin, and ciprofloxacin have been reported to potentially alleviate conditions like autism spectrum disorder (ASD) by impacting gut microbiota composition (Braakman and van Ingen, 2018). Additionally, antibiotics may have both positive and negative effects on the treatment of inflammatory bowel disease (IBD) and irritable bowel syndrome (IBS) (Lee et al., 2020).

Unlike fecal microbiota transplantation (FMT), which generally aims to restore healthy microbiota, antibiotics can disrupt gut microbiota balance. They often eradicate a significant portion of resident microbiota, creating an environment conducive to pathogen overgrowth and resulting in ecological imbalance (Duan et al., 2022). The use of antibiotics has been linked to disruptions in gut microbiota that can affect the nervous system, with evidence suggesting that penicillin and quinolones may increase the risk of depression and anxiety (Aygun, 2020).

The impact of antibiotics on gut microbiota varies by type. For example, macrolides can decrease the abundance of Actinobacteria, oral vancomycin can reduce Firmicutes while increasing Proteobacteria, and penicillin generally causes minimal changes to gut microbiota composition (Fan et al., 2022). Given the potential for antibiotics to disrupt gut microbiota homeostasis, careful consideration of dosage and choice of antibiotic is essential to mitigate adverse effects on gut health.

Probiotics

Bifidobacterium and Lactobacillus are prevalent probiotics found in foods like yogurt, which is rich in these beneficial microorganisms (Yan and Polk, 2020). Research indicates that probiotics containing Bifidobacterium and Lactobacillus can enhance cognitive function in Alzheimer's disease (AD) patients and reduce motor dysfunction in Parkinson's disease (PD) mouse models (Naomi et al., 2021). Sun et al., reported that Clostridium butyricum might exert neuroprotective effects on PD mice through the gut-brain axis (Sun et al., 2021). Additionally, Lactobacillus acidophilus, Bifidobacterium bifidum, Lactobacillus reuteri, and Lactobacillus fermentum have been shown to improve motor function in PD patients (Tamtaji et al., 2019).

In studies on depression, a decrease in Bifidobacterium and Lactobacillus abundance has been observed, while administering these probiotics from healthy feces has alleviated depressive symptoms in mice (Jang et al., 2019). Similarly, probiotics such as Eubacterium, Ruminococcaceae, Erysipelothrix, and Spirillaceae may benefit individuals with autism (Liu et al., 2019), and a three-month regimen of Lactobacillus acidophilus, Lactobacillus rhamnosus, and Bifidobacterium longum has been reported to reduce autism severity (Shaaban et al., 2018).

Overall, using probiotics to manage inflammatory diseases shows considerable promise. However, challenges remain regarding the composition, stability, and authenticity of probiotic strains, as well as the optimal dosage, duration, and type of probiotics. Furthermore, because probiotics generally do not permanently colonize the intestine, their beneficial effects require consistent daily intake.

Prebiotics

Prebiotics are defined as substances that are selectively used by host microorganisms to confer health benefits (Swanson et al., 2020). They include compounds such as inulin, fructo-oligosaccharides, galactooligosaccharides, and resistant starch. These prebiotics are found in everyday foods like fruits, vegetables, grains, and dairy products (Bamigbade et al., 2022). While prebiotics are not absorbed by the intestine, they are utilized by specific gut microbiota, which can promote beneficial changes in the microbiota composition (Verkhnyatskaya et al., 2019).

Similar to probiotics, prebiotics can influence CNS activity and modulate inflammatory responses by altering gut microbiota composition (Carlessi et al., 2021). For instance, an open study demonstrated that a combination of papain and pepsin could alleviate symptoms of autism spectrum disorder (ASD) (Song et al., 2022). Additionally, Dong et al. found that polymannuronic acid improved motor function in mice with Parkinson's disease, reduced inflammation in the intestine, brain, and systemically, and enhanced the integrity of the intestinal tract and blood-brain barrier (Dong et al., 2020).

These findings suggest that prebiotics can help mitigate inflammation, though further research is needed to fully understand their beneficial effects.

Synbiotics

Synbiotics are defined as a combination of prebiotics and probiotics, which together confer health benefits to the host (Swanson et al., 2020). By integrating live microorganisms with substrates that selectively support their growth, synbiotics can enhance the efficacy of both prebiotics and probiotics, allowing for synergistic effects that maximize their positive impact on health (Trone et al., 2023). Synbiotics are known to improve intestinal function by modulating gut microbiota (Li et al., 2021) and can also bolster immune system function, thereby reducing the risk of postoperative infections (Yadav et al., 2022).

For example, synbiotics containing Bifidobacterium infantis, Lactobacillus rhamnosus, Bifidobacterium lactis, and Lactobacillus paracasei combined with fructooligosaccharides have been shown to benefit patients with autism spectrum disorder (ASD) by increasing levels of beneficial bacteria and decreasing pathogenic bacteria (Wang Y. et al., 2020). Similarly, synbiotics that include GOSs, Lactobacillus helveticus, and Bifidobacterium longum have demonstrated positive effects on depression (Kazemi et al., 2019).

Despite these promising results, several issues remain to be addressed regarding the use of synbiotics. These include determining the optimal combinations, ratios, and dosages of prebiotics and probiotics to maximize their benefits. Further research is needed to refine these aspects and to fully understand the potential of synbiotics in therapeutic applications.

Postbiotics

Postbiotics are bioactive compounds produced by bacteria or through interactions between bacteria and the host, with common examples including short-chain fatty acids (SCFAs) and intestinal peptides. Although research on postbiotics for treating inflammatory diseases is still developing, recent studies have highlighted their potential effects on neurological disorders through the gut-brain axis (van de Wouw et al., 2018). Supplementation with PUFAs has been shown to alleviate some symptoms of autism (Cheng et al., 2017). These fatty acids might help regulate inflammatory processes and support neurological function. These compounds have been found to reduce colitis by lowering the accumulation of alpha-synaptic proteins (Perez-Pardo et al., 2018). They also demonstrate potential benefits in mitigating inflammatory responses and improving gut health. While propionic acid therapy has been associated with autism-like phenotypes in rats (Sharma et al., 2022), it also underscores the complexity of SCFA effects, suggesting that further research is needed to understand their full range of impacts. Intestinal peptides are known to influence neurological diseases, but their effectiveness may be limited by their interactions within the MGB axis signals (Lach et al., 2018). Research into how these peptides can be harnessed effectively remains ongoing.

Postbiotics represent a promising area of research, particularly in their potential to influence neurological and inflammatory diseases through the MGB axis. While the effects of SCFAs and intestinal peptides are increasingly recognized, there is still much to learn about their full therapeutic potential and mechanisms of action. The development of heat-killed probiotics as a stable and safe alternative to live probiotics highlights the ongoing innovation in utilizing gut microbiota for health benefits. Further studies are needed to optimize postbiotic applications and fully understand their impact on various health conditions.

Conclusion

The gut microbiota and the gut-brain axis play crucial roles in the development and management of neurological disorders in animals. Understanding these interactions opens new avenues for therapeutic interventions, potentially transforming the approach to treating and preventing neurological diseases.

References

Campbell, C., Kandalgaonkar, M. R., Golonka, R. M., Yeoh, B. S., Vijay-Kumar, M., and Saha, P. (2023). Crosstalk between gut microbiota and host immunity: impact on inflammation and immunotherapy. *Biomedicines* 11 (2), 274–334. doi: 10.3390/biomedicines11020294.

Flint H. J., Bayer E. A. Plant Cell Wall Breakdown by Anaerobic Microorganisms from the Mammalian Digestive Tract. *Ann N Y Acad Sci*. (2008) 1125:280–8. doi:10.1196/annals.1419.022.

Fulling C., Dinan T. G., Cryan J. F. Gut microbe to brain signaling: what happens in vagus. *Neuron.* 2019;101(6):998–1002.

Gilbert, J. A., Blaser, M. J., Caporaso, J. G., Jansson, J. K., Lynch, S. V., and Knight, R. (2018). Current understanding of the human microbiome. *Nat. Med.* 24 (4), 392–400. doi:10.1038/nm.45.

Human Microbiome Project C. Structure, function and diversity of the healthy human microbiome. *Nature.* 2012;486(7402):207–214.

Kraimi, N., Dawkins, M., Gebhardt-Henrich, S. G., Velge, P., Rychlik, I., Volf, J., Creach, P., Smith, A., Colles, F and Leterrier, C. Influence of the microbiota-gut-brain axis on behavior and welfare in farm animals: A review. *Physiol. Behav.* 2019, 210, 112658.

Krautkramer, K. A., Fan, J., and Backhed, F. (2021). Gut microbial metabolites as multi-kingdom intermediates. *Nat. Rev. Microbiol.* 19 (2), 77–94. doi: 10.1038/s41579-020-0438-4.

Li, H. J., Li, D. Q., Zhang, Y. L., Ding, X. F., Gao, H. T., Zhu, Y., et al. (2023). Modulation of gut microbiota alleviates cerebral ischemia/reperfusion injury in rats by inhibiting M1 polarization of microglia. *Front. Pharmacol.* 14. doi: 10.3389/ fphar.2023.1123387.

Long-Smith C., O'Riordan K. J., Clarke G., Stanton C., Dinan T. G., Cryan JF. Microbiota-gut-brain axis: new therapeutic opportunities. *Annu Rev Pharmacol Toxicol.* 2020;60:477–502.

Martyanov S. V., Botchkova E. A., Plakunov V. K., Gannesen A. V. The impact of norepinephrine on mono-species and dual-species staphylococcal biofilms. *Microorganisms.* 2021;9(4).

Nicholson, S. E., Watts, L. T., Burmeister, D. M., Merrill, D., Scroggins, S., Zou, Y., et al. (2019). Moderate traumatic brain injury alters the gastrointestinal microbiome in a time-dependent manner. *Shock* 52 (2), 240–248. do i : 10.1097/ SHK.0000000000001211.

Ozutsumi Y., Tajima K., Takenaka A., Itabashi H. The effect of protozoa on the composition of rumen bacteria in cattle using 16S rRNA gene clone libraries. *Biosci Biotechnol Biochem.* (2005) 69:499–506. doi: 10.1271/bbb.69.499.

Probiotics and fructo-oligosaccharide intervention modulate the microbiota-gut brain axis to improve autism spectrum reducing also the hyper-serotonergic state and the dopamine metabolism disorder. *Pharmacol. Res.* 157, 104784.

Vernocchi P., Del Chierico F., Putignani L. Gut microbiota profiling: metabolomics based approach to unravel compounds affecting human health. *Front Microbiol.* 2016;7:1144.

Wastyk, H. C., Fragiadakis, G. K., Perelman, D., Dahan, D., Merrill, B. D., Yu, F. B., et al. (2021). Gut-microbiota-targeted diets modulate human immune status. *Cell* 184 (16), 4137–4153.e14. doi: 10.1016/j.cell.2021.06.019.

Wei, G. Z., Martin, K. A., Xing, P. Y., Agrawal, R., Whiley, L., Wood, T. K., et al. (2021). Tryptophan-metabolizing gut microbes regulate adult neurogenesis via the aryl hydrocarbon receptor. *Proc. Natl. Acad. Sci. U.S.A.* 118 (27), e2021091118. doi: 10.1073/pnas.2021091118.

Wenzel, T. J., Gates, E. J., Ranger, A. L., and Klegeris, A. (2020). Short-chain fatty acids (SCFAs) alone or in combination regulate select immune functions of microglialike cells. *Mol. Cell. Neurosci.* 105, 103493. doi: 10.1016/j.mcn.2020.103493.

Woelfer, M., Kasties, V., Kahlfuss, S., and Walter, M. (2019). The role of depressive subtypes within the neuroinflammation hypothesis of major depressive disorder. *Neuroscience* 403, 93–110. doi: 10.1016/j.neuroscience.2018.03.034.

Wu, S. C., Chen, W. T., Muo, C. H., and Hsu, C. Y. (2020). A population-based cohort study examining the long-term risk of repeated surgery in non-helicobacter pylori-infected PPU patients who underwent simple closure. *J. Gastrointest. Surg.* 24 (11), 2587–2595. doi: 10.1007/s11605-019-04442-3.

Wu, Y., Wang, Y., Hu, A., Shu, X., Huang, W., Liu, J., et al. (2022). Lactobacillus plantarum-derived postbiotics prevent Salmonella-induced neurological dysfunctions by modulating gut-brain axis in mice. *Front. Nutr.* 9. doi: 10.3389/fnut.2022.946096.

Xiao, L., Yan, J., Yang, T., Zhu, J., Li, T., Wei, H., et al. (2021). Fecal microbiome transplantation from children with autism spectrum disorder modulates tryptophan and serotonergic synapse metabolism and induces altered behaviors in germ-free mice. *mSystems* 6 (2), e01343-20. doi: 10.1128/mSystems.01343-20.

Xiao, W., Su, J., Gao, X., Yang, H., Weng, R., Ni, W., et al. (2022). The microbiotagut-brain axis participates in chronic cerebral hypoperfusion by disrupting the metabolism of short-chain fatty acids. *Microbiome* 10 (1), 62. doi: 10.1186/s40168-022-01255-6.

Xu, F., Han, L., Wang, Y., Deng, D., Ding, Y., Zhao, S., et al. (2023). Prolonged anesthesia induces neuroinflammation and complement-mediated microglial synaptic elimination involved in neurocognitive dysfunction and anxiety-like behaviors. *BMC Med.* 21 (1), 7. doi: 10.1186/s12916-022-02705-6.

Xu, T., Liu, J., Li, X. R., Yu, Y., Luo, X., Zheng, X., et al. (2021). The mTOR/NFkappaB pathway mediates neuroinflammation and synaptic plasticity in diabetic encephalopathy. *Mol. Neurobiol.* 58 (8), 3848–3862. doi: 10.1007/s12035-021-02390-1.

Yadav, M. K., Kumari, I., Singh, B., Sharma, K. K., and Tiwari, S. K. (2022). Probiotics, prebiotics and synbiotics: Safe options for next-generation therapeutics. *Appl. Microbiol. Biotechnol.* 106 (2), 505–521. doi: 10.1007/s00253-021-11646-8.

Yan, F., and Polk, D. B. (2020). Probiotics and probiotic-derived functional factorsmechanistic insights into applications for intestinal homeostasis. *Front. Immunol.* 11. doi: 10.3389/fimmu.2020.01428.

Yao, C., Li, Y., Luo, L., Xie, F., Xiong, Q., Li, T., et al. (2023). Significant differences in gut microbiota between irritable bowel syndrome with diarrhea and healthy controls in southwest China. *Dig Dis. Sci.* 68 (1), 106–127. doi: 10.1007/s10620-022-07500-0.

Yarandi, S. S., Kulkarni, S., Saha, M., Sylvia, K. E., Sears, C. L., and Pasricha, P. J. (2020). Intestinal bacteria maintain adult enteric nervous system and nitrergic neurons via toll-like receptor 2-induced neurogenesis in mice. *Gastroenterology* 159 (1), 200–213.e8. doi: 10.1053/j.gastro.2020.03.050.

Yuan, M. J., Li, W., and Zhong, P. (2021). Research progress of ghrelin on cardiovascular disease. *Biosci. Rep.* 41 (1), BSR20203387. doi: 10.1042/BSR20203387.

Yuan, M., Jiang, Z., Bi, G., Nomura, K., Liu, M., Wang, Y., et al. (2021). Patternrecognition receptors are required for NLR-mediated plant immunity. *Nature* 592 (7852), 105–109. doi: 10.1038/s41586-021-03316-6.

Zaky, A., Glastras, S. J., Wong, M., Pollock, C. A., and Saad, S. (2021). The role of the gut microbiome in diabetes and obesity-related kidney disease. *Int. J. Mol. Sci.* 22 (17), 9641–9666. doi: 10.3390/ijms22179641.

Zhang, J., Xu, M., Zhou, W., Li, D., Zhang, H., Chen, Y., et al. (2020). Deficiency in the anti-apoptotic protein DJ-1 promotes intestinal epithelial cell apoptosis and aggravates inflammatory bowel disease via p53. *J. Biol. Chem.* 295 (13), 4237–4251. doi: 10.1074/jbc.RA119.010143.

Zhang, L., Wang, J., Wang, J., Yang, B., He, Q., and Weng, Q. (2020). Role of DJ-1 in immune and inflammatory diseases. *Front. Immunol.* 11. doi: 10.3389/fimmu.2020.00994.

Zhao, X., Wu, H., Zhu, R., Shang, G., Wei, J., Shang, H., et al. (2023). Combination of thalidomide and Clostridium butyricum relieves chemotherapy-induced nausea and vomiting via gut microbiota and vagus nerve activity modulation. *Front. Immunol.* 14. doi: 10.3389/fimmu.2023.1220165.

Zhao, Z., Ning, J., Bao, X. Q., Shang, M., Ma, J., Li, G., et al. (2021). Fecal microbiota transplantation protects rotenone-induced Parkinson's disease mice via suppressing inflammation mediated by the lipopolysaccharide-TLR4 signaling pathway through the microbiota-gut-brain axis. *Microbiome* 9 (1), 226. doi: 10.1186/s40168-021-01107-9.

Zheng, M., Karki, R., Williams, E. P., Yang, D., Fitzpatrick, E., Vogel, P., et al. (2021). TLR2 senses the SARS-CoV-2 envelope protein to produce inflammatory cytokines. *Nat. Immunol.* 22 (7), 829–838. doi: 10.1038/s41590-021-00937-x.

Zheng, S. Y., Li, H. X., Xu, R. C., Miao, W. T., Dai, M. Y., Ding, S. T., et al. (2021). Potential roles of gut microbiota and microbial metabolites in Parkinson's disease. *Ageing Res. Rev.* 69, 101347. doi: 10.1016/j.arr.2021.101347.

Zhong, J. G., Lan, W. T., Feng, Y. Q., Li, Y. H., Shen, Y. Y., Gong, J. H., et al. (2023). Associations between dysbiosis gut microbiota and changes of neurotransmitters and short-chain fatty acids in valproic acid model rats. *Front. Physiol.* 14. doi: 10.3389/fphys.2023.1077821.

Zhou, C., Chen, L. L., Lu, R. Q., Ma, W. W., and Xiao, R. (2021). Alteration of intestinal microbiota composition in oral sensitized C3H/HeJ mice is associated with changes in dendritic cells and T cells in mesenteric lymph nodes. *Front. Immunol.* 12. doi: 10.3389/fimmu.2021.631494.

Zhu, M., Liu, X., Ye, Y., Yan, X., Cheng, Y., Zhao, L., et al. (2022). Gut microbiota: A novel therapeutic target for parkinson's disease. Front. Immunol. 13. doi: 10.3389/fimmu.2022.937555.

Zou, R., Xu, F., Wang, Y., Duan, M., Guo, M., Zhang, Q., et al. (2020). Changes in the gut microbiota of children with autism spectrum disorder. *Autism Res.* 13 (9), 1614–1625. doi: 10.1002/aur.2358.

Zuo, T., and Ng, S. C. (2018). The gut microbiota in the pathogenesis and therapeutics of inflammatory bowel disease. *Front. Microbiol.* 9. doi: 10.3389/fmicb.2018.02247.

Chapter 9

Microbiomes as a Medicinal Approach

Akhter Rasool[1]
T. Sarath[2,*]
N. Arunmozhi[3]
and U. S. Kalyaan[4]
[1] National Dairy Research Institute, Karnal, India
[2]Veterinary Clinical Complex, Veterinary College and Research Institute, Salem, Tamil Nadu, India
[3]Department of Veterinary Gynaecology and Obstetrics, Tamil Nadu Veterinary and Animal Sciences University, India
[4]Resident Veterinary Services Section, Madras Veterinary College, Chennai, Tamil Nadu, India

Abstract

The chapter explores the pivotal role of microorganisms across various biological domains, emphasizing their influence on animal and human health, biotechnology, and therapeutic potentials. It discusses the complexity of microbial networks, particularly in the gut microbiome, which regulates metabolic functions, energy balance, and immune responses. Microbial dysbiosis, or imbalance, is linked to several diseases, highlighting the therapeutic promise of interventions like fecal microbial transplantation (FMT), phage therapy and probiotics. The role of microbial metabolites such as short-chain fatty acids (SCFAs), bile acids, and indole derivatives in health and disease is explored, alongside their potential as therapeutic agents.

Keywords: microbiome, dysbiosis, diseases, therapy, health

Introduction

Microorganisms play a crucial role in various biological fields, including animal and human health and biotechnology (Clemente et al., 2012). However, microbial networks remain poorly understood due to challenges in isolating many microbial species and the heterogeneous nature of their interactions (Muller et al., 2018). The gut microbiome is a complex ecosystem of

---

* Corresponding Author's Email: drsarathvet@gmail.com

In: Microbiomes in Animal Nutrition, Health, and Disease
Editor: Tanmoy Rana
ISBN: 979-8-89530-425-9

microorganisms that inhabit and critically maintain the homeostasis of the gastrointestinal (GI) tract (Thursby and Juge, 2017). Intestinal microbes regulate metabolic function and energy balance, and an altered microbial ecology is believed to contribute to the development of several metabolic diseases. Bacteria release fatty acids, peroxidases, proteases, and bacteriocins that prevent pathogen proliferation in the intestinal community (Cani, 2007). There is evidence that a healthy microbiota is essential for overall health and protects against disease, leading to interest in its therapeutic potential. Specific bacterial communities are promising clinical targets for treating inflammatory and infectious diseases. Fecal microbial transplantation (FMT) has proven to be an effective therapy for treating gastrointestinal disorders (Rasool et al., 2023).

Gut Microbial Metabolites

The gut microbiota is defined as the entire population of bacteria, viruses, parasites, and fungi colonizing the intestinal tract (Sender et al., 2016). Factors such as diet, age, stress, and diseases can increase or decrease the relative abundance and diversity of bacterial species in the GI tract and other body sites. Microbial metabolites can modulate the immune system by impacting the physiology and gene expression of host cells (Turnbaugh et al., 2007). Diverse metabolic activity allows the microbiota to fill available ecological niches maximally and competitively inhibit pathogen colonization at all sites (Sommer et al., 2017). However, metabolites from the microbiome can sometimes form toxic compounds that damage the host epithelium and cause inflammation (Fan et al., 2015). Factors like food particle form and size, food composition (essential nutrients, antinutrients such as α-amylase inhibitors), and transit time (Wong and Jenkins, 2007) can alter the composition of fecal microbial communities (Roager et al., 2016). Intestinal bacteria can endogenously synthesize essential co-factors for host energy metabolism and gene expression regulation, such as B vitamins (Biesalski, 2016). Gut microbes also harvest energy from remaining complex carbohydrates (Singh et al., 2017). Microbial end-products like butyrate, a preferred energy source, exhibit anti-inflammatory effects by stimulating both intestinal epithelial cells (IECs) and antigen-presenting cells (APCs) to produce cytokines TGF-β, IL-10, and IL-18 and inducing the differentiation of naïve T cells to T regulatory cells (Lee and Hase, 2014). Acetate and propionate also have anti-inflammatory effects, dampening pro-inflammatory cytokine production mediated by toll-like receptor (TLR) stimulation, and propionate, like butyrate, can induce T cell differentiation into T regulatory cells (Koh et al., 2016). In protein metabolism, the gut microbiota rapidly assimilates ammonia into microbial amino acid biosynthesis (Portune et al., 2016). Gut microbiota provides branched-chain fatty acids, modulating glucose, and lipid metabolism, often used as biomarkers of protein catabolism (Yao et al., 2016). Furthermore, branched-chain amino acids obtained from the gut microbiome, such as leucine, isoleucine, valine, and glycine, are required for synthesizing glutathione, a major antioxidant. Bacterial species within the gut microbiota can decarboxylate basic amino acids. The end products, agmatine, putrescine, and spermine, inhibit pro-inflammatory cytokines like IL-1, IL-6, and TNF-α, acting as anti-inflammatory agents through nitric oxide synthase inhibition (Kibe et al., 2014; August et al., 1995). Similarly, arginine can be converted to glutamate, which can be deaminated to produce 4-aminobutyrate (GABA), a major inhibitory neurotransmitter. Microbial dysbiosis leads to chronic GI inflammation, often associated with depression and anxiety, commonly comorbid with GI disorders, including irritable bowel syndrome (IBS) (Bravo et al., 2011). Histamine production by gut bacteria has been shown to inhibit pro-inflammatory cytokines TNF-α in vivo (Thomas et al., 2012) and IL-1 and IL-12 in vitro (Elenkov et al., 1998). Indole, a major bacterial metabolite of tryptophan produced by many Bacteroides and Enterobacteriaceae species (Gao et al., 2018), inhibits the colonization capabilities of pathogens such as Salmonella enterica (Li and Young, 2013). Microbial dysregulation is reported in many disorders, including inflammatory bowel disease (IBD) (Linden et al., 2005), IBS (Stasi et al., 2014), and cardiovascular disease (Ramage and Villalón, 2008). Tyramine, a neurotransmitter produced by certain gut bacteria like Enterococcus and Enterobacteriaceae (Pugin et al., 2017), facilitates norepinephrine release and serotonin synthesis (Shalaby, 1996; Yano et al., 2015). Gut microorganisms possess *lipases, lactobacilli, enterococci, clostridia,* and *Proteobacteria*, utilizing triglycerides and reducing glycerol to 1,3-propanediol (De Weirdt et al., 2010), forming 3-Hydroxypropanal (reuterin) as an intermediate with antimicrobial properties (Cleusix et al., 2007). The association between gut microbiota-mediated lipid metabolism and health consequences is still unclear. Intestinal microbiota are involved in producing vitamins, isoflavonoids, converting prodrugs to bioactive forms, and transforming bile acids and xenobiotics (Blaut and Clavel, 2007), all essential for animal health.

Gut Microbiome and Dysbiosis

The composition and functions of the gut microbiome may be influenced by various factors, including genetics, mode of delivery, age, diet, geographic location, and medical treatments (Clemente et al., 2012; Brown et al., 2013). Alteration of the microbiota composition, or dysbiosis, is defined as an imbalance in the microbiome structure resulting from an abnormal ratio of commensal and pathogenic bacterial species. Dysbiosis is associated with inflammatory bowel diseases (IBDs), metabolic disorders, cardiovascular disease, obesity, cancer, and infections, particularly with Clostridium difficile infection (CDI) (Wu & Lewis, 2013). Inflammatory diseases such as Crohn's disease and ulcerative colitis in humans are linked to the composition of the gut microbiome, and other microorganisms, including viruses and fungi, have been hypothesized to play a role in their pathogenesis (Iliev et al., 2012). High-protein diets are thought to result in the production of carcinogenic metabolites by the colonic microbiota, potentially inducing neoplasia in the colonic epithelium (Carding et al., 2015). Bacterial conversion of metabolites, which are low-molecular-weight signaling molecules synthesized by microorganisms, can influence disease risk, particularly for diseases affecting the liver, cardiovascular system, and kidneys (Nicholson et al., 2012). Individuals colonized by bacteria of the genera *Faecalibacterium, Bifidobacterium, Lactobacillus, Coprococcus, and Methanobrevibacter* have a significantly lower tendency to develop obesity-related diseases such as type-2 diabetes and ischemic cardiovascular disorders (Ley et al., 2006; Le Chatelier et al., 2013).

Skin and Reproductive Tract Health

Microbial colonization is influenced by the ecology and the epidermal layers of the skin surface. The *Actinobacteria* phylum is the most abundant on the skin. *Staphylococcus* spp., *Corynebacterium* spp., and the fungi *Candida* spp. and *Malassezia* spp. are also frequently associated with various skin diseases, including atopic dermatitis, psoriasis, eczema, and abnormal flaking and itching of the scalp (Grice et al., 2009), which occur due to derangement of the local cutaneous immune system.

Placentas from normal and preterm deliveries contain different populations of microbial species (Groer et al., 2014). A variety of microbes exist in amniotic fluid, umbilical cord blood, meconium (first stool), placental, and fetal membranes. Infants born vaginally have more diverse gut microbial communities similar to their mother's vaginal microbiota, while the microbiomes of newborns delivered by cesarean section resemble skin microbiota (Dominguez-Bello et al., 2010). The lack of exposure to the maternal vaginal microbiome might explain why cesarean section babies are at greater risk of developing type 1 diabetes, celiac disease, asthma, and obesity (DiGiulio et al., 2008; Dominguez-Bello et al., 2010).

The vagina has a low pH due to the secretion of lactic acid and hydrogen peroxide by Lactobacillus spp. If Lactobacillus decreases under the effects of antibiotics, *Gardnerella vaginalis, Peptostreptococcus anaerobius, Prevotella* spp., *Mobiluncus* spp., *Sneathia, Atopobium vaginae, Ureaplasma, Mycoplasma,* and numerous fastidious or uncultivated anaerobes can cause bacterial vaginosis (BV) (Belizário and Napolitano, 2015). The interplay of *Lactobacillus* species is crucial for maintaining the balance of the vaginal microbiota by producing lactic acid and potentially bacteriocins (Tachedjian et al., 2017; Maldonado-Barragán et al., 2016). These compounds inhibit microbes associated with dysbiosis, thus preserving homeostasis and reducing disease risk. Estrogen plays a vital role in sustaining the vaginal microbiome (Hickey et al., 2012). In the cervical flora, the presence of Gram-negative bacteria, such as *Chlamydia trachomatis* and *Gardnerella vaginalis*, combined with a lack of Lactobacillus, is associated with infertility. Additionally, the absence of a Lactobacillus-dominated flora in the endometrium is linked to recurrent implantation failures (Vitale et al., 2021).

There is a strong association between gut microbes and reproductive tract microbes. The gut microbial population can influence the reproductive tract microbiota; for example, a high-fat diet significantly impacts both the gut and reproductive tract microbiomes. Additionally, it has been suggested that the commensal microbiome may influence the signaling mechanisms of vitamins K and A, which are essential for fertility maintenance (Wang and Xie, 2022).

Probiotics and Probiotics

Probiotics are live microorganisms that improve the balance of the intestinal flora, thereby fostering healthy gut functions through a healthy gut microbiome (Gareau et al., 2010; Whelan and Quigley, 2013). Probiotic microorganisms are mainly lactic acid-producing bacteria from the *Lactobacillus* and *Bifidobacterium* genera. These probiotics can prevent or ameliorate clinical symptoms of irritable bowel syndrome, inflammatory and necrotizing enterocolitis, and acute diarrhea (Ng et al., 2009; Gareau et al., 2010; Whelan and Quigley, 2013). They regulate the balance of intestinal microbiota by increasing the production of a mucosal barrier by goblet epithelial cells (Etzold et al., 2014). Probiotic immune-modulatory effects are associated with a decrease in the production of pro-inflammatory cytokines, as well as microbial peptides called bacteriocins (Ng et al., 2009; Whelan and Quigley, 2013). The bacterial species *Bifidobacterium bifidum* has been used in the prevention and treatment of infantile eczema. Intra-vaginal administration of *Lactobacillus rhamnosus* GR-1 and *L. fermentum* RC-14 has been shown to have a positive effect on the prevention of recurrent bacterial vaginosis (BV) and candidiasis (Anukam et al., 2006; Vuotto et al., 2014). Systemic ammonia levels rise in cirrhosis patients and are linked to hepatic encephalopathy (HE). Probiotic *Escherichia coli* (E. coli) Nissle 1917 has been genetically modified to metabolize ammonia, converting it into L-arginine within the gut (Kurtz et al., 2019).

Prebiotics are functional food ingredients, including fructo-oligosaccharides (FOS) and trans-galacto-oligosaccharides (TOS), such as inulin (Roberfroid, 2000), that can change the composition and/or activity of the colonic flora (Roberfroid, 2000; Roberfroid, 2007; Brownawell et al., 2012). Prebiotics can promote the growth of beneficial bacteria such as *lactobacilli* and *bifidobacteria* strains (Roberfroid, 2000, 2007).

Microbial metabolites, including indole derivatives, bile acids, and short-chain fatty acids (SCFAs), fall under the category of postbiotics, which are preparations of non-viable microorganisms or their components that provide health benefits to the host. These metabolites have potential therapeutic and prognostic applications. For instance, bile acid modulators and SCFAs have been studied in clinical trials for their efficacy in treating liver diseases (Bajaj and Schnabl, 2022).

Fecal Microbial Transplantation

The use of antibiotics can disrupt and destabilize the normal bowel microbiota, predisposing patients to develop *Clostridium difficile* infections. Direct transfer of an entire microbial community from a healthy donor, a process known as fecal microbial transplantation (FMT), is highly effective but can increase the risks of transmitting other pathogens (Brandt and Reddy, 2011; Rasool et al., 2023). Dysbiosis of the gut microbiota in metabolic syndrome is linked to a proinflammatory state and impaired mucosal barrier function (De Groot et al., 2017). FMT is one such microbiota-modulating therapy that can help maintain microbial balance and manage associated gastrointestinal disorders. Fecal microbiota transplantation (FMT) aims to directly change the recipient's gut microbiota to normalize its composition and achieve therapeutic benefits. This procedure can be done via esophagogastroduodenoscopy, colonoscopy, retention enema, or oral capsule (Wang et al., 2019). FMT has been used to treat patients with inflammatory bowel disease (IBD) complicated by *Clostridium difficile* infection (CDI), and the outcome depends mainly on the metabolic and immune profile, as well as the genetic makeup of both the donor and the recipient (Debast et al., 2014).

Phage Therapy

Phages are viruses that infect bacteria, categorized as virulent or temperate based on their functional behavior. They were historically used as antibacterial agents, but their importance declined with the widespread use of antibiotics. However, phage therapy is gaining renewed interest as a safe alternative, even when administered intravenously in the form of phage cocktails (Schooley et al., 2017). Using phages to eliminate pathogenic bacteria can potentially reduce disease incidence. Challenges in implementing phage therapy in clinical practice include the limited host range of some phages, determining the optimal route of administration, the development of bacterial resistance to phages, and understanding the kinetics within the gastrointestinal tract (Bajaj and Schnabl, 2022).

Supplementation of probiotics, prebiotics, phage therapy, and fecal microbiota transplantation (FMT) for patients with gastrointestinal (GIT) disorders has shown promising biochemical improvements and beneficial effects. However, the mechanisms underlying their therapeutic potential in treating infertility remain unclear. Further research is needed to elucidate the role of these treatments in the management or prevention of infertility.

Conclusion

Microorganisms, particularly those inhabiting the gut, play critical roles in maintaining health through complex interactions with their hosts. The composition and activity of the gut microbiome are influenced by various factors including diet, age, and medical treatments, impacting host metabolism and immune function. Therapeutic approaches like FMT, probiotics, and prebiotics offer promising avenues for managing gastrointestinal disorders and improving overall health. Future research into microbial metabolites and phage therapy could further enhance our understanding and therapeutic capabilities in microbiome-related diseases.

References

Anukam, K. C., Osazuwa, E., Osemene, G. I., Ehigiagbe, F., Bruce, A. W., and Reid, G. (2006). Clinical study comparing probiotic Lactobacillus GR-1 and RC-14 with metronidazole vaginal gel to treat symptomatic bacterial vaginosis. *Microbes Infect.* 8, 2772–2776.

Auguet, M., Viossat, I., Marin, J. G., & Chabrier, P. E. (1995). Selective inhibition of inducible nitric oxide synthase by agmatine. *The Japanese Journal of Pharmacology*, 69(3), 285-287.

Bajaj, J. S., Ng, S. C., & Schnabl, B. (2022). Promises of microbiome-based therapies. *Journal of hepatology*, 76(6), 1379-1391.

Belizário, J. E., & Napolitano, M. (2015). Animal microbiomes and their roles in dysbiosis, common diseases, and novel therapeutic approaches. *Frontiers in microbiology*, 6, 1050.

Bravo, J. A., Forsythe, P., Chew, M. V., Escaravage, E., Savignac, H. M., Dinan, T. G., ... & Cryan, J. F. (2011). Ingestion of Lactobacillus strain regulates emotional behavior and central GABA receptor expression in a mouse via the vagus nerve. *Proceedings of the National Academy of Sciences*, 108(38), 16050-16055.

Brandt, L. J., and Reddy, S. S. (2011). Fecal microbiota transplantation for recurrent Clostridium difficile infection. *J. Clin. Gastroenterol.* 45(Suppl.), S159–S167. doi: 10.1097/MCG.0b013e318222e603.

Brown, C. T., Sharon, I., Thomas, B. C., Castelle, C. J., Morowitz, M. J., and Banfield, J. F. (2013). Genome resolved analysis of a premature infant gut microbial community reveals a Varibaculum cambriense genome and a shift towards fermentation-based metabolism during the third week of life. *Microbiome* 1:30. doi: 10.1186/2049-2618-1-30.

Brownawell, A. M., Caers, W., Gibson, G. R., Kendall, C. W., Lewis, K. D., Ringel, Y., et al. (2012). Prebiotics and the health benefits of fiber: current regulatory status, future research, and goals. *J. Nutr*. 142, 962–974.

Biesalski, H. K. (2016). Nutrition meets the microbiome: micronutrients and the microbiota. *Annals of the new York Academy of Sciences*, 1372(1), 53-64.

Cani, P. D., Amar, J., Iglesias, M. A., Poggi, M., Knauf, C., Bastelica, D., ... & Burcelin, R. (2007). Metabolic endotoxemia initiates obesity and insulin resistance. *Diabetes,* 56(7), 1761-1772.

Carding, S., Verbeke, K., Vipond, D. T., Corfe, B. M., & Owen, L. J. (2015). Dysbiosis of the gut microbiota in disease. *Microbial ecology in health and disease*, 26(1), 26191.

Cleusix, V., Lacroix, C., Vollenweider, S., Duboux, M., & Le Blay, G. (2007). Inhibitory activity spectrum of reuterin produced by Lactobacillus reuteri against intestinal bacteria. *BMC microbiology*, 7, 1-9.

Clemente, J. C., Ursell, L. K., Parfrey, L. W., and Knight, R. (2012). The impact of the gut microbiota on animal health: an integrative view. *Cell* 148, 1258–1270. doi: 10.1016/j.cell.2012.01.035.

Debast S. B., Bauer M. P., Kuijper E. J., *Eur. Soc. Clin. Microbiol. Infect. Dis*. (2014).

European Society of Clinical Microbiology and Infectious Diseases: update of the treatment guidance document for Clostridium difficile infection. *Clin. Microbiol. Infect*. 20:Suppl. 21–26.

De Groot P. F., Frissen M. N., de Clercq N. C., Nieuwdorp M. (2017). Fecal microbiota transplantation in metabolic syndrome: history, present and future. *Gut Microbes* 8:253–67.

De Weirdt R., Possemiers S., Vermeulen G., Moerdijk-Poortvliet T. C. W., Boschker H. T. S., Verstraete W., et al. (2010). Animal faecal microbiota display variable patterns of glycerol metabolism. *FEMS Microbiol Ecol.;*74:601–11.

DiGiulio, D. B., Romero, R., Amogan, H. P., Kusanovic, J. P., Bik, E. M., Gotsch, F., et al. (2008). Microbial prevalence, diversity and abundance in amniotic fluid during preterm labor: a molecular and culture-based investigation. *PLoS ONE* 3:e3056. doi: 10.1371/journal.pone.0003056.

Dominguez-Bello, M. G., Costello, E. K., Contreras, M., Magris, M., Hidalgo, G., Fierer, N., et al. (2010). Delivery mode shapes the acquisition and structure of the initial microbiota across multiple body habitats in newborns. *Proc. Natl. Acad. Sci. USA*. 107, 11971–11975. doi: 10.1073/pnas.1002 601107.

Elenkov I. J., Webster E., Papanicolaou D. A., Fleisher T. A., Chrousos G. P., Wilder R. L. (1998). Histamine potently suppresses animal IL-12 and stimulates IL-10 production via H2 receptors. *J Immunol*.;161:2586–93.

Etzold, S., Kober, O. I., Mackenzie, D. A., Tailford, L. E., Gunning, A. P., Walshaw, J., et al. (2014). Structural basis for adaptation of lactobacilli to gastrointestinal mucus. *Environ. Microbiol*. 16, 888–903. doi: 10.1111/1462 2920.12377.

Gao J., Xu K., Liu H., Liu G., Bai M., Peng C., et al. (2018) Impact of the gut microbiota on intestinal immunity mediated by tryptophan metabolism. *Front Cell Infect Microbiol*. 8:13.

Gareau, M. G., Sherman, P. M., and Walker, W. A. (2010). Probiotics and the gut microbiota in intestinal health and disease. *Nat. Rev. Gastroenterol. Hepatol*. 7, 503–514. doi: 10.1038/nrgastro.2010.117.

Grice, E. A., Kong, H. H., Conlan, S., Deming, C. B., Davis, J., Young, A. C., et al. (2009). Topographical and temporal diversity of the animal skin microbiome. *Science* 324, 1190–1192. doi: 10.1126/science.1171700.

Hickey R. J., Zhou X., Pierson J. D., Ravel J., Forney L. J. (2012). Understanding vaginal microbiome complexity from an ecological perspective. *Transl Res* 160:267–282.

Groer, M. W., Luciano, A. A., Dishaw, L. J., Ashmeade, T. L., Miller, E., and Gilbert, J. A. (2014). Development of the preterm infant gut microbiome: a research priority. *Microbiome* 2:38. doi: 10.1186/2049-2618-2-38.

Iliev I. D., Funari V. A., Taylor K. D., et al. (2012). Interactions between commensal fungi and the C type lectin receptor Dectin 1 influence colitis. *Science.* 336:1314 1317.

Kibe R., Kurihara S., Sakai Y., Suzuki H., Ooga T., Sawaki E., et al. (2014). Upregulation of colonic luminal polyamines produced by intestinal microbiota delays senescence in mice. *Sci Rep.* 4:4548.

Koh A., De Vadder F., Kovatcheva-Datchary P., Bäckhed F. (2016). From dietary fiber to host physiology: short-chain fatty acids as key bacterial metabolites. *Cell.* 165:1332–45.

Kurtz, C. B., Millet, Y. A., Puurunen, M. K., Perreault, M., Charbonneau, M. R., Isabella, V. M., ... & Miller, P. F. (2019). An engineered E. coli Nissle improves hyperammonemia and survival in mice and shows dose-dependent exposure in healthy humans. *Science translational medicine*, 11(475), eaau7975.

Lee W. J., Hase K. (2014). Gut microbiota-generated metabolites in animal health and disease. *Nat Chem Biol.* 10:416–24.

Le Chatelier, E., Nielsen, T., Qin, J., Prifti, E., Hildebrand, F., Falony, G., et al. (2013). Richness of animal gut microbiome correlates with metabolic markers. *Nature* 500, 541–546. doi: 10.1038/nature12506.

Ley, R. E., Turnbaugh, P. J., Klein, S., and Gordon, J. I. (2006). Microbial ecology: animal gut microbes associated with obesity. *Nature* 444, 1022–1023. doi: 10.1038/4441022a.

Linden D. R., Foley K. F., McQuoid C., Simpson J., Sharkey K. A., Mawe G. M. (2005). Serotonin transporter function and expression are reduced in mice with TNBS-induced colitis. *Neurogastroenterol Motil Off J Eur Gastrointest Motil Soc.* 17:565–74.

Li G., Young K. D. (2013). Indole production by the tryptophanase TnaA in Escherichia coli is determined by the amount of exogenous tryptophan. *Microbiol Read Engl.* 159:402–10.

Maldonado-Barragán A., Caballero-Guerrero B., Martín V., Ruiz-Barba J. L., Rodríguez J. M. (2016). Purification and genetic characterization of gassericin E, a novel co-culture inducible bacteriocin from Lactobacillus gasseri EV1461 isolated from the vagina of a healthy woman. *BMC Microbiol* 16:37.

Muller, E. E., Faust, K., Widder, S., Herold, M., Arbas, S. M., & Wilmes, P. (2018). Using metabolic networks to resolve ecological properties of microbiomes. *Current Opinion in Systems Biology*, 8, 73-80.

Nicholson J. K., Holmes E., Kinross J., Burcelin R., Gibson G., Jia W., et al. (2012). Host-gut microbiota metabolic interactions. *Science*. 336: 1262 7.

Ng, S. C., Hart, A. L., Kamm, M. A., Stagg, A. J., and Knight, S. C. (2009). Mechanisms of action of probiotics: recent advances. *Inflamm. Bowel Dis.* 15, 300–310. doi: 10.1002/ibd.20602.

Pugin B., Barcik W., Westermann P., Heider A., Wawrzyniak M., Hellings P., et al. (2017). A wide diversity of bacteria from the animal gut produces and degrades biogenic amines. *Microb Ecol Health Dis.* 28:1353881.

Portune K. J., Beaumont M., Davila A. M., Tomé D., Blachier F., Sanz Y. (2016). Gut microbiota role in dietary protein metabolism and health-related outcomes: the two sides of the coin. *Trends Food Sci Technol.* 57:213–32.

Ramage A. G., Villalón C. M. (2008). 5-hydroxytryptamine and cardiovascular regulation. *Trends Pharmacol Sci.* 29:472–81.

Rasool, A., Porteen, K., Sarath, T., & Joshy, K. (2023). Exploring the Potential of Probiotics and Fecal Microbiota in Maintaining a Healthy Gut Microbiota. *Journal of Scientific Research and Reports*, 29(7), 101-106.

Roager H. M., Hansen L. B. S., Bahl M. I., Frandsen H. L., Carvalho V., Gøbel R. J., et al. (2016). Colonic transit time is related to bacterial metabolism and mucosal turnover in the gut. *Nat Microbiol.* 1:16093.

Roberfroid, M. (2007). Prebiotics: the concept revisited. *J. Nutr.* 137(3 Suppl. 2), 830S–837S.

Roberfroid, M. B. (2000). Prebiotics and probiotics: are they functional foods? *Am. J. Clin. Nutr.* 71(6 Suppl.), 1682S–1687S.

Sender R., Fuchs S., Milo R. (2016). Revised estimates for the number of human and bacteria cells in the body. *PLOS Biol* 14:e1002533.

Shalaby A. R. (1996). Significance of biogenic amines to food safety and animal health. *Food Res Int.* 29:675–90.

Stasi C., Bellini M., Bassotti G., Blandizzi C., Milani S. (2014). Serotonin receptors and their role in the pathophysiology and therapy of irritable bowel syndrome. *Tech Coloproctol.* 18:613–21.

Singh R. K., Chang H. W., Yan D., Lee K. M., Ucmak D., Wong K., et al. (2017). Influence of diet on the gut microbiome and implications for animal health. *J Transl Med.* 15:73.

Sommer F., Anderson J. M., Bharti R., Raes J., Rosenstiel P. (2017). The resilience of the intestinal microbiota influences health and disease. *Nat Rev Microbiol.* 15:630–8.

Tachedjian G., Aldunate M., Bradshaw C. S., Cone R. A. (2017). The role of lactic acid production by probiotic Lactobacillus species in vaginal health. *Res Microbiol* 168:782–792.

Thomas C. M., Hong T., van Pijkeren J. P., Hemarajata P., Trinh D. V., Hu W., et al. (2012). Histamine derived from probiotic Lactobacillus reuteri suppresses TNF via modulation of PKA and ERK signaling. *PLoS One*. 7:e31951.

Turnbaugh P. J., Ley R. E., Hamady M., Fraser-Liggett C., Knight R., Gordon J. I. (2007). The animal microbiome project: exploring the microbial part of ourselves in a changing world. *Nature*. 449:804–10.

Thursby E., Juge N. (2017). Introduction to the animal gut microbiota. *Biochem J.* 474:1823–36.

Vuotto, C., Longo, F., and Donelli, G. (2014). Probiotics to counteract biofilm associated infections: promising and conflicting data. *Int. J. Oral Sci*. 6,189–194. doi: 10.1038/ijos.2014.52.

Vitale, S. G., Ferrari, F., Ciebiera, M., Zgliczyńska, M., Rapisarda, A. M. C., Vecchio, G. M., ... & Cianci, S. (2021). The role of genital tract microbiome in fertility: a systematic review. *International journal of molecular sciences*, 23(1), 180.

Wang, J. W., Kuo, C. H., Kuo, F. C., Wang, Y. K., Hsu, W. H., Yu, F. J., ... & Wu, D. C. (2019). Fecal microbiota transplantation: Review and update. *Journal of the Formosan Medical Association*, 118, S23-S31.

Wang, Y., & Xie, Z. (2022). Exploring the role of gut microbiome in male reproduction. *Andrology*, 10(3), 441-450.

Wong J. M. W., Jenkins D. J. A. (2007). Carbohydrate digestibility and metabolic effects. *J Nutr*. 137:2539S–46S.

Wu, G. D., & Lewis, J. D. (2013). Analysis of the animal gut microbiome and association with disease. *Clinical Gastroenterology and Hepatology*, 11(7), 774-777.

Whelan, K., and Quigley, E. M. (2013). Probiotics in the management of irritable bowel syndrome and inflammatory bowel disease. *Curr. Opin. Gastroenterol*. 29, 184–189. doi: 10.1097/MOG.0b013e32835d7bba.

Yao C. K., Muir J. G., Gibson P. R. (2016). Review article: insights into colonic protein fermentation, its modulation and potential health implications. *Aliment Pharmacol Ther*. 43:181–96.

Yano J. M., Yu K., Donaldson G. P., Shastri G. G., Ann P., Ma L., et al. (2015). Indigenous bacteria from the gut microbiota regulate host serotonin biosynthesis. *Cell.* 161:264–76.

Chapter 10

Microbiomes in the Innate Immunity of Animals

Khumtya Debbarma[1,*]
Mrinmoyee Sarma[2]
Monoshree Sarma[3]
and Das Adwitiya[4]

[1]Department of Veterinary Pharmacology and Toxicology, College of Veterinary Science, Guwahati, Assam, India
[2]Department of Veterinary Public Health and Epidemiology, Institute of Veterinary Science and Animal Husbandry, Siksha o Anusandhan, Bhubaneswar, Odisha, India
[3]Department of Veterinary Pharmacology and Toxicology, Institute of Veterinary Science and Animal Husbandry, Siksha o Anusandhan, Bhubaneswar, Odisha, India
[4]Division of Microbiology, Indian Veterinary Research Institute, Bengaluru, Karnataka, India

Abstract

The microbiome, a complex community of microorganisms residing on and within animals, plays a critical role in shaping innate immune responses. In animals, commensal microbes not only act as a physical barrier against infections but also modulate immune responses by interacting with pattern recognition receptors (PRRs) on immune cells. These interactions lead to the production of antimicrobial peptides, cytokines, and other immune mediators that fortify epithelial barriers and fine-tune inflammatory responses. Microbiome-derived metabolites, such as short-chain fatty acids (SCFAs), further influence immune cell differentiation and function, promoting a balanced immune environment. Disruption of this microbial community, often due to environmental stressors, diet changes, or antibiotic use, can result in dysbiosis, leading to impaired immune responses and increased susceptibility to infections and inflammatory diseases.

Introduction

The innate immune system serves as the first line of defense against pathogens in animals. One critical and often overlooked component of this system is the microbiome. The microbiome,

---

* Corresponding Author's Email: khumtyadebbarma999@gmail.com

In: Microbiomes in Animal Nutrition, Health, and Disease
Editor: Tanmoy Rana
ISBN: 979-8-89530-425-9

composed of a vast array of microorganisms including bacteria, fungi, viruses, and protozoa, resides on and within animal bodies. These microbial communities play a crucial role in maintaining homeostasis and enhancing the innate immune responses. This chapter explores the complex relationships between the microbiome and the innate immune system in animals, elucidating the mechanisms through which the microbiome influences immune responses, and the implications for animal health and disease.

The Microbiome and Its Components

The term "microbiome" refers to the collective genome of the microorganisms (bacteria, fungi, viruses, archaea, and other microbes) that inhabit various environments, including the human and animal body.

Composition of the Microbiome

The microbiome consists of diverse microbial populations (see Figure 1) that colonize various niches in the animal body, such as the skin, gut, respiratory tract, and reproductive organs. The composition of these communities can vary significantly depending on factors such as diet, environment, genetics, and health status.

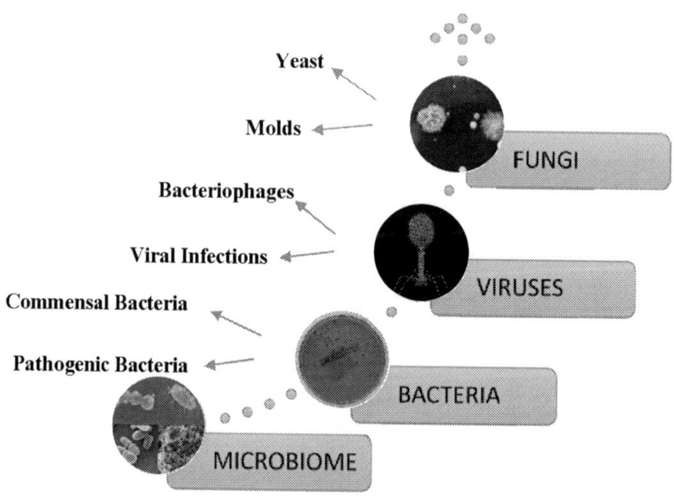

Figure 1. Components of Microbiome.

Function of the Microbiome

1. *Digestion and Nutrient Absorption*: The gut microbiome helps break down complex carbohydrates, proteins, and fats, facilitating digestion and absorption of nutrients. Certain bacteria in the gut produce enzymes that the human body cannot, aiding in the

breakdown of otherwise indigestible plant fibres into short-chain fatty acids, which provide energy and other health benefits.

2. *Immune System Modulation*: The microbiome interacts with the immune system, helping to train and regulate it. A healthy microbiome supports the immune system in distinguishing between harmful pathogens and harmless or beneficial microbes, thus preventing overreactions that could lead to allergies or autoimmune diseases.
3. *Protection against Pathogens*: The microbiome acts as a barrier against harmful pathogens by competing for nutrients and space, producing antimicrobial substances, and stimulating the immune system to fend off infections.
4. *Metabolism*: Microbes in the gut are involved in metabolizing drugs and xenobiotics, impacting their effectiveness and toxicity. They also produce metabolites that influence systemic metabolism, including lipid metabolism, glucose regulation, and the production of vitamins like vitamin K and certain B vitamins.
5. *Brain Function and Behaviour:* The gut-brain axis describes the complex communication network between the gut microbiome and the brain, influencing mood, behaviour, and cognitive functions. This interaction is mediated through the production of neurotransmitters, immune system modulation, and the regulation of inflammation.
6. *Development and Maintenance of the Gastrointestinal Tract*: The microbiome contributes to the development and maintenance of the gut lining and affects gut motility and the production of mucus, which protects the gut lining from damage.
7. *Detoxification*: Some gut bacteria help detoxify harmful compounds, including those found in food, environmental toxins, and metabolic byproducts, preventing their absorption and contributing to overall health.
8. *Influence on Weight and Obesity*: The composition of the gut microbiome can influence body weight by affecting the digestion of food, energy extraction, and fat storage. Imbalances in the microbiome have been linked to obesity and metabolic disorders.

The Innate Immune System in Animals

The innate immune system is the first line of defense in animals, providing immediate protection against infections and pathogens. Unlike the adaptive immune system, which requires time to develop a specific response, the innate immune system responds rapidly to threats using a broad range of mechanisms that are not specific to a particular pathogen.

Key Components of the Innate Immune System in Animals (Figure 2).

1. *Physical Barriers:*
   - Skin: Acts as a physical barrier preventing the entry of pathogens.
   - Mucous Membranes: Line the respiratory, gastrointestinal, and urogenital tracts, trapping pathogens and facilitating their removal.
   - Secretions: Such as saliva, tears, and mucus, contain antimicrobial substances like lysozyme that break down bacterial cell walls.

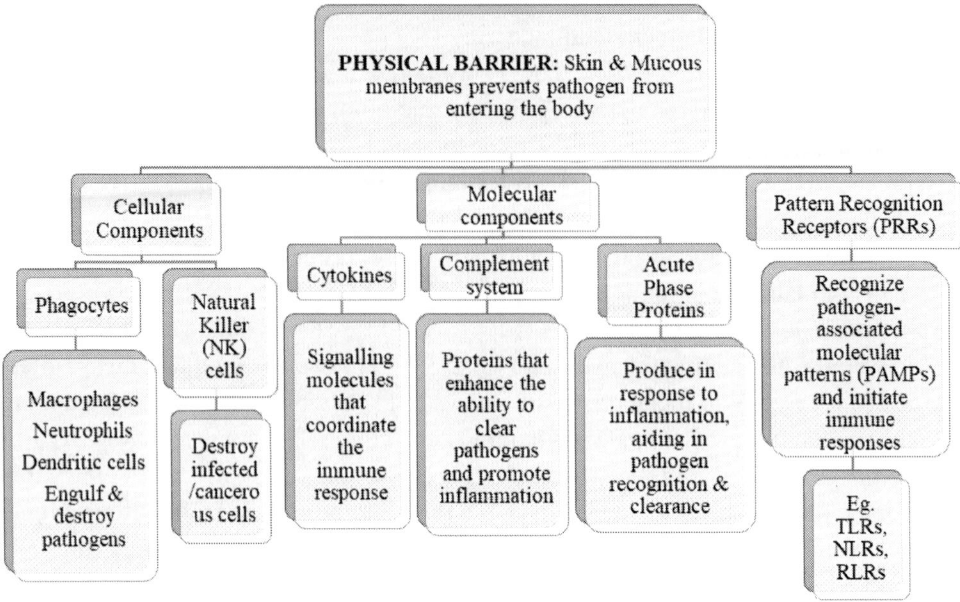

TLRs: Toll-like receptors; NLRs: NOD-like receptors; RLRs: RIG-I-like receptors

Figure 1. Key Components of the Innate Immunity System in Animals.

2. *Cellular Components:*
   - Phagocytes: Cells like macrophages, neutrophils, and dendritic cells that engulf and destroy pathogens through phagocytosis.
   - Natural Killer (NK) Cells: Recognize and kill infected or cancerous cells by inducing apoptosis.
   - Mast Cells: Play a role in inflammation and allergy by releasing histamine and other mediators.
3. *Molecular Components:*
   - Cytokines: Signalling molecules like interferons and interleukins that regulate immune responses, inflammation, and cell communication.
   - Complement System: A series of proteins that enhance the ability of antibodies and phagocytic cells to clear pathogens, promote inflammation, and attack the pathogen's cell membrane.
   - Acute Phase Proteins: Produced by the liver in response to inflammation, these proteins include C-reactive protein (CRP) and serum amyloid A, which help in pathogen recognition and clearance.
4. *Pattern Recognition Receptors (PRRs):*
   - Toll-like Receptors (TLRs): Recognize pathogen-associated molecular patterns (PAMPs) and initiate immune responses.
   - NOD-like Receptors (NLRs): Detect intracellular pathogens and initiate inflammatory responses.
   - RIG-I-like Receptors (RLRs): Detect viral RNA in the cytoplasm and trigger antiviral responses.

5. *Inflammatory Response:*
   - The innate immune system triggers inflammation in response to infection or injury, characterized by redness, heat, swelling, and pain. This process recruits immune cells to the site of infection, increases blood flow, and facilitates the healing process.

Mechanisms of Innate Immunity

Key mechanisms include:

- *Recognition of Pathogens*: Pattern recognition receptors (PRRs) on immune cells detect pathogen-associated molecular patterns (PAMPs).
- *Phagocytosis:* Immune cells engulf and destroy pathogens.
- *Inflammatory Response*: The release of cytokines and chemokines recruits additional immune cells to infection sites.
- *Antimicrobial Peptides:* Small proteins that directly kill or inhibit microbial growth.

Microbiome and Innate Immune Interactions

The interaction between the microbiome and the innate immune system is a critical aspect of health in animals. The microbiome, consisting of trillions of microorganisms (bacteria, viruses, fungi, and other microbes) residing primarily in the gut, plays a key role in regulating immune responses. Here's an overview of how these interactions occur:

Microbiome's Role in Immune System Development

The microbiome influences the maturation and function of the innate immune system, particularly in early life. The presence of beneficial microbes helps in the development of immune cells and the establishment of immune tolerance.

Microbial Colonization: The colonization of the gut by microbes after birth is essential for the proper development of immune cells such as macrophages and dendritic cells.

Pattern Recognition and Microbial Signalling

Pattern Recognition Receptors (PRRs)

Innate immune cells recognize microbes through PRRs such as Toll-like receptors (TLRs), which detect microbial molecules like lipopolysaccharides (LPS) from bacteria. These receptors can distinguish between commensal (beneficial) microbes and pathogenic ones, allowing the immune system to tolerate beneficial microbes while responding to harmful invaders.

Microbial Products as Immune Modulators

- *Short-Chain Fatty Acids (SCFAs):* Produced by the fermentation of dietary fibres by gut bacteria, SCFAs like butyrate and propionate have anti-inflammatory effects and enhance the barrier function of the gut lining.
- *Metabolites and Bacterial Products*: Certain microbial products can modulate immune responses, either promoting tolerance or triggering inflammation depending on the context.

Maintenance of Barrier Integrity

The microbiome plays a crucial role in maintaining the integrity of mucosal barriers (e.g., the gut lining), which is essential for preventing the invasion of pathogens.

- *Mucus Production:* Beneficial microbes stimulate the production of mucus, which acts as a physical barrier against pathogens.

Defense against Pathogens

- *Competition*: Commensal microbes compete with pathogens for resources and attachment sites, preventing colonization by harmful organisms.
- *Stimulation of Immune Responses*: The presence of the microbiome primes the innate immune system, enabling a rapid and effective response to pathogens.

Dysbiosis and Immune Dysregulation

- *Dysbiosis:* An imbalance in the microbiome (e.g., due to antibiotics, diet changes, or infections) can lead to immune dysregulation, contributing to conditions like chronic inflammation, allergies, and autoimmune diseases.
- *Leaky Gut:* Dysbiosis can compromise the gut barrier, leading to increased intestinal permeability ("leaky gut"), which allows harmful substances to enter the bloodstream and trigger an immune response.

Innate Immune System Feedback to Microbiome

The innate immune system can shape the microbiome composition through the secretion of antimicrobial peptides and other immune mediators, ensuring a balanced microbial environment.

Conclusion

The microbiome plays a pivotal role in the innate immunity of animals, influencing immune development, modulation, and responses to pathogens. Maintaining a healthy and balanced microbiome is essential for preventing diseases and enhancing overall health. Continued research into the interactions between the microbiome and the innate immune system will pave the way for innovative therapeutic approaches to improve animal health and welfare.

References

Belkaid, Y., & Hand, T. W. (2014). Role of the microbiota in immunity and inflammation. *Cell*, 157(1), 121-141.

Buffie, C. G., & Pamer, E. G. (2013). Microbiota-mediated colonization resistance against intestinal pathogens. *Nature Reviews Immunology*, 13(11), 790-801.

Honda, K., & Littman, D. R. (2016). The microbiota in adaptive immune homeostasis and disease. *Nature*, 535(7610), 75-84.

Hooper, L. V., Littman, D. R., & Macpherson, A. J. (2012). Interactions between the microbiota and the immune system. *Science*, 336(6086), 1268-1273.

Kamada, N., Seo, S. U., Chen, G. Y., & Núñez, G. (2013). Role of the gut microbiota in immunity and inflammatory disease. *Nature Reviews Immunology*, 13(5), 321-335.

Chapter 11

The Application of Bacteriophage Therapy to Improve Gut Health

Chayna Singha Mahapatra[1,*]
and Soubhik Sinhamahapatra[2]
[1]Department of Veterinary Microbiology, College of Veterinary and Animal Sciences, Bihar Animal Sciences University, Bihar, India
[2]Midnapore City College, Vidyasagar University, West Bengal, India

Abstract

The human body hosts a pool of microorganisms (microbiome) comprising bacteria, archaea, fungi, viruses, and protozoa that perform important functions that have a strong impact on our health in terms of the digestion of food and protective immunity. Virome is the viral fraction of the microbiome, and is generally dominated by phages. Bacteriophages are viruses that infect bacteria and they possess specific characteristics, including specificity for particular or closely related bacterial species. They have many applications, including as potential alternatives to antibiotics against multi-resistant bacterial pathogens and as control agents in bacteria-contaminated environments. Microbiota present in the gastrointestinal tract play a key role in the regulation of many gastrointestinal diseases. In particular, gastrointestinal diseases especially those induced by antibiotic-resistant bacterial strains bacteriophages, influence homeostasis by exerting a selective pressure on those bacterial communities.

Keywords: microbiome, bacteriophage, gastrointestinal diseases, antibiotics

Introduction

Phage therapy dates back to the first trials conducted by Felix d'Herèlle in 1921 in patients with dysentery (Moelling et al., 2018) when it was not widely accepted. Though at present, clinical trials are rising worldwide encouraging results are fostering phage therapy as a real alternative to combat pathogenic bacteria (Domingo-Calap and Delgado-Martínez, 2018), these treatments

---

* Corresponding Author's Email: chaynasmp@gmail.com

In: Microbiomes in Animal Nutrition, Health, and Disease
Editor: Tanmoy Rana
ISBN: 979-8-89530-425-9

are not yet regulated in the Western countries and are currently used as compassionate treatment. Regulation for the therapeutic use of phages pertain strict use of lytic phages (Young and Gill, 2015) as temperate have potentiality for acquisition of pathogenic traits like antibiotic resistance genes through horizontal gene transfer (Gordillo Altamirano and Barr, 2019). However, suppression of the transcription of bacterial virulence factors could be achieved by genetically engineered temperate phages (Hsu et al., 2020).

The microbial communities (microbiomes) present in the gastrointestinal tract (GIT) contribute to the overall well-being and health (Hou et al., 2022) as they are neither potentially beneficial nor can be eradicated. Host's well-being is attained through the mediation of nutrient absorption, synthesis of vitamins and neurotransmitters, the development and modulation of the immune system, and modifying resistance against pathogens, among others (Heintz-Buschart and Wilmes, 2018) and impact is connected to the bacterial component of the microbiome (Chow et al., 2010). Thus maintenance of the equilibrium between the healthy and harmful gut microorganisms is essential. One prominent and most abundant microorganisms of GIT are the bacteriophages (Mukhopadhya et al., 2019). These phages, are viruses that specifically infect bacteria and are the most abundant within the human biome and they have the ability to modulate human gut microbiome. This quality is attributed by the predation of bacteria or the promotion of bacterial survival by the transfer of genes that enhance bacterial fitness upon infection. Therefore, disturbances in this ecosystem, reflected by alterations in the microbial communities, may link to multiple diseases, from gut inflammation to neurological disorders (Mulak and Bonaz, 2015; Thursby and Juge, 2017).

The Gut Microbiome and Its Components

The human gut originates from the mouth, traversing through the esophagus, stomach, and small and large intestines, it terminates in the anus. These organs harbor a large number and numerous classes of microorganisms that make up the gut microbiota, including virus, bacteria, fungi, protozoa, and archaea of the following species (Table 1).

Table 1. List of different classes of microorganisms that make up the gut microbiota

Virus	Bacteria	Fungi	Protozoa	Archae
Bacteriophage	Faecalibacteria	Candida	Giardia	Methanobrevibacter
Adenovirus	Lactobacillus	Saccharomyces	Entamoeba	Methanomicrococcus
Picornavirus	Ruminococcus	Aspergillus	Dientamoeba	Methanospaera
Calicivirus	Enterobacteriaceae	Penicillium	Cryptosporidium	Methanospirrilum
Reovirus	Clostridium	Cryptococcus	Trichomonas	Halorubrum
Retrovirus	Streptococcus	Malassezia	Blastocystis	Sulfobales
Herpesvirus	Enterococcus	Trichosporon		Nitrosospaeralis
	Bifidobacterium			
	Helicobacter			
	Bacteroides			
	Proteobacteria			
	Fusobacteria			

Phages: An Overview

Phages are obligate parasites that require a host for reproduction and are the most abundant and diverse biological entities on earth (Suttle, 2005). Bacteriophages are the predating viruses on the bacterial population. Structurally, the majority of phages are composed of a nucleic-acid genome packaged inside a protein shell. Phage capsids are highly variable, both in size and morphology (i.e., polyhedral, filamentous, or pleomorphic). Some phages present an outer lipid membrane in addition to their protein capsid (Dion et al., 2020). Phage genomes are variable in size (ranging between ~3.5 kb and ~540 kb) and are composed of either single or double-stranded DNA or RNA (Yang et al., 2021; Zhai et al., 2021). Through the ages of co-evolution with bacteria, some of the phages possessed the ability to defeat bacterial defense systems by genetic evolution or even inventing new functions. Even after a single mutation in their tail fibers, phages can jump from susceptible to resistant and survive there (De Sordi et al., 2017).

Bacteriophages: Classification

They are classified based on their life cycle, genomic properties as well as morphology. Regarding genomic properties, DNA phages can be double-stranded DNA (dsDNA) or single-stranded DNA (ssDNA). Also, RNA phages can be double-stranded RNA (dsRNA) or single-stranded RNA (ssRNA). A specific group of RNA Phages are retroviruses. Circular replicating phages possess circular genomes and replicate independently of the host cell's replication machinery. Temperate, lysogenic, and lytic processes classify phages according to their life cycle (Mukhopadhya et al., 2019). Lytic phages replicate within the host cell, eventually release of daughter phage particles by cell lysis. All virulent phages are strictly lytic in nature. Temperate (pseudo-lysogenic) phages can be lytic, causing host cell death, or lysogenic, becoming dormant within the host after integration into the host DNA (Table 2).

Table 2. List of different classes of bacteriophages along with their morphology, genomic property and life cycle

Class	Morphology	Genomic properties	Life cycle
Caudovirales	Long, non-enveloped tails	DNA Phages	Temperate
Myoviridae,	Contractile, complex baseplate, long flexible tails	dsDNA	
Podoviridae	Short, non-contractile tails	dsDNA	
Siphoviridae	Non-contractile and rigid/non-flexible tails	ssDNA	
Filamentous	Long, flexible, and filament-like structures with the genetic material inside a protein coat	dsRNA and ssRNA	Lytic
Tectiviridae	Possess a "tectiviral"-appearing lipid envelope surrounding an icosahedral head	Retroviruses	Lysogenic
Inoviridae	Long, flexible, helical, and icosahedral heads (encapsulating the genetic material)	ssDNA	Circular Replicating
Leviviridae	Small-sized and possess a ssRNA genome in an icosahedral capsid	RNA phages	Temperate phage
Microviridae	Non-enveloped and small-sized and ss DNA genome in an icosahedral capsid	ssDNA	Virulent phage, Lytic
Pleolipoviridae	Pleomorphic; thus, they can be filamentous, spherical, or irregular and possesses a lipid-containing envelope	ssDNA and dsDNA	Archaeal Virus

Phages in the Human Gastrointestinal Tract

In spite of their presence throughout the gut, the density of phages is in an ascending order through the gastrointestinal tract from the small to the large intestine (Shkoporov and Hill, 2019). As measured by phage particles count (Kim et al., 2011) and estimation of viral genome numbers in feces of healthy adults (Shkoporov et al., 2019), the density of phages in the large intestine ranges between 10 and 8×10 phage virions per gram of feces, Phages appeared to make up the vast majority (97.7%) of gut viral genomes (Gregory et al., 2020) where DNA phages account the most and RNA phages are rare and negligible. The majority of viral RNA genomes within the gut originate from plant viruses acquired in diet (Lim et al., 2015). The two most abundant fecal phage clades, crAssphage (Cross Assembly phage) and Gubaphage (Gut Bacteroidales phage) are dsDNA phages that infect bacteria of the genus *Bacteroides* (Camarillo-Guerrero et al., 2021). The phage composition of the gut has been reported to remain stable for up to 1 year period in healthy adults (Shkoporov et al., 2019), and alterations have been associated with gastrointestinal diseases, such as Clostridiosis and inflammatory bowel diseases (IBD) (Clooney et al., 2019; Park et al., 2019). Similarly, Parkinson's disease patients were observed to have increased levels of lytic *Lactococcus* phages and depleted levels of *Lactococcus* bacteria, the latter of which regulate gut permeability and produce dopamine (Tetz et al., 2018). Furthermore, Crohn's disease patients presented an increased *Caudovirales* richness when compared to healthy controls, while ulcerative colitis patients did not (Norman et al., 2015), overall suggesting a connection between phage residence and the health state of the gut.

Phage Interactions in the Human GIT and Population Dynamics

The alterations of the gut phage composition in association with different diseases suggest a potential ecological influence of phages on the gut ecosystem. Phage interaction follows predator-prey-like dynamics (De Sordi et al., 2019) as, there are strong selective pressures for bacteria to evolve resistance mechanisms against lytic phages and also for the phages to develop strategies to evade those mechanisms (Hampton et al., 2020), respectively. *The Arms-race dynamic* (ARD) is characterized by competition for survival driven by predator-prey interactions, with the bacterial population developing counter-infection defenses, while the phage population develops methods to evade these defenses. *The fluctuating selection dynamic* (FSD) results from the bacterial fitness costs for developing phage-resistance mechanisms being disadvantageous in an environment in which multiple bacterial species are competing for resources. Population dynamics are strictly dependent on the environment in which they take place. In GIT, factors influencing the population dynamics include pH and bile acid levels, structural conformations, mucin layer, oxygen levels, and nutrient availability; all varying across the length of the GIT. This variability influences gut bacteria and, in certain cases, phage populations (Donaldson et al., 2016).

Role of Phages in the Human Gastrointestinal Tract

Classically, sequencing-based analysis suggested that the intestinal virome was mainly composed by single-stranded DNA (ssDNA) phages, mainly *Microviridae,* and double-stranded DNA (dsDNA) phages of the order Caudovirales (families *Myoviridae*, *Podoviridae* and *Siphoviridae*). Recent metagenomic studies identified a dsDNA phage genome, called *crAssphage* (crAss—cross-assembling), as a major fraction in the intestinal microbiota. They have two major role, one in intestinal dysbiosis and the other in immune modulation.

Phages and Intestinal Dysbiosis

Gut microbiome is fairly stable in the gastrointestinal tract as most of the phages in the human intestine are temperate phages. This component of temperate phages has led to the establishment of a global intestinal phagome, showing a correlation between phages and health status and their role in maintaining the structure and function of the intestinal microbiome (Manrique et al., 2016). This stability allows the maintenance of other microorganisms of the intestinal microbiota (Knowles et al., 2016) which is found to be altered in Ulcerative colitis (UC) and Chron's disease (CD). The inflamed gut in diseases condition is connected to increase in inhabitant intestinal pathogenic bacteria thus losing phage diversity over there. Whereas, bacteriophage richness, measured by the number of taxa per sample, was showed to be higher in patients undergoing these diseases, while bacterial richness decreased concomitantly (Norman et al., 2015). Several environmental conditions also trigger the stress response such as ultraviolet light and chemotherapeutic agents (Imamovic et al., 2016). Hence, most orally administered antibiotics alter the intestinal microbiota and promote dysbiosis.

Phages as Immunomodulators

Phages have the potential for regulation of the immune responses in the body. The immune system interacts with the microbiota by maintaining a non-inflammatory homeostasis. Multiple mechanisms like the physical mucosal barrier and the secretion of antimicrobial compounds help in maintaining homeostasis. The most immediate impact of phages on the immune system may be during sepsis, where lytic phages may reduce the bacterial load. Conversely, bacterial debris created by phages can also lead to sepsis. Besides the removal of invasive bacteria, intestinal phages can maintain immune homeostasis by reducing local immune and inflammatory reactions (Weiss and Hennet, 2017). Bacteria or lysed bacteria-induced inflammation may get partially neutralized by the buffering action of the immunomodulatory properties of phages (Van Belleghem et al., 2018). Cellular phage-mediated lysis appears to be involved in the production of pathogen-associated molecular patterns (PAMP) which may get translocated in case of increased intestinal permeability (Sinha and Maurice, 2019). As part of innate immunity, the gut environment favors obligate anaerobes, especially members of the Bacillota and Bifidobacteriaceae, while discouraging/suppressing the growth of facultative anaerobic pathogens. Even, Phages stimulate the removal of bacteria by phagocytosis through opsonization and pattern recognition receptors (PRRs) mediated recognition for immediate immunological response (den Besten et al., 2015; Martin-Gallausiaux et al., 2018). As adaptive immunity, the gut microbiota stimulates humoral immunity by inducing the B-cells to produce ample amounts of antibodies and cellular immunity by promoting the differentiation of T-cells (Bunker et al., 2015).

Phages in Cardio-Metabolic Diseases (CMDs)

Gut microbiome diversity and function have been consistently involved in CMDs including insulin resistance, diabetes, and obesity. For instance, a reduction in gut microbiome diversity and Ruminococcaceae, Roseburia, and Faecalibacteria and an increase in Enterobacteriaceae (*Escherichia, Shigella*, Klebsiella) Lactobacillus, and Streptococcus have been reported in individuals with CMDs. Microbiome promotes gut permeability which regulates excretion and absorption. Bacterial metabolites including short-chain fatty acids induce inflammation by deregulating the neuro-hormonal secretion through the production of pro-and anti-inflammatory signaling molecules from immune cells. Bacterial metabolites are released into the bloodstream and are responsible for several metabolic conditions, including insulin resistance, diabetes, and obesity, thus mediating microbial effects on human health. Thus phages attacking the bacteria responsible for CMDs can prevent the disease conditions.

Manipulation of the Gut Microbiome

Phage therapy and FVT are two different approaches that involve the use of viruses and phages to modify gut microbiome composition and potentially gain therapeutic benefits.

Phage Therapy

A key property of phages is their ability to target specific bacteria while leaving off-target species largely unaffected, which has led to their use as a treatment option known as phage therapy. Phage therapy involves the targeted approach to isolate and transplant phages that are effective against the specific bacterial strain causing the infection. Research findings suggest the use of phages as an alternative to antibiotics due to their narrower target range and fewer off-target effects on other bacterial species, minimal side effects on human hosts, and their ability to coevolve with their bacterial hosts. In the era of antimicrobial resistance, Phage therapy is becoming more accessible in Europe and the United States to patients in lack of other Solutions. Centre for Phage Research in Leicester, UK, which provides a biobank repository and plans to host a national library of phages to facilitate efforts in phage therapy. In comparison with conventional antibiotic therapy, safe specific, and retain the homeostatic nature of the gut microbiota makes preferred use of bacteriophages in the treatment of bacterial diseases. They play different roles in preventing and treating specific gut microbiota-related diseases via different mechanisms. For example, Microbiota transplantation is used to treat *Escherichia coli, Klebsiella pneumoniae,* and *Clostridioides difficile* recurrent infections where phages from the donors could colonize the recipient's microbiota for at least 12 months. A prophage sequence from the intestinal bacterium *Enterococcus hirae* triggers a specific immune response which enhances the cyclophosphamide treatment of cancer patients (Fluckiger et al., 2020). Though acidic pH and hydrolytic enzymes can reduce phage count while transiting through the gut, administration of a single high dose of phages can reduce the bacterial load to a certain level, providing an effective treatment.

If the problem is not due to the dominance of a particular species but a general dysbiosis and not by, there comes the need of development of a phage cocktail which still faces the challenges like mode of delivery, the dosage and the stability of phage preparations.

Fecal Virome Transplant (FVT)

Fecal matter transplantation (FMT) is the method of transferring a wide range of microorganisms which is in contrast to Fecal Virome Transplant (FVT) as it specifically transfers viral components/virome from fecal matter. FVT involves the transfer of viral components from the stool of a healthy donor to the gut of a recipient. To restore the microbiome, the fecal matter of the donor is first processed through a size exclusion filter to get a selected population specific to the population. Though gross bacteria are removed, bacterial cell walls and metabolites may remain in the fecal matrix. Thus, FVT has the advantage of reduced risks of bacterial infection associated with FMTs. Improvement in oral glucose tolerance, liver pathology, and adipose inflammation has been established in preclinical studies in mice. Success in obtaining enrichment in the phageome following FMTs have been reported for various conditions, including *Clostridium difficile* infection, IBD, autism spectrum disorder, and others.

Though FVTs pose potential advantages to FMTs, the use of biological agents in FVTs could remain active in the gut microbiome of the recipient indefinitely giving persistent co-occurring side effects due to the treatment. Another significant challenge to the FVT is the transfer of some unwanted viruses like herpes virus with unknown consequences upon transmission. Therefore, a thorough screening, and inactivation of enveloped (eukaryotic) viruses by solvent treatment, is needed. Besides, there should be thorough characterization for a better understanding of the gut phageome in human health.

Potential Therapeutic Applications

Many pathogenic and opportunistic bacteria are inherent in the gut system and have led to gut disease initiation, development, and advancement. Phage therapies are, however, developed and targeted at these bacteria. Potential applications of phages for preventive and curative purposes have produced single phages, cocktails, genetically modified phages, and even a combination with some other therapeutics, including antibiotics and probiotics (Nikolich and Filippoy, 2020). The extensive literature review revealed that studies are developing therapies against the most prevalent bacteria associated with gut system infections, including *Vibrio* spp., *Escherichia coli*, *Clostridioides difficile*, *Salmonella* spp., *Fusobacterium nucleatum*, *Shigella* spp., *Klebsiella pneumoniae*, *Listeria monocytogenes*, *Ruminococcus gnavus*, and *Campylobacter* spp.

1. *Lytic Phages against Ulcerative Colitis*
   Clostridioides difficile Ulcerative Colitis (UC) is one of the most common inflammations of the large intestine that causes diarrhea in asymptomatic population. Phages could be transferred from the donor to the recipient, as patients broadly comprising over 50% of children and 15% of healthy adults. A recent study has shown that phageome alone may be sufficient to eliminate *C. difficile* infection and restore a healthy microbial structure (Cepko et al., 2020).
2. *Lytic Phages against Crohn's Disease*
   E. coli causing diarrhoea is a major global burden especially in locations with reduced access to clean water and medical care. Among *E. coli* pathotypes, Invasive Adherent Escherichia coli (AIEC) is the most serious one, causing acute and persistent diarrhea

and there is emergence of multi-drug-resistant strains (Galtier et al., 2017). In 2018, Monteiro and co-workers demonstrated that a single phage cocktail administration has reduced the level of LF82 strain of AIEC colonization (Monteiro et al., 2018). Another example using phages against diarrhogenic *E. coli* was the PDX phage of *Myoviridae* family which could lyse the intestinal colonization of targeted bacteria in a murine model (Galtier et al., 2017).

3. *Temperate Phages to Suppress Virulence Factors*
   Temperate phages could be modified to deliver synthetic genes in order to suppress virulence genes of pathogenic bacteria. For example, use of genetically modified temperate phages was done in a clinical trial to suppress the Shiga toxin (Stx) from an established *E. coli* population (Sausset et al., 2020). Many research in this field based on in-vivo animal models or suitable in-vitro system has shown the following proven therapeutic activities (Table 3).

Table 3. List of proven activity of bacteriophages against their targeted bacteria

Targeted Bacteria	Targeting Phage with proven activity
Pathogenic Vibrio spp	V cholerae-specif ic phages, ICP1, ICP2, and ICP3, with specif ic effects against V cholerae domiciled in the small intestine.
Pathogenic Escherichia coli	Phage phPE42 against E. coli clinical isolates as well as T4 phages
Pathogenic Salmonella spp	Phage SE20 single dose (2 × 10 PFU/mL) of the phage isolate provided a targeted and prophylactic effect against Salmonella enteritidis
Pathogenic Fusobacterium nucleatum	F. nucleatum-binding M13-phage-loaded silver nanoparticles (AgNPs) achieved targeted clearance of F. nucleatum and remodeled the tumor-immune microenvironment.
Pathogenic Shigella spp.	Shigella-specif ic bacteriophages (vB_Sf lS-ISF001 and vB_SsoS-ISF002) and a cocktail of both. The phage preparations were investigated against multidrug-resistant Shigella sonnei and Shigella f lexneri isolates.
Pathogenic Klebsiella pneumoniae	Lytic cocktails of five-phage targeting strains
Pathogenic Listeria monocytogenes	Phage cocktail designated as a foodborne outbreak pill (FOP) targeted at the L. monocytogenes
Pathogenic Ruminococcus gnavus	Six phages that infect the R. gnavus JCM 6515T strain.
Pathogenic Campylobacter spp.	Two double-stranded phages (Φ 16-izsam and Φ 7-izsam) of Caudovirale order and bacteriophages φ4, φ44, φ22, φCj1, φ198, and φ287 of Myoviridae and Siphoviridae

Conclusion

The emerging role of the phageome gives rise to the possibility of using phages for diagnosis and treatment. The associations between phages and cardio-metabolic diseases (CMDs) suggest that phages can act as potential biomarker better predicting the diseases than traditional biomarkers. Host specificity and minimal off-target effects are making them attractive options in treatment. Therapy like phage based therapy and FVT are still in their infancy, but their translational potential has shown a promising result for CMD treatment. Technological advancements including artificial intelligence and modern bioinformatics techniques are guiding to better characterize of the gut phageome and metagenomic analysis and this can better predict bacterial host of phages. A more complete picture of the gut microbiome consisting all other microbes (archaea, fungi, protozoa) is needed to bridge the gap between basic research and clinical applications. Moreover, the development of animal models that mimic the intestinal community of humans will be useful to test the impact of phages. Lastly, synthetic biology or genetic engineering is allowing the use of temperate phages for therapeutic purposes, opening a new area of phage-based research.

References

Bunker J. J., Flynn T. M., Koval J. C., Shaw D. G., Meisel M., McDonald B. D., Ishizuka I. E., Dent A. L., Wilson P. C., Jabri B., et al. Innate and adaptive humoral responses coat distinct commensal bacteria with immunoglobulin A. *Immunity.* 2015;43:541–553. doi: 10.1016/j.immuni.2015.08.007.

Camarillo-Guerrero, L. F., Almeida, A., Rangel-Pineros, G., Finn, R. D., Lawley, T. D. (2021). Massive Expansion of Human Gut Bacteriophage Diversity. *Cell* 184, 1098– 1109.e9. doi: 10.1016/j.cell.2021.01.029.

Cepko L. C. S., Garling E. E., Dinsdale M. J., Scott W. P., Bandy L., Nice T., Faber-Hammond J., Mellies J. L. Myoviridae phage PDX kills enteroaggregative Escherichia coli without human microbiome dysbiosis. *J. Med. Microbiol.* 2020;69:309–323. doi: 10.1099/jmm.0.001162.

Chow, J., Lee, S. M., Shen, Y., Khosravi, A., Mazmanian, S. K. (2010). Chapter 8 - Host– Bacterial Symbiosis in Health and Disease. Advances in immunology, 107, pp. 243-274.

Clooney, A. G., Sutton, T. D. S., Shkoporov, A. N., Holohan, R. K., Daly, K. M., O'Regan, O., et al (2019). Whole-Virome Analysis Sheds Light on Viral Dark Matter in Inflammatory Bowel Disease. *Cell Host Microbe* 26, 764–778.e5. doi: 10.1016/j.chom.2019.10.009

De Sordi L., Khanna V., Debarbieux L. 2017. The gut microbiota facilitates drifts in the genetic diversity and infectivity of bacterial viruses. *Cell Host Microbe* 22:801–808.e3.

De Sordi L., Lourenco M., Debarbieux L. 2019. The battle within: interactions of bacteriophages and bacteria in the gastrointestinal tract. *Cell host Microbe,* 25(2), pp. 210-218.

den Besten G., Gerding A., van Dijk T. H., Ciapaite J., Bleeker A., van Eunen K., Havinga R., Groen A. K., Reijngoud D.-J., Bakker B. M. Protection against the Metabolic Syndrome by Guar Gum-Derived Short-Chain Fatty Acids Depends on Peroxisome Proliferator-Activated Receptor γ and Glucagon-Like Peptide-1. *PLoS ONE.* 2015;10:e0136364. doi: 10.1371/journal.pone.0136364.

Dion, M. B., Oechslin, F., Moineau, S. (2020). Phage Diversity, Genomics and Phylogeny. *Nat. Rev. Microbiol.* 18, 125–138. doi: 10.1038/s41579-019-0311-5.

Domingo-Calap P., Delgado-Martínez J. Bacteriophages: Protagonists of a Post-Antibiotic Era. *Antibiotics.* 2018;7:66. doi: 10.3390/antibiotics7030066.

Donaldson, G. P., Lee, S. M., Mazmanian, S. K. (2016). Gut Biogeography of the Bacterial Microbiota. *Nat. Rev. Microbiol.* 14, 20–32. doi: 10.1038/nrmicro3552.

Fluckiger A., Daillère R, Sassi M., Sixt BS, Liu P., Loos F., Richard C., Rabu C., Alou M. T., Goubet A. G., Lemaitre F., Ferrere G., Derosa L., Duong C. P. M., Messaoudene M., Gagné A., Joubert P., De Sordi L., Debarbieux L., Simon S, Scarlata C. M., Ayyoub M., Palermo B., Facciolo F., Boidot R., Wheeler R., Boneca I. G., Sztupinszki Z., Papp K., Csabai I., Pasolli E., Segata N., Lopez-Otin C., Szallasi Z., Andre F., Iebba V., Quiniou V., Klatzmann D., Boukhalil J., Khelaifia S., Raoult D., Albiges L., Escudier B., Eggermont A., MamiChouaib F., Nistico P., Ghiringhelli F., Routy B., Labarrière N., Cattoir V., et al. 2020. Cross-reactivity between tumor MHC class I-restricted antigens and an enterococcal bacteriophage. *Science* 369:936–942.

Galtier M., De Sordi L., Sivignon A., de Vallee A., Maura D., Neut C., Rahmouni O., Wannerberger K., Darfeuille-Michaud A., Desreumaux P., et al. Bacteriophages Targeting Adherent Invasive Escherichia coli Strains as a Promising New Treatment for Crohn's Disease. *J. Crohns Colitis.* 2017;11:840–847. doi: 10.1093/ecco-jcc/jjw224.

Gordillo Altamirano F. L., Barr J. J. Phage Therapy in the Postantibiotic Era. *Clin. Microbiol. Rev.* 2019;32 doi: 10.1128/CMR.00066-18.

Gregory, A. C., Zablocki, O., Zayed, A. A., Howell, A., Bolduc, B., Sullivan, M. B. (2020). The Gut Virome Database Reveals Age-Dependent Patterns of Virome Diversity in the Human Gut. *Cell Host Microbe* 28, 1–17. doi: 10.1016/j.chom.2020.08.003.

Hampton, H. G., Watson, B. N. J., Fineran, P. C. (2020). The Arms Race Between Bacteria and Their Phage Foes. *Nature* 577, 327–336. doi: 10.1038/s41586-019-1894-8.

Heintz-Buschart, A., Wilmes, P. (2018). Human Gut Microbiome: Function Matters. *Trends Microbiol.* 26, 563–574. doi: 10.1016/j.tim.2017.11.002

Hou K., Wu Z. X., Chen X. Y., Wang J. Q., Zhang D., Xiao C., Zhu D., Koya J. B., Wei L., Li J., et al. Microbiota in health and diseases. Sig. Transduct. Target. Ther. 2022;7:135. doi: 10.1038/s41392-022-00974-4.

Hsu B. B., Way J. C., Silver P. A. Stable Neutralization of a Virulence Factor in Bacteria Using Temperate Phage in the Mammalian Gut. *Msystems.* 2020;5 doi: 10.1128/mSystems.00013-20.

Imamovic L., Balleste E., Martinez-Castillo A., Garcia-Aljaro C., Muniesa M. Heterogeneity in phage induction enables the survival of the lysogenic population. *Environ. Microbiol.* 2016;18:957–969. doi: 10.1111/1462-2920.13151.

Kim, M. S., Park, E. J., Roh, S. W., Bae, J. W. (2011). Diversity and Abundance of Single Stranded DNA Viruses in Human Feces. *Appl. Environ. Microbiol.* 77, 8062–8070. doi: 10.1128/AEM.06331-11

Knowles B., Silveira C. B., Bailey B. A., Barott K., Cantu V. A., Cobian-Guemes A. G., Coutinho F. H., Dinsdale E. A., Felts B., Furby K. A., et al. Lytic to temperate switching of viral communities. *Nature.* 2016;531:466–470. doi: 10.1038/nature17193.

Lim, E. S., Zhou, Y., Zhao, G., Bauer, I. K., Droit, L., Ndao, I. M., et al. (2015). Early Life Dynamics of the Human Gut Virome and Bacterial Microbiome in Infants. *Nat. Med.* 21, 1228–1234. doi: 10.1038/nm.3950.

Manrique P., Bolduc B., Walk S. T., van der Oost J., de Vos W. M., Young M. J. Healthy human gut phageome. *Proc. Natl. Acad. Sci. USA.* 2016;113:10400–10405. doi: 10.1073/pnas.1601060113.

Martin-Gallausiaux C., Béguet-Crespel F., Marinelli L., Jamet A., Ledue F., Blottière H. M., Lapaque N. Butyrate produced by gut commensal bacteria activates TGF-beta1 expression through the transcription factor SP1 in human intestinal epithelial cells. *Sci. Rep.* 2018;8:9742. doi: 10.1038/s41598-018-28048-y.

Moelling K., Broecker F., Willy C. A Wake-Up Call: We Need Phage Therapy Now. *Viruses.* 2018;10:688. doi: 10.3390/v10120688.

Monteiro R., Pires D. P., Costa A. R., Azeredo J. Phage Therapy: Going Temperate? *Trends Microbiol.* 2019;27:368–378. doi: 10.1016/j.tim.2018.10.008.

Mukhopadhya I., Segal J. P., Carding S. R., Hart A. L., Hold G. L. The gut virome: The 'missing link' between gut bacteria and host immunity? *Ther. Adv. Gastroenterol.* 2019;12:1756284819836620. doi: 10.1177/1756284819836620.

Mukhopadhya I., Segal J. P., Carding S. R., Hart A. L., Hold G. L. The gut virome: The 'missing link' between gut bacteria and host immunity? *Ther. Adv. Gastroenterol.* 2019;12:1756284819836620. doi: 10.1177/1756284819836620.

Mulak, A., Bonaz, B. (2015). Brain-Gut-Microbiota Axis in Parkinson's Disease. *World J. Gastroenterol.* 21, 10609. doi: 10.3748/wjg.v21.i37.10609.

Nikolich M. P., Filippov A. A. Bacteriophage Therapy: Developments and Directions. *Antibiotics.* 2020;9:135. doi: 10.3390/antibiotics9030135.

Norman J. M., Handley S. A., Baldridge M. T., Droit L., Liu C. Y., Keller B. C., Kambal A., Monaco C. L., Zhao G., Fleshner P., et al. Disease-specif ic alterations in the enteric virome in inflammatory bowel disease. *Cell.* 2015;160:447–460. doi: 10.1016/j.cell.2015.01.002.

Norman, J. M., Handley, S. A., Baldridge, M. T., Droit, L., Liu, C. Y., Keller, B. C., et al. (2015). Disease-Specifi c Alterations in the Enteric Virome in Infl ammatory Bowel Disease. *Cell* 160, 447–460. doi: 10.1016/j.cell.2015.01.002.

Park, H., Laffi n, M. R., Jovel, J., Millan, B., Hyun, J. E., Hotte, N., et al. (2019). The Success of Fecal Microbial Transplantation in Clostridium Diffi cile Infection Correlates With Bacteriophage Relative Abundance in the Donor: A Retrospective Cohort Study. *Gut Microbes* 10, 676–687. doi: 10.1080/19490976.2019.1586037

Sausset R., Petit M. A., Gaboriau-Routhiau V., De Paepe M. New insights into intestinal phages. *Mucosal Immunol.* 2020;13:205–215. doi: 10.1038/s41385-019-0250-5.

Shkoporov A. N., Hill C. Bacteriophages of the Human Gut: The "Known Unknown" of the Microbiome. *Cell Host Microbe.* 2019;25:195–209. doi: 10.1016/j.chom.2019.01.017.

Shkoporov, A. N., Clooney, A. G., Sutton, T. D. S., Ryan, F. J., Daly, K. M., Nolan, J. A., et al. (2019). The Human Gut Virome Is Highly Diverse, Stable, and Individual Specifi c. *Cell Host Microbe* 26, 527–541.e5. doi: 10.1016/j.chom.2019.09.009.

Sinha A., Maurice C. F. Bacteriophages: Uncharacterized and Dynamic Regulators of the Immune System. *Mediat. Inf lamm.* 2019;2019:3730519. doi: 10.1155/2019/3730519.

Suttle, C. A. (2005). Viruses in the Sea. *Nature* 437, 356–361. doi: 10.1038/nature04160.

Tetz, G., Tetz, V. (2018). Bacteriophages as New Human Viral Pathogens. *Microorganisms* 6, 54. doi: 10.3390/microorganisms6020054.

Van Belleghem J. D., Dabrowska K., Vaneechoutte M., Barr J. J., Bollyky P. L. Interactions between Bacteriophage, Bacteria, and the Mammalian Immune System. *Viruses.* 2018;11 doi: 10.3390/v11010010.

Weiss G. A., Hennet T. Mechanisms and consequences of intestinal dysbiosis. *Cell Mol. Life Sci.* 2017;74:2959–2977. doi: 10.1007/s00018-017-2509-x.

Yang, M., Xia, Q., Sen, D., Zhang, Z., Qin, F., Zhao, Y. (2021). Genomic Characterization and Distribution Pattern of a Novel Marine OM43 Phage. *Front. Microbiol.* 12, 651326. doi: 10.3389/fmicb.2021.651326.

Young R., Gill J. J. Phage therapy redux—What is to be done? *Science.* 2015;350:1163–1164. doi: 10.1126/science.aad6791.

Zhai, Z., Zhang, Z., Zhao, G., Liu, X., Qin, F., Zhao, Y. (2021). Genomic Characterization of Two Novel RCA Phages Reveals New Insights Into the Diversity and Evolution of Marine Viruses. *Microbiol. Spectr.* 9, e0123921. doi: 10.1128/Spectrum.01239-21.

Chapter 12

The Interplay between Microbiome and Phytotherapeutics for Intestinal Health

Vijay Anand[1,*]
and Elamurugan Appavoo[2]
[1]Department of Animal Biotechnology, Madras Veterinary College (TANUVAS), Chennai, India
[2]Vaccine Research Centre - Viral Vaccines, Centre for Animal Health Studies (TANUVAS), Madhavaram Milk Colony, Chennai, India

Abstract

The dynamic interplay between the gut microbiome and phytotherapeutics offers significant potential for enhancing gut health. The gut microbiome, a complex ecosystem of microorganisms, is crucial for processes such as digestion, nutrient absorption, and immune system regulation. Disruptions to this balance, often caused by factors like diet or antibiotic use, can lead to dysbiosis and associated health issues. Phytotherapeutics, including polyphenols, terpenoids, and flavonoids, are recognized for selectively encouraging the growth of beneficial gut bacteria while inhibiting harmful ones, promoting a healthier intestinal environment. These compounds are metabolized by gut microbiota into bioactive metabolites, significantly impacting their health benefits. Additionally, therapeutic potential in treating gastrointestinal disorders is discussed, along with challenges related to bioavailability and the need for standardized formulations. The importance of further research to optimize the integration of these natural compounds into clinical practice is emphasized.

Keywords: gut microbiome, phytotherapeutics, intestinal health

Introduction

The gastrointestinal tract (GIT) is a complex and dynamic ecosystem essential for maintaining overall health. Central to this system is the gut microbiome, a diverse community of microorganisms, including bacteria, fungi, viruses, and archaea, that engage in a symbiotic

* Corresponding Author's Email: vijayanand.j@tanuvas.ac.in

In: Microbiomes in Animal Nutrition, Health, and Disease
Editor: Tanmoy Rana
ISBN: 979-8-89530-425-9

relationship with the host (Lozupone et al., 2012). This community is integral to numerous physiological processes such as digestion, nutrient absorption, and immune modulation (Hooper et al., 2001). Moreover, the gut microbiome is involved in the production of essential vitamins, the metabolism of bile acids, and the detoxification of xenobiotics (O'Keefe, 2016). In recent years, the gut microbiome has been increasingly recognized as a key factor in various health conditions, including obesity, diabetes, inflammatory bowel disease (IBD), and even neurological disorders like depression and anxiety (Sekirov et al., 2010; Dinan and Cryan, 2017). The delicate balance of microbial populations in the gut can be disrupted by factors such as diet, antibiotics, and stress, leading to dysbiosis—a state associated with various diseases (Tilg and Kaser, 2011). This growing understanding has led to a surge in interest in modulating the gut microbiome to promote health and prevent or treat disease.

Phytotherapeutics, plant-derived compounds, have emerged as promising agents for modulating the gut microbiome and promoting intestinal health. These compounds, which include polyphenols, alkaloids, terpenoids, and essential oils, are known for their antimicrobial, anti-inflammatory, and antioxidant properties (Vargas-Mendoza et al., 2019). These properties make them potent candidates for influencing the composition and function of the gut microbiome, thereby enhancing host health. Polyphenols, for example, are a diverse group of phytochemicals found in fruits, vegetables, tea, and wine, and are known to exert prebiotic-like effects, selectively promoting the growth of beneficial bacteria such as Bifidobacterium and Lactobacillus (Cardona et al., 2013). The metabolism of polyphenols by gut bacteria can also produce bioactive metabolites that contribute to their health benefits, including anti-inflammatory effects and protection against oxidative stress (Espín et al., 2017). Terpenoids, another class of phytochemicals, have been shown to modulate gut microbiota composition and exert protective effects against gastrointestinal diseases through their anti-inflammatory and antimicrobial activities (Singh et al., 2019). Furthermore, recent studies have highlighted the potential of phytotherapeutics in restoring gut microbiota balance and improving outcomes in conditions such as IBD and colorectal cancer (Gill et al., 2018; Nani et al., 2019). The integration of phytotherapeutics into clinical practice offers a promising approach to harnessing the gut microbiome's therapeutic potential, providing a natural and holistic avenue for maintaining and improving gut health. This chapter delves into the mechanisms by which phytotherapeutics influence the gut microbiome, the resulting health benefits, and the potential for integrating these approaches into clinical practice. By exploring these interactions, we can better understand how to leverage phytotherapeutics to enhance intestinal health and overall well-being.

The Gut Microbiome and Its Functions

The gut microbiome, composed of trillions of microorganisms including bacteria, archaea, fungi, and viruses, is predominantly dominated by the bacterial phyla Firmicutes and Bacteroidetes, with contributions from Actinobacteria, Proteobacteria, and Verrucomicrobia (Qin et al., 2010; Park et al., 2024). Its composition and diversity are influenced by factors such as diet, age, genetics, and environment (Palmer et al., 2007; Park et al., 2024; Radjabzadeh et al., 2020). The microbiome evolves from birth, with colonization patterns influenced by delivery mode and feeding practices, becoming more diverse and stable in adulthood (Palmer et al., 2007; Park et al., 2024). In adults, Firmicutes and Bacteroidetes predominate, whereas infants' microbiota is initially dominated by Actinobacteria and Proteobacteria (Qin et al., 2010; Park et al., 2024). Ecological interactions, such as mutualism and competition, shape microbial diversity, with predatory bacteria like Bdellovibrio playing a key role in maintaining balance (Park et al., 2024; Qin et al., 2010). Diet is a significant factor; high intake of fruits, vegetables, and whole grains enhances microbial diversity, while processed foods reduce it (Palmer et al., 2007; Qin et al., 2010). Specific dietary components, like fiber, promote beneficial bacteria, such as Bifidobacteria and Lactobacillus (Park et al., 2024). Genetic variation also influences microbiome composition, though environmental factors have a greater impact (Palmer et al., 2007; Qin et al., 2010). Differences in gut microbiome composition are observed between children and adults, with children showing lower diversity and distinct functional profiles, reflecting developmental and dietary shifts (Radjabzadeh et al., 2020). Micronutrients, including polyphenols and vitamins, also modulate gut microbiota, promoting beneficial bacteria and influencing functional pathways related to health (Tabatabaeizadeh et al., 2018).

Role in Digestion and Nutrient Absorption

Gut microbiota facilitate the breakdown of complex carbohydrates into short-chain fatty acids (SCFAs) like acetate, propionate, and butyrate, which serve as energy sources for colonocytes and have systemic health benefits (Flint et al., 2012). These SCFAs are produced through the fermentation of dietary fibers and resistant starches, which the human digestive enzymes cannot break down. The gut microbiome also synthesizes essential vitamins such as vitamin K and certain B vitamins (Thiamine, Riboflavin, and Folate), contributing to the host's nutritional status (LeBlanc et al., 2013). Additionally, gut bacteria play a role in bile acid metabolism and the deconjugation of bile acids, which aids in lipid digestion and absorption (Ridlon et al., 2016). Digestion and absorption of nutrients predominantly occur in the small intestine and stomach, where approximately 85% of carbohydrates, 66-95% of proteins, and nearly all fats are absorbed. This process leaves the colon and the bulk of the gut microbiome exposed to food residues that are largely indigestible by the host (Cummings and Macfarlane, 1997). After most of the digestible nutrients have been absorbed, the colon assumes a crucial role in processing the remaining indigestible components of food, especially dietary fibre. The gut microbiota in the colon ferments these fibres, producing short-chain fatty acids (SCFAs) such as butyrate, acetate, and propionate. These SCFAs are essential for the health of colonocytes—the cells lining the colon—and play significant roles in energy regulation, enhancing nutrient absorption, and maintaining the intestinal barrier function (Morrison and Preston, 2016; Topping and Clifton, 2001). Furthermore, SCFA production contributes to the local and systemic health of the host by improving insulin sensitivity, increasing energy expenditure, and potentially regulating appetite (Macfarlane and Macfarlane, 2011).

Effects of Dietary Components on Microbiome Composition

The gut microbiome, a complex community of microorganisms in the gastrointestinal tract, is significantly influenced by dietary components. Diets rich in fiber, polyphenols, and unsaturated fatty acids support a diverse and balanced microbiome, promoting gut health and reducing the risk of chronic diseases. Conversely, diets high in refined sugars, fats, and low in fiber can disrupt this balance, leading to dysbiosis, a condition associated with inflammation and other adverse health outcomes (Rinninella et al., 2019). Omega-3 polyunsaturated fatty acids (PUFAs) modulate the gut microbiome and enhance intestinal immunity by increasing beneficial bacteria like Bifidobacterium and Lactobacillus while reducing pro-inflammatory species. This shift promotes a healthier gut environment, highlighting the therapeutic potential of omega-3 PUFAs in managing gut-related disorders (Fu et al., 2021). A mechanistic perspective on the fermentation of amino acids by gut bacteria reveals that this process is influenced by factors such as pH and dietary starch availability. The presence of resistant starch modifies fermentation pathways, leading to the production of short-chain fatty acids (SCFAs), which are crucial for maintaining gut health. These insights suggest that strategic dietary choices, particularly those involving resistant starch, can significantly impact gut microbiome composition and function (Smith and Macfarlane, 1998). Protein metabolism produces L-carnitine, which bacterial fermentation converts to TMAO. TMAO has been linked to atherosclerosis and colorectal cancer. Red meat consumption, associated with higher colorectal cancer risk, contrasts with the protective effects of dietary fibre (Zhen et al., 2023). Collectively, these studies underscore the profound influence of dietary components on gut microbiome composition. They suggest that intentional dietary modifications can be a powerful tool for maintaining and improving gut health, with potential implications for preventing and managing various chronic diseases. As research continues to evolve, understanding the complex relationship between diet and the gut microbiome will be essential for developing targeted nutritional strategies to enhance health and well-being.

Modulation of Gut Microbiome by Phytotherapeutics

Phytochemicals are non-nutrient biologically active plant components that can modify the composition of gut microflora. They selectively stimulate the proliferation or inhibition of certain microbial communities in the intestine. These compounds often have low bioavailability and long retention times in the intestine, leading to beneficial impacts on gut microbiota populations. For example, polyphenols from fruits and vegetables can act as prebiotics, promoting beneficial bacteria such as Bifidobacteria and Lactobacilli, and inhibiting harmful bacteria like E. coli and Salmonella (Gibson et al., 2017; Pandey and Rizvi, 2009). This modulation can improve gut health, metabolic functions, and overall well-being (Del Rio et al., 2013; Scalbert et al., 2005).

Molecular Basis of Phytochemical-Gut Microbiota Interactions

Phytoconstituents, such as polyphenols, flavonoids, and alkaloids, interact with gut microbiota to influence host physiology. These interactions often involve complex molecular signaling pathways, such as AMPK, PPARs, and NF-κB, which are crucial in regulating inflammation, metabolism, and immune responses (Ozdal et al., 2016). For instance, saponins from legumes increase gut microbial diversity and SCFA production, which helps regulate glucose metabolism and reduce inflammation (Guo et al., 2020). Similarly, ginsenosides from ginseng modulate the gut microbiota by promoting beneficial bacteria and reducing harmful ones, thereby enhancing metabolic health (Ding et al., 2023).

Role of Nutraceuticals in Gut Microbiota Modulation

Nutraceuticals, including probiotics, prebiotics, and postbiotics, have shown significant potential in modulating gut microbiota. These compounds can improve gut health by enhancing beneficial bacterial populations and reducing pathogens. For example, essential oils (EOs) like oregano oil have antimicrobial properties that enhance gut barrier function and promote beneficial bacteria. Studies on animal models have shown that EOs can significantly increase bacterial diversity and the abundance of probiotic bacteria like Bifidobacterium and Lactobacillus, improving gut health and reducing disease susceptibility (Brenes and Roura, 2010; Hyldgaard et al., 2012).

Effects of Specific Phytochemicals on Gut Microbiota

Different phytochemicals have specific effects on gut microbiota:

- *Polyphenols:* Found in fruits, vegetables, tea, and wine, polyphenols can promote beneficial bacteria like Bifidobacteria and inhibit harmful ones. For example, resveratrol increases Bifidobacteria abundance, reduces inflammation, and enhances metabolic health (Cardona et al., 2013; Larrosa et al., 2009).
- Flavonoids: These include flavones, flavanones, and flavonols found in various fruits and vegetables. They act as antioxidants and prebiotics, promoting the growth of beneficial bacteria and reducing gut dysbiosis (Kawabata et al., 2019; Stevenson et al., 2019).
- Saponins: These compounds from legumes increase SCFA-producing bacteria, which are crucial for maintaining gut health and metabolic functions (Guo et al., 2020).
- Carotenoids: Compounds like beta-carotene and astaxanthin modulate gut microbiota by increasing beneficial bacteria and reducing inflammatory bacteria (Yonekura and Nagao, 2007; Liu et al., 2011).

Potential Therapeutic Applications

The modulation of gut microbiota by phytotherapeutics holds significant therapeutic potential for managing metabolic disorders such as obesity, diabetes, and cardiovascular diseases. By targeting specific microbial pathways and enhancing beneficial microbial populations, phytotherapeutics can improve metabolic outcomes and reduce disease risks (Lee et al., 2013; Tomas-Barberan et al., 2016). Additionally, understanding the molecular mechanisms underlying these interactions can aid in developing novel therapeutic strategies and nutraceutical interventions to combat metabolic dysregulation and promote overall health (Delzenne et al., 2011; Cani & Van Hul, 2015).

Impact of Phytotherapeutics Gut Microbiome

Cattle

The interplay between microbiome and phytotherapeutics in maintaining intestinal health, particularly in ruminants, is a dynamic and complex process. The research on various phytochemicals and plant-based additives highlights their significant role in modulating the gut microbiome, enhancing digestive efficiency, and mitigating environmental impacts such as methane emissions. Studies have shown that the use of specific phytochemicals, such as those derived from Ginkgo fruit and Capsicum oleoresin, can positively influence rumen fermentation and microbial populations. For instance, Ginkgo fruit extract, particularly from the Kyuju cultivar, contains antimicrobial compounds like anacardic acids. These compounds have been found to reduce methane production by up to 53%, while increasing propionate levels—a beneficial shift in fermentation processes that supports energy metabolism without adversely affecting the digestibility of essential nutrients. However, the presence of antinutritional compounds in Ginkgo necessitates further processing to ensure its safe incorporation into diets (Oh et al., 2017). In addition to individual phytochemicals, combined phytotherapeutic formulations like Farmatan™ (containing tannins, essential oils, sodium acetate, and organic zinc) and Pine Tree Energy Supplements (PTES) have shown promise in enhancing ruminal digestion. These supplements not only boost volatile fatty acid production, improving energy metabolism, but also reduce harmful bacterial populations and decrease the prevalence of pathogens in the gastrointestinal tract. This modulation of the gut microbiome is crucial for both the health and productivity of ruminants (Zaitsev et al., 2022). Capsaicin, another phytochemical of interest, demonstrates potential as an alternative to traditional antimicrobial growth promoters. Though the study results indicate a trend toward capsaicin influencing the fecal microbiome, further research is needed to confirm its efficacy in significantly altering the gut microbial structure in ruminants (Bierly et al., 2024). Tannins, another group of phytochemicals, exhibit dual roles in ruminant nutrition. They can inhibit the digestion of proteins and fibres, leading to increased nitrogen excretion. However, they also show potential in reducing methane emissions, a key environmental concern, by inhibiting methanogens in the rumen. The variability in the impact of tannins on milk and meat production, particularly regarding fatty acid composition, underscores the need for careful management in their use (Besharati et al., 2022).

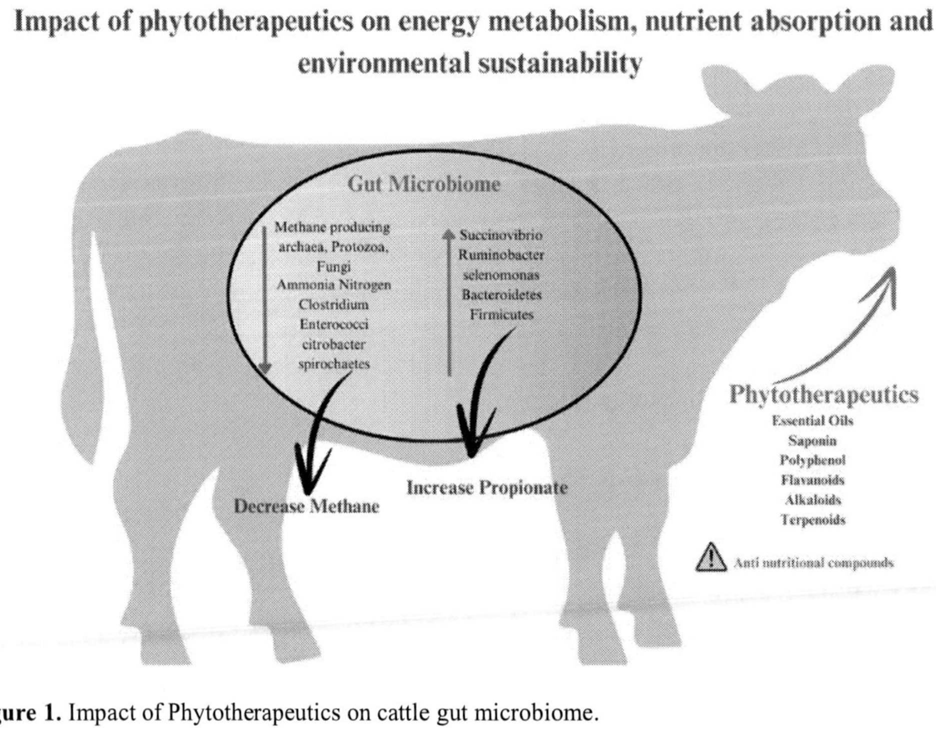

Figure 1. Impact of Phytotherapeutics on cattle gut microbiome.

Moreover, the exploration of phytotherapeutics for controlling gastrointestinal nematodes (GINs) in ruminants reveals their potential as alternatives to chemical anthelmintics. Phytochemicals such as tannins, flavonoids, and saponins have shown nematicidal activity, although their efficacy in vivo is often compromised due to complex ruminant digestive physiology. Advances such as nanotechnology may offer solutions to protect these compounds from degradation and improve their efficacy (Borges and Borges, 2016). Overall, phytogenic additives have the potential to selectively modulate ruminal bacterial communities, as evidenced by studies on grazing beef cattle. Such modulation can benefit nutrient utilization, animal productivity, and environmental sustainability (Figure 1).

Sheep and Goats

In ruminants such as sheep and goats, the interplay between diet and the ruminal microbiota has gained significant attention. Tannin-rich diets, such as those containing Sericea lespedeza, have been shown to significantly alter the composition of the ruminal microbiota, impacting both microbial diversity and abundance. These effects can vary depending on the host species, indicating that phytotherapeutics could be tailored to enhance specific microbial functions that are beneficial for ruminant health and productivity (Min et al., 2024). A metagenomic study cataloged over 5,000 metagenome-assembled genomes from the gut microbiota of Caprinae, many of which represent potentially new species. This catalog serves as a valuable resource for further exploring the effects of various phytotherapeutics on these novel microbial populations and their enzymatic functions (Zhang et al., 2022). The interaction between host species and their gut microbiota is crucial for understanding how phytotherapeutics influence not only microbial diversity but also metabolic pathways that affect overall animal health and feed efficiency. Integrating these findings could lead to more targeted use of phytotherapeutics to optimize the gut microbiome for improved health and productivity in sheep and goats.

Horses

In horses, recent research has provided detailed insights into the relationship between phytotherapeutics and gut health, particularly concerning behavioral and parasitic conditions. For example, a study on the gut microbiome characteristics of horses with a history of cribbing behavior—a behavior often associated with gastrointestinal dysfunction—revealed distinct differences in fecal microbiota between cribbing and non-cribbing horses. This suggests that cribbing behavior may be linked to a disrupted gut microbiome. However, a gastric health support supplement used in the study did not significantly alter the gut environment, indicating that further research is needed to identify effective interventions (Arias-Esquivel et al., 2024). Additionally, research has demonstrated that diet, particularly high-fibre or high-starch diets supplemented with polyphenol-rich sainfoin, can significantly affect the composition of intestinal microbiota and the severity of strongyle infections in horses. These findings suggest that phytotherapeutic supplementation can influence both gut health and parasitic burden in horses (Laroche et al., 2023).

Pigs

In pigs, the use of phytotherapeutics and herbal medicines in diets is increasingly seen as a natural alternative to antibiotics, aiming to enhance gut health and overall performance. Specific herbs like Filipendula ulmaria and Orthosiphon aristatus, traditionally used for urinary tract disorders, are metabolized by the pig gut microbiota into compounds such as quercetin and kaempferol, which possess anti-inflammatory and diuretic properties. This indicates their potential for managing urinary tract issues in pigs while reducing the reliance on antibiotics (Popowski et al., 2021). Furthermore, Lythrum salicaria (Purple Loosestrife), historically used for treating diarrhea, has been shown to modulate gut microbiota in post-weaning piglets, promoting the production of beneficial postbiotic metabolites like urolithins that support gut health without causing dysbiosis (Dadi et al., 2020). The combination of phytogenics, including essential oils and dry herbs, with encapsulated sodium butyrate, has also been effective in replacing antibiotics as growth promoters in piglets, improving feed conversion ratios, enhancing growth, and reducing diarrhea incidence (da Silva et al., 2022).

Dogs

In dogs, phytotherapeutics are being increasingly explored for their effects on the gut microbiome, particularly in managing gastrointestinal health and metabolic conditions. For instance, adding a polyphenol-rich fiber bundle to dog food has been shown to positively impact the gastrointestinal microbiome and metabolome, especially in dogs with chronic gastroenteritis. The prebiotic fibres promoted beneficial changes in microbial populations, leading to improved gut health by enhancing saccharolytic activities and reducing bacterial putrefaction processes. This dietary intervention also influenced systemic metabolic signatures, suggesting potential benefits beyond gut health, including modulating inflammation and enhancing gastrointestinal motility (Fritsch et al., 2023).

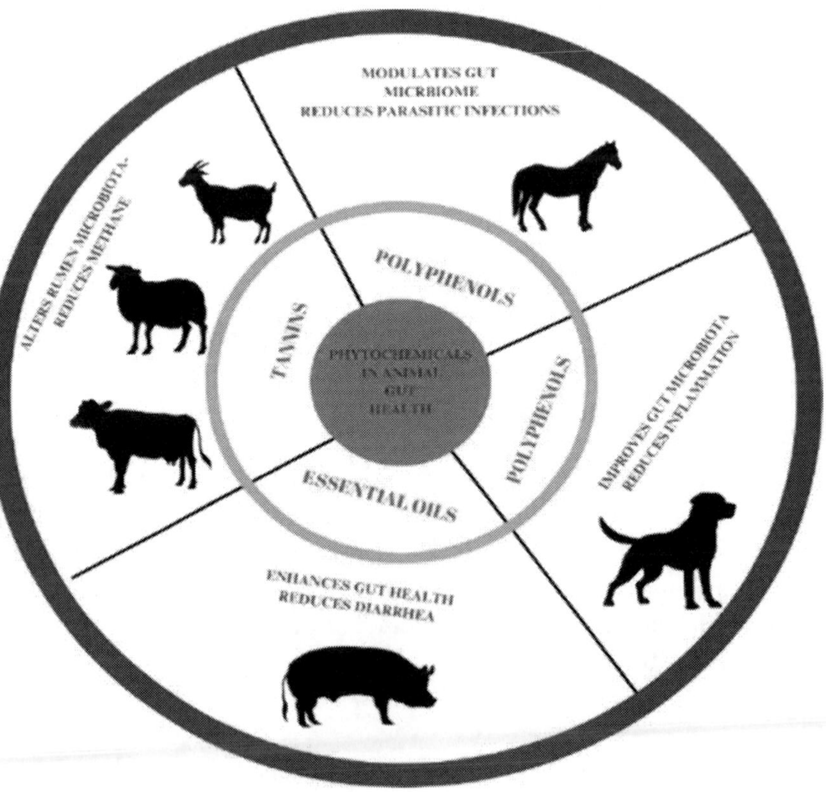

Figure 2. Role of phytochemicals in gut health of animals.

Phytotherapeutics offer promising potential for enhancing gut health across species by modulating the microbiome. Tailored dietary interventions can improve health outcomes and support sustainable animal management. Integrating these strategies may lead to more effective and holistic approaches in animal husbandry (Figure 2).

4. Bioavailability and Metabolism of Phytotherapeutics

Role of Gut Microbiota in Metabolizing Phytotherapeutics

Gut microbes can metabolize phytotherapeutics into bioactive metabolites with enhanced or altered biological activities compared to the parent compounds. This process highlights the bidirectional interaction between phytotherapeutics and the gut microbiome (Selma et al., 2009). For example, the gut microbiota can convert polyphenols such as ellagitannins from pomegranates into urolithins, which have been shown to possess anti-inflammatory and anticancer properties (Selma et al., 2014). Similarly, the microbial metabolism of flavonoids can produce metabolites with distinct biological activities, influencing the overall health benefits of these compounds (Williamson and Clifford, 2017).

Factors Influencing Bioavailability

The bioavailability of phytotherapeutics is influenced by factors such as their chemical structure, the presence of other dietary components, and individual differences in gut microbiota composition (Pérez-Jiménez et al., 2010). The absorption and metabolism of phytotherapeutics can vary significantly between individuals, leading to differences in their therapeutic efficacy. Strategies to enhance the bioavailability of phytotherapeutics include the use of nanoformulations, encapsulation, and combination with other bioactive compounds that can enhance their absorption and stability. Understanding the factors that influence the bioavailability of phytotherapeutics is crucial for optimizing their therapeutic potential.

Effects of Gut Microbiota on the Bioavailability of Phytotherapeutics

The gut microbiota profoundly influences the bioavailability and bioactivity of phytochemicals, including those found in ginkgo leaf extracts (GLE), which are rich in flavonoids and terpenoids. These compounds undergo extensive microbial metabolism, significantly impacting their pharmacokinetics and therapeutic efficacy. In studies focusing on GLE, antibacterial treatments in mice that disrupt gut microbiota have been shown to increase the bioavailability of isorhamnetin, a key flavonoid in GLE, suggesting that gut bacteria play a crucial role in the metabolism and absorption of these bioactive compounds (Choi et al., 2019). This phenomenon is consistent with broader research on the gastrointestinal biotransformation of phytochemicals, which emphasizes that gut bacteria metabolise dietary compounds and convert them into bioactive metabolites that may either enhance or diminish their health benefits (Singh et al., 2020). For instance, polyphenols—a significant class of phytochemicals—undergo microbial metabolism in the colon, producing metabolites with distinct biological activities compared to their parent compounds. This metabolic process is crucial for understanding the bioavailability and health impacts of polyphenols in the diet (Koudoufio et al., 2020).

Furthermore, the gut microbiome's role as a metabolic organ is increasingly recognized, particularly in the first-pass metabolism of botanical supplements. The microbial biotransformation often results in lower circulating levels of the original compounds, with health effects linked to specific active metabolites produced within the gut. This has been observed in the metabolism of various botanical supplements, including GLE, where the gut microbiota's metabolic activities significantly influence these compounds' pharmacokinetics and overall bioactivity (Possemiers et al., 2011). Additionally, specific microbial taxa have been identified as key players in this process. For example, certain Lactobacillus and Bifidobacterium strains are known to enhance the bioavailability of flavonoids by producing enzymes that facilitate their conversion into absorbable forms (Walsh et al., 2018). Moreover, dietary interventions aimed at modulating gut microbiota composition—such as prebiotics and probiotics—have been proposed as strategies to improve the bioavailability and therapeutic effects of GLE and other botanical supplements (Holscher, 2017).

Clinical Implications and Future Directions

Therapeutic Potential in Gastrointestinal Disorders

Phytotherapeutics hold promise for the prevention and treatment of gastrointestinal disorders such as IBD, IBS, and colorectal cancer. Clinical trials have shown the beneficial effects of compounds like resveratrol and curcumin in these conditions. For example, resveratrol, a polyphenol found in grapes and berries, has been shown to reduce inflammation and oxidative stress in patients with ulcerative colitis (Samsami-Kor et al., 2015). Similarly, curcumin has been found to induce remission in patients with mild to moderate ulcerative colitis, highlighting its potential as an adjunctive therapy (Lang et al., 2015).

Challenges and Considerations

Challenges in the use of phytotherapeutics include variability in bioavailability, potential interactions with medications, and the need for standardized formulations (Williamson, 2001). The therapeutic efficacy of phytotherapeutics can be influenced by factors such as the source of the plant material, extraction methods, and individual differences in metabolism and gut microbiota composition. Standardization of phytotherapeutic formulations is essential to ensure consistent therapeutic outcomes. This includes the identification and quantification of active compounds, as well as the use of validated analytical methods to ensure quality control.

Integration into Clinical Practice

Integrating phytotherapeutics into clinical practice requires collaboration between researchers, clinicians, and regulatory bodies to ensure safety, efficacy, and accessibility. Clinical guidelines and evidence-based recommendations are needed to guide the use of phytotherapeutics in the management of gastrointestinal disorders. Further research is needed to elucidate the mechanisms of action of phytotherapeutics, optimize their formulations, and conduct large-scale clinical trials to establish their therapeutic efficacy and safety profiles. Additionally, education and training for healthcare professionals on the use of phytotherapeutics can facilitate their integration into clinical practice.

Conclusion

The gut microbiome, a complex and dynamic ecosystem, plays a crucial role in digestion, nutrient absorption, and immune modulation. Disruptions to this microbial balance, often caused by poor diet, stress, and antibiotic use, can lead to dysbiosis, which is associated with various health disorders. Phytotherapeutics—natural compounds derived from plants—have emerged as a potential strategy to restore and maintain a healthy gut microbiome. Compounds such as polyphenols, terpenoids, and flavonoids have shown the ability to selectively promote beneficial gut bacteria while inhibiting harmful ones. They act through mechanisms like prebiotic effects, influencing short-chain fatty acid production, and modulating immune responses. The gut microbiota's role in metabolizing these compounds into bioactive metabolites further highlights the microbiome's importance in determining health outcomes. However, challenges like variability in bioavailability due to individual differences in gut microbiota composition and the need for standardized formulations remain. Addressing these issues is critical for optimizing the therapeutic use of phytotherapeutics. Continued research is necessary to fully understand these interactions and develop effective, evidence-based approaches for using phytotherapeutics in maintaining gut health and preventing gastrointestinal disorders.

References

Arias-Esquivel, A. M., Jeong, K. C., Fan, P., Lance, J., DeNotta, S., Wickens, C. 2024. Gut microbiome characteristics of horses with history of cribbing behavior: An observational study. *Journal of Veterinary Behavior*, 72:40-50.

Besharati, M., Maggiolino, A., Palangi, V., Kaya, A., Jabbar, M., Eseceli, H., De Palo, P., Lorenzo, J. M. 2022. Tannin in ruminant nutrition: Review. *Molecules*, 27(8273).

Bierly, S. A., Van Syoc, E. P., Westphalen, M. F., Miles, A. M., Gaeta, N. C., Felix, T. L., Hristov, A. N., Ganda, E. K. 2024. Alterations of rumen and fecal microbiome in growing beef and dairy steers fed rumen-protected capsicum oleoresin. *Journal of Animal Science*, 102: skae014.

Borges, D. G. L and Borges, F. A. 2016. Plants and their medicinal potential for controlling gastrointestinal nematodes in ruminants. *Nematoda*, 3.

Brenes, A and Roura, E. 2010. Essential oils in poultry nutrition: main effects and modes of action. *Animal Feed Science and Technology*, 158(1-2):1-14.

Cani, P. D and Van Hul, M. 2015. Novel opportunities for next-generation probiotics targeting metabolic syndrome. *Current Opinion in Biotechnology*, 32:21-27.

Cardona, F., Andrés-Lacueva, C., Tulipani, S., Tinahones, F. J., Queipo-Ortuño, M. I. 2013. Benefits of polyphenols on gut microbiota and implications in human health. *The Journal of Nutritional Biochemistry*, 24(8):1415-1422.

Choi, M. S., Kim, J. K., Kim, D. H., Yoo, H. H. 2019. Effects of gut microbiota on the bioavailability of bioactive compounds from Ginkgo leaf extracts. *Metabolites*, 9(7):132.

Cummings, J. H and Macfarlane, G. T. 1997. Role of intestinal bacteria in nutrient metabolism. JPEN *Journal of Parenteral and Enteral Nutrition*, 21(6):357-365.

da Silva, C. A., Dias, C. P., Callegari, M. A., Romano, G. S., de Souza, K. L., Jacob, D. V., Ulbrich, A. J., Goossens, T. 2022. Phytogenics and encapsulated sodium butyrate can replace antibiotics as growth promoters for lightly weaned piglets. *PLoS One*, 17(12).

Dadi, T. H., Vahjen, W., Zentek, J., Melzig, M. F., Granica, S., Piwowarski, J. P. 2020. Lythrum salicaria L. herb and gut microbiota of healthy post-weaning piglets: Focus on prebiotic properties and formation of postbiotic metabolites in ex vivo cultures. *Journal of Ethnopharmacology*, 261:113073.

Del Rio, D., Rodriguez-Mateos, A., Spencer, J. P. E., Tognolini, M., Borges, G., Crozier, A. 2013. Dietary (poly)phenolics in human health: Structures, bioavailability, and evidence of protective effects against chronic diseases. *Antioxidants and Redox Signaling*, 18(14):1818-1892.

Delzenne, N. M., Neyrinck, A. M., Cani, P. D. 2011. Modulation of the gut microbiota by nutrients with prebiotic properties: Consequences for host health in the context of obesity and metabolic syndrome. *Microbial Cell Factories*, 10(Suppl 1).

Dinan, T. G and Cryan, J. F. 2017. Gut instincts: Microbiota as a key regulator of brain development, ageing and neurodegeneration. *The Journal of Physiology*, 595(2):489-503.

Ding, H., Dong, J., Wang, Y., Huang, Q., Xu, J., Qiu, Z., Yao, F. 2023. Ginsenoside Rb1 interfered with macrophage activation by activating PPARγ to inhibit insulin resistance in obesity. *Molecules*, 28(7):3083.

Espín, J. C., González-Sarrías, A., Tomás-Barberán, F. A. 2017. The gut microbiota: A key factor in the therapeutic effects of (poly)phenols. *Biochemical Pharmacology*, 139:82-93.

Flint, H. J., Scott, K. P., Louis, P., Duncan, S. H. 2012. The role of the gut microbiota in nutrition and health. *Nature Reviews Gastroenterology & Hepatology*, 9(10):577-589.

Fritsch, D. A., Jackson, M. I., Wernimont, S. M., Feld, G. K., Badri, D. V., Brejda, J. J., Cochrane, C. Y., Gross, K. L. 2023. Adding a polyphenol-rich fiber bundle to food impacts the gastrointestinal microbiome and metabolome in dogs. *Frontiers in Veterinary Science*, 9:1039032.

Fu, Y., Wang, Y., Gao, H., Li, D., Jiang, R., Ge, L., Xu, K. 2021. Associations among dietary omega-3 polyunsaturated fatty acids, the gut microbiota, and intestinal immunity. *Mediators of Inflammation*, 2021:8879227.

Gibson, G. R., Scott, K. P., Rastall, R. A., Tuohy, K. M., Hotchkiss, A., Dubert-Ferrandon, A., Delzenne, N. M., et al. 2017. Dietary prebiotics: Current status and new definition. *Food Science and Technology Bulletin: Functional Foods*, 7(1):1-19.

Gill, S. K., Haskard, C. A., Lin, M. B., Gan, R. Y. 2018. Nutraceuticals, gut microbiota, and mucosal permeability—a new triad in the management of intestinal diseases. *Nutrients*, 10(11):1714.

Guo, S., Gillingham, T., Guo, Y., Wang, D., Yao, M. 2020. Dietary saponins improve animal health by modulating gut microbiota. *Journal of Animal Science and Biotechnology*, 11(1):1-8.

Holscher, H. D. 2017. Dietary fiber and prebiotics and the gastrointestinal microbiota. *Gut Microbes*, 8(2):172-184.

Hooper, L. V and Gordon, J. I. 2001. Commensal host-bacterial relationships in the gut. *Science*, 292(5519):1115-1118.

Hyldgaard, M., Mygind, T., Meyer, R. L. 2012. Essential oils in food preservation: mode of action, synergies, and interactions with food matrix components. *Frontiers in Microbiology*, 3:12.

Kawabata, K., Yoshioka, Y., Terao, J. 2019. Role of intestinal microbiota in the bioavailability and physiological functions of dietary polyphenols. *Molecules*, 24(2):370.

Koudoufio, M., Desjardins, Y., Feldman, F., Spahis, S., Delvin, E., Levy, E. 2020. Insight into polyphenol and gut microbiota crosstalk: Are their metabolites the key to understanding protective effects against metabolic disorders? *Antioxidants*, 9(10):982.

Lang, A., Salomon, N., Wu, J. C., Kopylov, U., Lahat, A., Har-Noy, O., Ching, J. Y., Cheong, P. K., Avidan, B., Gamus, D., Kaimakliotis, I., Eliakim, R., Ng, S. C., Ben-Horin, S. 2015. Curcumin in combination with mesalamine induces remission in patients with mild-to-moderate ulcerative colitis in a randomized controlled trial. *Clinical Gastroenterology and Hepatology*, 13(8):1444-1449.e1.

Laroche, N., Grimm, P., Julliand, S., Sorci, G. 2023. Disrupting hindgut microbiota through the di*et al*ters strongyles infections in horses. *Journal of Equine Veterinary Science*, 124:104350.

Larrosa, M., Luceri, C., Vivoli, E., Pagliuca, C., Lodovici, M., Moneti, G., Espín, J. C. 2009. Polyphenol metabolites from colonic microbiota exert anti-inflammatory activity on different inflammation models. *Molecular Nutrition & Food Research*, 53(8):1044-1054.

LeBlanc, J. G., Milani, C., de Giori, G. S., Sesma, F., van Sinderen, D., Ventura, M. 2013. Bacteria as vitamin suppliers to their host: a gut microbiota perspective. *Current Opinion in Biotechnology*, 24(2):160-168.

Lee, S. H and Jeon, Y. J. 2013. Anti-diabetic effects of brown algae derived phlorotannins, marine polyphenols through diverse mechanisms. *Fitoterapia*, 86:129-136.

Liu, C. F., Tseng, K. C., Chiang, S. S., Lee, B. H., Hsu, W. H., Pan, T. M. 2011. Immunomodulatory and antioxidant potential of Lactobacillus exopolysaccharides. *Journal of the Science of Food and Agriculture*, 91(12):2284-2291.

Lozupone, C. A., Stombaugh, J. I., Gordon, J. I., Jansson, J. K., Knight, R. 2012. Diversity, stability and resilience of the human gut microbiota. *Nature*, 489(7415):220-230.

Macfarlane, G. T and Macfarlane, S. 2011. Fermentation in the human large intestine: its physiologic consequences and the potential contribution of prebiotics. *Journal of Clinical Gastroenterology*, 45.

Min, B. R., Wang, W., Pitta, D. W., Indugu, N., Patra, A. K., Wang, H. H., Abrahamsen, F., Hilaire, M., Puchala, R. 2024. Characterization of the ruminal microbiota in sheep and goats fed different levels of tannin-rich Sericea lespedeza hay. *Journal of Animal Science*, ska198.

Morrison, D. J and Preston, T. 2016. Formation of short chain fatty acids by the gut microbiota and their impact on human metabolism. *Gut Microbes*, 7(3):189-200.

Nani, B. D., Lima, S. F., Figueiredo, R. C., Borges, M. C., Costa, M. C. 2019. Gut microbiota and probiotics in colorectal cancer. *World Journal of Gastroenterology*, 25(43):6445-6466.

O'Keefe, S. J. D. 2016. Diet, microorganisms and their metabolites, and colon cancer. *Nature Reviews Gastroenterology & Hepatology*, 13(12):691-706.

Oh, S., Shintani, R., Koike, S., Kobayashi, Y. 2017. Ginkgo fruit extract as an additive to modify rumen microbiota and fermentation and to mitigate methane production. *Journal of Dairy Science*, 100(3):1923-1934.

Ozdal, T., Sela, D. A., Xiao, J., Boyacıoğlu, D., Chen, F., Capanoglu, E. 2016. The reciprocal interactions between polyphenols and gut microbiota and effects on bioaccessibility. *Nutrients*, 8(2):78.

Palmer, C., Bik, E. M., DiGiulio, D. B., Relman, D. A., Brown, P. O. 2007. Development of the human infant intestinal microbiota. *PLoS Biology*, 5(7).

Pandey, K. B and Rizvi, S. I. 2009. Plant polyphenols as dietary antioxidants in human health and disease. *Oxidative Medicine and Cellular Longevity*, 2(5):270-278.

Park, G., Kim, S., Lee, W., Kim, G., Shin, H. 2024. Deciphering the impact of defecation frequency on gut microbiome composition and diversity. *International Journal of Molecular Sciences*, 25(9):4657.

Pérez-Jiménez, J., Neveu, V., Vos, F., Scalbert, A. 2010. Identification of the 100 richest dietary sources of polyphenols: an application of the Phenol-Explorer database. European Journal of Clinical Nutrition, 64 Suppl 3.

Popowski, D., Zentek, J., Piwowarski, J. P., Granica, S. 2022. Gut microbiota of pigs metabolizes extracts of Filipendula ulmaria and Orthosiphon aristatus—Herbal remedies used in urinary tract disorders. *Planta Medica*, 88(3-04):254-261.

Possemiers, S., Bolca, S., Verstraete, W., Heyerick, A. 2011. The intestinal microbiome: a separate organ inside the body with the metabolic potential to influence the bioactivity of botanicals. *Fitoterapia*, 82(1):53-66.

Qin, J., Li, R., Raes, J., Arumugam, M., Burgdorf, K. S., Manichanh, C., Nielsen, T., et al. 2010. A human gut microbial gene catalogue established by metagenomic sequencing. *Nature*, 464(7285):59-65.

Radjabzadeh, D., Khodorova, A. S. M., Kraaij, R. D., Zhernakova, M. T. L., Meurs, C. J. M. van, van Meurs, C. J. M. 2020. Diversity, compositional and functional differences between the human gut microbiota of children and adults. *Scientific Reports*, 10:14064.

Ridlon, J. M., Harris, S. C., Bhowmik, S., Kang, D. J., Hylemon, P. B. 2016. Consequences of bile salt biotransformations by intestinal bacteria. *Gut Microbes*, 7(1):22-39.

Rinninella, E., Cintoni, M., Raoul, P., Lopetuso, L. R., Scaldaferri, F., Pulcini, G., Miggiano, G. A. D., Gasbarrini, A., Mele, M. C. 2019. Food components and dietary habits: keys for a healthy gut microbiota composition. *Nutrients*, 11(10):2393.

Samsami-Kor, M., Daryani, N. E., Asl, P. R., Hekmatdoost, A. 2015. Anti-inflammatory effects of resveratrol in patients with ulcerative colitis: A randomized, double-blind, placebo-controlled pilot study. *Archives of Medical Research*, 46(4):280-285.

Scalbert, A., Manach, C., Morand, C., Rémésy, C., Jiménez, L. 2005. Dietary polyphenols and the prevention of diseases. *Critical Reviews in Food Science and Nutrition*, 45(4):287-306.

Sekirov, I., Russell, S. L., Antunes, L. C., Finlay, B. B. 2010. Gut microbiota in health and disease. *Physiological Reviews*, 90(3):859-904.

Selma, M. V., Beltrán, D., García-Villalba, R., Espín, J. C., Tomás-Barberán, F. A. 2014. Description of urolithin production capacity from ellagic acid of two human intestinal Gordonibacter species. *Food & Function*, 5(8):1779-1784.

Selma, M. V., Espín, J. C., Tomás-Barberán, F. A. 2009. Interaction between phenolics and gut microbiota: Role in human health. *Journal of Agricultural and Food Chemistry*, 57(15):6485-6501.

Singh, B., Mal, G., Sharma, D., Sharma, R., Antony, C. P., Kalra, R. S. 2020. Gastrointestinal biotransformation of phytochemicals: Towards futuristic dietary therapeutics and functional foods. *Trends in Food Science & Technology*, 106:64-77.

Singh, R., Chandrashekharappa, S. R., Bodduluri, S. R., Baby, B. V., Hegde, B., Kotla, N. G., Hiwale, A. A., Saiyed, T., Patel, P., Vijay-Kumar, M., et al. 2019. Enhancement of the gut barrier integrity by a microbial metabolite through the Nrf2 pathway. *Nature Communications*, 10:89.

Smith, E. A and Macfarlane, G. T. 1998. Enumeration of amino acid fermenting bacteria in the human large intestine: effects of pH and starch on peptide metabolism and dissimilation of amino acids. FEMS Microbiology Ecology, 25(4):355-368.

Stevenson, D. E and Hurst, R. D. 2007. Polyphenolic phytochemicals—just antioxidants or much more? *Cellular and Molecular Life Sciences*, 64(22):2900-2916.

Tabatabaeizadeh, S. A., Tafazoli, N., Ferns, G. A., Avan, A., Ghayour-Mobarhan, M. 2018. Vitamin D, the gut microbiome, and inflammatory bowel disease. *Journal of Research in Medical Sciences*, 23:75.

Tilg, H and Kaser, A. 2011. Gut microbiome, obesity, and metabolic dysfunction. *The Journal of Clinical Investigation*, 121(6):2126-2132.

Tomas-Barberan, F. A., Selma, M. V., Espín, J. C. 2016. Interactions of gut microbiota with dietary polyphenols and consequences to human health. *Current Opinion in Clinical Nutrition and Metabolic Care*, 19(6):471-476.

Topping, D. L and Clifton, P. M. 2001. Short-chain fatty acids and human colonic function: roles of resistant starch and nonstarch polysaccharides. *Physiological Reviews*, 81(3):1031-1064.

Vargas-Mendoza, N., Madrigal-Santillán, E., Morales-González, A., Esquivel-Soto, J., Esquivel-Chirino, C., Morales-González, J. A. 2019. Polyphenols as potential therapeutic agents for chronic diseases: a review. *Current Pharmaceutical Design*, 25(22):2372-2392.

Walsh, J., Griffin, B. T., Clarke, G., Hyland, N. P. 2018. Drug-gut microbiota interactions: implications for neuropharmacology. *British Journal of Pharmacology*, 175(24):4415-4429.

Williamson, E. M. 2001. Synergy and other interactions in phytomedicines. *Phytomedicine*, 8(5):401-409.

Williamson, G and Clifford, M. N. 2017. Role of the small intestine, colon, and microbiota in determining the metabolic fate of polyphenols. *Biochemical Pharmacology*, 139:24-39.

Yonekura, L and Nagao, A. 2007. Intestinal absorption of dietary carotenoids. *Molecular Nutrition & Food Research*, 51(1):107-115.

Zaitsev, V. V., Bogolyubova, N. V., Zaitseva, L. M., Emelyanova, I. S., Korotkiy, V. P., Ryzhov, V. A. 2022. Effect of dietary supplements Farmatan™ and Pine Tree Energy on the regulation of ruminal digestion and microbiocenosis of lactating black and white cows. *Basrah Journal of Agricultural Sciences*, 35(2):99-109.

Zhang, X. X., Lv, Q. B., Yan, Q. L., Zhang, Y., Guo, R. C., Meng, J. X., Ma, H., Qin, S. Y., Zhu, Q. H., Li, C. Q., Liu, R., Liu, G., Li, S. H., Sun, D. B., Ni, H. B. 2022. A catalog of over 5,000 metagenome-assembled microbial genomes from the Caprinae gut microbiota. *Microbiology Spectrum*, 10(6).

Zhen, J., Zhou, Z., He, M., Han, H., Lv, E., Wen, P., Liu, X., Wang, Y., Cai, X., Tian, J., Zhang, M., Kang, X. 2023. The gut microbial metabolite trimethylamine N-oxide and cardiovascular diseases. *Frontiers in Endocrinology*, 14:1085041.

Chapter 13

Microbiota in Reproductive Health

Rajesh Kumar[1,*]
H. Chauhan[2]
Sushant Srivastava[1]
and Pramod Kumar[3]

[1]Department of Veterinary Gynaecology and Obstetrics, College of Veterinary Science and Animal Husbandry, Acharya Narendra Deva University of Agriculture and Technology, Ayodhya, UP, India
[2]Department of Veterinary Microbiology, COSc, Kamdhenu University Sardarkrushinagar, Dantiwada, Banaaskantha, Gujarat
[3]Department of Veterinary Physiology and Biochemistry, College of Veterinary Science and Animal Husbandry, Acharya Narendra Deva University of Agriculture and Technology, Ayodhya, UP, India

Abstract

Microbiota plays vital roles in the host, including immunological, neuroendocrine, and metabolic programming and regulation. The host's health is significantly impacted in the short and long term by the proper acquisition of the microbiota throughout early life and its modifications throughout life. The various niches of microbiota are not a separate entity rather it is network of interconnected communities that are continuously exchanging. The microbiotas can reach to various parts of reproductive tract viz. vagina, cervix, uterus, uterine tubes as well as follicular fluid via various niches. The surroundings environment and mode of delivery shapes the microbial community of newborn and plays many vital roles like protection against pathogens and maintenance of reproductive health and Socio-chemical signaling etc. Thus, the dysbiosis can results battery of infertility problems. In addition, modulation of reproductive tract microbiota can be used to mitigate infertility issues like vaginitis, metritis, endometritis etc in animals. Moreover, comprehensive understanding of reproductive microbiota and its role in modulating physiological functions will aid in the development of management and therapeutic strategies to augment reproductive efficiency.

Keywords: microbiota, dysbiosis, reproductive health, infertility

---

* Corresponding Author's Email: drrajesh25@gmail.com

In: Microbiomes in Animal Nutrition, Health, and Disease
Editor: Tanmoy Rana
ISBN: 979-8-89530-425-9

Introduction

Animals are holobionts, or "superorganisms," made up of both the cells of the corresponding host species and the microorganisms known as the "microbiota," which live on their skin and mucosal surfaces. There is currently no agreed-upon meaning for the similarly related term "microbiome," which some writers use interchangeably with "microbiota." For instance, according to The Human Microbiome Project (Peterson et al., 2009), the human microbiome is the whole of the bacteria, viruses, eukaryotes, and other microorganisms that coexist with the human body. Some writers, however, interpret this term as the entirety of the genetic material that the microbiota contains, while others also include their metabolites (Cho and Blaser, 2012). Numerous alternative interpretations of the term "microbiome" have been put forth based on varying ecological, host, and scientific viewpoints (Berg et al., 2020). Although the phrases are frequently used synonymously, from now on we shall exclusively use "microbiome" to refer to the aggregate genomes of all the microorganisms living in a host. Over the last few decades, it has been clear that the microbiota plays vital roles in the host, including immunological, neuroendocrine, and metabolic programming and regulation (Power et al., 2017; Mills et al., 2019). Therefore, the host's health is significantly impacted in the short and long term by the proper acquisition of the microbiota throughout early life and its modifications throughout life (Milani et al., 2017). Moreover, dysbiosis, or an imbalance in the microbiota's composition, upsets homeostasis, triggering an inflammatory reaction and changing the organism's physiological metabolic profile (Turnbaugh et al., 2006). The majority of methods used in the study of the microbiota have been culture-based ones, ranging from traditional approaches and procedures to the introduction of novel substrates and growth environments known as "culturomics."This phrase refers to expanding a biological sample's repertoire of cultivable bacteria by the diversity of culture conditions and identification using matrix-assisted laser desorption/ionization mass spectrometry (MALDI-TOF). The proof of microbial viability is crucial in these methods. Conversely, the culture-independent methods (which include those that employ next-generation sequencing [NGS] platforms) that have been employed thus far are restricted to the identification of microbial nucleic acid sequences and are mostly utilized in relation to the prokaryotic elements of the microbiota. In this context, nearly every part of the female and male reproductive tracts, as well as several other mammalian host regions that were previously believed to be sterile, have been shown to contain microbial DNA (Reid et al., 2015). The phrase "reproductive microbiome" was thus created as a result. The presence of an extra-vaginal microbiota in the female reproductive tract is, at the very least, a contentious topic. It is widely known that the mammalian vagina has a site-specific microbiota that plays important functions in genital and reproductive health. Numerous findings have been reached without taking into account the NGS approach's inherent biases and technological limitations. Nonetheless, this subject is expected to get more attention from researchers due to its significant ramifications. In this review, we address the relationship between reproductive events and the function of the microbiotas of other body regions and present the current state of knowledge, limitations, and gaps in our understanding of the female reproductive microbiota, with a focus on the bacteria of the extra-vaginal genital tract. The majority of current research on the female reproductive tract's microbiota has been on humans. Based on the DNA discovered in samples of the female vaginal tract, many of these studies have only described the major bacterial phyla and species. More research on humans and animals is required to understand the shifting composition of this microbiota and, in particular, its functional activities. A few investigations examining various animal species have been published in the last ten years ((Santos et al., 2011; Jeon et al., 2015; Miller et al., 2017; Comizzoli et al., 2019; Galvao et al., 2019; Barba et al., 2020; Quereda et al., 2020[a]; Ong et al., 2021; Srinivasan et al., 2021; Adnane et al., 2022). There will be significant effects of this kind of knowledge on enhancing female health and reproductive success.

Origin of Reproductive Tract Microbiota

In a strict ecological sense, the various niches of microbiota are not separate entity rather it is a network of interconnected communities that are continuously exchanging (Neckovic et al., 2020). The microbiotas can reach to reproductive tract via various niches. The surroundings environment and mode of delivery shape the microbial community of newborns (Dominguez-Bello et al., 2010). In humans, *Lactobacillus, Prevotella,* or *Sneathia* spp. predominate in the general microbiota of the newborn delivered vaginally, which is comparable to the maternal vaginal microbiota. Like the microbiota of mother skin, the general microbiome of neonates delivered by cesarean section is mostly made up of *Staphylococcus, Corynebacterium*, and *Propionibacterium* species ((Dominguez-Bello et al., 2010). Consequently, an individual's microbial load and diversity are influenced by their early exposure to the maternal microbiome (Mazmanian et al., 2005; Dominguez-Bello et al., 2010). It is believed that bacteria like *Bacteroides* and *Enterobacteriaceae* that cause reproductive disorders in cattle are derived from the environment or the stomach (Kraipowich et al., 2000; Wexler, 2007). Furthermore, the gastrointestinal tract is the source of the vaginal microbiota (Laguardia-Nascimento et al., 2015), while other studies concluded that methanogen species are present in the vaginal microbial population and that, as a result, the vaginal microbiome aids in the formation of the digestive tract microbiota (Deng et al., 2019). Another significant potential pathway for uterine pathogen infection is the hematogenous route (Galvao et al., 2019).

Microbiota of Reproductive Tract at Different Niches

Vaginal Microbiota

The microbiota of vaginal vault is site (France et al., 2022) and species-specific and the composition depends upon a battery of factors like species, age, stage in the estrous (Quereda et al., 2020[a]) or menstrual cycle, diseases (Kudo et al., 2021), hormone treatments, stress (Stumpf et al., 2013; Yildirim et al., 2014; Miller et al., 2016; Uchihashi et al., 2015; Chen et al., 2020; France et al., 2022), parity (primiparous *vs* multiparous) (Quadros et al., 2020), postpartum periods (Kudo et al., 2021) and other external factors like presence spermatozoa in the vagina which has its peculiar microbiota (Baud et al., 2019; Cojkic et al., 2021; Molina et al., 2021). More than 70% of the vaginal microbiota of humans belongs to the genus Lactobacillus and this appears to unique to humans as the Lactobacillus population is far less (<1%) in other non-human primates and other mammals (Miller et al., 2016). By virtue of abundance of Lactobacillus in human vagina, the vaginal pH of humans is more acidic (pH 4.5) compared with vagina other mammals (pH5.4-7.8). Moreover, these pH values are reaches at nadir during highest estrogen dominance (Miller et al., 2016). Under estrogen dominance exfoliated and lysed epithelial cells produce more glycogen, which is ultimately used by Lactobacillus to produce lactic acid (Amabebe et al., 2018), the latter is responsible for low vaginal pH (O'Hanlon et al., 2013). The lactobacillus protects the vaginal mucosa from pathogenic microorganisms by adhering to vaginal mucosa (Clabaut et al., 2021). Unlike in humans, the protection against invading pathogens is provided by other bacteria, which has species specific interactions with immune systems (Heil et al., 2019; Koedooder et al., 2019; Lyman et al., 2019) but such interactions yet not completely deciphered. Differences in genital microbiota among healthy and diseased subjects might be due difference in mounted immunologic response (Li et al., 2022). In short, the main function of vaginal microbiota in human and animals are protection against microbial challenges and boost immunologic defense system, which in turn a important prerequisite for endometrial health, fertilization, implantation and pregnancy success (Bjurstr€om et al., 1993; Groppetti et al., 2012; Duijkeren et al., 2020; Golinska et al., 2021; Torres Luque et al., 2021; Zhang et al., 2021; France et al., 2022). In addition, certain bacterial metabolites in vagina namely benzopyrene and glycerophospholipids are positively correlated with success rate of implantation in women (Fu et al., 2020). Indeed, prostaglandin (PG) plays an important role in process of implantation; Benzopyrene being a Cyclooxygenase 2 (COX-2)-inhibitor, modulates PG biosynthesis (Fu et al., 2020). Furthermore, Glycerophospholipids being a precursor of arachidonic acid, which in turn act as a limiting factor PG biosynthesis (Smith et al., 1996; Zerani et al., 2021). Furthermore, vaginal microbiota may facilitate chemical communication (via pheromones) between opposite sexes of same species via production of odoriferous compounds (pheromones) in livestock (Ungerfeld et al., 2005; Srinivasan et al., 2021) and wild animals (Theis et al., 2013; Comizzoli et al., 2021).

In conclusion, vaginal dysbiosis has been associated with sub-fertility in mammals (Wang et al., 2016; Wang et al., 2017; Fu et al., 2020; Golinska et al., 2021; Ichiyama et al., 2021; Kiefer et al., 2021; Liang et al., 2022; Moreno et al., 2022). The vaginal microbiota differ significantly during different phases of estrus cycle (follicular and luteal phase) (Wang et al., 2019) and it is important to recognize vaginal microbiota to improve pregnancy success following artificial insemination (Srinivasan et al., 2021) and to predict the success of assisted reproductive technologies (ART) (Schoenmakersa et al., 2020; Diaz-martínez et al., 2021; Tsonis et al., 2021; Patel et al., 2022). Moreover, intravaginal administration of microflora has been proposed as a new approach to improve reproductive efficiency (Appiah et al., 2020).

Cervical Microbiota

The cervix act as an anatomical barriers prevent entry of pathogens into uterine milieu (Sheldon and Dobson, 2004; Azawi, 2008). Moreover, cervical mucus prevents ascending infections from lower reproductive tract (Sheldon and Dobson, 2004). Thus, in order to forecast the future fertility after parturition, cervix needs to be evaluated separately (Deguillaume et al., 2012). The cervix also called as "neck of the womb," being the part of uterus, the health of farmer included the anatomical and functional soundness of cervix as well (Appiah et al., 2020). The cervical microbiota differs between cows with the history of abortion than those of cows without any record of abortion (Galvao, 2018). In addition, profile of cervical microbiota is altered in cows suffering from uterine infections (Appiah et al., 2020). Furthermore, alterations in cervical microbiota can be used as biomarker to spot development of cervical cancers in woman (Curty et al., 2020). Similarly, nature of cervico-vaginal microbiota may be used to assess pathology of upper reproductive tract (Chen et al., 2017).

Uterine Microbiota

In contrast to previous view, recent evidence suggest that uterus is not sterile and has site specific microbiota (Chen et al., 2017; Moreno et al., 2017; Pelzer et al., 2017; Baker et al., 2018), which still prevails the organ even during pregnancy (Karstrup et al., 2017; Moore et al., 2017). The uterine microbes have a cross-talk with endometrial cells (Appiah et al., 2020) and have regulating effects on local endometrial immunity (Agostinis et al., 2019), the organisms mainly involves are *Fusobacteria specie*s, and *Prevotella* spp (Karstrup et al., 2017; Moore et al., 2017). Unlike in gut and vagina, the microbes are far less abundant in the uterus, and only small proportion of this load contributes to postpartum uterine infections (Appiah et al., 2020), but this view is not supported by few previous studies (Santos et al., 2011; Santos et al., 2012), which might be due to difference in technique used as well breed of cattles. Regardless of health status of cows, uterine microbiota has shown more diversity and is comprised of *Actinobacteria, Bacteroidetes*, *Firmicutes*, *Fusobacteria*, *Proteobacteria*, *Tenericutes*, and Unassigned taxa (Santos et al., 2012). The greater diversity of uterine microbiota might be due to breach in the barrier (loss of pregnancy seal) between interior to exterior environment at parturition, causes dispersal and mixing of vaginal and uterine microbiota (Miranda-CasoLuengo et al., 2019). Early postpartum uterus contaminated with variety of microbes (Galvao, 2018), which may be classified on the basis of their pathogenic potential as opportunistic contaminants, potential pathogens and uterine pathogens (Williams et al., 2005). In addition to vagina, skin, gut and environment might be a potential source of uterine microbiota. Thus, uterine microbes have a definite bearing on uterine health, and have a pronounced effect of subsequent fertility (Ryan et al., 2020). Furthermore, previous studies suggest a link between uterine health, luteal function, fertility sub index and reproductive function in dairy cattle (Canadas et al., 2020), whereas no such links were recorded by Rojas et al., (2020), and the difference in results was might be due to difference in study methods, breed and genotype of the host (Appiah et al., 2020). Some reports suggest improved fertility with *alpha-hemolytic streptococci* at 7 days postpartum (DPP) (Gilbert et al., 2016) and *E. coli* during lactation (LeBlanc et al., 2002) but results are not consistent (de Boer et al., 2015). In addition to varying effects of microbes on uterine health, host also contributed the same via different mechanisms like avoidance, tolerance and resistance towards the potential pathogens (Sheldon et al., 2020). In order to prevent uterine infections, we should breed a more resilient cattle, who has strong immunity and able to tolerate uterine pathogens fairly well (Kinig and May, 2019).

Oviductal Microbiota

Important events such as sperm capacitation, fertilization, and early embryo development take place in the oviduct within a complex signaling cascade involving Janus kinase/signal transducer and activator of transcription (JAK/STAT) and ERK pathways (Gentilini et al., 2007; Jin et al., 2017). The information available about the microorganisms that may inhabit or transit the oviduct is scarce, although interesting interactions between gametes and non-pathogenic oviductal bacteria have been proposed. Although this review does not deal with the microbiota of the male reproductive system, we should mention that semen usually contains a rich and diverse microbiota (Cojkic et al., 2021; Farahani et al., 2021; Gachet et al., 2022; Koziol et al., 2022). Consequently, fertilization may not occur under sterile conditions (Stsepetova et al., 2020). The bacterial communities that have been described in the human oviduct seem to reflect those that are particularly abundant in human semen (e.g., *Staphylococcus* spp., *Enterococcus* spp., *Cutibacterium* spp.) or in the human vagina (*Lactobacillus* spp.) (Pelzer et al., 2018). In addition, bacterial profiles appear to change between the fimbria and the proximal oviduct (Brewster et al., 2022), the isthmus and ampulla, and the right and left oviduct (Pelzer et al., 2018), but no correlation with ovarian follicular or luteal status has been detected so far (Pelzer et al., 2018). Additionally, the oviductal microbiota can be affected by factors such as hormone treatments or menopause (Pelzer et al., 2018; Brewster et al., 2022). Further studies are required to establish the reproductive roles of oviduct microorganisms in humans and other mammalian species.

Follicular Fluid Microbiota

The oocyte's development environment has a direct impact on its quality and fertilisation potential. There has been inconsistent information on the presence of a microbiome at this location. In addition to identifying bacterial nucleic acids (including *Lactobacillus* spp., *Actinomyces* spp., and *Cutibacterium* spp.) and cells in human follicular fluid (FF), several scientists have also reported variations between the right and left ovaries of the same host (Pelzer et al., 2011; Pelzer et al., 2013[a]; Pelzer et al., 2013[b]; Borges et al., 2020). On the other hand, a well-controlled research conducted recently was unable to identify any particular bacteria in ovarian cystic fluid, irrespective of the kind of cyst and endometriosis (Oishi et al., 2022). It is yet need to determine if an FF-related microbiome exists, although its makeup has been successfully connected to pregnancy outcomes. Both healthy women and those with reproductive issues showed a positive link between the presence of *Lactobacillus* spp. in the FF and the pregnancy rate following IVF and embryo transfer [Pelzer et al., 2013[a]]. The generation of flavonoids and the suppression of host eicosanoid secretions by lactobacilli, which lower oxidative stress and inflammation and increase oocyte competence, maybe the cause of this effect (Owens et al., 2020). On the other hand, oocyte quality may be compromised by certain bacteria present in the FF (Pelzer et al., 2013[b]). Oocyte DNA fragmentation was brought on by the in vitro co-cultivation of mouse oocytes lacking in *Cutibacterium acnes* or *Lactobacillus gasseri*, which had previously been isolated from FF (Pelzer et al., 2013[a], Campisciano et al., 2020). This might be the cause of the poor success rate of IVF. It was previously noted that the bacteria in porcine FF were likewise harmful to oocyte maturation because they secreted a compound that prevented follicle-stimulating hormone (FSH) from attaching to its receptor (Sluss and Reichert, 1983). Further research is necessary to validate this discovery and identify the specific bacterial species that may be responsible for this unfavorable result, as these bacteria have not yet been discovered. When uterine infection happens, extracellular signal-regulated kinase/nuclear factor-kB (ERK/NF-kB) pathway stimulates Toll-like receptor 4 (TLR4) and other cytokines like tumour necrosis factor a (TNF-a), interleukins IL-1b, and IL-6, and activates intracellular signaling cascades such as the mitogen-activated protein kinase (MAPK) pathway (Wu et al., 2016; Zhao et al., 2019). Overstimulation of these pathways can cause FF to produce inflammatory mediators and increase oxidative stress in the follicular environment (Gilbert, 2019), both of which can cause an inflammatory response that hinders oocyte development. Other diseases of the female genital tract, including endometritis, have also been linked to lipopolysaccharides (LPS) found in the outer membrane of Gram-negative bacteria (Wu et al., 2016; Magata and Shimizu, 2017). Increased follicular apoptosis and atresia are brought on by the presence of LPS in the FF (Besnard et al., 2001), most likely as a result of LPS's suppression of steroidogenic enzyme transcription, which includes beta-catenin (Ferranti et al., 2021), P450, and CYP17 (Magata et al., 2014). When cows receive subcutaneous injections of LPS, both the FF concentration of LPS in the dominant follicles and the FF level of estradiol rise (Ferranti et al., 2021). Rats' progesterone response to follicular stimulation hormone (FSH) is, however, diminished by LPS (Besnard et al., 2001). Furthermore, LPS itself has an impact on the capability of oocytes to mature (Magata and Shimizu, 2017), disrupting processes such as meiotic resumption and cytoplasmic maturation, cytoskeletal dynamics, and oxidative stress (Magata and Shimizu, 2017). It also modifies the epigenetic markings of oocytes in sheep (Zhao et al., 2017) and cows (Rasekhi et al., 2020). Blastocystrates and the quantity of cells at the trophoblast stage are reduced (Magata and Shimizu, 2017), even in oocytes taken from heifers challenged in vivo with LPS and in vitro matured (IVM), IVF, and in vitro cultured (IVC) (Rincon et al., 2019). Embryo development is similarly impacted. Regarding the impact of the metabolites generated by the bacteria present in the FF, some conjecture is possible. For instance, increased hyaluronan concentrations have been linked to increased granulosa cell death (Baka et al., 2006). Certain hyaluronidase-producing bacteria that were isolated from human FF, including as *Cutibacterium* spp., *Streptococcus* spp., and *Escherichia coli*, have been linked to cumulus cells of low quality (Pelzer et al., 2012). An alternative theory suggests that extracellular vesicles (EVs) carrying immunomodulatory substances involved in cell-to-cell communication may be released by Gram-negative bacteria (Jan, 2017). These EVs may function as a signal for follicular cells to communicate with one another if they are released in the FF. In fact, it is known that FF contains exosomes carrying bioactive compounds that may enhance follicular and embryonic development (Sohel et al., 2013; da Silveira et al., 2017). How FF-related bacteria connect to follicle growth, oocyte maturation, and reproductive outcomes, yet not completely elucidated.

Factors Affecting Diversity of Reproductive Tract Microbiota

It's interesting to note that between 10 and 35 DPP, cows with subclinical endometritis and cows without uterine inflammation had comparable uterine microbiota; in contrast, the microbiota of the two other groups and the cows with clinical endometritis were not similar in composition (Pascottini et al., 2020). Compared to cows without uterine inflammation and those with subclinical endometritis, cows affected with clinical endometritis had a lower bacterial diversity, as evidenced by a higher prevalence of *Fusobacterium* and *Trueperella* and a lower abundance of *Escherichia, Lactobacillus, Prevotella, Schlegelella, Shigella,* and

Streptococcus (Miranda-CasoLuengo et al., 2019; Pascottini et al., 2020). Compared to cows with clinical endometritis, the latter group had a comparatively higher incidence of *Anaerococcus, Corynebacterium,* and *Staphylococcus*. The probability of developing clinical endometritis after 21 days postpartum is correlated with the microbial population of the uterus and vagina at 7 days postpartum (Miranda-CasoLuengo et al., 2019). Cows with endometritis have a vaginal microbiota at 7 DPP that is greatly enriched in *Bacteroides, Fusobacterium*, and *Helcococcus,* in contrast to cows without uterine inflammation, where *Firmicutes* predominate. The uterine and vaginal microbiota of these animals were compared, and the results showed that at 7 DPP, there was a comparable microbial makeup between the two compartments; however, between 21 and 50 DPP, there was less similarity (Miranda-CasoLuengo et al., 2019). These results demonstrate that, in contrast to cows without uterine inflammation, whose uterine and vaginal microbiota are well established by day 7 postpartum, the genital microbiota of cows at risk of developing endometritis is not fully established during the first month postpartum (Miranda-CasoLuengo et al., 2019).

A marked inflammation of the endometrium and myometrium soon after calving and before day 21 postpartum is indicative of metritis. The vaginal microbiota is relatively less abundant, and *Bacteroides, Porphyromonas*, and *Fusobacterium* are more abundant relative to one another (Sheldon et al., 2008; Galvao et al., 2019). Similarly, compared to cows without metritis, cows with metritis exhibited a uterine microbiota that was noticeably enriched in *F. necrophorum, Porphyromonas levii*, and *Prevotella melaninogenica* (Cunha et al., 2018). Another intriguing bacterium found early postpartum in the majority of cows is *E. coli.* This microbe is critical for the growth of *F. necrophorum*, which causes metritis, whereas the latter is frequently found in conjunction with *T. pyogenes* in endometritis cases (Bicalho et al., 2012; Galvao et al., 2019). According to Brotman et al. (2014), the vaginal microbial community in women with mild to moderate vaginal atrophy is often composed of invasive pathogens, primarily *Streptococcus* and *Prevotella*. When comparing the microbiota of cattle with and without necrotic vulvovaginitis, it is shown that the former have a higher abundance of *Bacateroidetes* and a comparatively higher microbial diversity (Shpigel et al., 2017).

Internal Factors

Presence of different bacterial species in the bovid vaginal tract is believed to impact reproductive outcomes (Laguardia-Nascimento et al., 2015). This variation may elucidate the mechanisms by which certain animals acquire resilience while others get infected to uterine illnesses. These variations are believed to be prevalent in other mammals, including humans (HMP, 2012).

Species

The vaginal microbiota exhibits significant diversity both across different species and within individuals of the same species, leading to disruptions in the control of reproductive hormones (Swartz et al., 2014). Lower estradiol concentration and subsequently delayed ovulation has been reported in endometrioses cattle or cattle subjected to bacterial lipopolysaccharide (LPS) (Sheldon et al., 2002; Williams et al., 2007). The bovine vaginal microbiota has abundant population of Enterococcus spp., Staphylococcus spp., and Streptococcus spp., whereas the ovine vaginal microbiota is mainly dominated by Bacillus spp., Corynebacterium spp., Escherichia spp., Staphylococcus spp., and Streptococcus spp. (Sawyer, 1977; de Ruiz Holgado et al., 1999; Otero et al., 2000; Manes et al., 2010). The vaginal microbiomes of both cows and ewes were primarily consisted of Proteobacteria, Fusobacteria, and Bacteroidetes at the phylum level. At the genus level, Aggregatibacter spp. and Streptobacillus spp. were the most common species (Swartz et al., 2014). Although Lactobacilli are found in 80% of sheep and 90% of cows' population, the overall microbial community is less plentiful, resulting in a nearly neutral vaginal pH. In contrast, women vaginal milieu is acidic in nature due to significant population of Lactobacilli spp. In addition to Lactobacilli spp., cows and ewes have several species in common like Sneathia spp., Porphyromonas spp., and Prevotella spp. (Swartz et al., 2014; Miller et al., 2016). A low ratio of Firmicutes to Bacteroidetes is premonitory sign of postpartum endometritis (Miranda-CasoLuengo et al., 2019).

Breed

The Gir or Gyr cattle is a famous Indian breed of zebuine cattle, is characterized by a high abundance of bacteria and fungus, with a limited presence of archaea (Giannattasio-Ferraz et al., 2019). The Gir vaginal microbiota is characterized by high abundance of *Firmicutes, Bacteroidetes, Proteobacteria,* and *Actinobacteria.* Moreover, among the fungal genera, *Mycosphaerella* and *Cladosporium* were the most commonly identified. Although archaea were present in small numbers, the *Methanobrevibacter* genus was the most prevalent. The vaginal microbiome of Nellore beef cattle is mainly comprised of *Firmicutes, Bacteroidetes, Proteobacteria,* and a portion of unclassified bacteria, which can account for up to 20% (Laguardia-Nascimento et al., 2015). *Mycosphaerella* was the most prevalent fungal genus, whereas *Methanobrevibacter* was the dominating archaeal genus. The vaginal microbiome of Holstein Friesian cow, which is the most common dairy breed in North Africa, Europe, and the USA, is primarily made up of the *Firmicutes, Tenericutes, Proteobacteria,* and *Bacteriodetes* phyla. Furthermore, the *Actinobacteria* and *Spirochaetae* are identified in lesser amounts (Quadros et al., 2020).

Mode of Delivery (Natural birth vs Caesarean Section)

The mode of delivery of offspring (i.e., natural birth or caesarean section) is a significant determinant of the microbial diversity the vaginal milieu (Dominguez-Bello et al., 2010). If the newborn is delivered by cesarean section, its general microbiota closely resembles that of the mother skin microbiota. However, in vaginal delivery (natural birth), its general microbiome is more comparable to the vaginal microbiota (Dominguez-Bello et al., 2010). Pre-existing bacterial communities offer protection against harmful microorganisms that could potentially infect the newborn. The maternal vagina serves as the primary origin of the natural microbiota, although a distinct neonatal microbiota is formed immediately after birth, including microorganisms from various additional sources. Neonates born through caesarean section are more likely to have methicillin-resistant Staphylococcus aureus (MRSA) skin infections (64–82%) compared to neonates born without any problems during natural birth (Dominguez-Bello et al., 2010).

Estrous Cycle and Menstrual Cycle

The taxonomic composition of the vaginal microbiota was different during the follicular and luteal phases in cattle (Quereda et al., 2020[a]). During proestrus and estrus, there are substantially greater levels of estradiol, while metestrus and diestrus are characterized by comparatively higher levels of progesterone. These hormonal variations have significant impacts on the vaginal pH in mammals (Gorodeski et al., 2005; Perry et al., 2008). Microorganisms have high sensitivity to acidic environments. Consequently, it is hypothesized that higher concentrations of estrogen will have an impact on the microbiome in the vaginal environment of some species. The specific impacts will vary depending on the concentration of estradiol. The populations of *Bacteroidetes* spp., *Histophilus somni*, *Actinobacillus seminist*, and unclassified *Fusobacterium* increase when there are lower levels of estrogen in the vaginal environment (during metestrus and diestrus phase) (Messman et al., 2020). When the vaginal mucus has higher levels of estradiol, unclassified *Pasteurellaceae* become the most common microorganisms in the vagina. Similarly, an increase in progesterone levels leads to a proportionate increase in the abundance of vaginal bacteria (De Ruiz Holgado et al., 1999; Laguardia-Nascimento et al., 2015). Furthermore, inseminating a heifer did not lead to a large change in the reproductive tract microbiota when assessed by traditional methods of bacterial culture or metataxonomic (16S rRNA gene) sequencing (Moraes et al., 2024).

In women, lower stability of the bacterial communities was indicated by the increased variety of bacteria and quantity of non-Lactobacillus species in the vagina during menses (Eschenbach et al., 2000). Similarly, Gajer et al., (2012) found that hormone concentrations were correlated with the microbiome, which was the least stable during menses. According to Gajer et al., (2012), peak progesterone production was followed by peak oestrogen secretion, which showed the highest stability. Stability during ovulation and the development and preparation of pregnancy is therefore probably essential for creating the ideal reproductive tract environment. According to Chen et al., (2017), more research has revealed variations in the functional pathways of bacteria throughout the proliferative and secretory phases. Pathways including pyrimidine and purine metabolism were enriched during the proliferative phase, which may have indicated greater bacterial cell multiplication. During the secretory phase, many pathways—including several processes related to the metabolism of amino acids—were enhanced. Furthermore, based on the phase of the menstrual cycle, the abundance of several bacterial species varied in the uterus and vagina (Chen et al., 2017). Therefore, bacterial populations and their corresponding roles are probably changing during the menstrual cycle as a result of the physiological changes that are occurring, particularly menses.

Gestation

Throughout gestation, the vaginal microbiota experiences reduced microbial diversity and an increased presence of archaeal organisms (Laguardia-Nascimento et al., 2015). The vaginal microbial population during pregnancy is less diversified due to the greater concentrations of progesterone, which limit the growth of bacterial species in the vagina (De Ruiz Holgado et al., 1999). The decreased microbial community in the vagina during pregnancy may increase the likelihood of dysbiosis and miscarriage. Similarly, the number of microorganisms in the vagina reduces as the length of the pregnancy grows (Aagaard et al., 2012; Gajer et al., 2012). In contrast to humans, the vaginal microbiota of cattle remains largely consistent during the gestational period (Deng et al., 2019). Nonpregnant heifers exhibited a high presence of

Pasteurellaceae spp. and Fusobacterium spp. in their vagina, while pregnant heifers had a higher occurrence of Pasteurella multocida (Messman et al., 2020).

Postpartum

Significant diversity has been noted in the uterine microbiome of cows without uterine infections during the initial month postpartum (Miranda-CasoLuengo et al., 2019; Pascottini et al., 2020). The uterine microbiota of cows without uterine inflammation is primarily comprised of *Porphyromonas, Bacillus, Schlegelella, Paracoccus,* and *Fusobacterium* (Pascottini et al., 2020). Curiously, the vaginal and uterine microbiomes of cows that did not have inflammation in their uterus during the first 50 days postpartum (DPP) were comparable and had a significant presence of *Firmicutes* spp. (Miranda-CasoLuengo et al., 2019). This observation can be attributed to the fact that shortly after giving birth, the cervical opening is less narrow, causing a blending of the vaginal and uterine environments and the dispersion of these substances throughout the reproductive system. A retrospective analysis comparing the genital microbiome of pregnant cows before giving birth with those that develop endometritis after 21 days postpartum (DPP) and those that maintain a healthy uterus showed that the pre-calving genital microbiome was similar to the vaginal microbiome of healthy cows that did not develop uterine infections after 21 DPP (Miranda-CasoLuengo et al., 2019).

External Factors

Plane of Nutrition

The nutrient composition of the food, particularly the energy content, during the postpartum period was found to impact the microbial community in the uterus of dairy cows, namely around the time of calving (Esposito et al., 2020). Cows fed 80% of their energy requirement experienced inflammation in their uterus, and the most common types of bacteria found in their uterine microbiome were Bacteroidetes and Fusobacteria. A study comparing the uterine microbiome before and after calving in cows that developed metritis and those that did not, found that cows with metritis had a higher presence of Bacteroides and Fusobacteria, and a lower presence of Proteobacteria and Tenericutes, which was significantly different from cows without metritis (Galvao et al., 2019). Hence, nutrition impacts the vaginal microbiome by influencing overall metabolic and immunological activities, consequently influencing the occurrence of dysbiosis and genital infections.

Disorders of the Reproductive System

The cows with subclinical endometritis exhibited a comparable uterine microbiota to cows without uterine inflammation during the period of 10 to 35 days postpartum (DPP), whereas cows with clinical endometritis had a distinct microbiota composition compared to the other two groups (Pascottini et al., 2020). Cows suffering from clinical endometritis exhibited reduced bacterial diversity, with a higher presence of *Fusobacterium* and *Trueperella*, and a lower abundance of *Escherichia coli, Shigella, Lactobacillus, Prevotella, Schlegelella,* and *Streptococcus*, as compared to cows without uterine inflammation and those with subclinical endometritis (Miranda-CasoLuengo et al., 2019; Pascottini et al., 2020). The second group had a higher occurrence of *Anaerococcus, Corynebacterium,* and *Staphylococcus* in comparison to cows suffering from clinical endometritis. The composition of microorganisms in the uterus and vagina at 7 days postpartum (DPP) is linked to the likelihood of developing clinical endometritis by 21 DPP (Miranda-CasoLuengo et al., 2019). The vaginal microbiome of cows with endometritis at 7 days postpartum (DPP) exhibits a significant abundance of *Bacteroides, Helcococcus,* and *Fusobacterium,* in contrast to cows without uterine inflammation, where *Firmicutes* are the major bacterial group. Analysis of the uterine and vaginal microbiomes in these animals revealed that at 7 days postpartum (DPP), the microbial makeup was comparable in both compartments. However, this similarity decreased between 21 and 50 DPP (Miranda-CasoLuengo et al., 2019). These data validate that the genital microbiome of cows who are at risk of developing endometritis is not fully established during the first month after giving birth, unlike in cows without uterine inflammation. In cows without inflammation, the uterine and vaginal microbiomes are well established by day 7 after giving birth (Miranda-CasoLuengo et al., 2019). Metritis, a condition characterized by significant inflammation of the endometrium and myometrium shortly after calving and within the first 21 days after giving birth, is associated with a decreased presence of vaginal microbiome and an increased presence of *Bacteroides, Porphyromonas,* and *Fusobacterium* (Sheldon et al., 2008; Galvão et al., 2019). In addition, cows diagnosed with metritis exhibited a significant increase in the presence of *F. necrophorum, Porphyromonaslevii,* and *Prevotella melaninogenica* in their uterine microbiota, as compared to cows without metritis (Cunha et al., 2018). E. coli is a bacterium that is commonly found in cows shortly after giving birth. Its presence is crucial for the growth of *F. necrophorum*, which can cause tometritis. On the other hand, *F. necrophorum* is often found alongside *T. pyogenes* in cases of endometritis. Women with mild or moderate vaginal atrophy typically have a predominance of invasive pathogens in their vaginal microbial population, particularly *Streptococcus* and *Prevotella* (Brotman et al., 2014). Cattle diagnosed with bovine necrotic vulvovaginitis (BNVV) exhibit a higher microbial diversity and a significant abundance of Bacateroidetes, in comparison to animals not diagnosed with BNVV (Shpigel et al., 2017).

Function of Reproductive Tract Microbiota

Protection against Pathogens and Maintenance of Fertility

Genital microbiota of animals and humans protect against invading pathogens. Lactic acid, which is produced by *Lactobacilli* in women, controls the pH of the vagina and prevents the growth of infections (O'Hanlon et al., 2013). Conversely, mucus, sugar, and proteins secreted by the genital tract are used by symbiotic bacteria as a source of vital nutrients (Tachedjian et al., 2017). Additionally, lactic acid causes the vaginal environment to become more acidic, interfering with intracellular processes and ultimately leading to the removal of microorganisms (Alakomi et al., 2000). The findings of in vitro investigations show that normal lactic acid concentrations inactivate *Neisseria gonorrhoeae*, *Escherichia coli* (*E. coli*), and *Chlamydia trachomatis* (Valore et al., 2002; Graver et al., 2011; Nardini et al., 2016). Protection against Human immunodeficiency virus type 1 (HIV-1) is one of the possible advantages of commensal bacteria in women. It appears that *Lactobacilli* in the vagina shield women against HIV-1 infection when they engage in sexual activity. HIV is less likely to be shed when the amount of free virions is limited by lactic acid and the ensuing biofilm (Taha et al., 1998; Hoang et al., 2020). When HIV-1 is exposed to typical lactic acid concentrations *in vitro*, it becomes permanently inactivated (Aldunate et al., 2013). This might be the way lactobacilli provide vaginal defence against infections.

Reproductive tract microbiota likely to affect fertility by maintaining a healthy genital tract milieu. Reproductive tract microbiota is more diverse in infertile women as compared to fertile women (Campisciano et al., 2017). Subjects with non-lactobacillus dominated genital tract had significantly lower implantation and pregnancy rates (Moreno et al., 2016). Furthermore, invasion of pathogens in the uterus may cause abortion or premature birth (Payne and Bayatibojakhi, 2014; Moreno et al., 2016; Al-Memar et al., 2020). Thus, uterine bacteria likely have a greater influence over pregnancy success and minor differences in potentially pathogenic bacterial abundances may affect the uterine environment and impact fertility (Ault-Seay et al., 2023).

Soscio-Chemical Signaling

Through the synthesis of pheromones, the vaginal microbiome is also connected to sociochemical signaling in several species (Archunan et al., 2014; Archunan, 2020; Srinivasan et al., 2021). The microbiota generates pheromones by direct synthesis of pheromonal cues or the fermentation hypothesis, which involves microorganisms breaking down endogenous organic components to form extremely volatile chemicals (LeClaire et al., 2014). According to research on buffalo, males exposed to female vaginal mucous during estrus had a recurring Flehmen reaction (Sankar et al., 2004). It's interesting to note that the pheromones generated change depending on how diverse the vaginal microbiota is (LeClaire et al., 2017; Srinivasan et al., 2021).

Microbial Dysbiosis and Reproductive tract Dysfunctions

Reproductive Dysfunction in Women

Bacterial vaginosis in women has a clear link with vaginal microbiota dysbiosis, which results in loss of *Lactobacillus* dominance and an elevated population of *Prevotella, Mobiluncus, Gardnerella, Ureaplasma* and *Mycoplasma* in women (Srinivasan et al., 2015). Furthermore, dysbiosis in vaginal microbiota is also associated with problems with pregnancy onset and infertility in women (Venneri et al., 2022). Furthermore, dysbiosis in vaginal and cervical microbiota may be associated with the development of uterine fibroids in women (Mao et al. 2023). Many cases of unexplained infertility (idiopathic infertilities) in women are because of dysbiosis of reproductive tract microbiota (Chopra et al., 2024).

Reproductive dysfunction in Animals

Disruptions in normal reproductive tract microbiota (dysbiosis) cause inflammation of the reproductive tract, which plays an important role in pathomechanisms of reproductive tract dysfunctions (Chase and Kaushik, 2019). Intrauterine invasion of pathogens during or after calving causes endometritis, metritis, and pyometra, which have a significant impact on fertility outcomes, and a major impact on profitability of dairy farms due to economic and welfare concerns (Galvao, 2018; Sheldon et al., 2018; Helfrich et al., 2020).

Subclinical and Clinical Endometritis

Although endometritis regarded as a multifactorial disease, E. coli is one of the significant pathogens that causes endometritis in cattle (Ricci et al., 2015; Salah et al., 2017; de Cássia Bicudo et al., 2019). Recently, the higher relative abundance of *Bacteroidetes* and *Fusobacterium* also plays a part in the pathogenesis of endometritis (Wang et al., 2018; Miranda-Casoluengo et al., 2019). In addition, there is the relationship between endometritis and cystic ovarian diseases (Yamamoto et al., 2020).

Metritis

Available evidences suggest that endometritis and metritis in not separate diseases rather it is a two stages of same entity different level of progression. Metritis is associated with dysbiosis of uterine microbiota, which is characterized by abundance of *Bacteroidetes* and *Fusobacteri*a, particularly *Bacteroides*, *Porphyromonas*, and *Fusobacterium* (Galvao et al., 2019), moreover *Helcococcus ovis is* recently revealed as an emerging pathogen in the pathogenesis of metritis in Holstein Dairy Cows (Cunha et al., 2019).

Modulation of Reproductive Tract Microbiota to Mitigate Infertility Issues

According to the Food and Agriculture Organization of United Nation (FAO), The Probiotics are live microorganisms that insure health benefits to the subject, when administered in optimum amount (Fuller, 1989). Changes in the female genital tract microbiome are consistently correlated to gynecological and obstetrical pathologies, and tract dysbiosis can impact reproductive outcomes during fertility treatment (Toson et al., 2022). Unlike the vaginal tract of human female subjects, the microbiota of the vaginal tract of dairy cows is not predominated by *Lactobacillus* strains (Miller et al., 2016). Additionally, the vaginal pH of dairy cows and humans differs, with humans having an acidic vaginal pH (3.5-4.5) and cows having a slightly acidic to slightly alkaline pH (pH 6.8-7.35) (Bicalho et al., 2019). As only humans have acidic vaginal pH levels, *Lactobacilli* are known to create lactic acid by hydrolyzing the glycogen found in human vaginal mucosa (Miller et al., 2016). The prevention of bacterial infection is significantly aided by the acidic vaginal pH (Bicalho et al., 2019). In order to make lactic acid, *Lactobacilli* utilize glycogen produced by vaginal epithelial cells. Surprisingly, glycogen has also been found in the uterus of dairy cows (Wrenn et al., 1968). Numerous researches in dairy cows investigated the possibility of employing intravaginal *Lactobacilli* to prevent vaginal and uterine infections given that lowering vaginal pH is a physical barrier to pathogenic bacterial infection across the globe. Administration of intravaginally a mixture of three strains of *Lactobacilli* (*Lactobacillus sakei*, *Pediococcus acidilactici* FUA 3140, and *P. acidilactici* FUA 3138 at 1010–1012 CFU of LAB/wk/cow) in transition dairy cows around calving could improve milk yield, lower plasma Haptoglobin (an indicator of uterine infections) and lower the occurrence of purulent vaginal discharges (Ametaj et al., 2014), but pregnancy rate did not improved than those of untreated cows. Similarly, few other researchers observed promising results using intravaginal probiotics as evidenced by declined systemic serum amyloid, elevated vaginal mucus IgA (Deng et al., 2014), lower blood NEFA (Ametaj, 2005) and reduced incidence of uterine infections (Genis et al., 2017) as compared to control cows.

The various mechanism of action of probiotics has been ascribed like improved function of epithelial barriers (Bermudez-Brito et al., 2012; Hummel et al., 2012), adhesion on epithelial layer (Juntunen et al., 2001), competing for essential nutrients (Mezzasalma et al., 2017), production of organic acids (Russell and Diez-Gonzalez, 1998), production of antimicrobials (Makras et al., 2006) and modulation of immune system (Deng et al., 2014) and modifying vaginal pH (Otero et al., 2006). In contrast, Quereda et al. (2020[b]) found no effect of intravaginal administration of probiotics on general health or fertility in sheep. Furthermore, comprehensive understanding of reproductive microbiota and its role in modulating physiological functions will aid in the development of management and therapeutic strategies to improve reproductive efficiency (Poole et al., 2023).

Conclusion

The reproductive tract microbiota is essential part of animal's reproductive health. Microbiota plays vital roles in the host, including immunological, neuroendocrine, and metabolic programming and regulation. The host's health is significantly impacted in the short and long term by the proper acquisition of the microbiota throughout early life and its modifications throughout life. The surroundings environment and mode of delivery shapes the microbial community of newborn and plays many vital roles like protection against pathogens and maintenance of reproductive health and Socio-chemical signaling etc. Thus, the dysbiosis can results battery of infertility problems. The surroundings environment and mode of delivery shapes the microbial community of newborn and plays many vital roles like protection against pathogens and maintenance of reproductive health and Socio-chemical signaling etc. Thus, the dysbiosis can results battery of infertility problems. In conclusion, comprehensive understanding of reproductive microbiota and its role in modulating physiological functions will aid in the development of management and therapeutic strategies to augment reproductive efficiency.

References

Aagaard, K., Riehle, K., Ma, J., Segata, N., Mistretta, T. A., Coarfa, C., ... & Versalovic, J. (2012). A metagenomic approach to characterization of the vaginal microbiome signature in pregnancy. *PloS one*, *7*(6), e36466.

Adnane, M., & Chapwanya, A. (2022). A review of the diversity of the genital tract microbiome and implications for fertility of cattle. *Animals*, *12*(4), 460.

Agostinis, C., Mangogna, A., Bossi, F., Ricci, G., Kishore, U., & Bulla, R. (2019). Uterine immunity and microbiota: a shifting paradigm. *Frontiers in immunology*, *10*, 2387.

Alakomi, H. L., Skytta, E., Saarela, M., Mattila-Sandholm, T., Latva-Kala, K., & Helander, I. M. (2000). Lactic acid permeabilizes gram-negative bacteria by disrupting the outer membrane. *Applied and environmental microbiology*, *66*(5), 2001-2005.

Aldunate, M., Tyssen, D., Johnson, A., Zakir, T., Sonza, S., Moench, T., ... & Tachedjian, G. (2013). Vaginal concentrations of lactic acid potently inactivate HIV. *Journal of Antimicrobial Chemotherapy*, *68*(9), 2015-2025.

Al-Memar, M., Bobdiwala, S., Fourie, H., Mannino, R., Lee, Y. S., Smith, A., Marchesi, J. R., Timmerman, D., Bourne, T., Bennett, P. R., & MacIntyre, D. A. (2020). The association between vaginal bacterial composition and miscarriage: a nested case-control study. *BJOG : an international journal of obstetrics and gynaecology*, *127*(2), 264–274. https://doi.org/10.1111/1471-0528.15972.

Amabebe, E., & Anumba, D. O. (2018). The vaginal microenvironment: the physiologic role of lactobacilli. *Frontiers in medicine*, *5*, 181.

Ametaj, B. N. (2005). A New understanding of the causes of fatty liver in dairy cows. Advanced. *Dairy Science and Technology, 17*, 97–112.

Appiah, M. O., Wang, J., & Lu, W. (2020). Microflora in the reproductive tract of cattle: a review. *Agriculture*, *10*(6), 232.

Archunan, G. (2020). Reproductive enhancement in buffalo: Looking at urinary pheromones and hormones. *Iranian Journal of Veterinary Research*, *21*(3), 163.

Archunan, G., Rajanarayanan, S., & Karthikeyan, K. (2014). 16 Cattle pheromones. *Neurobiology of Chemical Communication. 1st edn. CRC Press, New York*, 461-479.

Ault-Seay, T. B., Moorey, S. E., Mathew, D. J., Schrick, F. N., Pohler, K. G., McLean, K. J., & Myer, P. R. (2023). Importance of the female reproductive tract microbiome and its relationship with the uterine environment for health and productivity in cattle: A review. *Frontiers in Animal Science*, *4*, 1111636.

Azawi, O. I. (2008). Postpartum uterine infection in cattle. *Animal reproduction science*, *105*(3-4), 187-208.

Baka, S., & Malamitsi-Puchner, A. (2006). Novel follicular fluid factors influencing oocyte developmental potential in IVF: a review. *Reproductive biomedicine online*, *12*(4), 500-506.

Baker, J. M., Chase, D. M., & Herbst-Kralovetz, M. M. (2018). Uterine Microbiota: Residents, Tourists, or Invaders?. *Frontiers in immunology*, *9*, 208. https://doi.org/10.3389/fimmu.2018.00208.

Barba, M., Martínez-Boví, R., Quereda, J. J., Mocé, M. L., Plaza-Dávila, M., Jiménez-Trigos, E., ... & García-Roselló, E. (2020). Vaginal microbiota is stable throughout the estrous cycle in Arabian mares. *Animals*, *10*(11), 2020.

Baud, D., Pattaroni, C., Vulliemoz, N., Castella, V., Marsland, B. J., & Stojanov, M. (2019). Sperm microbiota and its impact on semen parameters. *Frontiers in microbiology*, *10*, 234.

Berg, G., Rybakova, D., Fischer, D., Cernava, T., Vergès, M. C. C., Charles, T., ... & Schloter, M. (2020). Microbiome definition re-visited: old concepts and new challenges. Microbiome 8: 103.

Bermudez-Brito, M., Plaza-Díaz, J., Muñoz-Quezada, S., Gómez-Llorente, C., & Gil, A. (2012). Probiotic mechanisms of action. *Annals of nutrition & metabolism*, *61*(2), 160–174. https://doi.org/10.1159/000342079.

Besnard, N., Horne, E. A., & Whitehead, S. A. (2001). Prolactin and lipopolysaccharide treatment increased apoptosis and atresia in rat ovarian follicles. *Acta physiologica Scandinavica*, *172*(1), 17–25. https://doi.org/10.1046/j.1365-201X.2001.00813.x.

Bicalho, M. L. S., Machado, V. S., Oikonomou, G., Gilbert, R. O., & Bicalho, R. C. (2012). Association between virulence factors of Escherichia coli, Fusobacterium necrophorum, and Arcanobacterium pyogenes and uterine diseases of dairy cows. *Veterinary microbiology*, *157*(1-2), 125-131.

Bjurström L. (1993). Aerobic bacteria occurring in the vagina of bitches with reproductive disorders. *Acta veterinaria Scandinavica*, *34*(1), 29–34. https://doi.org/10.1186/BF03548220.

Borges, E. D., Berteli, T. S., Reis, T. F. D., Silva, A. S. D., & Vireque, A. A. (2020). Microbial contamination in assisted reproductive technology: source, prevalence, and cost. *Journal of assisted reproduction and genetics*, *37*(1), 53-61.

Brewster, W. R., Burkett, W. C., Ko, E. M., Bae-Jump, V., McCoy, A. N., & Keku, T. O. (2022). An evaluation of the microbiota of the upper reproductive tract of women with and without epithelial ovarian cancer. *Gynecologic Oncology Reports*, *42*, 101017.

Brotman, R. M., Shardell, M. D., Gajer, P., Fadrosh, D., Chang, K., Silver, M. I., ... & Gravitt, P. E. (2014). Association between the vaginal microbiota, menopause status, and signs of vulvovaginal atrophy. *Menopause*, *21*(5), 450-458.

Campisciano, G., Florian, F., D'eustacchio, A., Stanković, D., Ricci, G., De Seta, F., & Comar, M. (2017). Subclinical alteration of the cervical–vaginal microbiome in women with idiopathic infertility. *Journal of cellular physiology*, *232*(7), 1681-1688.

Campisciano, G., Iebba, V., Zito, G., Luppi, S., Martinelli, M., Fischer, L., De Seta, F., Basile, G., Ricci, G., & Comar, M. (2020). *Lactobacillus iners* and *gasseri*, *Prevotella bivia* and HPV Belong to the Microbiological Signature Negatively Affecting Human Reproduction. *Microorganisms*, *9*(1), 39. https://doi.org/10.3390/microorganisms9010039.

Canadas, E. R., Herlihy, M. M., Kenneally, J., Grant, J., Kearney, F., Lonergan, P., & Butler, S. T. (2020). Associations between postpartum phenotypes, cow factors, genetic traits, and reproductive performance in seasonal-calving, pasture-based lactating dairy cows. *Journal of dairy science*, *103*(1), 1016-1030.

Chase, C., & Kaushik, R. S. (2019). Mucosal immune system of cattle: all immune responses begin here. *Veterinary Clinics of North America: Food Animal Practice*, *35*(3), 431-451.

Chen, C., Song, X., Wei, W., Zhong, H., Dai, J., Lan, Z., ... & Jia, H. (2017). The microbiota continuum along the female reproductive tract and its relation to uterine-related diseases. *Nature communications*, *8*(1), 875.

Chen, H., Fu, K., Pang, B., Wang, J., Li, H., Jiang, Z., ... & Cao, R. (2020). Determination of uterine bacterial community in postpartum dairy cows with metritis based on 16S rDNA sequencing. *Veterinary and Animal Science*, *10*, 100102.

Cho, I., & Blaser, M. J. (2012). The human microbiome: at the interface of health and disease. *Nature Reviews Genetics*, *13*(4), 260-270.

Chopra, C., Kumar, V., Kumar, M., & Bhushan, I. (2024). Role of vaginal microbiota in idiopathic infertility: a prospective study. *Microbes and Infection*, *26*(4), 105308.

Clabaut, M., Suet, A., Racine, P. J., Tahrioui, A., Verdon, J., Barreau, M., ... & Feuilloley, M. G. (2021). Effect of 17β-estradiol on a human vaginal Lactobacillus crispatus strain. *Scientific Reports*, *11*(1), 7133.

Cojkic, A., Niazi, A., Guo, Y., Hallap, T., Padrik, P., & Morrell, J. M. (2021). Identification of bull semen microbiome by 16S sequencing and possible relationships with fertility. *Microorganisms*, *9*(12), 2431.

Comizzoli, P., & Power, M. (2019). Reproductive microbiomes in wild animal species: a new dimension in conservation biology. *Reproductive sciences in animal conservation*, 225-240.

Comizzoli, P., Power, M. L., Bornbusch, S. L., & Muletz-Wolz, C. R. (2021). Interactions between reproductive biology and microbiomes in wild animal species. *Animal microbiome*, *3*(1), 87. https://doi.org/10.1186/s42523-021-00156-7.

Cunha, F., Jeon, S. J., Daetz, R., Vieira-Neto, A., Laporta, J., Jeong, K. C., Barbet, A. F., Risco, C. A., & Galvão, K. N. (2018). Quantifying known and emerging uterine pathogens, and evaluating their association with metritis and fever in dairy cows. *Theriogenology*, *114*, 25–33. https://doi.org/10.1016/j.theriogenology.2018.03.016.

Cunha, F., Jeon, S. J., Kutzer, P., Jeong, K. C., & Galvão, K. N. (2019). Draft Genome Sequences of Helcococcus ovis Strains Isolated at Time of Metritis Diagnosis from the Uterus of Holstein Dairy Cows. *Microbiology resource announcements*, *8*(22), e00402-19. https://doi.org/10.1128/MRA.00402-19.

Curty, G., de Carvalho, P. S., & Soares, M. A. (2019). The role of the cervicovaginal microbiome on the genesis and as a biomarker of premalignant cervical intraepithelial neoplasia and invasive cervical cancer. *International journal of molecular sciences*, *21*(1), 222.

da Silveira, J. C., Andrade, G. M., Del Collado, M., Sampaio, R. V., Sangalli, J. R., Silva, L. A., Pinaffi, F. V. L., Jardim, I. B., Cesar, M. C., Nogueira, M. F. G., Cesar, A. S. M., Coutinho, L. L., Pereira, R. W., Perecin, F., & Meirelles, F. V. (2017). Supplementation with small-extracellular vesicles from ovarian follicular fluid during in vitro production modulates bovine embryo development. *PloS one*, *12*(6), e0179451.

de Boer, M., Buddle, B. M., Heuer, C., Hussein, H., Zheng, T., LeBlanc, S. J., & McDougall, S. (2015). Associations between intrauterine bacterial infection, reproductive tract inflammation, and reproductive performance in pasture-based dairy cows. *Theriogenology*, *83*(9), 1514–1524. https://doi.org/10.1016/j.theriogenology.2015.01.032.

de Cássia Bicudo, L., Oba, E., Bicudo, S. D., da Silva Leite, D., Siqueira, A. K., de Souza Monobe, M. M., ... & Ribeiro, M. G. (2019). Virulence factors and phylogenetic group profile of uterine Escherichia coli in early postpartum of high-producing dairy cows. *Animal Production Science*, *59*(10), 1898-1905.

de Ruiz Holgado, A. A. P., & Nader-Macias, M. E. (1999). Lactobacilli and enterococci isolated from the bovine vagina during the estrous cycle. *Anaerobe*, *5*, 305-307.

Deguillaume, L., Geffré, A., Desquilbet, L., Dizien, A., Thoumire, S., Vornière, C., ... & Chastant-Maillard, S. (2012). Effect of endocervical inflammation on days to conception in dairy cows. *Journal of dairy science*, *95*(4), 1776-1783.

Deng, F., McClure, M., Rorie, R., Wang, X., Chai, J., Wei, X., ... & Zhao, J. (2019). The vaginal and fecal microbiomes are related to pregnancy status in beef heifers. *Journal of Animal Science and Biotechnology*, *10*, 1-13.

Diaz-Martínez, M. D. C., Bernabeu, A., Lledó, B., Carratalá-Munuera, C., Quesada, J. A., Lozano, F. M., ... & Bernabeu, R. (2021). Impact of the vaginal and endometrial microbiome pattern on assisted reproduction outcomes. *Journal of Clinical Medicine*, *10*(18), 4063.

Dominguez-Bello, M. G., Costello, E. K., Contreras, M., Magris, M., Hidalgo, G., Fierer, N., & Knight, R. (2010). Delivery mode shapes the acquisition and structure of the initial microbiota across multiple body habitats in newborns. *Proceedings of the National Academy of Sciences*, *107*(26), 11971-11975.

Eschenbach, D. A., Thwin, S. S., Patton, D. L., Hooton, T. M., Stapleton, A. E., Agnew, K., ... & Stamm, W. E. (2000). Influence of the normal menstrual cycle on vaginal tissue, discharge, and microflora. *Clinical Infectious Diseases*, *30*(6), 901-907.

Esposito, G., Raffrenato, E., Lukamba, S. D., Adnane, M., Irons, P. C., Cormican, P., ... & Chapwanya, A. (2020). Characterization of metabolic and inflammatory profiles of transition dairy cows fed an energy-restricted diet. *Journal of animal science*, *98*(1), skz391.

Farahani, L., Tharakan, T., Yap, T., Ramsay, J. W., Jayasena, C. N., & Minhas, S. (2021). The semen microbiome and its impact on sperm function and male fertility: a systematic review and meta-analysis. *Andrology*, *9*(1), 115-144.

Ferranti, E. M., Aloqaily, B. H., Gifford, C. A., Forrest, K. K., Löest, C. A., Wenzel, J. C., & Gifford, J. H. (2022). Effects of lipopolysaccharide on beta-catenin, aromatase, and estrogen production in bovine granulosa cells in vivo and in vitro. *Domestic Animal Endocrinology*, *78*, 106652.

France, M., Alizadeh, M., Brown, S., Ma, B., & Ravel, J. (2022). Towards a deeper understanding of the vaginal microbiota. *Nature microbiology*, *7*(3), 367-378.

Fu, M., Zhang, X., Liang, Y., Lin, S., Qian, W., & Fan, S. (2020). Alterations in Vaginal Microbiota and Associated Metabolome in Women with Recurrent Implantation Failure. *mBio*, *11*(3), e03242-19. https://doi.org/10.1128/mBio.03242-19.

Gachet, C., Prat, M., Burucoa, C., Grivard, P., & Pichon, M. (2022). Spermatic microbiome characteristics in infertile patients: Impact on sperm count, mobility, and morphology. *Journal of clinical medicine*, *11*(6), 1505.

Gajer, P., Brotman, R. M., Bai, G., Sakamoto, J., Schütte, U. M., Zhong, X., Koenig, S. S., Fu, L., Ma, Z. S., Zhou, X., Abdo, Z., Forney, L. J., & Ravel, J. (2012). Temporal dynamics of the human vaginal microbiota. *Science translational medicine*, *4*(132), 132ra52. https://doi.org/10.1126/scitranslmed.3003605.

Galvao, K. N. (2018). Postpartum uterine diseases in dairy cows. *Animal Reproduction (AR)*, *9*(3), 290-296.

Galvao, K. N., Bicalho, R. C., & Jeon, S. J. (2019). Symposium review: The uterine microbiome associated with the development of uterine disease in dairy cows. *Journal of dairy science*, *102*(12), 11786-11797.

Genis, S., Sanchez-Chardi, A., Bach, A., Fabregas, F., & Aris, A. (2017). A combination of lactic acid bacteria regulates Escherichia coli infection and inflammation of the bovine endometrium. *Journal of dairy science*, *100*(1), 479–492. https://doi.org/10.3168/jds.2016-11671.

Gentilini, D., Busacca, M., Di Francesco, S., Vignali, M., Viganò, P., & Di Blasio, A. M. (2007). PI3K/Akt and ERK1/2 signalling pathways are involved in endometrial cell migration induced by 17beta-estradiol and growth factors. *Molecular human reproduction*, *13*(5), 317–322. https://doi.org/10.1093/molehr/gam001.

Giannattasio-Ferraz, S., Laguardia-Nascimento, M., Gasparini, M. R., Leite, L. R., Araujo, F. M. G., de Matos Salim, A. C., ... & Barbosa-Stancioli, E. F. (2019). A common vaginal microbiota composition among breeds of Bos taurus indicus (Gyr and Nellore). *Brazilian Journal of Microbiology*, *50*, 1115-1124.

Gilbert, R. O. (2019). Symposium review: Mechanisms of disruption of fertility by infectious diseases of the reproductive tract. *Journal of dairy science*, *102*(4), 3754-3765.

Gilbert, R. O., & Santos, N. R. (2016). Dynamics of postpartum endometrial cytology and bacteriology and their relationship to fertility in dairy cows. *Theriogenology*, *85*(8), 1367-1374.

Golińska, E., Sowińska, N., Tomusiak-Plebanek, A., Szydło, M., Witka, N., Lenarczyk, J., & Strus, M. (2021). The vaginal microflora changes in various stages of the estrous cycle of healthy female dogs and the ones with genital tract infections. *BMC veterinary research*, *17*(1), 8. https://doi.org/10.1186/s12917-020-02710-y.

Gorodeski, G. I., Hopfer, U., Liu, C. C., & Margles, E. (2005). Estrogen acidifies vaginal pH by up-regulation of proton secretion via the apical membrane of vaginal-ectocervical epithelial cells. *Endocrinology*, *146*(2), 816-824.

Graver, M. A., & Wade, J. J. (2011). The role of acidification in the inhibition of Neisseria gonorrhoeae by vaginal lactobacilli during anaerobic growth. *Annals of Clinical Microbiology and Antimicrobials*, *10*, 1-5.

Groppetti, D., Pecile, A., Barbero, C., & Martino, P. A. (2012). Vaginal bacterial flora and cytology in proestrous bitches: role on fertility. *Theriogenology*, *77*(8), 1549–1556. https://doi.org/10.1016/j.theriogenology.2011.11.022.

Heil, B. A., Paccamonti, D. L., & Sones, J. L. (2019). Role for the mammalian female reproductive tract microbiome in pregnancy outcomes. *Physiological genomics*, *51*(8), 390–399. https://doi.org/10.1152/physiolgenomics.00045.2019.

Hoang, T., Toler, E., DeLong, K., Mafunda, N. A., Bloom, S. M., Zierden, H. C., ... & Ensign, L. M. (2020). The cervicovaginal mucus barrier to HIV-1 is diminished in bacterial vaginosis. *PLoS pathogens*, *16*(1), e1008236.

Human Microbiome Project Consortium (2012). Structure, function and diversity of the healthy human microbiome. *Nature*, *486*(7402), 207–214. https://doi.org/10.1038/nature11234.

Hummel, S., Veltman, K., Cichon, C., Sonnenborn, U., & Schmidt, M. A. (2012). Differential targeting of the E-Cadherin/β-Catenin complex by gram-positive probiotic lactobacilli improves epithelial barrier function. *Applied and environmental microbiology*, *78*(4), 1140–1147. https://doi.org/10.1128/AEM.06983-11.

Ichiyama, T., Kuroda, K., Nagai, Y., Urushiyama, D., Ohno, M., Yamaguchi, T., ... & Tanaka, A. (2021). Analysis of vaginal and endometrial microbiota communities in infertile women with a history of repeated implantation failure. *Reproductive Medicine and Biology*, *20*(3), 334-344.

Jan, A. T. (2017). Outer membrane vesicles (OMVs) of gram-negative bacteria: a perspective update. *Frontiers in microbiology*, *8*, 1053.

Jeon, S. J., Vieira-Neto, A., Gobikrushanth, M., Daetz, R., Mingoti, R. D., Parize, A. C. B., ... & Galvão, K. N. (2015). Uterine microbiota progression from calving until establishment of metritis in dairy cows. *Applied and environmental microbiology*, *81*(18), 6324-6332.

Jin, S. K., & Yang, W. X. (2017). Factors and pathways involved in capacitation: how are they regulated?. *Oncotarget*, *8*(2), 3600–3627. https://doi.org/10.18632/oncotarget.12274.

Juntunen, M., Kirjavainen, P. V., Ouwehand, A. C., Salminen, S. J., & Isolauri, E. (2001). Adherence of probiotic bacteria to human intestinal mucus in healthy infants and during rotavirus infection. *Clinical and diagnostic laboratory immunology*, *8*(2), 293–296. https://doi.org/10.1128/CDLI.8.2.293-296.2001.

Karstrup, C. C., Klitgaard, K., Jensen, T. K., Agerholm, J. S., & Pedersen, H. G. (2017). Presence of bacteria in the endometrium and placentomes of pregnant cows. *Theriogenology*, *99*, 41–47. https://doi.org/10.1016/j.theriogenology.2017.05.013.

Kiefer, Z. E., Koester, L. R., Studer, J. M., Chipman, A. L., Mainquist-Whigham, C., Keating, A. F., ... & Ross, J. W. (2021). Vaginal microbiota differences associated with pelvic organ prolapse risk during late gestation in commercial sows. *Biology of Reproduction*, *105*(6), 1545-1561.

Koedooder, R., Mackens, S., Budding, A., Fares, D., Blockeel, C., Laven, J., & Schoenmakers, S. (2019). Identification and evaluation of the microbiome in the female and male reproductive tracts. *Human reproduction update*, *25*(3), 298–325. https://doi.org/10.1093/humupd/dmy048.

König, S., & May, K. (2019). Invited review: Phenotyping strategies and quantitative-genetic background of resistance, tolerance and resilience associated traits in dairy cattle. *Animal*, *13*(5), 897-908.

Koziol, J. H., Sheets, T., Wickware, C. L., & Johnson, T. A. (2022). Composition and diversity of the seminal microbiota in bulls and its association with semen parameters. *Theriogenology*, *182*, 17-25.

Kraipowich, N. R., Morris, D. L., Thompson, G. L., & Mason, G. L. (2000). Bovine abortions associated with Bacteroides fragilis fetal infection. *Journal of veterinary diagnostic investigation*, *12*(4), 369-371.

Kudo, H., Sugiura, T., Higashi, S., Oka, K., Takahashi, M., Kamiya, S., ... & Usui, M. (2021). Characterization of reproductive microbiota of primiparous cows during early postpartum periods in the presence and absence of endometritis. *Frontiers in Veterinary Science*, *8*, 736996.

Laguardia-Nascimento, M., Branco, K. M. G. R., Gasparini, M. R., Giannattasio-Ferraz, S., Leite, L. R., Araujo, F. M. G., ... & Barbosa-Stancioli, E. F. (2015). Vaginal microbiome characterization of Nellore cattle using metagenomic analysis. *PLoS One*, *10*(11), e0143294.

LeBlanc, S. J., Duffield, T. F., Leslie, K. E., Bateman, K. G., Keefe, G. P., Walton, J. S., & Johnson, W. H. (2002). Defining and diagnosing postpartum clinical endometritis and its impact on reproductive performance in dairy cows. *Journal of dairy science*, *85*(9), 2223-2236.

Leclaire, S., Jacob, S., Greene, L. K., Dubay, G. R., & Drea, C. M. (2017). Social odours covary with bacterial community in the anal secretions of wild meerkats. *Scientific Reports*, *7*(1), 3240.

Leclaire, S., Nielsen, J. F., & Drea, C. M. (2014). Bacterial communities in meerkat anal scent secretions vary with host sex, age, and group membership. *Behavioral Ecology*, *25*(4), 996-1004.

Li, J., Zhu, Y., Mi, J., Zhao, Y., Holyoak, G. R., Yi, Z., Wu, R., Wang, Z., & Zeng, S. (2022). Endometrial and vaginal microbiome in donkeys with and without clinical endometritis. *Frontiers in microbiology*, *13*, 884574. https://doi.org/10.3389/fmicb.2022.884574.

Liang, H., Cai, R., Li, C., Glendon, O. H. M., Chengcheng, H., & Yan, H. (2022). High-throughput sequencing of 16S rRNA gene analysis reveals novel taxonomic diversity among vaginal microbiota in healthy and affected sows with endometritis. *Research in veterinary science*, *143*, 33-40.

Lyman, C. C., Holyoak, G. R., Meinkoth, K., Wieneke, X., Chillemi, K. A., & DeSilva, U. (2019). Canine endometrial and vaginal microbiomes reveal distinct and complex ecosystems. *PloS one*, *14*(1), e0210157. https://doi.org/10.1371/journal.pone.0210157.

Magata, F., & Shimizu, T. (2017). Effect of lipopolysaccharide on developmental competence of oocytes. *Reproductive toxicology (Elmsford, N.Y.)*, *71*, 1–7. https://doi.org/10.1016/j.reprotox.2017.04.001.

Magata, F., Horiuchi, M., Miyamoto, A., & Shimizu, T. (2014). Lipopolysaccharide (LPS) inhibits steroid production in theca cells of bovine follicles in vitro: distinct effect of LPS on theca cell function in pre- and post-selection follicles. *The Journal of reproduction and development*, *60*(4), 280–287. https://doi.org/10.1262/jrd.2013-124.

Makras, L., Triantafyllou, V., Fayol-Messaoudi, D., Adriany, T., Zoumpopoulou, G., Tsakalidou, E., Servin, A., & De Vuyst, L. (2006). Kinetic analysis of the antibacterial activity of probiotic lactobacilli towards Salmonella enterica serovar Typhimurium reveals a role for lactic acid and other inhibitory compounds. *Research in microbiology*, *157*(3), 241–247. https://doi.org/10.1016/j.resmic.2005.09.002.

Manes, J., Fiorentino, M. A., Kaiser, G., Hozbor, F., Alberio, R., Sanchez, E., & Paolicchi, F. (2010). Changes in the aerobic vaginal flora after treatment with different intravaginal devices in ewes. *Small Ruminant Research*, *94*(1-3), 201-204.

Mao, X., Chen, H., Peng, X., Zhao, X., Yu, Z., & Xu, D. (2023). Dysbiosis of vaginal and cervical microbiome is associated with uterine fibroids. *Frontiers in cellular and infection microbiology*, *13*, 1196823. https://doi.org/10.3389/fcimb.2023.1196823.

Mazmanian, S. K., Liu, C. H., Tzianabos, A. O., & Kasper, D. L. (2005). An immunomodulatory molecule of symbiotic bacteria directs maturation of the host immune system. *Cell*, *122*(1), 107–118. https://doi.org/10.1016/j.cell.2005.05.007.

Messman, R. D., Contreras-Correa, Z. E., Paz, H. A., Perry, G., & Lemley, C. O. (2020). Vaginal bacterial community composition and concentrations of estradiol at the time of artificial insemination in Brangus heifers. *Journal of animal science*, *98*(6), skaa178.

Mezzasalma, V., Manfrini, E., Ferri, E., Boccarusso, M., Di Gennaro, P., Schiano, I., Michelotti, A., & Labra, M. (2017). Orally administered multispecies probiotic formulations to prevent uro-genital infections: a randomized placebo-controlled pilot study. *Archives of gynecology and obstetrics*, *295*(1), 163–172. https://doi.org/10.1007/s00404-016-4235-2.

Milani, C., Duranti, S., Bottacini, F., Casey, E., Turroni, F., Mahony, J., ... & Ventura, M. (2017). The first microbial colonizers of the human gut: composition, activities, and health implications of the infant gut microbiota. *Microbiology and molecular biology reviews*, *81*(4), 10-1128.

Miller, E. A., Beasley, D. E., Dunn, R. R., & Archie, E. A. (2016). Lactobacilli dominance and vaginal pH: why is the human vaginal microbiome unique?. *Frontiers in microbiology*, *7*, 1936.

Miller, E. A., Beasley, D. E., Dunn, R. R., & Archie, E. A. (2016). Lactobacilli Dominance and Vaginal pH: Why Is the Human Vaginal Microbiome Unique?. *Frontiers in microbiology*, *7*, 1936. https://doi.org/10.3389/fmicb.2016.01936.

Miller, E. A., Livermore, J. A., Alberts, S. C., Tung, J., & Archie, E. A. (2017). Ovarian cycling and reproductive state shape the vaginal microbiota in wild baboons. *Microbiome*, *5*, 1-14.

Mills, S., Stanton, C., Lane, J. A., Smith, G. J., & Ross, R. P. (2019). Precision nutrition and the microbiome, part I: current state of the science. *Nutrients*, *11*(4), 923.

Miranda-CasoLuengo, R., Lu, J., Williams, E. J., Miranda-CasoLuengo, A. A., Carrington, S. D., Evans, A. C. O., & Meijer, W. G. (2019). Delayed differentiation of vaginal and uterine microbiomes in dairy cows developing postpartum endometritis. *PloS one*, *14*(1), e0200974. https://doi.org/10.1371/journal.pone.0200974.

Molina, N. M., Plaza-Díaz, J., Vilchez-Vargas, R., Sola-Leyva, A., Vargas, E., Mendoza-Tesarik, R., ... & Altmäe, S. (2021). Assessing the testicular sperm microbiome: a low-biomass site with abundant contamination. *Reproductive BioMedicine Online*, *43*(3), 523-531.

Moore, S. G., Ericsson, A. C., Poock, S. E., Melendez, P., & Lucy, M. C. (2017). Hot topic: 16S rRNA gene sequencing reveals the microbiome of the virgin and pregnant bovine uterus. *Journal of dairy science*, *100*(6), 4953-4960.

Moraes, J. G. N., Gull, T., Ericsson, A. C., Poock, S. E., Caldeira, M. O. & Lucy, M. C. (2024). Establishment of the uterine microbiomefollowing artificial insemination invirgin heifers. *Frontiers in Microbiology*, 15:1385505.doi: 10.3389/fmicb.2024.1385505.

Moreno, C. G., Luque, A. T., Galvão, K. N., & Otero, M. C. (2022). Bacterial communities from vagina of dairy healthy heifers and cows with impaired reproductive performance. *Research in Veterinary Science*, *142*, 15-23.

Moreno, I., & Franasiak, J. M. (2017). Endometrial microbiota—new player in town. *Fertility and sterility*, *108*(1), 32-39.

Moreno, I., Codoñer, F. M., Vilella, F., Valbuena, D., Martinez-Blanch, J. F., Jimenez-Almazán, J., ... & Simon, C. (2016). Evidence that the endometrial microbiota has an effect on implantation success or failure. *American journal of obstetrics and gynecology*, *215*(6), 684-703.

Nardini, P., Ñahui Palomino, R. A., Parolin, C., Laghi, L., Foschi, C., Cevenini, R., Vitali, B., & Marangoni, A. (2016). Lactobacillus crispatus inhibits the infectivity of Chlamydia trachomatis elementary bodies, in vitro study. *Scientific reports*, *6*, 29024. https://doi.org/10.1038/srep29024.

Neckovic, A., van Oorschot, R. A., Szkuta, B., & Durdle, A. (2020). Investigation of direct and indirect transfer of microbiomes between individuals. *Forensic Science International: Genetics*, *45*, 102212.

O'Hanlon, D. E., Moench, T. R., & Cone, R. A. (2013). Vaginal pH and microbicidal lactic acid when lactobacilli dominate the microbiota. *PloS one*, *8*(11), e80074.

Oishi, S., Mekaru, K., Tanaka, S. E., Arai, W., Ashikawa, K., Sakuraba, Y., ... & Aoki, Y. (2022). Microbiome analysis in women with endometriosis: Does a microbiome exist in peritoneal fluid and ovarian cystic fluid? *Reproductive Medicine and Biology*, *21*(1), e12441.

Ong, C. T., Turni, C., Blackall, P. J., Boe-Hansen, G., Hayes, B. J., & Tabor, A. E. (2021). Interrogating the bovine reproductive tract metagenomes using culture-independent approaches: A systematic review. *Animal microbiome*, *3*(1), 41.

Otero, C., Saavedra, L., Silva de Ruiz, C., Wilde, O., Holgado, A. R., & Nader-Macías, M. E. (2000). Vaginal bacterial microflora modifications during the growth of healthy cows. *Letters in Applied Microbiology*, *31*(3), 251-254.

Otero, M. C., Morelli, L., & Nader-Macías, M. E. (2006). Probiotic properties of vaginal lactic acid bacteria to prevent metritis in cattle. *Letters in applied microbiology*, *43*(1), 91–97. https://doi.org/10.1111/j.1472-765X.2006.01914.x.

Owens, C. E., Daniels, K. M., Ealy, A. D., Knowlton, K. F., & Cockrum, R. R. (2020). Graduate Student Literature Review: Potential mechanisms of interaction between bacteria and the reproductive tract of dairy cattle. *Journal of dairy science*, *103*(11), 10951-10960.

Pascottini, O. B., Van Schyndel, S. J., Spricigo, J. W., Rousseau, J., Weese, J. S., & LeBlanc, S. J. (2020). Dynamics of uterine microbiota in postpartum dairy cows with clinical or subclinical endometritis. *Scientific reports*, *10*(1), 12353.

Patel, N., Patel, N., Pal, S., Nathani, N., Pandit, R., Patel, M., ... & Parekh, B. (2022). Distinct gut and vaginal microbiota profile in women with recurrent implantation failure and unexplained infertility. *BMC Women's Health*, *22*(1), 113.

Payne, M. S., & Bayatibojakhi, S. (2014). Exploring preterm birth as a polymicrobial disease: an overview of the uterine microbiome. *Frontiers in immunology*, *5*, 595.

Pelzer, E. S., Allan, J. A., Cunningham, K., Mengersen, K., Allan, J. M., Launchbury, T., ... & Knox, C. L. (2011). Microbial colonization of follicular fluid: alterations in cytokine expression and adverse assisted reproduction technology outcomes. *Human reproduction*, *26*(7), 1799-1812.

Pelzer, E. S., Allan, J. A., Theodoropoulos, C., Ross, T., Beagley, K. W., & Knox, C. L. (2012). Hormone-dependent bacterial growth, persistence and biofilm formation–a pilot study investigating human follicular fluid collected during IVF cycles. *PloS one*, *7*(12), e49965.

Pelzer, E. S., Allan, J. A., Waterhouse, M. A., Ross, T., Beagley, K. W., & Knox, C. L. (2013[a]). Microorganisms within human follicular fluid: effects on IVF. *PloS one*, *8*(3), e59062.

Pelzer, E. S., Harris, J. E., Allan, J. A., Waterhouse, M. A., Ross, T., Beagley, K. W., & Knox, C. L. (2013[b]). TUNEL analysis of DNA fragmentation in mouse unfertilized oocytes: the effect of microorganisms within human follicular fluid collected during IVF cycles. *Journal of reproductive immunology*, *99*(1-2), 69-79.

Pelzer, E. S., Willner, D., Buttini, M., Hafner, L. M., Theodoropoulos, C., & Huygens, F. (2018). The fallopian tube microbiome: implications for reproductive health. *Oncotarget*, *9*(30), 21541.

Pelzer, E., Gomez-Arango, L. F., Barrett, H. L., & Nitert, M. D. (2017). Review: Maternal health and the placental microbiome. *Placenta*, *54*, 30–37. https://doi.org/10.1016/j.placenta.2016.12.003.

Perry, G. A., & Perry, B. L. (2008). Effect of preovulatory concentrations of estradiol and initiation of standing estrus on uterine pH in beef cows. *Domestic Animal Endocrinology*, *34*(3), 333-338.

Peterson, J., Garges, S., Giovanni, M., McInnes, P., Wang, L., Schloss, J. A., ... & NIH HMP Working Group. (2009). The NIH human microbiome project. *Genome research*, *19*(12), 2317-2323.

Poole, R. K., Soffa, D. R., McAnally, B. E., Smith, M. S., Hickman-Brown, K. J., & Stockland, E. L. (2023). Reproductive Microbiomes in Domestic Livestock: Insights Utilizing 16S rRNA Gene Amplicon Community Sequencing. *Animals : an open access journal from MDPI*, *13*(3), 485. https://doi.org/10.3390/ani13030485.

Power, M. L., Quaglieri, C., & Schulkin, J. (2017). Reproductive microbiomes: a new thread in the microbial network. *Reproductive Sciences*, *24*(11), 1482-1492.

Quadros, D. L., Zanella, R., Bondan, C., Zanella, G. C., Facioli, F. L., da Silva, A. N., & Zanella, E. L. (2020). Study of vaginal microbiota of Holstein cows submitted to an estrus synchronization protocol with the use of intravaginal progesterone device. *Research in veterinary science*, *131*, 1-6.

Quereda, J. J., Barba, M., Mocé, M. L., Gomis, J., Jiménez-Trigos, E., García-Muñoz, Á., Gómez-Martín, Á., González-Torres, P., Carbonetto, B., & García-Roselló, E. (2020[a]). Vaginal Microbiota Changes During Estrous Cycle in Dairy Heifers. *Frontiers in veterinary science*, *7*, 371. https://doi.org/10.3389/fvets.2020.00371.

Quereda, J. J., García-Roselló, E., Barba, M., Mocé, M. L., Gomis, J., Jiménez-Trigos, E., ... & Gómez-Martín, Á. (2020[b]). Use of probiotics in intravaginal sponges in sheep: a pilot study. *Animals*, *10*(4), 719.

Rasekhi, M., Mohammadi-Sangcheshmeh, A., Daliri, M., Bakhtiarizadeh, M., Shariati, V., Rahimi, M., ... & Tvrdá, E. (2020). Transcriptional profile of ovine oocytes matured under lipopolysaccharide treatment in vitro. *Theriogenology*, *157*, 70-78.

Reid, G., Brigidi, P., Burton, J. P., Contractor, N., Duncan, S., Fargier, E., ... & Ansell, J. (2015). Microbes central to human reproduction. *American Journal of Reproductive Immunology*, *73*(1), 1-11.

Ricci, A., Gallo, S., Molinaro, F., Dondo, A., Zoppi, S., & Vincenti, L. (2015). Evaluation of subclinical endometritis and consequences on fertility in Piedmontese beef cows. *Reproduction in Domestic Animals*, *50*(1), 142-148.

Rincón, J. A. A., Gindri, P. C., Mion, B., de Ávila, F. G., Barbosa, A. A., Maffi, A. S., ... & Schneider, A. (2019). Early embryonic development of bovine oocytes challenged with LPS in vitro or in vivo. *Reproduction*, *158*(5), 453-463.

Rojas Canadas, E. R., Herlihy, M. M., Kenneally, J., Grant, J., Kearney, F., Lonergan, P., & Butler, S. T. (2020). Associations between ovarian cyclicity, uterine health, indicators of bioenergetic status, and genetic traits during early lactation in seasonal-calving, pasturebased lactating dairy cows. *J. Dairy Sci*, *103*, 1002-1015.

Russell, J. B., & Diez-Gonzalez, F. (1998). The effects of fermentation acids on bacterial growth. *Advances in microbial physiology*, *39*, 205–234. https://doi.org/10.1016/s0065-2911(08)60017-x.

Ryan, N. J., Meade, K. G., Williams, E. J., O'Farrelly, C., Grant, J., Evans, A. C., & Beltman, M. E. (2020). Purulent vaginal discharge diagnosed in pasture-based Holstein-Friesian cows at 21 days postpartum is influenced by previous lactation milk yield and results in diminished fertility. *Journal of Dairy Science*, *103*(1), 666-675.

Salah, N., & Yimer, N. (2017). Cytological endometritis and its agreement with ultrasound examination in postpartum beef cows. *Veterinary world*, *10*(6), 605–609. https://doi.org/10.14202/vetworld.2017.605-609.

Sankar, R., & Archunan, G. (2004). Flehmen response in bull: role of vaginal mucus and other body fluids of bovine with special reference to estrus. *Behavioural processes*, *67*(1), 81-86.

Santos, T. M. A., Gilbert, R. O., & Bicalho, R. C. (2011). Metagenomic analysis of the uterine bacterial microbiota in healthy and metritic postpartum dairy cows. *Journal of dairy science*, *94*(1), 291-302.

Santos, T. M., & Bicalho, R. C. (2012). Diversity and succession of bacterial communities in the uterine fluid of postpartum metritic, endometritic and healthy dairy cows. *PloS one*, *7*(12), e53048.

Sawyer, G. J. (1977). Observations on the bacterial population of the os cervix of the ewe before and after embryo death. *Australian Veterinary Journal*, *53*(11), 542-544.

Schoenmakers, S., & Laven, J. (2020). The vaginal microbiome as a tool to predict IVF success. *Current Opinion in Obstetrics and Gynecology*, *32*(3), 169-178.

Sheldon, I. M., & Dobson, H. (2004). Postpartum uterine health in cattle. *Animal reproduction science*, *82*, 295-306.

Sheldon, I. M., Molinari, P. C., Ormsby, T. J., & Bromfield, J. J. (2020). Preventing postpartum uterine disease in dairy cattle depends on avoiding, tolerating and resisting pathogenic bacteria. *Theriogenology*, *150*, 158-165.

Sheldon, I. M., Noakes, D. E., Rycroft, A. N., Pfeiffer, D. U., & Dobson, H. (2002). Influence of uterine bacterial contamination after parturition on ovarian dominant follicle selection and follicle growth and function in cattle. *Reproduction (Cambridge, England)*, *123*(6), 837–845.

Sheldon, I. M., Williams, E. J., Miller, A. N., Nash, D. M., & Herath, S. (2008). Uterine diseases in cattle after parturition. *Veterinary journal (London, England : 1997)*, *176*(1), 115–121. https://doi.org/10.1016/j.tvjl.2007.12.031.

Shpigel, N. Y., Adler-Ashkenazy, L., Scheinin, S., Goshen, T., Arazi, A., Pasternak, Z., & Gottlieb, Y. (2017). Characterization and identification of microbial communities in bovine necrotic vulvovaginitis. *The Veterinary Journal*, *219*, 34-39.

Sluss, P. M., & Reichert Jr, L. E. (1983). Presence of bacteria in porcine follicular fluid and their ability to generate an inhibitor of follicle-stimulating hormone binding to receptor. *Biology of reproduction*, *29*(2), 335-341.

Smith, W. L., Garavito, R. M., & DeWitt, D. L. (1996). Prostaglandin endoperoxide H synthases (cyclooxygenases)-1 and -2. *The Journal of biological chemistry*, *271*(52), 33157–33160. https://doi.org/10.1074/jbc.271.52.33157.

Sohel, M. M. H., Hoelker, M., Noferesti, S. S., Salilew-Wondim, D., Tholen, E., Looft, C., ... & Tesfaye, D. (2013). Exosomal and non-exosomal transport of extra-cellular microRNAs in follicular fluid: implications for bovine oocyte developmental competence. *PloS one*, *8*(11), e78505.

Srinivasan, M., Adnane, M., & Archunan, G. (2021). Significance of cervico-vaginal microbes in bovine reproduction and pheromone production–A hypothetical review. *Research in Veterinary Science*, *135*, 66-71.

Srinivasan, S., Morgan, M. T., Fiedler, T. L., Djukovic, D., Hoffman, N. G., Raftery, D., ... & Fredricks, D. N. (2015). Metabolic signatures of bacterial vaginosis. *MBio*, *6*(2), 10-1128.

Štšepetova, J., Baranova, J., Simm, J., Parm, Ü., Rööp, T., Sokmann, S., ... & Mändar, R. (2020). The complex microbiome from native semen to embryo culture environment in human in vitro fertilization procedure. *Reproductive Biology and Endocrinology*, *18*, 1-13.

Stumpf, R. M., Wilson, B. A., Rivera, A., Yildirim, S., Yeoman, C. J., Polk, J. D., ... & Leigh, S. R. (2013). The primate vaginal microbiome: comparative context and implications for human health and disease. *American journal of physical anthropology*, *152*, 119-134.

Swartz, J. D., Lachman, M., Westveer, K., O'Neill, T., Geary, T., Kott, R. W., ... & Yeoman, C. J. (2014). Characterization of the vaginal microbiota of ewes and cows reveals a unique microbiota with low levels of lactobacilli and near-neutral pH. *Frontiers in veterinary science*, *1*, 19.

Tachedjian, G., Aldunate, M., Bradshaw, C. S., & Cone, R. A. (2017). The role of lactic acid production by probiotic Lactobacillus species in vaginal health. *Research in microbiology*, *168*(9-10), 782-792.

Taha, T. E., Hoover, D. R., Dallabetta, G. A., Kumwenda, N. I., Mtimavalye, L. A., Yang, L. P., ... & Miotti, P. G. (1998). Bacterial vaginosis and disturbances of vaginal flora: association with increased acquisition of HIV. *Aids*, *12*(13), 1699-1706.

Theis, K. R., Venkataraman, A., Dycus, J. A., Koonter, K. D., Schmitt-Matzen, E. N., Wagner, A. P., Holekamp, K. E., & Schmidt, T. M. (2013). Symbiotic bacteria appear to mediate hyena social odors. *Proceedings of the National Academy of Sciences of the United States of America*, *110*(49), 19832–19837. https://doi.org/10.1073/pnas.1306477110.

Torres Luque, A., Fontana, C., Pasteris, S. E., Bassi, D., Cocconcelli, P. S., & Otero, M. C. (2021). Vaginal bacterial diversity from healthy gilts and pregnant sows subjected to natural mating or artificial insemination. *Research in veterinary science*, *140*, 26–37. https://doi.org/10.1016/j.rvsc.2021.07.023.

Toson, B., Simon, C., & Moreno, I. (2022). The Endometrial Microbiome and Its Impact on Human Conception. *International journal of molecular sciences*, *23*(1), 485. https://doi.org/10.3390/ijms23010485.

Tsonis, O., Gkrozou, F., & Paschopoulos, M. (2021). Microbiome affecting reproductive outcome in ARTs. *Journal of Gynecology Obstetrics and Human Reproduction*, *50*(3), 102036.

Turnbaugh, P. J., Ley, R. E., Mahowald, M. A., Magrini, V., Mardis, E. R., & Gordon, J. I. (2006). An obesity-associated gut microbiome with increased capacity for energy harvest. *Nature*, *444*(7122), 1027-1031.

Uchihashi, M., Bergin, I. L., Bassis, C. M., Hashway, S. A., Chai, D., & Bell, J. D. (2015). Influence of age, reproductive cycling status, and menstruation on the vaginal microbiome in baboons (Papio anubis). *American Journal of Primatology*, *77*(5), 563-578.

Ungerfeld, R., & Silva, L. (2005). The presence of normal vaginal flora is necessary for normal sexual attractiveness of estrous ewes. *Applied Animal Behaviour Science*, *93*(3-4), 245-250.

Valore, E. V., Park, C. H., Igreti, S. L., & Ganz, T. (2002). Antimicrobial components of vaginal fluid. *American journal of obstetrics and gynecology*, *187*(3), 561-568.

Venneri, M. A., Franceschini, E., Sciarra, F., Rosato, E., D'ettorre, G., & Lenzi, A. (2022). Human genital tracts microbiota: dysbiosis crucial for infertility. *Journal of Endocrinological Investigation*, *45*(6), 1151-1160.

Wang, J., Li, C., Nesengani, L. T., Gong, Y., Zhang, S., & Lu, W. (2017). Characterization of vaginal microbiota of endometritis and healthy sows using high-throughput pyrosequencing of 16S rRNA gene. *Microbial pathogenesis*, *111*, 325–330. https://doi.org/10.1016/j.micpath.2017.08.030.

Wang, J., Liu, C., Nesengani, L. T., Gong, Y., Yang, Y., Yang, L., & Lu, W. (2019). Comparison of vaginal microbial community structure of beef cattle between luteal phase and follicular phase. *Indian Journal of Animal Research*, *53*(10), 1298-1303.

Wang, J., Sun, C., Liu, C., Yang, Y., & Lu, W. (2016). Comparison of vaginal microbial community structure in healthy and endometritis dairy cows by PCR-DGGE and real-time PCR. *Anaerobe*, *38*, 1–6. https://doi.org/10.1016/j.anaerobe.2015.11.004.

Wang, M. L., Liu, M. C., Xu, J., An, L. G., Wang, J. F., & Zhu, Y. H. (2018). Uterine microbiota of dairy cows with clinical and subclinical endometritis. *Frontiers in microbiology*, *9*, 2691.

Wexler, H. M. (2007). Bacteroides: the good, the bad, and the nitty-gritty. *Clinical microbiology reviews*, *20*(4), 593-621.

Williams, E. J., Fischer, D. P., Noakes, D. E., England, G. C., Rycroft, A., Dobson, H., & Sheldon, I. M. (2007). The relationship between uterine pathogen growth density and ovarian function in the postpartum dairy cow. *Theriogenology*, *68*(4), 549-559.

Williams, E. J., Fischer, D. P., Pfeiffer, D. U., England, G. C., Noakes, D. E., Dobson, H., & Sheldon, I. M. (2005). Clinical evaluation of postpartum vaginal mucus reflects uterine bacterial infection and the immune response in cattle. *Theriogenology*, *63*(1), 102-117.

Wu, H., Zhao, G., Jiang, K., Li, C., Qiu, C., & Deng, G. (2016). Engeletin alleviates lipopolysaccharide-induced endometritis in mice by inhibiting TLR4-mediated NF-κB activation. *Journal of agricultural and food chemistry*, *64*(31), 6171-6178.

Yamamoto, N., Nishimura, R., Gunji, Y., & Hishinuma, M. (2020). Research of postpartum endometritis in Japanese Black cattle with cystic ovarian disease by vaginal mucus test and endometrial cytology. *Archives Animal Breeding*, *63*(1), 1-8.

Yildirim, S., Yeoman, C. J., Janga, S. C., Thomas, S. M., Ho, M., Leigh, S. R., ... & Stumpf, R. M. (2014). Primate vaginal microbiomes exhibit species specificity without universal Lactobacillus dominance. *The ISME journal*, *8*(12), 2431-2444.

Zerani, M., Polisca, A., Boiti, C., & Maranesi, M. (2021). Current Knowledge on the Multifactorial Regulation of Corpora Lutea Lifespan: The Rabbit Model. *Animals : an open access journal from MDPI*, *11*(2), 296. https://doi.org/10.3390/ani11020296.

Zhang, J., Liu, M., Ke, S., Huang, X., Fang, S., He, M., Fu, H., Chen, C., & Huang, L. (2021). Gut and Vagina Microbiota Associated With Estrus Return of Weaning Sows and Its Correlation With the Changes in Serum Metabolites. *Frontiers in microbiology*, *12*, 690091. https://doi.org/10.3389/fmicb.2021.690091.

Zhao, G., Zhang, T., Wu, H., Jiang, K., Qiu, C., & Deng, G. (2019). MicroRNA let-7c Improves LPS-Induced Outcomes of Endometritis by Suppressing NF-κB Signaling. *Inflammation*, *42*(2), 650–657. https://doi.org/10.1007/s10753-018-0922-4.

Zhao, S. J., Pang, Y. W., Zhao, X. M., Du, W. H., Hao, H. S., & Zhu, H. B. (2017). Effects of lipopolysaccharide on maturation of bovine oocyte in vitro and its possible mechanisms. *Oncotarget*, *8*(3), 4656.

Chapter 14

Microbiomes of Poultry Health, Disease, and Production

Naveen Swaroop Murikipudi[1,*]
Vasantha SKI[2]
Susmitha T.[3]
Varij Nayan[4]
Rajesh Kumar[5]
and Aswani Kumar K.[1]
[1]Department of Veterinary Biochemistry, Sri Venkateswara Veterinary University, Tirupati, India
[2]Department of Veterinary Physiology, Sri Venkateswara Veterinary University, Tirupati, India
[3]Poultry Science, Sri Venkateswara Veterinary University, Tirupati, India
[4]Animal Biochemistry Division, ICAR- National Diary Research Institute, Karnal, India
[5]Acharya Narendra Deva University of Agriculture and Technology, Ayodhya, India

Abstract

The poultry industry has evolved into a major source of protein worldwide, making it a crucial aspect of food security and economic development. To ensure optimal health and productivity in poultry, understanding the role of the microbiome the diverse community of microorganisms inhabiting the gastrointestinal tract (GIT) is essential. This chapter explores the impact of the poultry microbiome on health and reproduction, providing insights into how it affects the well-being and productivity of poultry.

Keywords: poultry, health, gastrointestinal tract (GIT), reproduction

Introduction

The microbiome encompasses a complex and diverse array of bacteria, archaea, fungi, protozoa, and viruses residing in the GI tract of poultry. This microbial community plays a crucial role in maintaining the host's physiological functions. The balance and composition of this microbiome can be influenced by factors such as diet, environment, genetic makeup, and

---

* Corresponding Author's Email: mnaswp@gmail.com

In: Microbiomes in Animal Nutrition, Health, and Disease
Editor: Tanmoy Rana
ISBN: 979-8-89530-425-9

management practices. The microbiome affects the overall performance and is essential for preserving intestinal health. The majority of research on gut microbiota has been done on broilers, whose microbial communities differ from those of layers. At the phylum level, Proteobacteria, Firmicutes, Bacteroidetes, Fusobacteria, and Actinobacteria predominate in the normal gut microbiota of laying chickens. Age, genotype, and production system of chickens all affect the intestinal microbiota's makeup. Short-chain fatty acids, indole, tryptamine, vitamins, and bacteriocins are among the gut microbiota's metabolites that are involved in immunological homeostasis, barrier maintenance, and host-microbiota cross-talk. Members of the resident gut microbiota also prevent and regulate the colonization of foodborne pathogens (Khan et al., 2020).

Microbial Diversity and Function

The microbiome varies significantly across different sections of the poultry GI tract, reflecting the specialized functions of each region. The crop, for instance, harbors a mixture of lactic acid bacteria that help in fermenting dietary carbohydrates, while the caecum contains a diverse and dense community of microbes involved in nutrient absorption and immune system modulation. The microbiome of poultry is an intricate and dynamic ecosystem composed of diverse microorganisms, including bacteria, archaea, fungi, protozoa, and viruses. This microbial community is crucial for the physiological and metabolic functions of poultry, influencing digestion, immune responses, and overall health. The composition and diversity of the poultry microbiome vary significantly across different sections of the gastrointestinal (GI) tract, reflecting the specific functional roles of each region.

Microbial Composition across the GI Tract

The GI tract of poultry consists of several distinct sections, each hosting a unique microbial community:

- *Crop:* The crop, a storage organ located at the base of the esophagus, plays a role in temporarily storing and moistening ingested food. The microbial composition in the crop primarily includes lactic acid bacteria such as *Lactobacillus* and *Enterococcus* species. These bacteria are adept at fermenting dietary carbohydrates, producing lactic acid and other metabolites. The crop's microbial composition may also vary based on dietary factors, feeding habits, and environmental influences.
- *Gizzard:* The gizzard, a muscular organ that mechanically grinds food, has limited microbial diversity compared to other sections of the GI tract. The harsh, acidic environment, along with mechanical grinding, restricts microbial colonization. Nonetheless, certain acid-tolerant bacteria, such as *Lactobacillus* species, may persist in this region.
- *Small Intestine*: The small intestine comprises three parts: the duodenum, jejunum, and ileum. This region serves as a crucial site for nutrient absorption and enzymatic digestion. The microbial community in the small intestine consists of a moderate diversity of bacteria, including *Lactobacillus*, *Bifidobacterium*, and *Clostridium*

species. These bacteria assist in breaking down complex carbohydrates, proteins, and lipids into simpler compounds, facilitating nutrient absorption by the host.

- *Caecum:* The caecum, located at the junction of the small and large intestines, harbors the densest microbial community in the poultry GI tract. This region is particularly important for microbial fermentation of dietary fibers and non-digestible carbohydrates. The bacterial composition in the caecum includes diverse genera such as *Bacteroides*, *Ruminococcus*, *Firmicutes*, and *Bacteroidetes*. The fermentation process in the caecum produces short-chain fatty acids (SCFAs) such as butyrate, propionate, and acetate, which serve as energy sources for the host and as signaling molecules for metabolic regulation.
- *Large Intestine*: The large intestine, including the colon and rectum, has a microbial community that plays a role in the final stages of nutrient absorption and waste elimination. The microbial diversity in this region is intermediate between the small intestine and caecum, and it includes species such as *E. coli*, *Enterococcus*, and *Lactobacillus*. The fermentation of residual nutrients and fiber continues in the large intestine, producing additional SCFAs and other metabolites.

Factors Influencing Microbial Composition and Diversity

Several factors influence the composition and diversity of the poultry microbiome:

- *Diet:* Dietary composition plays a significant role in shaping the poultry microbiome. Fiber-rich diets favor the growth of beneficial bacteria such as *Bifidobacterium* and *Lactobacillus*, while protein- and fat-rich diets can alter the balance of gut bacteria, potentially leading to dysbiosis.
- *Age:* The microbial community in the poultry GI tract changes over time, with the initial colonization influenced by environmental exposure and vertical transmission from parent to offspring. The microbiome stabilizes as the bird matures, with the diversity and composition evolving based on diet, genetics, and other factors.
- *Genetics:* Genetic makeup influences the microbial composition in poultry. Certain genetic traits may predispose poultry to specific microbial communities, affecting the overall health and productivity of the bird.
- *Environment:* Environmental factors such as housing conditions, hygiene practices, and exposure to antibiotics can also affect the poultry microbiome. For example, antibiotic use can lead to a reduction in microbial diversity, resulting in dysbiosis.

Microbiome and Poultry Health

Nutrient Absorption and Metabolism

The microbiome significantly influences nutrient absorption and metabolism. Microbial enzymes break down complex carbohydrates, fibers, and proteins into simpler compounds, facilitating nutrient absorption in the intestines. Short-chain fatty acids (SCFAs), such as butyrate, propionate, and acetate, are produced by bacterial fermentation of dietary fibers in the caecum, providing energy to the host and serving as signaling molecules for metabolic regulation.

Immune Function

The microbiome also plays a key role in modulating the poultry immune system. It acts as a barrier against pathogenic microorganisms by competing for nutrients and adhering to the intestinal lining, preventing harmful pathogens from establishing. Moreover, the microbiome influences the development and function of the host's immune system, including the production of antibodies and modulation of inflammatory responses.

Disease Prevention

An imbalance in the poultry microbiome (dysbiosis) can lead to various health issues. For instance, dysbiosis may contribute to conditions such as necrotic enteritis caused by Clostridium perfringens, and avian influenza can exacerbate dysbiosis, weakening the immune system and making poultry more susceptible to secondary infections.

Microbiome and Poultry Reproduction

Impact on Fertility

Emerging research suggests that the microbiome may affect reproductive functions in poultry. Dysbiosis can alter hormone levels and reproductive organ function, potentially reducing fertility. Additionally, SCFAs produced by microbial fermentation can influence reproductive hormone synthesis, indirectly affecting fertility.

Vertical Transmission

The microbiome also plays a role in vertical transmission, where microbial communities are transferred from parent to offspring, influencing chick development. Vertical transmission can occur through the egg's shell, cloacal contact, or parental care behaviors.

Enhancing Poultry Microbiome for Health and Productivity

Probiotics and Prebiotics

Probiotics and prebiotics can be used to optimize the poultry microbiome as they have potential roles of prebiotics and probiotics in the reshaping of the gut microbiota of layer chickens to enhance gut health and food safety. These supplements improve gut health, nutrient absorption, and immune function, leading to enhanced growth and productivity. Prebiotic and probiotic supplementation increase host performance by fortifying the gut microbiota against colonization by gut pathogens like Salmonella and Campylobacter. The generation of organic acids, stimulation of the host immune system, and synthesis of antimicrobial compounds are the processes by which prebiotics and probiotics act. Probiotic candidates that have demonstrated encouraging results in improving food safety and gut health include isolates of Lactobacillus, Bifidobacterium, Bacillus, Saccharomyces, and Faecalibacterium. Furthermore, inulin and a variety of complex carbohydrates, such as fructose, galactose and mannose oligosaccharides, and show promise as means of enhancing gut health (Khan et al., 2020). Attia et al., 2020 observed improved egg shell quality which might be related to the ability of probiotics to enhance calcium absorption and retention in the shell glands of laying hens.

Dietary Management

Dietary adjustments can shape the poultry microbiome. For example, incorporating fibrous ingredients and fermentable substrates into feed can enhance the growth of beneficial bacterial populations, contributing to improved digestion and immune function.

Antibiotic-Free Production

With the growing concern over antibiotic resistance, the poultry industry is transitioning towards antibiotic-free production systems. This shift necessitates alternative approaches to managing the microbiome, including probiotic and prebiotic supplementation and improved hygiene practices.

Gut Health and Competitive Exclusion

The gastrointestinal tract (GIT) of poultry is a complex ecosystem, home to a diverse array of microorganisms. Maintaining a balanced microbiome is essential for gut health and overall well-being. Probiotics, particularly lactic acid bacteria (such as *Lactobacillus* and *Enterococcus*) and bifidobacteria, play a crucial role in this balance by promoting competitive exclusion:

1. *Nutrient Competition:* Probiotic bacteria compete with pathogenic bacteria for nutrients in the gut, including sugars, amino acids, and other organic compounds. This competition limits the availability of nutrients for harmful microbes, reducing their growth and ability to colonize the GI tract.
2. *Adhesion Site Competition:* Probiotics also compete with pathogenic bacteria for adhesion sites on the intestinal epithelium. By adhering to the intestinal lining, probiotic bacteria prevent harmful microbes from attaching and establishing themselves in the gut, reducing the risk of infection.
3. *Production of Antimicrobial Compounds:* Some probiotic strains produce antimicrobial compounds, including organic acids (such as lactic acid and acetic acid), bacteriocins, and hydrogen peroxide. These compounds create an acidic environment and inhibit the growth of pathogenic bacteria, further promoting a balanced microbiome.
4. *Immune Modulation:* Probiotics help modulate the poultry immune system, contributing to gut health. They stimulate the production of antibodies, particularly IgA, which bind to pathogens and prevent them from establishing infections. Probiotics also influence cytokine production, modulating inflammatory responses and further supporting immune function.

Preventing Necrotic Enteritis

Necrotic enteritis is a significant disease affecting poultry, caused by the bacterium *Clostridium perfringens*. Probiotics play a vital role in preventing this disease by maintaining a balanced gut microbiome:

1. *Competitive Exclusion* of *Clostridium perfringens*: Probiotic bacteria, particularly lactic acid bacteria, and bifidobacteria, compete with *C. perfringens* for nutrients and adhesion sites, limiting its growth and ability to establish infections.
2. *Reduction of Toxin Production*: Probiotic strains such as *Lactobacillus* spp. and *Bifidobacterium* spp. can influence the expression of virulence factors in *C. perfringens*, reducing its production of toxins that cause necrotic enteritis. This reduction in toxin production minimizes tissue damage and disease severity.
3. *Gut Barrier Integrity:* Probiotics help maintain the integrity of the intestinal barrier, reducing the risk of pathogen translocation and systemic infections. This barrier function is particularly important in preventing necrotic enteritis, as it minimizes the spread of *C. perfringens* and its toxins throughout the bird's body.

Microbiome in Poultry Diseases

The gut microbiome plays a crucial role in modulating poultry health, and an imbalance in this microbial community, known as dysbiosis, can lead to various diseases. Here's a closer look at some key poultry diseases influenced by the microbiome:

1. *Necrotic Enteritis*: This disease, caused by the bacterium *Clostridium perfringens*, leads to significant losses in the poultry industry. An imbalance in the gut microbiome, characterized by a decrease in beneficial bacteria such as *Lactobacillus* and *Bifidobacterium*, can create an environment conducive to the overgrowth of *C. perfringens*, resulting in inflammation, necrosis, and sometimes mortality (Abd El-Hack et al., 2022). Probiotic supplementation helps prevent necrotic enteritis by promoting competitive exclusion, inhibiting the growth of *C. perfringens*, and modulating immune responses Lutful Kabir, 2009).
2. *Salmonellosis:* This disease is caused by *Salmonella* species, including *Salmonella enteritidis* and *Salmonella typhimurium*, which can colonize the GI tract, leading to gastrointestinal infections and potential vertical transmission (Pérez-Cobas et al., 2015). A balanced microbiome can protect against *Salmonella* infections by promoting competitive exclusion and enhancing immune function, reducing the risk of infection.
3. *Colibacillosis:* This condition results from pathogenic strains of *Escherichia coli*, causing symptoms such as diarrhea, enteritis, and sometimes systemic infections, leading to reduced productivity and mortality (Pan and Yu, 2014). Probiotic supplementation helps prevent colibacillosis by maintaining a balanced microbiome and stimulating immune responses, protecting against *E. coli* infections.
4. *Avian Influenza*: This viral disease affects poultry worldwide, leading to significant economic losses. Avian influenza can exacerbate dysbiosis, weakening the immune system and making poultry more susceptible to secondary infections (Korver, 2012). A balanced microbiome can support immune function, aiding recovery from avian influenza and reducing the severity of secondary infections.
5. *Coccidiosis:* This disease is caused by protozoan parasites from the genus *Eimeria*, leading to enteritis, diarrhea, and reduced growth rates in poultry. Dysbiosis can exacerbate coccidiosis severity by weakening immune responses and reducing gut barrier integrity. Probiotics can mitigate the effects of coccidiosis by promoting a

balanced microbiome, supporting immune function, and maintaining gut barrier integrity.

Microbiome in Egg Production

The microbiome, particularly the gastrointestinal (GI) tract microbiome, plays a significant role in various aspects of poultry health and productivity, including egg production. Here's a detailed look at how the microbiome influences egg production:

Nutrient Absorption and Metabolism

The microbiome influences egg production by impacting nutrient absorption and metabolism as illustrated in Figure 1.

1. *Nutrient Availability*: Microbial communities in the GI tract break down complex carbohydrates, proteins, and fibers into simpler compounds, facilitating the absorption of essential nutrients such as vitamins, minerals, and amino acids (Salim et al., 2012). These nutrients are critical for egg formation, including yolk and eggshell production.
2. *SCFAs:* The fermentation of dietary fibers by gut bacteria produces short-chain fatty acids (SCFAs), including butyrate, propionate, and acetate (Stanley et al., 2014). These SCFAs provide energy to the host and serve as signaling molecules, influencing metabolic processes that impact egg production.
3. *Cholesterol and Fat Metabolism*: The microbiome also affects cholesterol and fat metabolism, which are crucial for yolk production. Beneficial bacteria such as *Bifidobacterium* and *Lactobacillus* help modulate cholesterol levels, contributing to optimal egg yolk composition (Kogut & Arsenault, 2016).

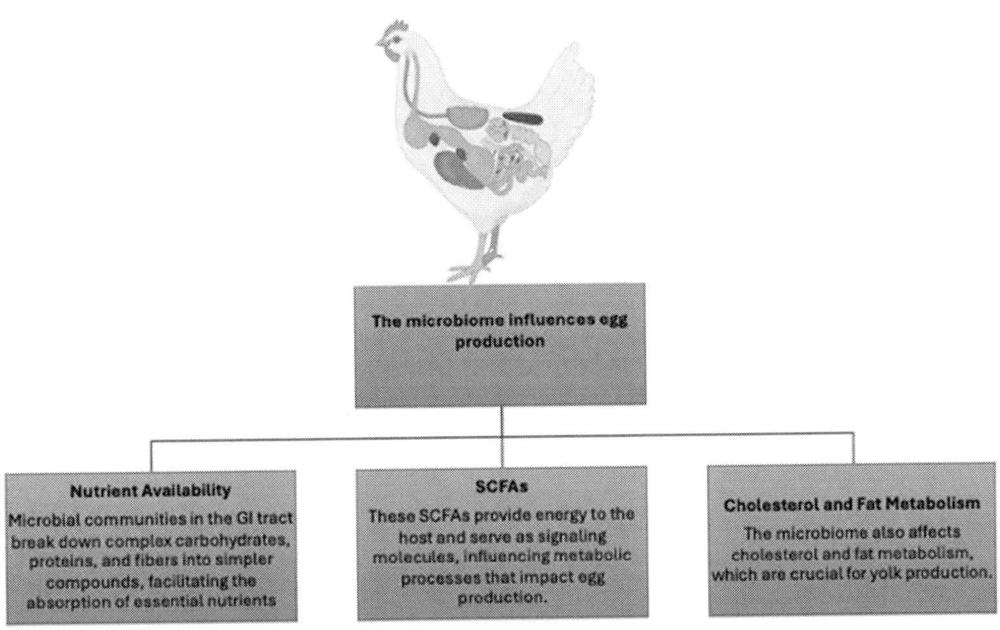

Figure 1. An illustration that depicts the influence of microbiome on egg production.

Immune Function

A balanced microbiome contributes to immune function, indirectly supporting egg production:

1. *Immune Modulation:* Probiotics and beneficial bacteria in the microbiome modulate immune responses by stimulating antibody production, reducing inflammation, and influencing cytokine activity (Shini et al., 2017). This immune modulation helps protect reproductive organs from infections and inflammation that could impair egg production.

   The development of the intestinal immune system depends on the microbiota, but immune responses also control the composition and structure of the intestinal microbiota by developing special immunological adaptations to deal with the high bacterial load. Together, the immune systems make sure that commensal bacteria rarely cross the intestinal barrier. The communication between microbiota and the immune system is mediated by the interaction of bacterial components with pattern recognition receptors expressed by intestinal epithelium and various antigen-presenting cells which result in innate and adaptive immune response activation. Mucosal homeostasis and health status of the host are reflection of an interaction between the microbial community and the host (Kogut et al., 2020).

   The host-microbiota interaction affects the host's metabolism, immunity, and health (Marchesi et al., 2016). This crosstalk is mediated by dietary nutrients, host and microbiota metabolites, as well as antimicrobial compounds (Kogut et al., 2020). Defensins are non-specific antimicrobial peptides produced by the host that regulate microbiota growth and anatomical location (Den Hartog et al., 2016), and miRNAs that regulate bacterial transcripts and bacterial growth (Liu et al., 2016). In response to the local environmental, commensal microorganisms in the GIT initiate metabolic pathways that activate bacterial metabolism, enabling them to evade, modify, or outlive host innate immune destruction. Furthermore, certain compounds derived from microbes might encourage particular commensal processes that are advantageous to the microbe and the host. Similar to this, the microbes are recognized by the host when they produce particular molecules or components with distinctive chemical patterns, which trigger both innate and acquired immune responses. Thus, the adaptation of the commensal bacteria/viruses/ fungi residing in the host intestine resulted in a mutually beneficial coexistence for both microbiota and host during homeostasis (Broom and Kogut, 2018).

   According to Bene et al. (2017), the host immune response is controlled by the interdependent interaction between the microbiota and the host, which results in immunological tolerance to commensal microorganisms while maintaining susceptibility to invasive pathogens. This immune balance is upset when the gut microbial communities are changed, which results in immunological deregulation and increased vulnerability to illness.

   Antimicrobial peptides (AMPs) are produced when T cells produce IL-17 and IL-22 in response to pro-inflammatory cytokines produced by macrophages and DCs in response to microbial signals. Microbiota antigens are transported by dendritic cells to the Peyer's patches and small lymphoid follicles in the avian gut, where they stimulate the development of Th17 and regulatory T cells, which in turn stimulate the development of IgA-producing plasma B cells that secrete more IgA.

2. *Gut Barrier Integrity:* The microbiome supports the integrity of the intestinal barrier, reducing the risk of pathogen translocation and systemic infections. A healthy gut barrier minimizes the likelihood of diseases such as Colibacillosis and Salmonellosis that could negatively impact egg production (Pan and Yu, 2014).

Reproductive Hormones

The microbiome also affects reproductive hormones that play a role in egg production:

1. *Hormonal Regulation:* The microbiome influences the synthesis and metabolism of reproductive hormones such as estrogen and progesterone, which are essential for ovulation and egg production (Pérez-Cobas et al., 2015). Dysbiosis can disrupt hormonal balance; potentially reducing egg production. GnRH is a primary reproductive hormone synthesized in the hypothalamus and regulates gonadotropin secretion from the anterior pituitary (Ubuka et al., 2008). The peptide hormone is released from nerve terminals in the median eminence into portal capillary vessels (Ni et al., 2011). It plays a vital role in the control of ovulation rate, sexual maturity and reproductive aging. Two isoforms of GnRH are presented in the domestic chicken cGnRH-I and cGnRH-II GnRH plays a significant role in the activation of the reproductive axis and release of reproductive hormones in non-laying hens to increase ovarian activity and facilitate laying rate. It is also associated with an increase in other hormones FSH, LH, estrogen, and progesterone, and also induces ovary morphology and signs of maturity (Hanafy and Elnesr, 2021). The secretion and regulation of gonadotropins, FSH, and LH are critical for normal reproductive function (Jalob & Al-Noori, 2010). Changes in levels of FSH, LH, and progesterone have been found to affect reproductive efficiency in layers (Onagbesan et al., 2006). LH is responsible for ovulation, high levels of the same indicate a rise in egg development (Tanabe et al., 1981).
2. *SCFAs and Hormones:* SCFAs produced by microbial fermentation can also impact reproductive hormone synthesis, indirectly influencing fertility and egg production (Pan & Yu, 2014). In a work done by Wang et al., 2022 it has been concluded that dietary supplementation of combined effect of 600 μg/kg MCE (Macleaya cordata extract) and 5 × 108 CFU/kg compound Bacillus improved laying performance by improving microbiota with enhanced antioxidant capacity and intestinal barrier, regulate reproductive hormones and the concentration of cecal short-chain fatty acids of laying hens.

Probiotic and Prebiotic Supplementation

Probiotic and prebiotic supplementation can enhance the poultry microbiome, supporting egg production. The inclusion of probiotics and prebiotics could enhance the production performance as well immune status of birds by augmenting the T cell-mediated immune response (Khan et al., 2020). An important application of prebiotics and probiotics in layers is to reduce the colonization of pathogenic bacteria like Salmonella and Campylobacter in the gut. Salmonella internalizes into enterocytes and survives within macrophages and M cells and triggers an inflammatory response. Hence, the positive manipulation of the gut microbiota is a useful approach to improve food safety and control avian diseases such as spotty liver disease (Khan et al., 2020). Mitigation of food pathogens like Salmonella continues to be a focus in poultry production, whether on shell eggs or meat products. Targeted in-feed additives are a component of a food safety program and might help to reduce *Salmonella* infection. The addition of a combination probiotic and probiotic product (Fulfill Plus) reduced *Salmonella* colonization in both the intestines and ovaries of layers (Kimminau et al., 2021).

Probiotics

Use of probiotics is one of the promising strategies for maintaining egg production levels throughout laying hens'welfare (Lutful Kabir., 2009). Prebiotics are specific nutrients that resist digestive processes in the GIT, selectively ferment, and interact with one or a limited number of microorganisms in the intestinal microbiota (Callaway and Ricke, 2012). Probiotics, particularly lactic acid bacteria and bifidobacteria help maintain a balanced microbiome by promoting competitive exclusion, inhibiting the growth of harmful bacteria, and supporting nutrient absorption. This balance supports reproductive health and egg production. Numerous bacterial strains have been investigated and their effects on immune system stimulation, animal performance, egg production, and gut microbiome modifications have all been evaluated (Peralta-Sanchez et al., 2019). Rhodobacter capsulatus improved hen health and egg quality during the last period of the laying. Bacillus subtilis has been tested successfully against Salmonella infection and increased egg quality and production or B. licheniformis acted as an immune system enhancer and a hormone regulator (Wang et al., 2022). The use of E. faecalis UGRA10 showed some advantages as enterococci are common bacteria in warm blood animals and have a beneficial interaction with the immune system, hormones and metabolism, and gut microbiota (Han et al., 2016).

Prebiotics

Prebiotics, such as fructooligosaccharides (FOS) and mannanoligosaccharides (MOS), promote the growth of beneficial bacteria, supporting nutrient absorption, metabolic regulation, and immune function (Stanley et al., 2013). This enhancement indirectly benefits egg production. Swiatkiewicz et al., (2010) reported a positive effect of prebiotic supplementation on eggshell quality parameters in laying hens; which is related to the availability of minerals. Scholz-Ahrens et al., (2007) stated that the high solubility of minerals is attributed to increased production of short-chain fatty acids.

Goblet cells proliferate and the mucus layer's chemical characteristics develop as a result of interactions between the epithelium, nutrition, and microbiota that are impacted by prebiotics and probiotics. As a result, there may be an increase in the quantity, viscosity, and acidity of mucus secretions, which can improve the integrity of the gut barrier, strengthen resistance to digestive infections, and promote GIT health (Bansil and Turner, 2006).

The inclusion of prebiotics could also lead to increased mucus layer thickness, the number of goblet cells in intestinal villi, and epithelial mucin activity (Teng and Kim, 2018). The mucus layer is essential for shielding the intestinal epithelium from the many aggressors that are present in the gut lumen, such as bacteria, chemicals, enzymes, and physical elements (Bansil and Turner, 2006). In both humans and chickens, Mucin 2 (MUC2) is a secreted glycoprotein produced by goblet cells, and it constitutes a major component of the gut mucosa (Shani and Irani M., 2024). It has been demonstrated by Van Hemme et al., (2019) that supplementation with immune system modulators like prebiotics, including cell walls of yeast ingredients, is a valuable tool for enhancing immune system response.

Conclusion

The microbiome plays a vital role in the health, productivity, and overall well-being of poultry. Understanding the diverse and dynamic microbial community residing in the gastrointestinal (GI) tract is crucial for optimizing poultry production. This document has outlined the significant impact of the microbiome on nutrient absorption, immune function, disease prevention, and reproductive health in poultry. The composition and diversity of the microbiome vary across different sections of the GI tract, each serving unique functional roles. Factors such as diet, age, genetics, and environment significantly influence microbial balance. Maintaining a balanced microbiome is essential for enhancing nutrient absorption, promoting immune function, and preventing diseases such as necrotic enteritis, salmonellosis, and colibacillosis. Probiotics and prebiotics have emerged as effective strategies to support and enhance the poultry microbiome. These supplements promote the growth of beneficial bacteria, improve gut health, and reduce the reliance on antibiotics, addressing concerns over antibiotic resistance. In addition to health benefits, the microbiome influences reproductive functions and egg production in poultry. By modulating nutrient absorption, immune responses, and reproductive hormones, a balanced microbiome contributes to optimal fertility and productivity. In conclusion, the microbiome is integral to poultry health and production. Future research and practical applications should continue to focus on manipulating the microbiome through dietary management, probiotic and prebiotic supplementation, and improved management practices to achieve sustainable and efficient poultry production. By leveraging the microbiome, the poultry industry can enhance animal welfare, productivity, and economic viability.

References

Abd El-Hack ME, El-Saadony MT, Elbestawy AR, El-Shall NA, Saad AM, Salem HM, El-Tahan AM, Khafaga AF, Taha AE, AbuQamar SF, El-Tarabily KA. Necrotic enteritis in broiler chickens: disease characteristics and prevention using organic antibiotic alternatives - a comprehensive review. *Poultry Science*. 2022 Feb;101(2):101590.

Attia, Y. A., Al-Harthi, M. A., & Abo El-Maaty, H. M. (2020). Calcium and cholecalciferol levels in late-phase laying hens: effects on productive traits, egg quality, blood biochemistry, and immune responses. *Frontiers in Veterinary Science*, *7*, 389.

Bansil, R., & Turner, B. S. (2006). Mucin structure, aggregation, physiological functions and biomedical applications. *Current opinion in colloid & interface science*, *11*(2-3), 164-170.

Bene, K., Varga, Z., Petrov, V. O., Boyko, N., &Rajnavolgyi, E. (2017). Gut microbiota species can provoke both inflammatory and tolerogenic immune responses in human dendritic cells mediated by retinoic acid receptor alpha ligation. *Frontiers in Immunology*, *8*, 427.

Broom, L. J., & Kogut, M. H. (2018). Inflammation: friend or foe for animal production?. *Poultry Science*, *97*(2), 510-514.

Callaway, T. R., & Ricke, S. C. (2012). *Direct-fed microbials and prebiotics for animals: science and mechanisms of action. Springer.*

Den Hartog, G., De Vries-Reilingh, G., Wehrmaker, A. M., Savelkoul, H. F. J., Parmentier, H. K., & Lammers, A. (2016). Intestinal immune maturation is accompanied by temporal changes in the composition of the microbiota. *Beneficial microbes*, *7*(5), 677-685.

Han, Geon Goo, Eun Bae Kim, Jinyoung Lee, Jun-Yeong Lee, Gwideuk Jin, Jongbin Park, Chul-Sung Huh et al. "Relationship between the microbiota in different sections of the gastrointestinal tract, and the body weight of broiler chickens." *Springerplus* 5 (2016): 1-9.

Hanafy, A. M., &Elnesr, S. S. (2021). Induction of reproductive activity and egg production by gonadotropin-releasing hormone in non-laying hens. *Reproduction in Domestic Animals*, *56*(9), 1184-1191.

Jalob, Z. K., & Al-Noori, M. A. (2010). The effect of Gonadotroph in releasing hormone (GnRH) in the reproductive traits of layer chickens. *Kufa Journal for Veterinary Medical Sciences*, 1(2), 97–103.

Khan, S., Moore, R. J., Stanley, D., &Chousalkar, K. K. (2020). The gut microbiota of laying hens and its manipulation with prebiotics and probiotics to enhance gut health and food safety. *Applied and environmental microbiology*, *86*(13), e00600-20.

Kimminau, E. A., Karnezos, T. P., Berghaus, R. D., Jones, M. K., Baxter, J. A., & Hofacre, C. L. (2021). Combination of probiotic and prebiotic impacts Salmonella Enteritidis infection in layer hens. *Journal of Applied Poultry Research*, *30*(4), 100200.

Kogut, M. H., & Arsenault, R. J. (2016). Gut health: The new paradigm in food animal production. *Frontiers in Veterinary Science, 3*, 71.

Kogut, M. H., Lee, A., & Santin, E. (2020). Microbiome and pathogen interaction with the immune system. *Poultry science*, *99*(4), 1906-1913.

Liu, S., Da Cunha, A. P., Rezende, R. M., Cialic, R., Wei, Z., Bry, L., ... & Weiner, H. L. (2016). The host shapes the gut microbiota via fecal microRNA. *Cell host & microbe*, *19*(1), 32-43.

Lutful Kabir, S. M. (2009). The role of probiotics in the poultry industry. *International journal of molecular sciences*, *10*(8), 3531-3546.

Marchesi, J. R., Adams, D. H., Fava, F., Hermes, G. D., Hirschfield, G. M., Hold, G., ... & Hart, A. (2016). The gut microbiota and host health: a new clinical frontier. *Gut*, *65*(2), 330-339.

Ni, Y., Lu, L., Chen, J., & Zhao, R. (2011). Changes of hypothalamic GnRH-I, POMC and NPY mRNA expression and serum IGF-I and leptin concentrations during maturation of shaoxing ducks (*Anas platyrhynchos*). *Asian-Australasian Journal of Animal Sciences*, 24(9), 1211–1216. https://doi.org/10.5713/ajas.2011.11024.

Onagbesan, O. M., Metayer, S., Tona, K., Williams, J., Decuypere, E., & Bruggeman, V. (2006). Effects of genotype and feed allowance on plasma luteinizing hormones, follicle-stimulating hormones, progesterone, estradiol levels, follicle differentiation, and egg production rates of broiler breeder hens. *Poultry Science*, 85(7), 1245–1258. https://doi.org/10.1093/ps/85.7.1245.

Pan, D., & Yu, Z. (2014). Intestinal microbiome of poultry and its interaction with host health. *Poultry Science, 93*(2), 282–288.

Peralta-Sánchez, J. M., Martín-Platero, A. M., Ariza-Romero, J. J., Rabelo-Ruiz, M., Zurita-González, M. J., Baños, A., ... & Martínez-Bueno, M. (2019). Egg production in poultry farming is improved by probiotic bacteria. *Frontiers in microbiology*, *10*, 1042.

Pérez-Cobas, A. E., et al. (2015). *Direct-fed microbials and prebiotics for animals: science and mechanisms of action. Poultry Science, 94*(5), 1000–1006.

Scholz-Ahrens, K. E., Ade, P., Marten, B., Weber, P., Timm, W., Asil, Y., ... &Schrezenmeir, J. (2007). Prebiotics, Probiotics, and Synbiotics Affect Mineral Absorption, Bone Mineral Content, and Bone Structure1. *The Journal of nutrition*, *137*(3), 838S-846S.

Shani, M. A., & Irani, M. (2024). Feeding strategy and prebiotic supplementation: Effects on immune responses and gut health in the early life stage of broiler chickens. *Research in Veterinary Science*, *171*, 105226.

Korver D R. (2012) Implications of changing immune function through nutrition in poultry. *Animal Feed Science and Technology, 173*(1-2), 54–64.

Pan D, Yu Z. Intestinal microbiome of poultry and its interaction with host and diet. *Gut Microbes.* 2014 Jan-Feb;5(1):108-19.

Świątkiewicz, S., Koreleski, J., &Arczewska, A. (2010). Laying performance and eggshell quality in laying hens fed diets supplemented with prebiotics and organic acids. *Czech Journal of Animal Science*, *55*(7), 294-306.

Tanabe, Y., Nakamura, T., Tanase, H., & Doi, O. (1981). Comparisons of plasma LH, progesterone, testosterone and estradiol concentrations in male and female chickens (*Gallus domesticus*) from 28 to 1141 days of age. *Endocrinologia Japonica*, 28(5), 605–613. https://doi.org/10.1507/endocrj1954.28.605.

Teng, P. Y., & Kim, W. K. (2018). Roles of prebiotics in intestinal ecosystem of broilers. *Frontiers in Veterinary Science*, *5*, 245.

Ubuka, T., McGuire, N. L., Calisi, R. M., Perfito, N., & Bentley, G. E. (2008). The control of reproductive physiology and behavior by gonadotropin-inhibitory hormone. *Integrative and Comparative Biology*, 48(5), 560–569. https://doi.org/10.1093/icb/icn019.

Van Hamme, V., Spaepen, R., & Smeets, N. (2019). Algal beta-(1, 3)-glucan and its effect on infectious bursal disease vaccination in poultry. *VlaamsDiergeneeskundigTijdschrift*, *88*(6).

Wang, F., Zou, P., Xu, S. *et al.* Dietary supplementation of *Macleaya cordata* extract and *Bacillus* in combination improve laying performance by regulating reproductive hormones, intestinal microbiota and barrier function of laying hens. *J Animal Sci Biotechnol* 13, 118 (2022). https://doi.org/10.1186/s40104-022-00766-4.

Chapter 15

Microbiomes of Pig and Horse Health, Disease, and Production

Amitava Roy[1,*]
and Tanmoy Rana[2]
[1]Department of Livestock Farm Complex, W.B. University of Animal & Fishery Sciences, Kolkata, India
[2]Department of Veterinary Clinical Complex, W.B. University of Animal & Fishery Sciences, Kolkata, India

Abstract

It is often recognized that pigs are an important livestock animal that contribute to the production of meat. However, compared to other non-primate animals, pigs provide good biological models for human health, which is why the gut microbiota of pigs has garnered a lot of attention recently. Numerous microorganisms, including bacteria, fungi, viruses, protozoa, and bacteriophages, are present in the gastrointestinal system of pigs. These microorganisms coexist in a complex symbiotic ecosystem with the host and are essential to the preservation of the pigs' many physiological, nutritional, and immune systems. However a variety of factors, such as management and rearing practices, environmental conditions, and the use of antibiotics, alters the bacteria population in the pig's gut, which impacts growth and health. So modulation of gut microbiota is required which not only helps in improving the health of pigs leading to higher production but also generates more revenue for the pig industry. As for other animals and humans, the horse gut microbiome is sensitive to diet, especially consumption of starch, fiber, and fat. Age, breeds, stress during competitions, transportation, and exercise may also impact the microbiome. Because of its size and its complexity, the equine GIT microbiota is prone to perturbations caused by external or internal stressors that may result in digestive diseases like gastric ulcer, diarrhea, colic, or colitis, and that are thought to be linked with systemic diseases like laminitis, equine metabolic syndrome or obesity. Therefore, the purpose of this chapter is to comprehend the structure and function of the common core microbiome found in each segment of the gastrointestinal tract and to identify putative microbial biomarkers of health or disease. These markers are important for predicting potential perturbations, optimizing global practices, and creating customized and adapted nutritional strategies.

Keywords: gut microbiota, health, modulation, pig, equine, production

---

* Corresponding Author's Email: amitavanutri07@gmail.com

In: Microbiomes in Animal Nutrition, Health, and Disease
Editor: Tanmoy Rana
ISBN: 979-8-89530-425-9

Introduction

Numerous microorganisms, including bacteria, fungi, viruses, protozoa, and bacteriophages, are present in the gastrointestinal system of pigs. More than 98% of the isolated genetic sequences found in the gut, out of roughly 10^{10}–10^{14} cells, are derived from bacteria (Turner, 2018). According to Lee and Mazmanian (2010) and Brestoff and Artis (2013), these bacteria live in a complex symbiotic environment with their host and are essential to the maintenance of the pigs' diverse physiological, nutritional, and immunological processes. These systems support the pigs' productivity while also assisting in their health management in farms. The gut microbiome is home to a variety of dangerous bacteria, including Salmonella, Clostridium, Escherichia coli, and other opportunistic pathogens, in addition to helpful microbes. Pathogens in the gut exert pathogenicity upon any kind of movement towards abnormalities, which accounts for illness and mortality in the pig population. Antimicrobials are therefore frequently employed as a preventative and health-management measure to minimize the loss in the cattle and animal husbandry industries to tackle that. In addition, the industry uses antimicrobials as growth factors and throughout the weaning process for piglets to achieve output targets. This leads to post-weaning diarrhea as well as detrimental effects on the gut microbiota of the weaned piglets and disruption of the gut lumen environment (Konstantinov et al., 2006). Overuse of antimicrobials damages good bacteria and can make it more difficult to get the pigs healthy. Antimicrobials that promote growth also increase the number of bacteria that are resistant to antibiotics, which is extremely concerning for public health (Dibner and Richards, 2005).

Horses are large non-ruminant herbivores and rely on microbial fermentation for energy, with more than half of their maintenance energy requirement coming from microbial fermentation occurring in their enlarged caecum and colon. To achieve that, the gastro-intestinal tract (GIT) of horses harbors a broad range of various microorganisms, differing in each GIT segment, which are essential for efficient utilization of feed, especially to use nutrients that are not or little degraded by endogenous enzymes. In addition, like in other animal species, the GIT microbiota is in permanent interplay with the host's cells and is involved in a lot of functions among which inflammation, immune homeostasis, and energy metabolism. In-depth research on gut microbiota, probiotic supplementation with various prebiotics, feeding highly fermentable carbohydrates, and microbial transplantation are strategies to lessen reliance on antibiotics for pig and equine health maintenance and growth to increase productivity. This chapter focuses on several variables that impact the intestinal microbiota of pigs along with equine and how they contribute to the preservation of health and productive pig farming and equine health.

Intestinal Microbiota of Pig

There are many different bacterial and non-bacterial components that make up the pig gut microbiome. The population of these constituents is influenced by a variety of factors, including management styles, feeding patterns, and environments in which they are raised.

Bacterial Components of the Microbiome

Soon after birth, bacteria begin to colonize the piglets' guts (Fouhse et al., 2016). Escherichia coli and Streptococcus spp. are used for initial colonisation, which creates an anaerobic environment that aids in the colonization of *Bacteroides* spp., *Bifidobacterium* spp., *Clostridium* spp., and *Lactobacillus* spp. (Petri et al., 2010). The predominant bacteria in the piglet's digestive system during the pre-weaning stage are *Lactobacilli.* (Petri et al., 2010). The bacteria discovered in pig feces during the post-weaning phase were categorized into five groups by Kim et al. (2011): Firmicutes, Bacteroidetes, Proteobacteria, Actinobacteria, and Spirochaetes. Of them, the phylum Firmicutes makes up the greatest fraction (together with Bacteroidetes), collectively making up 90% of the total bacterial population, mostly in pigs that are 3 weeks old. The population of Firmicutes grows with time about the Bacteroidetes phylum. Unexpectedly, unclassified bacterial groupings proliferate as well. When the pigs are 10 weeks old, *Prevotella* spp. account for 30% of all classifiable bacteria; nevertheless, their concentration only drops to 3.5–4% at 22 weeks old (Kim et al., 2011). *Prevotella* spp. are the most common bacteria. *Prevotella* species, *Anaerobacter* species, *Streptococcus* species, *Lactobacillus* species, *Coprococcus* species, *Megasphaera* species, *Subdoligranulum* species, *Blautia* species, *Oscillibacter* species, *Faecalibacterium* species, *Dialister* species, *Sarcina* spp., and *Butyricoccus* spp. are the most prevalent species (Table 1).

Table 1. The predominant population of bacteria in certain intestinal segments

Part of the Intestine	Predominant bacteria
Jejunum	Cyanobacteria, Firmicutes, Actinobacteria, Proteobacteria
Ilium	Firmicutes, Proteobacteria
Colon and caecum	Bacteroidetes , Firmicutes

Non-Bacterial Components of the Microbiome

Urubschurov et al. (2011) screened pig feces using the 26S rRNA gene in conjunction with denaturing gradient gel electrophoresis (DGGE) and DNA sequencing in an attempt to identify the non-bacterial components in the pig gut. Using a cultured approach, the researcher's isolated yeast including Candida catenulate, Kazachtania slooffiae, Galactomyces geotrichum, and Candidaglabrata (Urubschurov et al., 2011). Shan et al. (2011) demonstrated in a different study that 99 percent of the RNA viruses found in pig feces belonged to the families Picornaviridae, Astroviridae, Coronaviridae, and Caliciviridae. The DNA viruses that are found belong to the families Parvoviridae and Circoviridae.

Features of the Horse Gastrointestinal Tract Microbiota and Its Time Evolution throughout Life

The majority of the data used in this chapter came from studies that used DNA sequencing techniques or other omics techniques. Over the past 15 years, these methods have greatly expanded the scientific understanding of complex microbial ecosystems, particularly digestive ecosystems. The term "microbiota" describes a diverse group of harmful, commensal, and symbiotic microorganisms, such as fungi, bacteria, archaea, and protozoa. According to a new definition (Gresse et al., 2019), the microbiome is the culmination of the microbiota and its milieu of activity, which is made up of plasmids, viruses, dead cell-extracellular DNA, and microbial structural components including proteins, polysaccharides, nucleic acids, etc. Its composition is the outcome of the host's long-term evolutionary adaptation to its food. Molecular markers for bacteria and archaea have been identified by focusing on variable sections of the 16S rRNA gene in the analysis of the gut microbiota in horses by DNA sequencing techniques. Even though eukaryotic communities like ciliate protozoa and anaerobic fungi have been described in the hindgut of horses and are thought to be involved in digestion (Yeoman et al., 2018), there is a dearth of information about the true ecological roles of these communities because there hasn't been much published literature on how these communities have been evaluated by DNA sequencing techniques. Therefore, the Archaea and Bacteria domains of the microbiota will be the primary focus of this review.

Is There a True "Common Core Microbiota" among Healthy Adult Horses, or Is It Just a Myth?

Given its significance for both healthy and sick horses, the digestive site should be carefully considered while examining the horse's GIT microbiota. The distal portion of the GIT has more richness and evenness (Lyte, 2011), indicating that this environment is more complicated. The different GIT segments may have a "common core microbiota," or a collection of microbial species that are shared by all or most horses (Lyte, 2013), with notable variances between the foregut and hindgut. According to Lyte (2013), the definition of the core community in feces at the Operational Taxonomic Unit (OTU) level is "being present in all samples included in the study at 0.1% relative abundance (or greater)". According to several studies, fecal bacterial communities differ from those in the upper tract but are not substantially different from those in the colon (Mach et al., 2017) or even the caecum (Mura et al., 2019). Therefore, it is intriguing to consider using feces as a partial non-invasive marker of what occurs in the hindgut (particularly the colon) but not in the foregut.

Because the definitions of a "normal" GIT microbiota and a "common core microbiota" differ from study to study, it is still challenging to define these concepts in horses. One intriguing point to note is that, unlike ruminants and other animal species where a small number of OTUs are highly dominant, no single OTU dominates the microbial core community of horses. The large intestine and faeces still have a significant percentage of unidentified read assignments at the genus level, even though we are beginning to understand the common core microbiota of horses. This suggests that the genera contained in these samples may be different from those isolated from other animals. Just 30% of the caecal microbiota has been cultured, indicating a clear absence of cultured representatives of those new bacteria (Schoster et al., 2020). Future research is necessary to thoroughly identify and comprehend the horse's GIT microbiota. Furthermore, we don't know much about the "functional core microbiota," or the roles that the microbiota plays.

Evolution of the Horse Microbiota with Age

The first week of a foal's life is regarded as crucial since respiratory conditions, enteritis, and sepsis all increase morbidity and death during this time. Therefore, from a nutritional and health perspective, it is crucial to optimize the colonization of the foal's gut from birth, with a balanced balance of functionally relevant microbial communities.

The mare plays a crucial part in the newborn's acquisition of gut colonization because microbial transmission happens as a result of the neonate and dam's recurrent interactions. The meconium of foals is home to a diverse community of bacteria, including Aerococcus (of the

Clostridiales order), an anaerobic fermenter of the mammalian gut ecosystem, and other members of the common gut microbiota such as *Streptococcus, Enterococcus,* and *Enterobacterioaceae* (Clark et al., 2018) *Acinetobacter, Stenotrophomonas,* and *Sanguibacter* have all been identified as opportunistic bacteria from various animal hosts.

The makeup of the mare's gut microbiota was found to be similar to that of the foal's rectal microbiota at birth (Clark et al., 2018). This suggests that the gut microbiota of the dam particularly provides microbial components from the gut environment to the meconium community. Furthermore, the fetus may receive a diverse microbial subset from the amniotic fluid due to its low bacterial DNA load (Clark et al., 2018). Thus, the mare's dendritic cells, which pierce the host intestinal epithelia and sample luminal bacteria or bacterial antigens that would subsequently be released into the placenta via the bloodstream, could be the internal transmission route of microbial components, as has also been hypothesized in other animal species and humans (Peachey et al., 2019). This method of supplying the foal with bacterial components would prepare its immune system for the development of the gut microbiome. Since these current findings were collected on samples with relatively low biomass utilizing DNA sequencing-based methods, it is still unknown if the bacteria involved in this microbial translocation are fragmented, dead, or living.

In addition to early life, as horses live longer, the microbiome may change with age, as other species have shown. It is acknowledged that alterations in the fecal microbiota are linked to senescence and inflammation-aging in both people and canines. There are currently few available data on how age affects a horse's microbiome. While Paul et al. (2021) found no differences between health old (median 22.6 1.8 y old) and adult horses (median 6.8 3.7 y old), Venable et al. (2017) observed a decrease in richness but no differences in the structure and composition of the fecal microbiota in healthy old horses (19–28 y old) vs. adult horses (5–12 y old). However, Mach et al. (2020) found that adult ponies (mean 9.83 3.21 yr) had a lower relative abundance of Fibrobacteres and a greater alpha-diversity of Proteobacteria compared to old ponies (mean 21.55 2.94 yr). This suggests that adult ponies have a better ability to use fibre than old equines. Nevertheless, despite the study's elegance, the scientists were unable to identify any distinct biomarkers in the metabolome or fecal microbiota, highlighting the challenge of establishing links between the GIT microbiota and host phenotype based just on feces, which are not representative of the entire digestive tract.

Factors Affecting the Gut Microbiota in Pigs

The microbial population in the pig's gut can be influenced by several factors, including the breed of the animal, the management techniques employed, the impact of host genetics and environmental factors (such as pens and kinship), age, body weight, obesity, diet, types of enrichment, and the use of antimicrobials (Wang et al., 2019).

Breed

Pig breed may potentially have an impact on the makeup of the gut microbiota. Bian et al. (2016) observed that the bacterial community structure in the feces of obese Meishan (a Chinese domestic breed) and slim Yorkshire (a breed of domestic piglets) differed on day 14 and day 49. Compared to Yorkshire sows, the Meishan sows had comparatively larger

Firmicutes but lower Bacteroidetes abundances; this could have an impact on piglets' postnatal development.

Birth

Research has indicated that the microbial community of pigs is predominantly impacted by microbial contacts that occur during the early stages of life, specifically during the act of birth (Patil et al., 2020). In actuality, the mode of delivery—vaginal/normal delivery or cesarean section—as well as the piglets' diet have a significant impact on the microbial composition of the swine intestine (Patil et al., 2020).

Age

The age of the pigs has an impact on the microbial colonization in their guts as well (Holman et al., 2017). According to Dowarah et al. (2017), the bacteria that colonize the earliest (between birth and two days of age) are Escherichia sp., Clostridium sp., Fusobacterium sp., Streptococcus sp., and Enterococcus sp. At six hours of age, Clostridium spp. makeup around 34% of the entire microbial population in the pigs' gut; by the time they are 20 days old that number drops to 1% (Petri et al., 2010).

Diet and Feed Efficiency

Food has a big influence on the population of gut microbes, and feeding pigs a healthy, balanced diet at different phases of life also helps prevent dysbiosis, which is the cause of many chronic diseases (Doré and Blottière, 2015). However, any dietary modifications or imbalanced rations combined with low feed digestibility can alter the abundance of pathogenic bacteria in the gut, which could negatively impact the health and productivity of the animal. Feed efficiency is a crucial economic and environmental factor that has been thoroughly examined to determine whether farming is feasible, and the swine business is no exception.

Dietary Fibre

Dietary fibres are plant polysaccharides that may be fermented by microorganisms in the large intestines of animals with single stomas (Bindelle et al., 2008). Dietary fibre, which is included in the feed, has been shown to improve intestinal health and animal welfare by reducing the incidence of diarrhea in pigs at the weaning stage. In actuality, short-chain organic acids (acetate, lactate, propionate, and butyrate) are produced when bacteria breakdown food fibres and these acids stimulate the growth of the digestive tract by strengthening the gut epithelia (Montagne et al., 2003). *Salmonella, E. coli*, and *Clostridium* development are also inhibited by these acids (Pickard et al., 2017). Non-starch polysaccharides reduce the pH of the colon, lengthen intestinal villi, promote the proliferation of commensals in the gut, and boost the generation of short-chain organic acids (Lindberg, 2014).

Probiotics and Prebiotics

When taken in sufficient quantities, probiotics, a class of microorganisms, improve a person's general health (Linares et al., 2017). Probiotics enhance the digestion and metabolism of nutrients, which leads to improved animal growth and, ultimately, a notable improvement in the animal's production performance (Wang and Donovan, 2015). According to Galdeano et al., (2010), probiotic supplementation increases the number of beneficial bacteria in the pig's gut, such as Lactobacilli, and reduces the population of harmful bacteria, such as E. coli, Clostridia spp., etc. This results in a significantly higher growth rate and body weight gain, which helps the pig industries meet their production targets (Wells et al., 2005). Prebiotics are substances that support the increase of advantageous microorganisms in the gut population (Hutkins et al., 2016).

Dietary Copper

When added to animal feed, copper, an important micronutrient, can support a variety of animal biological processes. To mention a few benefits, it enhances the activity of digestive enzymes, aids in iron metabolism, and shields the animal from oxidative stress (Zhang et al., 2019). Weaned piglets on copper-enriched diets are said to grow more body weight, have higher feed conversion ratios, and experience fewer diarrhea episodes. Because copper has anti-microbial and growth-promoting qualities, the pig business has traditionally supplemented feed with copper (Patil et al., 2020). According to Højberg et al. (2005), this procedure aids in the reduction of Lactobacillus and Streptococcus colonies in the pig gut. Increased unsaturated fatty acid content from high dietary copper intake results in the development of softer pork fat (Dêbski, 2016).

Stressors and Weaning

It is commonly known that stress in any form, including heat stress and weaning, impacts livestock's ability to develop, reproduce, produce, and be susceptible to illness (Parkunan et al., 2015). A piglet's life is divided into six periods, with weaning being the most important one. Weaning-related stress modifies the microbial metabolic patterns in the gut in addition to microbial composition and function (Li et al., 2018). When the host animal experiences stressful situations like confinement, weaning, heat, etc. during this time, functional alterations in the gut microbiota are seen (Patil et al., 2020). Pig production standards are hampered by the ileal Lactobacilli population's significant changes brought on by the stress of weaning (Urubschurov et al., 2011).

Antibacterials

In the pig business, antibiotics such as tetracyclines, penicillins, tylosin, and sulfamethazine are frequently utilized as growth promoters (Kim et al., 2016). However, these substances negatively affect the commensals that inhabit the pig's stomach (Dibner and Richards, 2005). Overuse of antibiotics increases immunological tolerance, which makes the body less able to fight against infections from pathogenic species that are resistant to antibiotics in the future (Schokker et al., 2014). Antibacterials have been shown to have detrimental impacts on host health and gut flora in recent times. Moreover, prolonged use of antibiotics leads to the development of antibiotic resistance in addition to a decrease in the diversity and relative abundance of bacteria. Therefore, individualised antibiotic treatment is frequently recommended to reduce the effects of antibiotics on the gut microbiota and prevent this side effect.

Horse GIT Microbiota Variations Based on Abiotic Elements

Impact of Diet Composition

Similar to other mammals, horses' nutrition has a significant impact on the makeup of their gut bacteria. Because of the existence of active fungal and bacterial populations in their hindgut that break down fibre (Morgan et al., 2015) and digest high-forage diets (Mach et al., 2020), horses are physiologically well suited to handle them. However, to boost energy density—which is typically necessary for horses who exercise—this forage diet is frequently substituted with high-starch diets. Numerous investigations have contrasted the effects on equine fecal microbiota of high starch and high forage diets. The first research on viable cultivable functional bacterial groups, the investigations by Millinovich et al. (2007) and Jouany et al. (2009) were the first to show a shift in bacterial populations with a high starch diet. Because of the lower pH, higher lactic acid concentration, and decreased fibre digestibility, these bacterial alterations were linked to decreased polysaccharide-degrading and glycoside hydrolase bacterial activity. Sub-acute ruminal acidosis (SARA) is a well-defined term used to describe this form of dysbiosis in ruminants' rumens (Bailey et al., 2003). Changes in the makeup and/or activity of the hindgut microbiota in horses result in altered digestive health. Horses are more prone to laminitis or colic pain due to the buildup of lactate and pH decrease in the caecum-colon that causes subclinical acidosis (Al Jassim et al., 2005). Laminitis in horses fed an oligofructose load has been linked to lactate-producing Gramme-positive bacteria, including Streptococcus bovis-Streptococcus equinus species (Kaneko et al., 1997).

Transportation

It is well known that stress from transportation plays a significant role in a number of horse health-related problems, including digestive diseases (Nagano and Yano, 2020). During transit, a number of variables, including humidity, temperature, air quality, feed, and water allowance, may have a substantial impact on gut health. Gastrointestinal diseases accounted for less than 30% of health issues associated to transportation, according to an Australian survey on transport management techniques (Zak et al., 2020). Stress related to transportation may have a negative impact on the composition and activities of the hindgut microbiota because the gut microbiota-brain axis communication pathway might be altered in stressful situations. In fact, a two-hour truck ride could already cause a shift in the populations of faecal bacteria and cause a dysbiotic state (Litvak et al., 2017).

High Performance

Stresses of a physical and mental nature may also impact the gut microbiota-brain axis in endurance horses. Physical and psychological stress during exercise was strongly associated with alterations in the composition of the gastrointestinal microbiota in mice subjected to six weeks of forced treadmill running. These alterations included shifts in the balance of bacterial populations implicated in the breakdown of intestinal mucus and immune function (Abot et al., 2022). Interruption of the intestinal barrier during high-intensity exercise might result in an inflammatory response and leaky gut due to the redirection of blood flow away from the intestines and heat injury to the intestinal mucosa (Rath and Haller, 2022). The gut microbiota composition was linked to equitation conditions in a study by Coleman et al. (2019) with a cohort of 185 healthy horses classified according to equitation discipline, specialty, and level of performance. Specifically, the Gala and Cadre Noir specialties were most closely linked to fecal microbiota composition, notably with a decrease in the relative abundance of the butyrate-producing Lachnospiraceae AC2044 group and the Clostridiales family XIII. These decreases were likely caused by the highly stressful training and show events, as compared to other specialties. A subset of mitochondria-related differentially expressed genes implicated in processes like energy metabolism, oxidative stress, and inflammation in sportive vs. resting horses were described by Mach et al. (2017) in their study of the gut-mitochondria crosstalk in endurance horses.

Heat Stress Conditions

The five climatic variables of a horse's microclimate are ambient temperature, relative humidity, precipitation, wind velocity, and sun radiation, according to Cymbaluk and Christison (1990). The most significant single climatic stressor is ambient temperature. In hot weather, horses' intake of feed might drop by 15% to 20% (Comtet-Marre et al., 2018). It has also been shown that rising temperatures and humidity indexes (THI) cause ruminants to consume less. Reduced rumen motility under heat stress results in feed having a longer residence time within the rumen, increasing dry matter digestibility (Turroni et al., 2014). Additionally, changes in the composition of the microbiota community and metabolic activity (Harlow et al., 2017) raise the risk of ruminal acidosis and reduce milk production (Magne et al., 2020).

Conclusion

It is often recognized that pigs are an important livestock animal that contribute to the production of meat. However, compared to other non-primate animals, pigs provide good biological models for human health, which is why the gut microbiota of pigs has garnered a lot of attention recently. Pigs and humans share a great degree of similarities in terms of their immune systems, genomes, and protein sequences. Pork is therefore a better model than other species when examining the relationship between the early gut microbiota and health hazards.

By keeping a balance between beneficial and potentially harmful bacteria, will not only guarantee that the gastrointestinal microflora of piglets is at its ideal composition, but it will also assist the industry in removing its reliance on antibiotics and other antibacterial to fight sickness or spur growth. Modulating the microbiome to improve horse health seems like a promising open route, but it's also a technique that's part of a full, holistic approach to guarantee high levels of athletic performance, healthy colon fermentation and digestion, and ultimately well-being. In this context, novel microbial-derived products like non-viable bacteria or yeast or bacterial/fungal compounds, which are referred to as postbiotics, as well as pre-and probiotics, are microbiome-focused treatment and prevention methods for gastrointestinal disorders or even more severe systemic syndrome. With the use of these instruments, it will be possible to create sophisticated consortia, specialized products, and next-generation probiotics. They will also provide previously unheard-of insights into the functioning of all pre-, pro-, post-, and para-probiotics.

References

Abot, A., Fried, S., Cani, P. D., Knauf, C. Reactive Oxygen Species/Reactive Nitrogen Species as Messengers in the Gut: Impact on Physiology and Metabolic Disorders. *Antioxid. Redox Signal.* 2022, 37, 394–415.

Al Jassim, R. A. M., Scott, P. T., Trebbin, A. L., Trott, D., Pollitt, C. C. The genetic diversity of lactic acid producing bacteria in the equine gastrointestinal tract. *FEMS Microbiol. Lett.* 2005, 248, 75–81.

Bailey, S. R., Marr, C. M., Elliott, J. Identification and quantification of amines in the equine caecum. *Res. Vet. Sci.* 2003, 74, 113–118.

Bian G., Ma S., Zhu Z., Su Y., Zoetendal E. G. et al. 2016. Age, introduction of solid feed and weaning are more important determinants of gut bacterial succession in piglets than breed and nursing mother as revealed by a reciprocal cross-fostering model. *Environ Microbiol*, 18(5): 1566-1577, doi: 10.1111/1462-2920.13272.

Bindelle J., Buldgen A. and Leterme P., 2008. Nutritional and environmental consequences of dietary fibre in pig nutrition: A review. *Biotechnol Agron Soc Environ*, 12(1): 3247-3256.

Brestoff J. R. and Artis D., 2013. Commensal bacteria at the interface of host metabolism and the immune system. *Nat Immunol*, 14(7): 676-684, doi: 10.1038/ni.2640.

Clark, A., Sallé, G., Ballan, V., Reigner, F., Meynadier, A., Cortet, J., Koch, C., Riou, M., Blanchard, A., Mach, N. Strongyle Infection and Gut Microbiota: Profiling of Resistant and Susceptible Horses Over a Grazing Season. *Front. Physiol.* 2018, 9, 272.

Coleman, M. C., Whitfield-Cargile, C. M., Madrigal, R. G., Cohen, N. D. Comparison of the microbiome, metabolome, and lipidome of obese and non-obese horses. *PLoS ONE* 2019, 14, e0215918.

Comtet-Marre, S., Chaucheyras-Durand, F., Bouzid, O., Mosoni, P.; Bayat, A. R., Peyret, P., Forano, E. FibroChip, a Functional DNA Microarray to Monitor Cellulolytic and Hemicellulolytic Activities of Rumen Microbiota. *Front. Microbiol.* 2018, 9, 215.

Cymbaluk, N. F., Christison, G. I. Environmental Effects on Thermoregulation and Nutrition of Horses. *Vet. Clin. N. Am. Equine Pract.* 1990, 6, 355–372.

Dêbski B., 2016. Supplementation of pigs diet with zinc and copper as alternative to conventional antimicrobials. *Pol J Vet Sci,* 19(4): 917-924, doi: 10.1515/pjvs-2016-0113

Dibner J. J. and Richards J. D., 2005. Antibiotic growth promoters in agriculture: history and mode of action. *Poult Sci,* 84(4): 634-643, doi: 10.1093/ ps/84.4.634.

Dibner J. J. and Richards J. D., 2005. Antibiotic growth promoters in agriculture: history and mode of action. *Poult Sci*, 84(4): 634-643, doi: 10.1093/ ps/84.4.634.

Doré J. and Blottière H., 2015. The influence of diet on the gut microbiota and its consequences for health. *Curr Opin Biotechnol*, 32: 195-199, doi: 10.1016/j.copbio.2015.01.002.

Dowarah R., Verma A. K., Agarwal N., Patel B. H. M. and Singh P., 2017. Effect of swine based probiotic on performance, diarrhoea scores, intestinal microbiota and gut health of grower-finisher crossbred pigs. *Livest Sci,* 195: 74-79, doi: 10.1016/j.livsci.2016.11.006.

Fouhse J. M., Zijlstra R. T. and Willing B. P., 2016. The role of gut microbiota in the health and disease of pigs. *Anim Front*, 6(3): 30-36, doi: 10.2527/ af.2016-0031.

Galdeano C. M., de Moreno de Leblanc A., Dogi C. and Perdigón G., 2010. Lactic Acid Bacteria as Immunomodulators of the Gut-Associated Immune System. In: *Biotechnology of Lactic Acid Bacteria: Novel Applications* eds., Wiley-Blackwell, Oxford, UK, pp124-140.

Gresse, R., Chaucheyras Durand, F., Dunière, L., Blanquet-Diot, S., Forano, E. Microbiota Composition and Functional Profiling Throughout the Gastrointestinal Tract of Commercial Weaning Piglets. *Microorganisms* 2019, 7, 343.

Harlow, B. E., Lawrence, L. M., Harris, P. A., Aiken, G. E., Flythe, M. D. Exogenous lactobacilli mitigate microbial changes associated with grain fermentation (corn, oats, and wheat) by equine fecal microflora ex vivo. *PLoS ONE* 2017, 12, e0174059.

Højberg O., Canibe N., Poulsen H. D., Hedemann M. S. and Jensen B. B., 2005. Influence of dietary zinc oxide and copper sulfate on the gastrointestinal ecosystem in newly weaned piglets. *Appl Environ Microbiol*, 71(5): 2267-2277, doi: 10.1128/AEM.71.5.2267-2277.2005.

Holman D. B., Brunelle B. W., Trachsel J. and Allen H. K., 2017. Meta-analysis to define a core microbiota in the swine gut. *mSystems*, 2(3): e00004-17, doi: 10.1128/mSystems.00004-17.

Hutkins R. W., Krumbeck J. A., Bindels L. B., Cani P. D., Fahey Jr. G. et al. 2016. Prebiotics: why definitions matter. *Curr Opin Biotechnol*, 37: 1-7, doi: 10.1016/j.copbio.2015.09.001

Jouany, J. P., Medina, B., Bertin, G., Julliand, V. Effect of live yeast culture supplementation on hindgut microbial communities and their polysaccharidase and glycoside hydrolase activities in horses fed a high-fiber or high-starch diet. *J. Anim. Sci.* 2009, 87, 2844–2852.

Kaneko, T., Bando, Y., Kurihara, H., Satomi, K., Nonoyama, K., Matsuura, N. Fecal microflora in a patient with short-bowel syndrome and identification of dominant lactobacilli. *J. Clin. Microbiol.* 1997, 35, 3181–3185.

Kim H. B., Borewicz K., White B. A., Singer R. S., Sreevatsan S. et al. 2011. Longitudinal investigation of the age-related bacterial diversity in the feces of commercial pigs. *Vet Microbiol,* 153(1-2): 124-133, doi: 10.1016/j.vetmic.2011.05.021.

Kim J., Guevarra R. B., Nguyen S. G., Lee J. H., Jeong D. K. et al. 2016. Effects of the antibiotics growth promoter tylosin on swine gut microbiota. *Microbiol Biotechnol*, 26(5): 876-882, doi: 10.4014/jmb.1512.12004.

Konstantinov S. R., Awati A. A., Williams B. A., Miller B. G., Jones P. et al. 2006. Post-natal development of the porcine microbiota composition and activities. *Environ Microbiol*, 8(7): 1191-1199, doi: 10.1111/ j.1462-2920.2006.01009.x.

Lee Y. K. and Mazmanian SK, 2010. Has the microbiota played a critical role in the evolution of the adaptive immune system? *Science*, 330(6012): 1768-1773, doi: 10.1126/science.1195568.

Li Y., Guo Y., Wen Z., Jiang X., Ma X. et al. 2018. Weaning stress perturbs gut microbiome and its metabolic profile in piglets. *Sci Rep*, 8: 18068, doi: 10.1038/s41598-018-33649-8.

Linares D. M., Gómez C., Renes E., Fresno J. M., Tornadijo M. E. et al. 2017. Lactic acid bacteria and Bifidobacteria with potential to design natural biofunctional health-promoting dairy foods. *Front Microbiol*, 8: 846, doi: 10.3389/ fmicb.2017.00846.

Lindberg J. E., 2014. Fiber effects in nutrition and gut health in pigs. J *Anim Sci Biotechnol*, 5: 15, doi: 10.1186/2049-1891-5-15.

Litvak, Y., Byndloss, M. X., Tsolis, R. M., Bäumler, A. J. Dysbiotic Proteobacteria expansion: A microbial signature of epithelial dysfunction. *Curr. Opin. Microbiol.* 2017, 39, 1–6.

Lyte, M. Microbial endocrinology in the microbiome-gut-brain axis: How bacterial production and utilization of neurochemicals influence behavior. *PLoS Pathog*. 2013, 9, e1003726.

Lyte, M. Probiotics function mechanistically as delivery vehicles for neuroactive compounds: Microbial endocrinology in the design and use of probiotics. *BioEssays News Rev. Mol. Cell.* Dev. Biol. 2011, 33, 574–581.

Mach, N., Foury, A., Kittelmann, S., Reigner, F., Moroldo, M., Ballester, M., Esquerré, D., Rivière, J., Sallé, G., Gérard, P., et al. The Effects of Weaning Methods on Gut Microbiota Composition and Horse Physiology. *Front. Physiol* 2017, 8, 535.

Mach, N., Foury, A., Kittelmann, S., Reigner, F., Moroldo, M., Ballester, M., Esquerré, D., Rivière, J., Sallé, G., Gérard, P., et al. The Effects of Weaning Methods on Gut Microbiota Composition and Horse Physiology. *Front. Physiol* 2017, 8, 535.

Mach, N., Ruet, A., Clark, A., Bars-Cortina, D., Ramayo-Caldas, Y., Crisci, E., Pennarun, S., Dhorne-Pollet, S., Foury, A., Moisan, M. P., et al. Priming for welfare: Gut microbiota is associated with equitation conditions and behavior in horse athletes. *Sci. Rep*. 2020, 10, 8311.

Mach, N., Ruet, A., Clark, A., Bars-Cortina, D., Ramayo-Caldas, Y., Crisci, E., Pennarun, S., Dhorne-Pollet, S., Foury, A., Moisan, M. P., et al. Priming for welfare: Gut microbiota is associated with equitation conditions and behavior in horse athletes. *Sci. Rep*. 2020, 10, 8311.

Magne, F., Gotteland, M., Gauthier, L., Zazueta, A., Pesoa, S., Navarrete, P., Balamurugan, R. The Firmicutes/ Bacteroidetes Ratio: A Relevant Marker of Gut Dysbiosis in Obese Patients? *Nutrients* 2020, 12, 1474.

Millinovich, G. J., Trott, D. J., Burrell, P. C., Croser, E. L., Al Jassim, R. A. M., Morton, J. M., van Eps, A. W., Pollitt, C. C. Fluorescence in situ hybridization analysis of hindgut bacteria associated with the development of equine laminitis. *Environ. Microbiol.* 2007, 9, 2090–2100.

Montagne L., Pluske J. R. and Hampson D. J., 2003. A review of interactions between dietary fibre and the intestinal mucosa, and their consequences on digestive health in young non-ruminant animals. *Anim Feed Sci Technol*, 108(1-4): 95-117, doi: 10.1016/S0377-8401(03)00163-9.

Morgan, R., Keen, J., McGowan, C. Equine metabolic syndrome. *Vet. Rec.* 2015, 177, 173–179.

Mura, E., Edwards, J., Kittelmann, S., Kaerger, K., Voigt, K., Mrázek, J., Moniello, G., Fliegerova, K. Anaerobic fungal communities differ along the horse digestive tract. *Fungal Biol.* 2019, 123, 240–246.

Nagano, T., Yano, H. Dietary cellulose nanofiber modulates obesity and gut microbiota in high-fat-fed mice. *Bioact. Carbohydr. Diet. Fibre* 2020, 22, 100214.

Parkunan T., Das B. C., Banerjee D., Mohanty N., Das P. K. et al. 2015. A comparative study on the expression profile of MCTs and HSPs in Ghungroo and Large White Yorkshire breeds of pigs during dif-ferent seasons. *Cell Stress Chaperones*, 20(3): 441-449, doi: 10.1007/s12192-014-0569-5.

Patil Y., Gooneratne R. and Ju X. H., 2020. Interactions between host and gut microbiota in domestic pigs: A review. *Gut Microbes*, 11(3): 310-334, doi: 10.1080/19490976.2019.1690363.

Patil Y., Gooneratne R. and Ju X. H., 2020. Interactions between host and gut microbiota in domestic pigs: A review. *Gut Microbes*, 11(3): 310-334, doi: 10.1080/19490976.2019.1690363.

Paul, L. J., Ericsson, A. C., Andrews, F. M., Keowen, M. L., Morales Yniguez, F., Garza, F., Banse, H. E. Gastric microbiome in horses with and without equine glandular gastric disease. *J. Vet. Intern. Med.* 2021, 35, 2458–2464.

Peachey, L. E., Castro, C., Molena, R. A., Jenkins, T. P., Griffin, J. L., Cantacessi, C. Dysbiosis associated with acute helminth infections in herbivorous youngstock—Observations and implications. *Sci. Rep.* 2019, 9, 11121.

Petri D., Hill J. E. and Van Kessel A. G., 2010. Microbial succession in the gastrointestinal tract (GIT) of the preweaned pig. *Livest Sci,* 133(1-3): 107-109, doi: 10.1016/j.livsci.2010.06.037.

Pickard J. M., Zeng M. Y., Caruso R. and Núñez G., 2017. Gut microbiota: role in pathogen colonization, immune responses, and inflammatory disease. *Immunol Rev*, 279(1): 70-89, doi: 10.1111/imr.12567.

Rath, E., Haller, D. Intestinal epithelial cell metabolism at the interface of microbial dysbiosis and tissue injury. *Mucosal Immunol.* 2022, 15, 595–604.

Schokker D., Zhang J., Zhang L., Vastenhouw S. A., Heilig G. H. J. et al. 2014. Early-life environmental variation affects intestinal microbiota and immune development in new-born piglets. *PLoS One*, 9: e100040, doi: 10.1371/ journal.pone.0100040.

Schoster, A., Weese, J. S., Gerber, V., Nicole Graubner, C. Dysbiosis is not present in horses with fecal water syndrome when compared to controls in spring and autumn. *J. Vet. Intern.* Med. 2020, 34, 1614–1621.

Shan T., Li L., Simmonds P., Wang C., Moeser A. et al. 2011. The fecal virome of pigs on a high-density farm. *J Virol*, 85(22): 11697-11708, doi: 10.1128/JVI.05217-11.

Turner P. V., 2018. The role of the gut microbiota on animal model reproducibility. *Anim Model Exp Med*, 1(2): 109-115, doi: 10.1002/ame2.12022.

Turroni, F., Ventura, M., Buttó, L. F., Duranti, S., O'Toole, P. W., Motherway, M. O.'C., van Sinderen, D. Molecular dialogue between the human gut microbiota and the host: A Lactobacillus and Bifidobacterium perspective. *Cell. Mol. Life Sci.* 2014, 71, 183–203.

Urubschurov V., Janczyk P., Souffrant W. B., Freyer G., Zeyner A. et al. 2011. Establishment of intestinal microbiota with focus on yeasts of unweaned and weaned piglets kept under different farm conditions. *FEMS Microbiol Ecol,* 77(3): 493-502, doi: 10.1111/j.1574-6941.2011.01129.x.

Urubschurov V., Janczyk P., Souffrant W. B., Freyer G., Zeyner A. et al. 2011. Establishment of intestinal microbiota with focus on yeasts of unweaned and weaned piglets kept under different farm conditions. *FEMS Microbiol Ecol,* 77(3): 493-502, doi: 10.1111/j.1574-6941.2011.01129.x.

Venable, E. B., Fenton, K. A., Braner, V. M., Reddington, C. E., Halpin, M. J., Heitz, S. A., Francis, J. M., Gulson, N. A., Goyer, C. L., Bland, S. D., et al. Effects of Feeding Management on the Equine Cecal Microbiota. *J. Equine Vet. Sci.* 2017, 49, 113–121.

Wang M. and Donovan S. M., 2015. Human microbiota-associated swine: current progress and future opportunities. *ILAR J*, 56(1): 63-73, doi: 10.1093/ilar/ilv006.

Wang X., Tsai T., Deng F., Wei X., Chai J. et al. 2019. Longitudinal investigation of the swine gut microbiome from birth to market reveals stage and growth performance associated bacteria. *Microbiome*, 7: 109, doi: 10.1186/s40168-019-0721-7.

Wells J. E., Yen J. T. and Miller D. N., 2005. Impact of dried skim milk in production diets on *Lactobacillus* and pathogenic bacterial shedding in growing-finishing swine1. *J Appl Microbiol,* 99(2): 400-407, doi: 10.1111/j.1365-2672.2005.02629.x.

Yeoman, C. J., Ishaq, S. L., Bichi, E.; Olivo, S. K., Lowe, J., Aldridge, B. M. Biogeographical Differences in the Influence of Maternal Microbial Sources on the Early Successional Development of the Bovine Neonatal Gastrointestinal tract. *Sci. Rep*. 2018, 8, 3197.

Zak, A., Siwinska, N.; Elzinga, S., Barker, V. D., Stefaniak, T., Schanbacher, B. J., Place, N. J., Niedzwiedz, A., Adams, A. A. Effects of equine metabolic syndrome on inflammation and acute-phase markers in horses. *Domest. Anim. Endocrinol*. 2020, 72, 106448.

Zhang Y., Zhou J., Dong Z., Li G., Wang J. et al. 2019. Effect of dietary copper on intestinal microbiota and antimicrobial resistance profiles of *Escherichia coli* in weaned piglets. *Front Microbiol,* 10: 2808, doi: 10.3389/ fmicb.2019.02808.

Chapter 16

Microbiomes of Sheep and Goat Health, Disease, and Production

Athira Karunakaran[1,*]
and Arunraj Rajendrakumar[2]
[1]Animal Parasitic Diseases Laboratory, USDA, Beltsville, Maryland, USA
[2]Department of Veterinary Medicine, Virginia-Maryland College of Veterinary Medicine, University of Maryland, College Park, USA

Abstract

The microbiome of sheep and goats is influenced by several interconnected factors. Understanding and addressing these factors is crucial for managing and optimizing the microbiomes of these livestock species, which can significantly impact their overall health, productivity, and resilience to environmental challenges. The digestive tract of sheep and goats holds paramount significance in animal production due to its pivotal role in nutrient absorption, energy metabolism, and overall performance. Consequently, this region has been the primary subject of extensive research into the microbiome. However, the microbiome of other body systems, such as the respiratory, reproductive, skin, and immune systems, also plays a vital role in the overall health and well-being of these animals. In this chapter, we aim to summarize the current knowledge on the composition and function of the microbial community in these various body systems and discuss the factors that affect the overall health of sheep and goat populations.

Keywords: sheep, microbiome, dysbiosis, goat, second genome

Introduction

Sheep (*Ovis aries)* and goats (*Capra hircus*) are globally recognized as significant domestic livestock species, with sheep possessing 54 chromosomes and goats possessing 60, characterized by the gradual development of unique physiological and behavioral traits that facilitate their adaptation to diverse ecological environments. This adaptability is attributed to

* Corresponding Author's Email: dr.athirack@gmail.com

In: Microbiomes in Animal Nutrition, Health, and Disease
Editor: Tanmoy Rana
ISBN: 979-8-89530-425-9

an extended period of natural selection and artificial domestication (Chen et al. 2021), (Daly et al. 2021). Sheep are classified as grazers, predominantly subsisting on low-growing grass, while goats are categorized as browsers, primarily consuming vegetation discarded by other livestock. This distinction demonstrates the significant adaptive survival capabilities of these species. The differentiation in ecological niches, influenced by the evolutionary habits and environments during their domestication, led to the development of distinct genomic, behavioral, and metabolic adaptations between sheep and goats. The adaptive capacity of animal species depends not solely on the host genome but also on the extensive genetic diversity of the microbiome. The gut microbiota, acknowledged as the "second genome" of the host, plays a pivotal role in nutrient metabolism, immune response, host physiology, and evolution. Hence, a comprehensive understanding of the differences in the composition and function of the gut microbiota in sheep and goats is imperative for gaining profound insight into these two species' adaptability during evolution (Raoult et al. 2021). It is necessary to delineate the distinction between the terms microbiota and microbiome. Microbiota is defined as the collective microorganisms present in a specific environment, whereas microbiome encompasses the microorganisms within a given habitat that exhibit diverse physicochemical properties, interact with their environment, and can bestow microbiological resistance to their host. Furthermore, the microbiome entails and considers the interactions among Bacteria (bacteriome), viruses (virome), fungi (mycobiome), archaea (archaeome), and helminth, as well as the interplay between biotic and abiotic factors. The perturbation of the microorganisms implicated in the microbiome can give rise to dysbiosis, characterized by the disruption of a healthy microbiome, thereby reducing the symbiotic relationships of the microbiota, impacting productivity, and precipitating health issues. The microbiome's composition is influenced by various factors, and understanding these factors helps in managing and optimizing the microbiome of sheep and goats, which can impact their overall health, productivity, and resilience to environmental challenges.

The diversity of the microbiome in sheep and goats, like in other animals, is influenced by various factors. Here are some key factors that play a role:

1. *Dietary Composition:* The type and quality of food animals consume significantly impact their microbiome diversity. Different diets (e.g., hay, grass, grains) provide varying nutrients that support different microbial populations in the gut. For example, goats eat more crude protein, while sheep prefer mature fiber forage. As a result, their rumens have distinct microbial communities. Goats have more bacteria that break down crude protein and produce volatile fatty acids (VFA), such as Saccharofermentans and Lachnospiraceae_XPB1014. On the other hand, sheep have higher levels of fungi and protozoa that break down fiber, such as Neocallimastigaceae and Metadinium (Langda et al. 2020).
2. *Grazing vs. Confined Feeding:* Grazing generally leads to a more diverse and potentially beneficial microbial community, while confined feeding can improve growth performance and certain meat quality traits but may reduce microbial diversity. For instance, a study on Mongolian sheep found that barn-confined sheep had higher levels of certain bacteria like Succinivibrio, Treponema, and Prevotella, whereas free-grazing sheep had higher levels of Bacteroides, RC9_gut_group, and Alistipes. These differences in microbial communities are linked to differences in metabolic profiles, including butyric acid and various amino acids (Jize et al. 2022).

3. *Geographical Location:* Regional differences in soil composition, vegetation, and climate can affect the microbial diversity in the environment, which in turn can influence the animal's microbiome.
4. *Age and Developmental Stage*: The microbiome of sheep and goats evolves as they mature from birth through adulthood. Early diet and environmental exposures shape the establishment and diversity of gut microbiota.
5. *Health and Disease:* The presence of infections, parasites, or chronic diseases can alter the microbial composition in the gut. Conversely, a healthy animal is more likely to have a stable and diverse microbiome.
6. *Genetics:* Host genetics can influence the composition of the microbiome to some extent. Certain genetic traits may predispose animals to host specific microbial communities.
7. *Stress and Management Practices:* Stress from transportation, handling, or changes in management practices (e.g., weaning) can affect the gut microbiota composition temporarily or long-term. The main biological materials used to measure cortisol levels are serum, saliva, urine, and feces. However, cortisol levels in these substances only reflect a short period (from a few minutes to 1-2 days). On the other hand, the concentration of cortisol in hair is an indicator of cortisol secretion and stress over longer periods. Measuring hair cortisol is a useful way to assess animal chronic stress. However, specific reference values for different species are not currently available (Möstl and Palme 2002), (Meyer and Novak 2012), (Keckeis et al. 2012).
8. *Antibiotic Use:* Antibiotics can disrupt the balance of gut microbiota, potentially reducing diversity and altering microbial populations.
9. *Social Interactions:* Group housing and social interactions among animals can influence the transmission of microbial species, affecting individual microbiome composition.
10. *Microbial Transmission:* The mode of birth (vaginal birth vs. cesarean section) and maternal microbial transmission (via milk or direct contact) can shape the initial colonization and diversity of the microbiome in young animals.

The digestive tract holds paramount significance in animal production due to its pivotal role in the overall performance. Consequently, this region has been the primary subject of extensive research into the microbiome. Followed by respiratory, reproductive, Skin, and immune systems.

Microbiome of Gastrointestinal Tract (GIT)

The oral, gut, and fecal microbiomes are interconnected, their specific compositions and correlations vary along the GIT and are influenced by factors like age, breed, and physiological state in small ruminants like sheep and goats. A well-maintained and balanced gut microbiome plays a crucial role in the growth, development, and metabolism of ruminants. The microorganisms within the rumen function as an organ that effectively produces short-chain fatty acids through the decomposition of plant polysaccharides such as starch, cellulose, and hemicellulose. These short-chain fatty acids play a role in thermogenesis and immune function. Maintaining homeostasis between symbiotic and pathogenic organisms in the gut is imperative for overall health. The initial colonization of the gut microbiome, influenced by maternal sources and diet, significantly contributes to the proper development and maturation of the immune system in young ruminants. Disruption of this process may result in dysbiosis and an increased susceptibility to disease in sheep and goats.

Oral Microbiome

A healthy oral microbiome is crucial for preventing oral diseases and sustaining overall health in small ruminants. Compared to sheep, goats have higher levels of *Pasteurella multocida*, *Cutibacterium avidum*, *Streptococcus equinus*, *Streptomyces gilvigriseus*, and *Corynebacterium kroppenstedtii* in their oral microbiome (Figure 1). Sheep have higher abundances of *Corynebacterium freneyi,* Campylobacter sp. NCTC (number 1), *Bifidobacterium pseudocatenulatum*, *Fibrobacter intestinalis*, and *Bifidobacterium choerinum* in their oral microbiome compared to goats.

In sheep, bacteria such as Petrimonas, Acinetobacter, Porphyromonas, and Aerococcus, with Firmicutes pathogens multiplying the fastest, are associated with periodontitis (broken mouth). In goats, periodontitis is linked to common bacteria such as *Fusobacterium nucleatum*, *Tannerella forsythia*, *Fusobacterium necrophorum*, *Campylobacter rectus*, *Eikenella corrodens*, *Prevotella buccae*, *Porphyromonas gingivalis*, Treponema denticola, *Porphyromonas endodontalis*, and *Treponema maltophilum* (Figure 2) (Grzeczka et al. 2023).

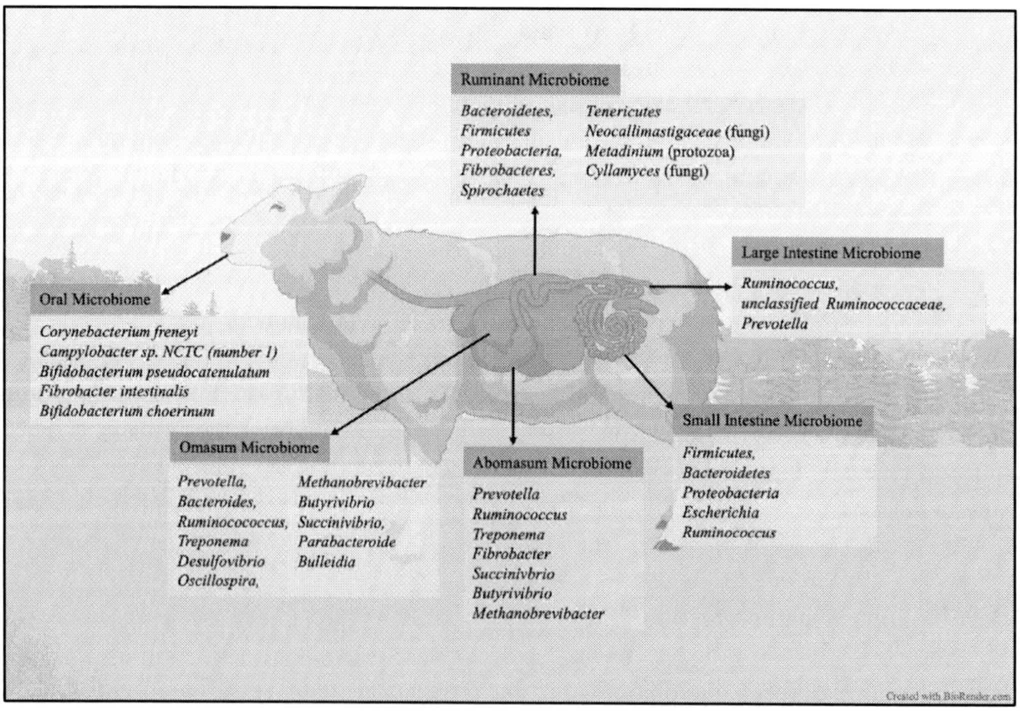

Figure 1. Most prevalent microbes present in the gastrointestinal tract of the Sheep.

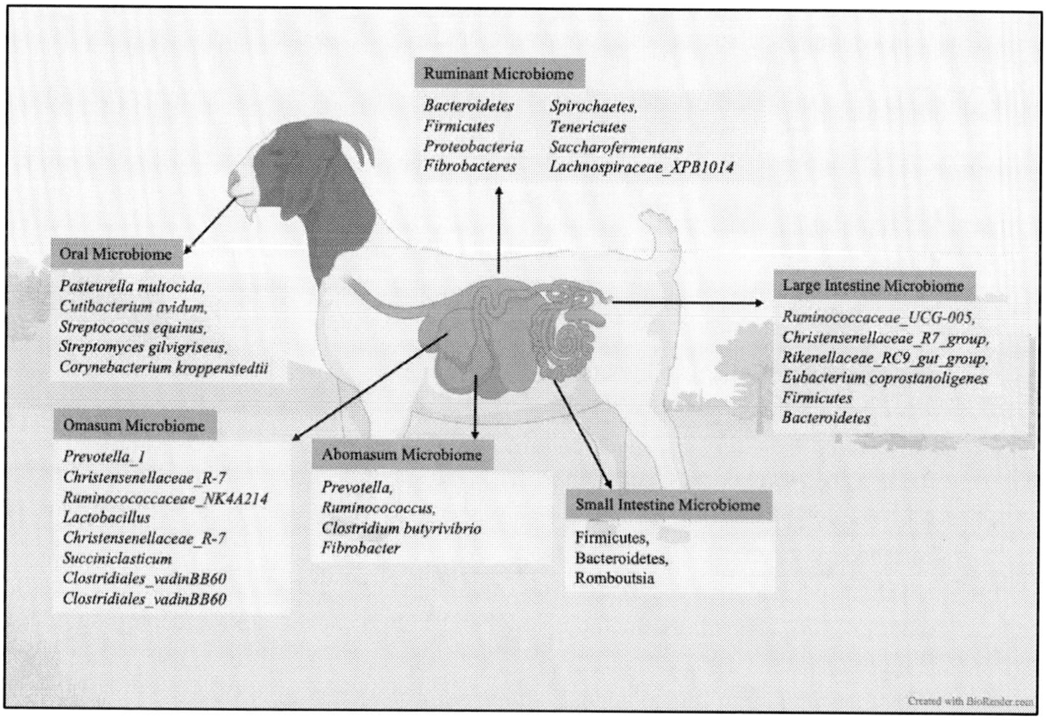

Figure 2. Most prevalent microbes present in the gastrointestinal tract of the Goat.

Microbiome of Stomach of Sheep and Goat

The microbiome of the four stomach compartments (rumen, reticulum, omasum, and abomasum) in sheep and goats shows distinct characteristics and variations. The microbiome in these compartments develops in stages, with the first stage (0-14 days) involving access to external microorganisms, followed by a transition stage (14-28 days), and finally, exogenous and endogenous microbial colonization beyond 28 days of age. While the rumen, reticulum, and omasum are primarily involved in the fermentation and digestion of plant materials, the abomasum functions more like a true stomach, producing hydrochloric acid and digestive enzymes. Although sheep and goats show similar overall patterns in their stomach microbiomes, there may be species-specific differences in the relative abundances of specific microbial groups.

Rumen and Reticulum

The rumen contains various microorganisms that produce volatile fatty acids and proteins through fermentation, with archaea playing a key role. This process releases nitrogen into the environment as methane gas (CH_4). Studies have identified specific microbial communities, including *Fibrobacter spp*., *Kandleria vitulina, Olsenella spp*., *Prevotella bryantii*, and *Sharpea azabuensis*, associated with reduced methane production in lambs (Kittelmann et al. 2014). The environmental adaptation of ruminants is strongly correlated with the microbiota present in the rumen. In both sheep and goats, the dominant bacterial phyla in the rumen and reticulum are *Bacteroidetes* and *Firmicutes*, which make up over 80% of the bacterial community. Other common phyla include *Proteobacteria*, *Fibrobacteres*, *Spirochaetes*, and *Tenericutes* (Wang et al. 2016).

Common dominant genera in both species include:

- Prevotella
- Bacteroides
- Butyrivibrio
- Ruminococcus
- Clostridium
- Oscillospira

Differences between sheep and goats:

- The family *Neocallimastigaceae* (fungi) and the genus *Metadinium* (protozoa) were found to be enriched in sheep (Figure 1).
- The bacterial genera *Saccharofermentans* and *Lachnospiraceae*_XPB1014 were enriched in goats.
- Goats were found to have higher levels of volatile fatty acids (VFA) producing bacteria like *Saccharofermentans*, including higher levels of acetic acid and propionic acid (Figure 2).
- The fiber-degrading fungal genus *Cyllamyces* was increased in sheep.

The differences in microbial composition suggest that goats may be better adapted to high-altitude pastures compared to sheep (Forcina et al. 2022).

Omsaum

The omasum microbiome is similar to that of the rumen and reticulum but with some distinctions. It has a lower microbial density compared to the rumen. It is dominated by the *Firmicutes* and *Bacteroidetes* phyla. It is important for water absorption and further breakdown of partially digested feed. In sheep omasum, the main bacteria genera include *Prevotella*, *Bacteroides*, *Ruminocococcus*, *Treponema*, *Desulfovibrio*, *Oscillospira*, *Methanobrevibacter*, *Butyrivibrio*, *Succinivibrio*, *Parabacteroide*, and *Bulleidia* (Figure 1). In case of goat, the omasum comprises of *Prevotella_1*, *Christensenellaceae_R-7*, Ruminococococcaceae_ NK4A214, *Lactobacillus*, *Christensenellaceae_R-7*, *Succiniclasticum*, *Clostridiales*_ vadinBB60, and *Clostridiales*_vadinBB60 (Figure 2) (Huaiquipán et al. 2023).

Abomasum

The abomasum has a distinct microbiome profile. It has lower microbial diversity compared to other compartments. In early life, it is dominated by Proteobacteria, which gradually decreases with age. It functions more like a true stomach, with a pH range of 3.5 to 4.0. In sheep abomasum, prevalent bacterial genera include *Prevotella, Ruminococcus, Treponema, Fibrobacter, Succinivbrio, Butyrivibrio, and Methanobrevibacter* (Figure 1). In the case of goats, the abomasum is composed of *Prevotella*, *ruminocococcus*, *Clostridium butyrivibrio*, and *fibrobacter* (Figure 2) (Miao et al. 2023).

Microbiome Small intestine of sheep and goat

The microbiomes of sheep and goats' small intestines are complex and exhibit similarities and differences in composition. In the small intestines of both sheep and goats, the dominant phyla are *Firmicutes* and *Bacteroidetes*, with the presence of *Proteobacteria* notably significant in sheep. In sheep, the prominent genera include *Escherichia, unclassified Lachnospiraceae* and *Ruminococcus*, (Figure 1) while in goats, *Romboutsia* is a key genus in the jejunum (Figure 2). The occurrence of *Methanobrevibacter* in the jejunum of goats underscores the involvement of archaea in the microbial ecosystem of the small intestine.

The small intestine plays a critical role in the absorption of nutrients, with the microbiome in this region playing a supportive role in this function. In goats, the microbiome of the small intestine is associated with vital adaptive functions, including the two-component system, ABC transporter, and transcription factors, all of which are essential for effective nutrient absorption.

(Two-component systems enable bacteria to detect and respond to changes in their environment, such as nutrient availability. These systems typically consist of a sensor kinase that detects environmental signals and a response regulator that mediates cellular responses. In terms of nutrient absorption, two-component systems can: detect the presence of specific nutrients in the environment, trigger the expression of genes encoding nutrient transporters and metabolic enzymes, and regulate cellular processes to optimize nutrient uptake and utilization

ABC transporters, or ATP-binding cassette transporters, are membrane proteins that use energy from ATP hydrolysis to transport various substrates across cell membranes. Specifically for nutrient absorption, ABC transporters:

- Actively import essential nutrients such as sugars, amino acids, peptides, and vitamins into the cell
- Can transport nutrients against concentration gradients, allowing cells to accumulate nutrients even when environmental concentrations are low
- Often have high specificity for particular nutrients, enabling efficient uptake

Transcription factors are proteins that regulate gene expression by binding to specific DNA sequences. In the context of nutrient absorption, transcription factors control the expression of genes involved in nutrient uptake and utilization.

Microbiome Large Intestine of Sheep and Goat

The microbiome present in the large intestine plays a significant role in the further breakdown of plant materials, particularly cellulose and other complex carbohydrates that have not undergone complete digestion in the rumen and small intestine. The large intestine typically contains a higher variety and abundance of microorganisms compared to the small intestine in both sheep and goats. This is probably because food stays in the large intestine for a longer time, creating an environment favorable for the growth of anaerobic microorganisms.

In the large intestine of sheep, the three most dominant genera are *Ruminococcus*, *unclassified Ruminococcaceae*, and *Prevotella* (Figure 1). In goats, the dominant genera in the large intestine include *Ruminococcaceae*_UCG005, *Bacteroides*, *Christensenellaceae*_R7_ group, *Rikenellaceae*_RC9_gut_group, and *Eubacterium coprostanoligenes* group (Figure 2). Both sheep and goats have *Firmicutes* and *Bacteroidetes* as dominant bacterial phyla in their large intestine microbiomes (Langda et al. 2020).

Lung Microbiome

Until recently it was assumed that the healthy mammalian lung did not harbor a microbiota, unlike other body sites. The sheep respiratory microbiota is of interest from both an animal health perspective and due to the potential use of the sheep as a large animal model for studying the lung microbiota. The predominant bacterial orders identified in sheep lungs are *Bacillales*, *Actinomycetales*, *Clostridiales*, and *Pseudomonadales*. Among the common genera detected are *Staphylococcus*, *Streptococcus*, *Corynebacterium*, *Jeotgalicoccus*, *Bacillus*, *Enterococcus*, *Klebsiella*, *Mannheimia*, *Micrococcus*, *Moraxella*, *Pasteurella*, and *Pseudomonas*. The microbiome of sheep lungs differs from that of the upper respiratory tract, distinguishing it from the situation in humans where similarity is observed. Furthermore, sheep lungs contain rumen-associated bacteria such as *Prevotella* and *Butyrivibrio*, likely due to inhalation. Considerable inter-individual variation characterizes the sheep lung microbiome, with each animal housing a distinct microbial community. Spatial variability is also notable, with distinct microbial profiles across different lung sites. Of particular note, *Staphylococcus equorum* is significantly more abundant in sheep lung fluids than in the upper respiratory tract. Likewise, a Brazilian study reveals *Bacillus* sp. as the predominant bacterium in both healthy and pneumonic sheep lungs. In summary, the sheep lung microbiome exhibits diversity and distinctness from the upper respiratory tract, demonstrating notable inter-individual and spatial variability. Core genera such as *Staphylococcus*, *Streptococcus*, and *Bacillus* are consistently identified across samples.

The diversity of the sheep lung microbiota has notable implications for sheep health management: For Example

1. *Spatial Variability within the Lungs*: Research indicates significant variations in bacterial communities among different lung segments in a single sheep at a given time. This spatial heterogeneity necessitates consideration when sampling the lung microbiome for health assessment.
2. *Individual Variability*: Each sheep possesses a distinct lung microbiome composition, even among healthy animals. This individuality likely impacts disease susceptibility and progression. It is essential to establish a baseline of "normal" microbiota variability in healthy sheep.
3. *Temporal Variability*: The lung microbiome can change over time within an individual sheep. Monitoring temporal shifts in the microbiome may aid in the early detection of respiratory disease development.
4. *Differences from Upper Respiratory Tract*: Unlike in humans, the sheep lung microbiome differs from that of the upper respiratory tract. Oropharyngeal swabs are inadequate for assessing the lower airway microbiota in sheep. Lung sampling is imperative for evaluating lung health.
5. *Presence of Rumen-associated Bacteria*: Sheep commonly inhale rumen bacteria such as *Prevotella* and *Butyrivibrio*. Overgrowth of these bacteria may indicate aspiration and pneumonia risk.
6. *Identification of Potential Pathogens*: Certain genera, such as *Mannheimia*, *Bibersteinia*, and *Pasteurella*, are linked to sheep pneumonia. Monitoring their abundance may aid in predicting disease risk.

Microbiome of the Reproductive System

The microbiome plays a crucial role in the reproductive health and fertility of small ruminants such as sheep and goats. The presence and abundance of specific types of microbiota in the vagina, uterus, placental tissues and fluids, semen, and milk can have an impact on fertility. The significance of this microbiome in influencing animal physiology is currently the subject of extensive research.

Female Reproductive Tract Microbiome

Vaginal and Uterine Microbiota

The earlier culture-based method identified *Staphylococcus*, *Corynebacterium*, *Bacillus*, *Streptococcus*, and *Escherichia coli* as the common bacteria present in the vaginal flora with ewes and does having a similar profile (Poole et al. 2023). The 16S rRNA gene-based study by the amplification of the V3-V4 region found *Aggregatibacter* and *Streptobacillus* as the prominent bacteria present in the vagina flora of ewes, with the earlier culture-based bacteria being less abundant. Predominately colonized by the bacterial phyla *Bacteroidetes*, *Fusobacteria*, and *Proteobacteria*. The presence of *lactobacillus spp* is less in the vaginal samples, and the alpha diversity of microbiota is less as compared to that of cattle but higher than in humans (Swartz et al. 2014). The vaginal and uterine microbiomes are crucial for maintaining a healthy environment for conception, embryo development, and pregnancy establishment in ewes and do. Dysbiosis in the reproductive tract can lead to conditions like endometritis, which negatively impacts fertility. Specific bacterial taxa like Porphyromonas, Fusobacterium, and Bacteroides have been associated with uterine infections and poor reproductive outcomes in small ruminants. A diverse and balanced microbiome in the reproductive tract is generally associated with better fertility and pregnancy rates. The abundance of microbiota is greater in non-pregnant Eves, with *Histophilus*, *Actinobacillus*, *Streptococcus*, *Mannheimia*, and *Fusobacterium* being more abundant in non-pregnant Eves (Serrano et al. 2020). *Histophilus* is a known livestock pathogen causing respiratory diseases and reproductive disorders such as abortion and epididymitis. An isolated study showed that the uterine flora of non-pregnant ewe is abundant in the genus *Finegoldia* (Smith et al. 2016). Moreover, during a normal estrus cycle, the bacterial load will be low in ewes and consist of *Achromobacter* spp, *Alcaligenes* spp, *Corynebacterium* spp, *Bacillus* spp, and *Escherichia coli*. Pathogenic bacteria such as *Pasteurella multocida, Corynebacterium pyogenes, Staphylococcus* spp, and *Streptococcus* spp were isolated after embryos were killed (Sawyer 1977).

In pregnant ewes, the vaginal microbial community changes gestation. *Actinobacillus*, *Histophilus*, and unclassified *Leptotrichiaceae* are the dominant species in early gestation. As gestation progresses, there is a shift towards increased presence of *Streptococcus* and *Staphylococcus*. *Staphylococcus, Streptococcus, and Ureaplasma* become the most abundant by mid to late gestation. These changes in the microbial population are likely linked to the hormonal fluctuations associated with the developing conceptus, particularly the rise in progesterone levels in the bloodstream (Cassas et al. 2024).

Placental Microbiome

Culture-based methods have traditionally been employed to identify specific pathogenic bacteria linked to placentitis and abortion in sheep and goats, including *Brucella, Campylobacter, Chlamydia, Listeria,* and *Coxiella*. Thus far, there have been no direct characterizations of the microbiome in the placentas of sheep or goats using 16S rRNA sequencing. However, a few studies have indicated the presence of a microbiome in the prenatal gut of lambs and kids, suggesting microbial seeding during gestation. The prenatal fetal gut consists mainly of phyla Proteobacteria, Actinobacteria, and Firmicutes, with *Escherichia coli* being the most abundant species (Bi et al. 2021).

Milk Microbiome

The 16S rRNA sequencing analysis of microbes present in healthy sheep milk has revealed the abundance of 5 genera: *Staphylococcus, Lactobacillus, Corynebacterium, Streptococcus,* and *Escherichia/Shigella*. A correlation was established between the microbial diversity and the milk's somatic cell count (SSC) to assess subclinical mastitis. In sheep without mastitis, the prevalent microbiota in the milk comprises *Staphylococcus, Lactobacillus,* and *Corynebacterium*. Sheep diagnosed with subclinical mastitis also exhibited this core microbiota in their milk samples, but they were also exposed to bacteria genera such as *Escherichia, Mannheimia,* and *Enterococcus*. The composition of the milk microbiota in small ruminants does not exhibit a clear consensus, as it depends on various variables including breed, environment, feed, and antibiotic usage (Esteban-Blanco et al. 2020; Toquet, Gómez-Martín, and Bataller 2021).

In goat milk samples without mastitis *Bacillus* spp., *Enterococcus* spp., and *Staphylococcus* spp. were found to be prevalent. In colostrum, Proteobacteria is dominant and as in goat milk Actinobacteria is dominant. Using the 16S rRNA sequencing method *Mycoplasma*, *Escherichia*, and *Enterococcus*, were found to be responsible for mastitis in goat.

Male Reproductive Tract Microbiome

The microbiome of the male reproductive tract, including semen, has been found to have an impact on sperm quality, motility, and fertility potential in rams and bucks. Some beneficial bacteria, such as *Lactobacillus* and *Pediococcus*, positively influence sperm function, while detrimental effects have been associated with bacteria like *Pseudomonas*. Culture methods have revealed the presence of prevalent bacteria such as *Escherichia coli*, *Staphylococcus*, *Streptococcus*, and *Bacillus* in ram seminal fluid (Tvrdá et al. 2022). However, a 16s RNA-based study demonstrated that *Corynebacterium* and *Pseudomonas* are the abundant species in ram semen (Serrano et al. 2020).

For goats, being seasonal breeders, their semen microbiota changes depending on the season. Research on the Murciano-Granadina breed indicated differences in microbiota diversity between breeding and non-breeding seasons. During the breeding season, the dominant phyla were *Firmicutes*, *Proteobacteria*, *Fusobacteria, Actinobacteria, and Bacteroidetes*, while during the non-breeding season, the dominant phyla were *Firmicutes, Proteobacteria, Actinobacteria, Bacteroidetes,* and *Cyanobacteria* (Mocé et al. 2022). The dominant genus during both seasons was *Ureaplasma*. *Sphingomonas* have been associated with good sperm quality and motility, while *Faecalibacterium* has been linked to reduced sperm quality. It is suggested that modulating the semen microbiome through probiotics or other interventions could potentially improve reproductive performance in males.

Potential Applications

Understanding the reproductive tract microbiome could lead to strategies like probiotics, prebiotics, or targeted antimicrobials to promote a healthy microbiome and improve fertility. Monitoring thc microbiome could help identify animals at risk of reproductive issues and guide management practices. In summary, the microbiome plays a crucial role in the reproductive health and fertility of small ruminants, and further research in this area could lead to interventions to enhance reproductive efficiency in these livestock species.

The Skin Microbiome of Sheep and Goat

The skin microbiome of sheep and goats plays an important role in protecting the host and can modulate the innate and adaptive immune response of the skin. Significant differences were found in the levels of certain bacterial phyla like *Firmicutes*, *Actinobacteria,* and *Proteobacteria* between sheep breeds (Świniarka, Uhruska, BCP line) and goat breeds (Boer, Saenian) living in the same environment and fed the same feed. These differences may be due to physicochemical differences in the skin surface like humidity, temperature, antimicrobial peptides, and lipid content.

Individual differences in skin microbiome composition were also observed, indicating that an individual animal's microbiome can be influenced by factors beyond just environment and diet. The results suggest the height of the animals may also impact the skin microbiome, as the level of *Proteobacteria* decreased with increasing height (Cholewińska et al. 2021).

Microbiome and Immune System

The microbial compositions of sheep and goats exhibit both similarities and key distinctions in their interactions with the host immune system. In young goats, the rumen and fecal microbiota are linked to the development and maturation of their immune system, with particular bacterial genera correlating with serum immunoglobulin G (IgG) and immunoglobulin M (IgM) levels as the goats age. In sheep, the rumen microbiome composition undergoes significant changes during the first year of life, predominantly affecting *Bacteroidetes* and *Proteobacteria* populations. This suggests the significant role of the rumen microbiome in the immune system development of young sheep. Specific rumen and fecal bacteria such as *Alistipes*, *Bacteroides*, *Blautia*, and *Prevotellaceae*_UCG.004 have been found to positively correlate with interleukin-2 (IL-2) and IL-6 in young goats under cold stress, suggesting their potential immunomodulatory properties. Furthermore, a comprehensive database of 5,810 microbial genomes from sheep and goat gut microbiomes identified 91 bacterial taxa that specifically colonize the sheep gut and encode enzymes for degrading complex glycans and mucins. This indicates that the sheep gut microbiome may possess unique adaptations for interacting with the host immune system compared to goats. Interestingly, the rearing environment appears to exert a greater impact on shaping the composition and function of the gut microbiome in sheep and goats than the host species itself. This implies that microbiome-immune interactions are influenced more by environmental factors than innate differences between the two ruminant species.

Microbiome and Disease

The microbiome plays a crucial role in sheep and goats' health and disease susceptibility. Among the common microbial diseases affecting these animals are Johne's disease, ovine progressive pneumonia, caseous lymphadenitis, enterotoxemia, coccidiosis, and respiratory diseases caused by *Pasteurella* and *Mannheimia*. These illnesses can have significant implications for animal health and productivity. One important factor to consider is gut dysbiosis, an imbalance in the microbiome, which can exacerbate diseases by promoting inflammatory responses in goats. For example, the disruption of gut microbiota worsened orf, a contagious disease caused by the epitheliotropic orf virus (ORFV). This disease primarily affects goats and sheep and is globally prevalent in asymptomatic goats. The disruption induced IL-17 production in the affected tissue, leading to excessive neutrophil accumulation in diseased lips, potentially contributing to tissue pathology (Tong et al. 2020). This highlights the intricate relationship between the gut microbiome and overall health in these animals. Additionally, the presence of apicomplexan parasites in goat kids has been associated with specific microbial changes. This includes a reduction in butyrate-producing bacteria, which are important for gut health, as well as an increase in mucosal inflammation and impaired tissue repair. These findings underscore the intricate interplay between microbial composition, parasitic infections, and the overall health of goat populations.

Table 1. Advantages of diverse microbial communities in the sheep and goats (Forcina et al. 2022)

Advantage	Description
Nutrient metabolism	The gut microbiota plays a crucial role in regulating energy metabolism and promoting nutrient absorption.
Immune function	The microbiome helps improve immune function in sheep and goats
Ecological adaptation	The ruminant gut microbiota is critical in helping sheep and goats adapt to different ecological environments.
Glycan degradation	Certain bacteria in the sheep gut encode enzymes for breaking down complex carbohydrates like glycans and mucins
Rumen development	Inoculation with rumen fluids early in life can boost rumen microbiome development and accelerate weaning in goat kids
Pathogen resistance	A healthy, diverse microbiome helps protect against colonization by pathogens like parasites as well as harmful bacteria such as Clostridium perfringens, which can cause severe gastrointestinal diseases.
Fermentation	Microbes ferment plant material in the rumen, producing important metabolites like short-chain fatty acids
Detoxification	The microbiome can help detoxify some plant compounds and other substances ingested by the animals

Table 2. Disadvantages of dysbiosis in the sheep and goat (Tong et al. 2020)

Disadvantage	Description
Exacerbation of Diseases	Worsen diseases such as orf promote inflammatory immune responses, leading to greater disease severity and higher clinical symptoms.
Digestive Issues	Digestive problems like diarrhea are linked to decreased beneficial bacteria and increased pathogenic bacteria.
Nutrient Absorption	Changes in the gut microbiome can affect the digestibility of nutrients such as phosphorus, impacting overall nutrient absorption and health.
Immune System Modulation	An imbalance in immune responses potentially causes increased inflammation and tissue damage.
Growth and Development	The gut microbiome plays a role in the growth and development of young animals. Dysbiosis can hinder proper development and reduce productive performance.
Microbial Diversity	A diverse gut microbiome is crucial for health. Dysbiosis reduces microbial diversity, which can negatively impact health and disease resistance.
Inflammation	Dysbiosis can lead to increased mucosal inflammation, which can impair gut function and overall health.

Conclusion

The study of the microbiome is complex and influenced by various factors. Changes in the population of microorganisms in the intestinal tract of production animals are primarily attributed to diet composition and the resulting biochemical processes. Additionally, an animal's life stage, external conditions, social behavior, and indigenous microorganisms in the intestinal tract can contribute to microbial diversity. Research indicates that the proportions of microorganisms differ significantly among various types of feed and across different animal species. Ruminants, in particular, exhibit similarities in their microbial communities, which are involved in the breakdown of fiber and carbohydrates through ruminal fermentation, including the presence of methanogenic archaea. *Firmicutes* and *Bacteroidetes* are the predominant phyla in the gastrointestinal tracts of ruminants such as sheep and goats, playing crucial roles in the degradation of high-fiber plant material and polysaccharides essential for their digestive processes.

The dominance of the *Firmicutes* phylum in the fecal bacterial communities of sheep and goats, especially at 6 months of age, suggests a significant role in gut health and maturation. Beyond these, *Actinobacteria* and *Proteobacteria* are also identified on the skin of sheep and goats, with varying levels based on breed, potentially contributing to the skin microbiome. *Lactobacillaceae*, a family within the Firmicutes phylum, has been the subject of specific research in sheep and their offspring, given the association of *Lactobacillus* bacteria with probiotic effects and gastrointestinal health. It's essential to note that a balanced and diverse distribution of bacterial phyla, rather than the predominance of a specific group, is likely most advantageous for the well-being of sheep and goats.

With the growing understanding of the functions and variables of the microbiome, microbiome-based studies are gaining increasing relevance. Current research on production animals has primarily focused on comprehending microbiome changes under defined feeding strategies and their impact on performance. It is essential to understand how diet components and various supplements affect microbial populations in the gastrointestinal tract of production animals, as modulating these populations could greatly enhance the health and quality of livestock products, improve feed efficiency, and optimize available resources.

References

Bi, Yanliang, Yan Tu, Naifeng Zhang, Shiqing Wang, Fan Zhang, Garret Suen, Dafu Shao, Shengli Li, and Qiyu Diao. 2021. "Multiomics Analysis Reveals the Presence of a Microbiome in the Gut of Fetal Lambs." *Gut* 70 (5): 853–64. https://doi.org/10.1136/gutjnl-2020-320951.

Cassas, Mackenzie S., Lucille C. Jonas, Chiron J. Anderson, Stephan Schmitz-Esser, and Curtis R. Youngs. 2024. "Temporal Changes in Ewe Vaginal Microbiota throughout Gestation." *Frontiers in Microbiology* 15 (February). https://doi.org/10.3389/fmicb.2024.1359678.

Chen, Ze-Hui, Ya-Xi Xu, Xing-Long Xie, Dong-Feng Wang, Diana Aguilar-Gómez, Guang-Jian Liu, Xin Li, et al. 2021. "Whole-Genome Sequence Analysis Unveils Different Origins of European and Asiatic Mouflon and Domestication-Related Genes in Sheep." *Communications Biology* 4 (1): 1307. https://doi.org/10.1038/s42003-021-02817-4.

Cholewińska, Paulina, Paulina Nazar, Andrzej Junkuszew, Jakub Smoliński, Katarzyna Czyż, and Anna Wyrostek. 2021. "The Level of Selected Bacterial Phyla on the Skin Surface of Small Ruminants According to the Breed and Species." *Animals* 11 (9): 2734. https://doi.org/10.3390/ani11092734.

Daly, Kevin G., Valeria Mattiangeli, Andrew J. Hare, Hossein Davoudi, Homa Fathi, Sanaz Beizaee Doost, Sarieh Amiri, et al. 2021. "Herded and Hunted Goat Genomes from the Dawn of Domestication in the Zagros Mountains." *Proceedings of the National Academy of Sciences* 118 (25): e2100901118. https://doi.org/10.1073/pnas.2100901118.

Esteban-Blanco, Cristina, Beatriz Gutiérrez-Gil, Fernando Puente-Sánchez, Héctor Marina, Javier Tamames, Alberto Acedo, and Juan José Arranz. 2020. "Microbiota Characterization of Sheep Milk and Its Association with Somatic Cell Count Using 16s rRNA Gene Sequencing." *Journal of Animal Breeding and Genetics* 137 (1): 73–83. https://doi.org/10.1111/jbg.12446.

Forcina, Giovanni, Lucía Pérez-Pardal, Júlio Carvalheira, and Albano Beja-Pereira. 2022. "Gut Microbiome Studies in Livestock: Achievements, Challenges, and Perspectives." *Animals* 12 (23): 3375. https://doi.org/10.3390/ani12233375.

Grzeczka, Arkadiusz, Marianna Lech, Gracjan Wozniak, Szymon Graczyk, Pawel Kordowitzki, Małgorzata Olejnik, Marek Gehrke, and Jędrzej Maria Jaśkowski. 2023. "Periodontitis Disease in Farmed Ruminants—Current State of Research." *International Journal of Molecular Sciences* 24 (11): 9763. https://doi.org/10.3390/ijms24119763.

Huaiquipán, Rodrigo, John Quiñones, Rommy Díaz, Carla Velásquez, Gastón Sepúlveda, Lidiana Velázquez, Erwin A. Paz, Daniela Tapia, David Cancino, and Néstor Sepúlveda. 2023. "Review: Effect of Experimental Diets on the Microbiome of Productive Animals." *Microorganisms* 11 (9): 2219. https://doi.org/10.3390/microorganisms11092219.

Jize, Zhang, Deqing Zhuoga, Zhang Xiaoqing, Ta Na, Gesang Jiacuo, Luosang Cuicheng, and Pingcuo Bandan. 2022. "Different Feeding Strategies Can Affect Growth Performance and Rumen Functions in Gangba Sheep as Revealed by Integrated Transcriptome and Microbiome Analyses." *Frontiers in Microbiology* 13 (August):908326. https://doi.org/10.3389/fmicb.2022.908326.

Keckeis, Karin, Michael Lepschy, Hanna Schöpper, Lukas Moser, Josef Troxler, and Rupert Palme. 2012. "Hair Cortisol: A Parameter of Chronic Stress? Insights from a Radiometabolism Study in Guinea Pigs." *Journal of Comparative Physiology B* 182 (7): 985–96. https://doi.org/10.1007/s00360-012-0674-7.

Kittelmann, Sandra, Cesar S. Pinares-Patiño, Henning Seedorf, Michelle R. Kirk, Siva Ganesh, John C. McEwan, and Peter H. Janssen. 2014. "Two Different Bacterial Community Types Are Linked with the Low-Methane Emission Trait in Sheep." Edited by Hauke Smidt. *PLoS ONE* 9 (7): e103171. https://doi.org/10.1371/journal.pone.0103171.

Langda, Suo, Chenguang Zhang, Ke Zhang, Ba Gui, De Ji, Ciren Deji, Awang Cuoji, Xiaolong Wang, and Yujiang Wu. 2020. "Diversity and Composition of Rumen Bacteria, Fungi, and Protozoa in Goats and Sheep Living in the Same High-Altitude Pasture." *Animals* 10 (2): 186. https://doi.org/10.3390/ani10020186.

Meyer, Jerrold S., and Melinda A. Novak. 2012. "Minireview: Hair Cortisol: A Novel Biomarker of Hypothalamic-Pituitary-Adrenocortical Activity." *Endocrinology* 153 (9): 4120–27. https://doi.org/10.1210/en.2012-1226.

Miao, Yongqiang, Xueliang Zhao, Jianlin Lei, Jingru Ding, Hang Feng, Ke Wu, Jiaohu Liu, et al. 2023. "Characterization of Lung Microbiomes in Pneumonic Hu Sheep Using Culture Technique and 16S rRNA Gene Sequencing." *Animals* 13 (17): 2763. https://doi.org/10.3390/ani13172763.

Mocé, María Lorena, Inés Carolina Esteve, Sara Pérez-Fuentes, Ernesto A. Gómez, and Eva Mocé. 2022. "Microbiota in Goat Buck Ejaculates Differs Between Breeding and Non-Breeding Seasons." *Frontiers in Veterinary Science* 9 (May):867671. https://doi.org/10.3389/fvets.2022.867671.

Möstl, E, and R Palme. 2002. "Hormones as Indicators of Stress." *Domestic Animal Endocrinology* 23 (1–2): 67–74. https://doi.org/10.1016/S0739-7240(02)00146-7.

Poole, Rebecca K., Dallas R. Soffa, Brooke E. McAnally, Molly S. Smith, Kyle J. Hickman-Brown, and Erin L. Stockland. 2023. "Reproductive Microbiomes in Domestic Livestock: Insights Utilizing 16S rRNA Gene Amplicon Community Sequencing." *Animals* 13 (3): 485. https://doi.org/10.3390/ani13030485.

Raoult, C. M. C., B. Osthaus, A. C. G. Hildebrand, A. G. McElligott, and C. Nawroth. 2021. "Goats Show Higher Behavioural Flexibility than Sheep in a Spatial Detour Task." *Royal Society Open Science* 8 (3): rsos.201627, 201627. https://doi.org/10.1098/rsos.201627.

Sawyer, G. J. 1977. "OBSERVATIONS ON THE BACTERIAL POPULATION OF THE *OS CERVIX* OF THE EWE BEFORE AND AFTER EMBRYO DEATH." *Australian Veterinary Journal* 53 (11): 542–44. https://doi.org/10.1111/j.1751-0813.1977.tb07942.x.

Serrano, Malena, Eric Climent, Fernando Freire, Juan F. Martínez-Blanch, Carmen González, Luis Reyes, M. Carmen Solaz-Fuster, Jorge H. Calvo, M. Ángeles Jiménez, and Francisco M. Codoñer. 2020. "Influence of the Ovine Genital Tract Microbiota on the Species Artificial Insemination Outcome. A Pilot Study in Commercial Sheep Farms." *High-Throughput* 9 (3): 16. https://doi.org/10.3390/ht9030016.

Smith, K. E., A. L. Garza, C. Robinson, R. L. Ashley, and S. L. Ivey. 2016. "1039 WS Influence of Sampling Location and Pregnancy on Composition of the Microbiome Associated with the Reproductive Tract of the Ewe." *Journal of Animal Science* 94 (suppl_5): 498–498. https://doi.org/10.2527/jam2016-1039.

Swartz, Jeffrey D., Medora Lachman, Kelsey Westveer, Thomas O'Neill, Thomas Geary, Rodney W. Kott, James G. Berardinelli, et al. 2014. "Characterization of the Vaginal Microbiota of Ewes and Cows Reveals a Unique Microbiota with Low Levels of Lactobacilli and Near-Neutral pH." *Frontiers in Veterinary Science* 1 (October). https://doi.org/10.3389/fvets.2014.00019.

Tong, Juanzhen, Wentao Ma, Rui Yang, Tianxing Wang, Xi Chen, Xinyue Zhang, Xidian Tang, Ying Wen, Jianjun Chang, and Dekun Chen. 2020. "Dysbiosis of the Gut Microbiota Maybe Exacerbate Orf Pathology by Promoting Inflammatory Immune Responses." *Veterinary Microbiology* 251 (December):108884. https://doi.org/10.1016/j.vetmic.2020.108884.

Toquet, Marion, Ángel Gómez-Martín, and Esther Bataller. 2021. "Review of the Bacterial Composition of Healthy Milk, Mastitis Milk and Colostrum in Small Ruminants." *Research in Veterinary Science* 140 (November):1–5. https://doi.org/10.1016/j.rvsc.2021.07.022.

Tvrdá, Eva, Miroslava Kačániová, Andrej Baláži, Jaromír Vašíček, Jakub Vozaf, Rastislav Jurčík, Michal Ďuračka, Jana Žiarovská, Ján Kováč, and Peter Chrenek. 2022. "The Impact of Bacteriocenoses on Sperm Vitality, Immunological and Oxidative Characteristics of Ram Ejaculates: Does the Breed Play a Role?" *Animals* 12 (1): 54. https://doi.org/10.3390/ani12010054.

Wang, Jin, Huan Fan, Ye Han, Jinzhao Zhao, and Zhijiang Zhou. 2016. "Characterization of the Microbial Communities along the Gastrointestinal Tract of Sheep by 454 Pyrosequencing Analysis." *Asian-Australasian Journal of Animal Sciences* 30 (1): 100–110. https://doi.org/10.5713/ajas.16.0166.

Chapter 17

Microbiomes of Cattle Health, Disease, and Production

T. Sarath[1,*]
Akhter Rasool[2]
N. Arunmozhi[3]
and U. S. Kalyaan[4]

[1]Veterinary Clinical Complex, VCRI, Salem, Tamil Nadu, India
[2]Department of Veterinary Gynaecology and Obstetrics, Madras Veterinary College, Chennai, India
[3]Department of Clinics, Madras Veterinary College, Chennai, India
[4]Resident Veterinary Services Section, Madras Veterinary College, Tamil Nadu Veterinary and Animal Sciences University, Chennai, India

Abstract

The intricate relationship between the microbiome and cattle health, diseases, and production is a subject of growing importance in contemporary research. This comprehensive review synthesizes current knowledge and research findings, shedding light on the multifaceted roles of microbial communities in various aspects of bovine well-being. The cattle microbiome, present in organs such as the gastro-intestinal tract, respiratory and reproductive tracts, has a profound impact on health and disease dynamics. Bovine Respiratory Disease (BRD), Bovine Digital Dermatitis (DD), Mastitis, Johne's disease, Metritis, Endometritis, Ruminal Acidosis, Acetonemia, and their connections to dysbiosis are explored in depth.

The reproductive tract microbiome's variability, influenced by factors such as diet, estrous cycle, and puberty attainment, is highlighted, emphasizing its crucial role in maintaining the structural integrity of the genital mucosal barrier and protecting against pathogens. Hormonal fluctuations, extrinsic factors like nutrition, and genital pathologies contribute to the complexity of the genital microbiome. Dysbiosis induced by intra-vaginal antibiotic treatment, changes in estrus behavior, and gestational hormonal variations underscore the delicate balance required for optimal reproductive health.

Optimizing bacterial communities in various organs emerges as a key focus, reflecting the shift towards proactive interventions to enhance animal health and productivity. The exploration of microbial dynamics offers innovative possibilities for interventions, emphasizing the potential for tailored approaches to manipulate microbial populations for

* Corresponding Author's Email: drsarathvet@gmail.com

In: Microbiomes in Animal Nutrition, Health, and Disease
Editor: Tanmoy Rana
ISBN: 979-8-89530-425-9

maximal success. This chapter encapsulates the evolving understanding of the microbiome's pivotal role in cattle health, disease prevention, and production efficiency, underscoring the need for ongoing research to refine strategies and advance the field.

Keywords: cattle health, microbiome, production, nutrition

Introduction

Ruminant husbandry practices have undergone substantial transformations throughout the past century. Despite enhancements in the efficiency of these systems, the global agricultural landscape is confronted with significant challenges in the approaching decades. Projections indicate a nearly twofold increase in global meat production and consumption, reaching an estimated 455 million tonnes annually by 2050 (Alexandratos and Bruinsma, 2012). This surge is attributed to both population growth, anticipated to reach 9 billion by 2050, and evolving dietary patterns fueled by rising incomes in developing nations, elevating the demand for animal protein (Alexandratos and Bruinsma, 2012).

Given the anticipated surge in global food demands, it is imperative to bolster the efficiency of both animal and crop-derived food production (Berry and Crowley 2013). The bacterial communities in various organs have indicated importance in the health and productivity of the animal; therefore, research is moving toward developing methods to optimize these bacterial communities for maximal success. Manipulating the microbiome of ruminants holds promise for enhancing animal performance and health, particularly mitigating the adverse effects of high-grain feeding on health and production. Various supplements, including buffers, yeasts, yeast culture products, direct-fed microbials, probiotics, and microbiota engineering, can attenuate these effects (Rasool et al., 2023).

In production species, microbial communities play pivotal roles in growth efficiency, yield, reproduction, and early-life development. Ruminants maintain a symbiotic relationship with microorganisms that are capable of producing enzymes to degrade polysaccharide plant fibrils. In bovines, gastrointestinal bacteria process feedstuffs into volatile fatty acids, microbial protein, and vitamins, fulfilling the host's requirements for maintenance, growth, production, and health (Newbold et al., 2020). The end products of cellulolytic bacterial interaction act as substrates for different microflora that start further degradation of such compounds. The composition of microbiota readily gets tuned to the changes in the feed habits of the dairy cow (Krause et al., 2003). The shift in the microbial composition from a dietary change can affect specific bacterial species. Analyzing the microbiota is crucial for refining animal nutrition strategies and promoting animal health.

Globally, over 3.5 billion domesticated ruminants, including cattle, sheep, and goats, constitute a significant source of food products for humans. These animals harbor a complex gut microbiome comprising approximately 10^{10} bacteria, 10^7 archaea, 10^8 protozoa, and 10^3 fungal spores per ml of rumen fluid (Mackie, 2000). This microbiome provides essential enzymes for breaking down plant fibers into volatile fatty acids and microbial crude protein. The metabolites produced by this microbiota play an important role in host physiology by maintaining homeostasis with the host's immune system (Das, and Ranjan, 2020). The expression of immune genes has been linked to the abundance of luminal bacteria and bacteria correlated with GIT in health in dairy calves (Okumura, and Takeda, 2018). Recently, studies have identified the most prevalent genera associated with a healthy cattle oral cavity, including Arcobacter, Gastranaerophilales, Planifilum, Escherichia-Shigella, Actinobacteria, Burkholderia, and Pseudomonas. Additionally, a diverse array of microflora is present in the reproductive tract, respiratory tract, mammary glands, and integumentary system (Figure 2). These microflora play a crucial role in protecting the host from various diseases and provide essential metabolites (Figure 1).

Rumen microbiology profoundly influences meat and milk production by optimizing nutrient uptake for growth. Understanding the cattle microbiome is paramount for maintaining cattle health and productivity. Notably, the composition of the newborn calf microbiomes initially reflects that of the oral microbiome more closely than the fecal or vaginal microbiome of their dams. By one week post-birth, the calf microbiomes shift to more closely resemble the fecal microbiome of their dams, accompanied by changes in bacterial numbers (Peixoto et al., 2021).

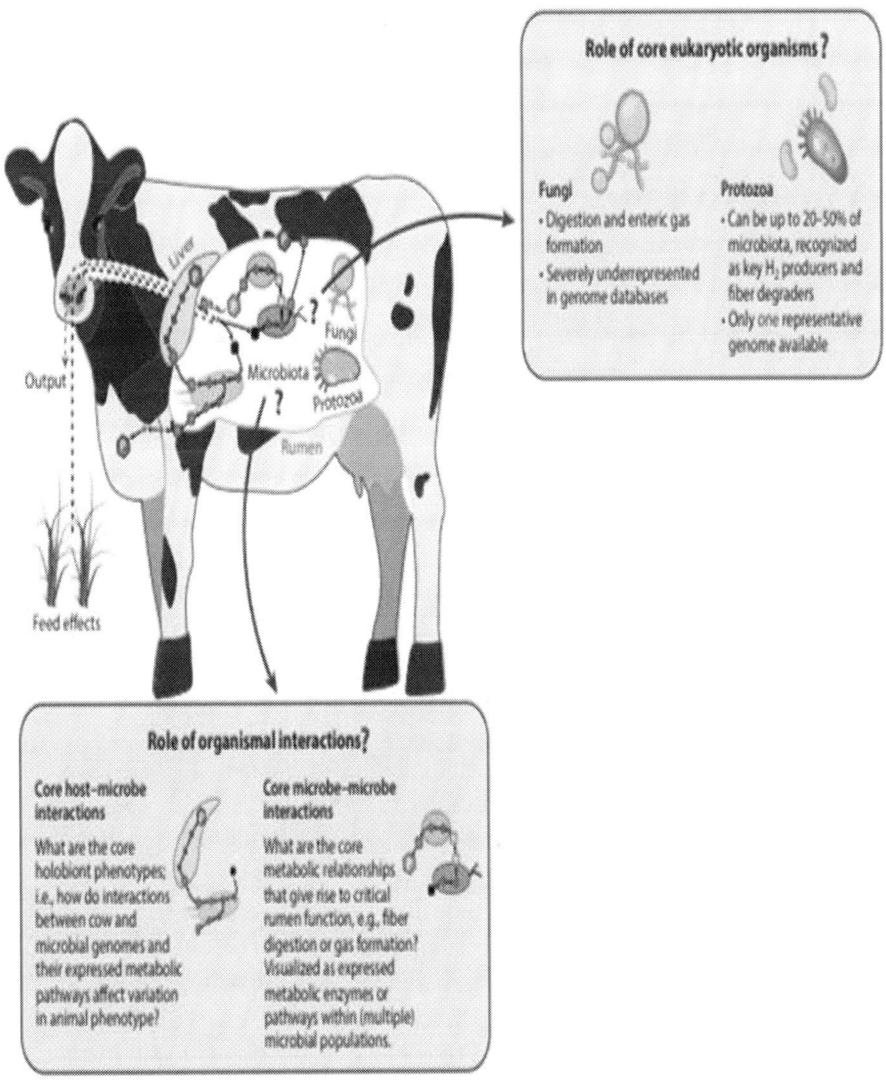

Figure 1. Ruminal Microbiome; Role of microbial communities and their interaction with the host (Perlman et al., 2022).

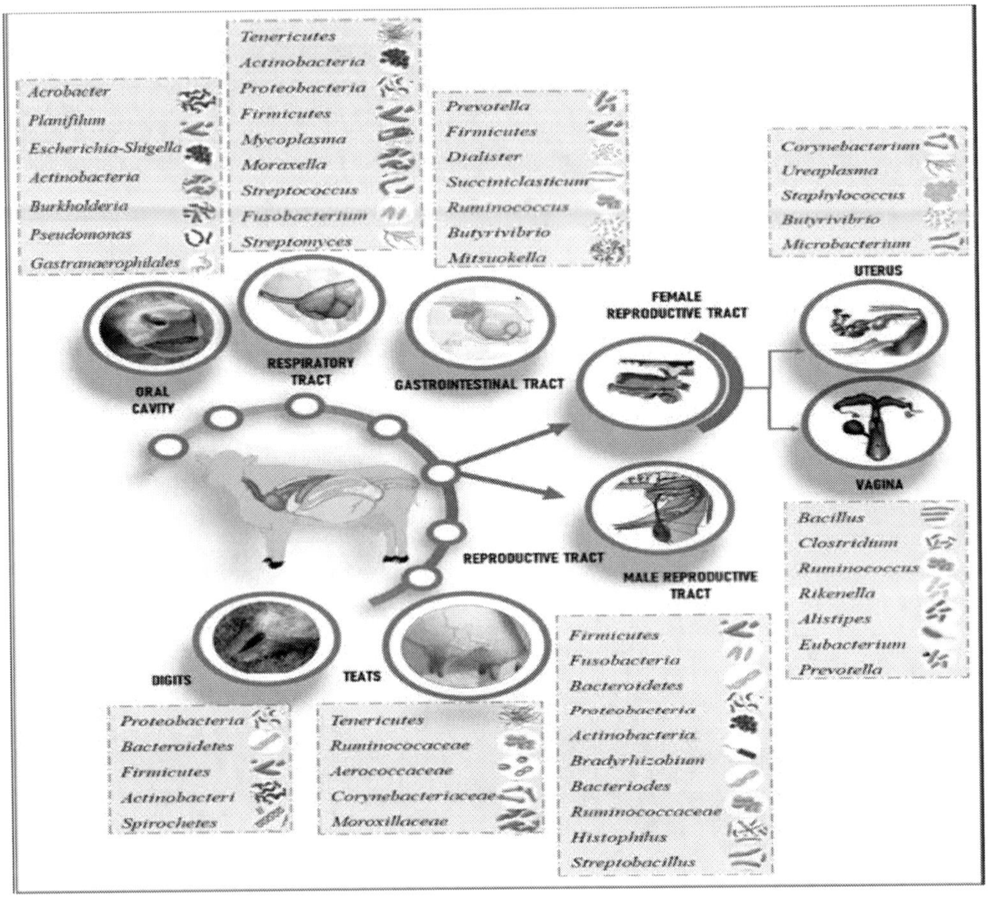

Figure 2. Healthy cattle microbiome in skin, reproductive organs, oral cavity, gastrointestinal tract, and respiratory tract (Khalil et al., 2022).

The microbiota serves as a physical barrier between pathogens and immune cells, engaging in competition with pathogens for adhesion sites and nutrients (Karstrup et al., 2017; Moore et al., 2017). Additionally, these microbes hinder pathogenic growth by synthesizing antimicrobial compounds, including organic acids, hydrogen peroxide, bacteriocins, and biosurfactants (Krishnan et al., 2020; Laguardia-Nascimento et al., 2015). Dysbiosis, characterized by an imbalance in microbial composition, has the potential to cause or exacerbate various disease phenotypes in cattle, such as Johne's disease, uterine, and metabolic diseases (e.g., ruminant acidosis) (Derakhshani et al., 2016; Jeon et al., 2016). To comprehend the therapeutic and prophylactic potential of the cattle microbiome, a thorough understanding of microbial composition, particularly in association with health and disease phenotypes, is essential.

Recent investigations have identified significant differences in the microbiome between modes of delivery, with vaginal delivery resulting in higher species richness, evenness, and diversity in both human and cattle infants compared to C-section delivery (Furman et al., 2020; Greenhalgh et al., 2016). After vaginal delivery, the rumen of calves exhibited a higher abundance of species from the phylum Proteobacteria, along with an increased relative abundance of gastrointestinal tract-related genera such as Bifidobacteria and Bacteroides (Cunningham et al., 2018), as well as vaginal tract-associated genera including Prevotella and

Butyrivibrio spp. (Furman et al., 2020). Conversely, calves delivered by C-section were more likely to have their rumen colonized by species from the phylum Firmicutes, and the genera Peptostreptococcus and Dorea. The reduced colonization and low maternal microbial persistence over time in C-section-delivered calves render them more susceptible to infections (Furman et al., 2020).

Reproductive Tract Microbiota

The reproductive tract microbiome in cattle exhibits high variability influenced by factors such as diet, estrous cycle phases, and puberty attainment, consequently impacting the uterine environment and endometrial tissue activity (Laguardia-Nascimento et al., 2015; Ault et al., 2019). This microbial community, resident in females from a young age, plays a crucial role in maintaining the structural integrity of the genital mucosal barrier, immunomodulation, and protection against pathogens. The gastrointestinal tract (GIT) microbiota, intricately linked to the functioning of the reproductive system, requires an optimal balance in abundance and diversity to promote female genital tract health. Circulating levels of follicle-stimulating hormone (FSH) and luteinizing hormone (LH) have been correlated with systemic lipopolysaccharide (LPS) and specific bacterial genera, influencing the delicate balance within the reproductive microbiome (Wang et al., 2020).

Specific bacterial families (Lachnospiraceae and Rikenellaceae) and genera (Acinetobacter, Bacillus, and Oscillospira) have been identified as indicative of a healthy vaginal microbiota, suggesting their potential as therapeutic targets (Moreno et al., 2020). Fluctuations in hormone levels throughout the estrous cycle and specific factors like estrous cyclicity and pregnancy contribute to the diversity of the genital microbiome (Barba et al., 2020; Deng et al., 2019). Extrinsic factors, including nutrition and genital pathologies, further shape the microbial landscape (Esposito et al., 2020; Mahalingam et al., 2019).

During gestation, higher progesterone concentrations lead to decreased bacterial diversity in the vagina. Dysbiosis induced by intravaginal antibiotic treatment and changes in estrus behavior can be associated with an overpopulation of *S. aureus*, increasing the risk of abortion. Modulating the vaginal microbiome may present an alternative to antibiotic therapies in preventing uterine diseases. Estrogen stimulates glycogen production by vaginal epithelial cells, serving as a substrate for Lactobacillus fermentation, producing lactic acid and hydrogen peroxide to maintain a low pH and defend against pathogen colonization (Rönnqvist et al., 2006).

The microbial diversity within cervical vaginal mucus (CVM) is a crucial factor, intricately linked to the health status of bovines, as highlighted by Knudsen et al. (2016). While the utility of CVM is extensively studied for diagnosing uterine diseases like metritis and endometritis, its impact on estrus detection and pheromone production in livestock species remains an underexplored avenue. The bovine vagina, constituting a dynamic mixture of aerobic, facultative anaerobic, and anaerobic bacteria, features prominent species such as Streptococcus sp., Staphylococcus sp., Enterococci, and Enterobacteriaceae. These organisms actively work to prevent the proliferation of pathogenic microorganisms under natural conditions, forming biofilms that stabilize the microbial environment during cyclicity and gestation (Otero et al., 2000; Tang et al., 2008; Swartz et al., 2014).

CVM microbes play a pivotal role in inhibiting infection during the follicular phase of the estrous cycle, producing reactive oxygen species (ROS) and organic acids. Lactic acid bacteria, in particular, contribute to maintaining a stable pH throughout the genital tract and CVM, thereby protecting the vagina from infection (Eva et al., 2014). Notably, the prevalence of specific bacteria, such as Citrobacter sp., in aborted buffalo suggests their potential use in predicting premature delivery and abortion in this species (Raza et al., 2013).

Understanding the microflora within CVM is critical for influencing pheromone production, as these microbes metabolize and ferment the contents of vaginal fluid. This process is exemplified in dogs, where vaginal microbes are responsible for producing the characteristic odor during estrus (Dzieciol et al., 2013). The attractiveness of vaginal pheromones to male animals, as observed in buffalo and cows, further underscores the significance of identifying estrus-specific microbes for accurate estrus detection (Rajanarayanan and Archunan, 2004).

Exploring the involvement of specific bacteria in CVM in the context of pregnancy diagnosis and abortion prediction holds immense potential for advancing pregnancy management in bovine and other livestock species. This multifaceted interaction between microbial communities and reproductive processes not only deepens our understanding of bovine reproductive health but also presents opportunities for targeted interventions to improve overall reproductive outcomes.

Probiotic management of postpartum uterine infections through intra-vaginal inoculation, including Streptococcus spp. and alpha-hemolytic streptococci, has shown promise in improving uterine health (LeBlanc et al., 2002). Dysbiosis during the postpartum period has been associated with the development of endometritis (Miranda-Casoluengo et al., 2019). The uterus harbors unique bacterial communities, distinct from the vagina, likely contributing to nutrient provision, metabolite production, and influencing the immunological environment of the reproductive tract (Clemmons et al., 2017).

Microbial dysbiosis during gestation in the uterus or vagina can impact pregnancy maintenance, potentially leading to preterm birth or miscarriage. Lactobacillus species, in conjunction with the common postpartum pathogen E. coli, has shown a reduced ability to infect endometrial cells, suggesting a potential strategy for reducing the prevalence of postpartum uterine diseases (Genıś et al., 2017). Notably, the association between bacteria present in the blood and the uterus suggests a potential origin from the gastrointestinal tract (Jeon et al., 2017). The presence of nutrients, particularly protein amino acids, and cytokines in the uterine environment may contribute to the relationship with the uterine microbiome, ultimately influencing reproductive efficiency.

Microbiota Dysbiosis and Associated Diseases

Internal and external factors can disrupt the delicate balance of this dynamic and complex niche, affecting the animal's health status. Dysbiosis, characterized by a decrease in abundance or loss of key symbiotic relationships within the metaorganism, can impact animal productivity and pose a threat to food production (Wilkins et al., 2019).

Dysbiosis, the disruption of microbial communities, plays a pivotal role in exacerbating various diseases. The commensal microbes comprising the microbiota of the respiratory tract, mucosal epithelium, and the immune system communicate to promote respiratory health, reduce inflammation, and maintain a functioning microbiota (Zeineldin et al., 2019). Commensal microbes produce bioactive molecules such as lactic acid and reactive oxygen species that inhibit the proliferation of pathogens (Tachedjian, et al., 2017). Bovine Respiratory Disease (BRD), a complex respiratory ailment, experiences heightened severity due to dysbiosis, compromising respiratory microbiota balance and fostering pathogenic growth (Alexander, 2018). Bovine Digital Dermatitis (DD), a significant cause of lameness, is influenced by dysbiosis affecting microbial communities, potentially involving opportunistic pathogens such as Treponema spp. and Dichelobacter nodosus, intensifying disease pathogenesis (Palmer et al., 2015). Mastitis, an inflammatory condition of the mammary gland, shows dysbiosis altering mammary microbiota, impacting microbial load and diversity, and influencing disease severity (Hoque et al., 2019; Ganda et al., 2016). Johne's disease, caused by Mycobacterium avium subsp. Paratuberculosis (MAP), demonstrates that dysbiosis in the gastrointestinal microbiome may facilitate MAP infection, worsening disease severity (Derakhshani et al., 2016; Fecteau et al., 2016). Metritis and endometritis, prevalent postpartum uterine diseases, exhibit dysbiosis characterized by altered microbial diversity, influencing the development and severity of these conditions (Galvão et al., 2019; Jeon et al., 2015; Jeon et al., 2016). Ruminal acidosis, resulting from a high carbohydrate diet, showcases dysbiosis involving reduced protozoan communities, changes in bacterial diversity, and an altered Firmicutes to Bacteroidetes ratio (Cholewi ́nska et al., 2020; Khafipour et al., 2016; Clemmons et al., 2019). Ruminal tympany, or bloat, stands as a significant contributor to morbidity and mortality in feedlot cattle (Horn, 2006). The occurrence of bloat and its impact on rumen microbial communities exhibit variability among individual animals and can be linked to factors such as the rate of fermentation and ruminal gas production (Bartley and Bassette, 1961). Notably, bloated cattle demonstrate increases in the relative abundance of certain bacterial genera, including Clostridium, Eubacterium, and Butyrivibrio. Conversely, there are decreases in the relative abundance of other genera such as Prevotella and Ruminococcus (Pitta et al., 2016). This shift in microbial composition highlights the intricate interplay between ruminal health and the dynamic balance of microbial communities, underscoring the potential role of specific bacterial populations in contributing to or mitigating the onset of bloat in feedlot cattle. Acetonemia, a metabolic disorder, experiences dysbiosis associated with changes in microbial phyla, contributing to the disorder's manifestation and impacting cattle health, fertility, and milk production (Rodriguez-Jimenez et al., 2018; Huang et al., 2019). Understanding these intricate disease-microbiome interactions is pivotal for developing targeted interventions to manage and prevent these conditions in cattle populations. Further research is warranted to unravel the complexities of these relationships and refine strategies for maintaining optimal bovine health.

A nuanced understanding of these microbial communities holds the potential to revolutionize approaches aimed at enhancing reproduction in cattle. Innovative strategies, including the application of probiotics, antibiotics, and bacteriophages, present promising avenues for manipulating the microbiota to foster the emergence of microflora communities that positively contribute to reproductive tract health. Probiotics play a crucial role in maintaining intestinal microbial homeostasis (Belguesmia et al., 2016). Their benefits extend to reducing the frequency of neonatal diarrhea (Agarwal et al., 2002), regulating ruminal pH (Whitley et al., 2009), and stimulating immunity (Bidarkar et al., 2014). In the realm of veterinary probiotic supplements for ruminants, Lactobacillus and Saccharomyces are among the most widely used, along with Bacillus, Bifidobacterium, Enterococcus, Propionibacterium, Megasphaera elsdenii, and Prevotella bryantii species (Krehbiel et al., 2003). This diverse array of probiotic strains underscores their versatility in promoting various aspects of digestive health and immune function in ruminant animals. As research progresses, the near future may witness the deliberate manipulation of microbiota to orchestrate a balanced immune response, thereby augmenting disease resistance and overall animal performance, as suggested by Gomez et al. (2019). This foresight underscores the transformative impact that microbiome manipulation could have on shaping the landscape of reproductive management in cattle, opening doors to novel and targeted interventions for improved reproductive outcomes and enhanced livestock health.

Conclusion

In conclusion, the bacterial communities throughout the body play a major role in the animal's overall performance, impacting health, disease susceptibility, and production efficiency in cattle. The intricate interplay between the microbiome and various physiological systems, such as the respiratory and reproductive tracts, underscores the need for a holistic understanding to optimize animal well-being. Future research in this field holds significant promise for advancing our comprehension of microbial dynamics in cattle. The ongoing exploration of dysbiosis-associated diseases, such as Bovine Respiratory Disease, Bovine Digital Dermatitis, and Metritis, provides a foundation for developing targeted interventions. Further investigations into the reproductive tract microbiome, considering factors like hormonal influences, nutrition, and extrinsic variables, offer opportunities to refine strategies for maintaining optimal reproductive health. As the field of microbiome research continues to evolve, collaborations between veterinary scientists, microbiologists, and livestock practitioners will be pivotal in translating research findings into practical applications on the farm. Harnessing the full potential of the microbiome in cattle holds promise not only for mitigating diseases but also for optimizing production practices, contributing to sustainable and resilient livestock systems. The integration of cutting-edge technologies and interdisciplinary approaches will undoubtedly shape the future of microbiome-related research, fostering advancements in cattle health, disease management, and production outcomes.

References

Abdela, N. Sub-acute ruminal acidosis (SARA) and its consequence in dairy cattle: A review of past and recent research at global prospective. *Achiev. Life Sci.* 2016, 10, 187–196.

Alexander, T. 145. The role of the bovine respiratory microbiota in health and disease. *J. Anim. Sci.* 2018, 96, 341.

Appiah, M. O., Wang, J., Lu, W. Microflora in the Reproductive Tract of Cattle: A Review (Running Title: The Microflora and Bovine Reproductive Tract). *Agriculture* 2020, 10, 232.

Ault, T. B., Clemmons, B. A., Reese, S. T., Dantas, F. G., Franco, G. A., Smith, T. P. L., et al. (2019). Bacterial taxonomic composition of the postpartum cow uterus and vagina prior to artificial insemination. *J. Anim. Sci.* 97 (10), 4305–4313. doi: 10.1093/jas/skz212.

Agarwal N., D. N. Kamra, L. C. Chaudhary, I. Agarwal, A. Sahoo, N. N. Pathak, Microbial status and rumen enzyme profile of crossbred calves fed on different microbial feed additives, *Lett. Appl. Microbiol.* 34 (2002) 329–336. doi:10.1046/j.1472- 674 765X.2002.01092.x.

Alexandratos, N., and J. Bruinsma. 2012. World agriculture towards 2030/2050: the 2012 revision. *ESA Working paper No*. 12-03. FAO, Rome.

Al-Memar, M., Bobdiwala, S., Fourie, H., Mannino, R., Lee, Y., Smith, A., et al. (2020). The association between vaginal bacterial composition and miscarriage: A nested case– control study. *International Journal of Obstetrics and Gynaecology* 127, 2, 264–274. doi: 10.1111/1471-0528.15972.

Barba, M., Martínez-Boví, R., Quereda, J., Mocé, M., Plaza-Dávila, M., Jiménez-Trigos, E., Gómez-Martín, Á., González-Torres, P., Carbonetto, B., García-Roselló, E. Vaginal Microbiota Is Stable throughout the Estrous Cycle in Arabian Maress. *Animals* 2020, 10, 2020.

Bartley E. E, R. Bassette, Bloat in Cattle. III. Composition of Foam in Legume Bloat, *J. Dairy Sci.* 44 (1961) 1365–1366. doi:10.3168/jds.S0022-0302(61)89890-1.

Bidarkar V. K., P. S. Swain, S. Ray, G. Dominic, Probiotics: Potential Alternative to 657 *Antibiotics in Ruminant Feeding*, 1 (2014) 1–4.

Belguesmia Y., Domenger, Caron, Dhulster, et al. Novel probiotic evidence of lactobacilli on immunomodulation and regulation of satiety hormones release in intestinal cells, *J. Funct. Foods*. 24 (2016) 276–286. doi:10.1016/j.jff.2016.04.014.

Berry D. P., Crowley J. J. 2013. Cell biology symposium: genetics of feed efficiency in dairy and beef cattle. *J. Anim. Sci.* 91(4):1594–613.

Clemmons, B. A., Voy, B. H., Myer, P. R. Altering the gut microbiome of cattle: Considerations of host-microbiome interactions for persistent microbiome manipulation. *Microb. Ecol.* 2019, 77, 523–536.

Cunningham, H. C., Austin, K. J., Cammack, K. M. Influence of maternal factors on the rumen microbiome and subsequent host performance. *Transl. Anim*. Sci. 2018, 2, S101–S105.

Clemmons, B. A., Mulon, P. Y., Anderson, D. E., Ault-Seay, T. B., Henniger, M. T., Schneider, L. G., et al. (2022). Ruminal bacterial communities and metabolome variation in beef heifers divergent in feed efficiency. *Ruminants* 2 (2), 282–296. doi: 10.3390/ ruminants2020019.

Cholewi´nska, P., Czyz, K., Nowakowski, P., Wyrostek, A. The microbiome of the digestive system of ruminants–A review. *Anim. Health Res. Rev.* 2020, 21, 3–14. (CrossRef).

Derakhshani, H., De Buck, J., Mortier, R., Barkema, H. W., Krause, D. O., Khafipour, E. The features of fecal and ileal mucosaassociated microbiota in dairy calves during early infection with Mycobacterium avium subspecies paratuberculosis. *Front. Microbiol*. 2016, 7, 426.

Das, P., Ranjan, R. Role of Gut Microbiome in Improving Animal Health and Productivity. *Indian J. Anim. Health* 2020, 59, 146–155.

Dzi˛ecioł, M., Niza´nski, W., Sta´nczyk, E., Kozdrowski, R., Najder-Kozdrowska, L., Twardo´n, J. The influence of antibiotic treatment˙ of bitches in oestrus on their attractiveness to males during mating. *Pol. J. Vet. Sci*. 2013, 16, 509–516.

Deng, F., McClure, M., Rorie, R., Wang, X., Chai, J., Wei, X., Lai, S., Zhao, J. The vaginal and fecal microbiomes are related to pregnancy status in beef heifers. *J. Anim. Sci. Biotechnol*. 2019, 10, 92.

Dzieciol, M., Nizanski, W., Stanczyk, E., Kozdrowski, R., Najder-Kozdrowska, L., Twardon, J., 2013. The influence of antibiotic treatment of bitches in oestrus on their attractiveness to males during mating. *Pol. J. Vet. Sci*. 16, 509–516.

Eva, S., Radomra, N., Sona, ˇG., Igor, V., Andrea, L., 2014. Bovine vaginal lactobacilli and their adherence to mucus in different phases of the estrous cycle. *Afr. J. Microbiol. Res*. 8, 3017–3024.

Esposito, G., Raffrenato, E., Lukamba, S. D., Adnane, M., Irons, P. C., Cormican, P., Tasara, T., Chapwanya, A. Characterization of metabolic and inflammatory profiles of transition dairy cows fed an energy-restricted diet. *J. Anim. Sci.* 2020, 98, 98.

Fecteau, M.-E., Pitta, D. W., Vecchiarelli, B.; Indugu, N., Kumar, S., Gallagher, S. C., Fyock, T. L., Sweeney, R. W. Dysbiosis of the fecal microbiota in cattle infected with Mycobacterium avium subsp. paratuberculosis. *PLoS ONE* 2016, 11, e0160353. (CrossRef).

Furman, O., Shenhav, L., Sasson, G., Kokou, F., Honig, H., Jacoby, S., Hertz, T., Cordero, O. X., Halperin, E., Mizrahi, I. Stochasticity constrained by deterministic effects of diet and age drive rumen microbiome assembly dynamics. *Nat. Commun*. 2020, 11, 1–13.

Genıs, S., Sa´´nchez-Chardi, A., Bach, À., Fàbregas, F., and Arıs, A. (2017). A´ combination of lactic acid bacteria regulates escherichia coli infection and inflammation of the bovine endometrium. *J. Dairy Sci.* 100 (1), 479–492. doi: 10.3168/ jds.2016-11671.

Gomez, D. E., Galvão, K. N., Rodriguez-Lecompte, J. C., Costa, M. C. The Cattle Microbiota and the Immune System: An Evolving Field. Vet. *Clin. Food Anim. Pract.* 2019, 35, 485–505.

Gregoire, A. T., and Parakkal, P. F. (1972). Glycogen content in the vaginal tissue of normally cycling and estrogen and progesterone-treated rhesus monkeys. *Biol. Reprod.* 7 (1), 9–14. doi: 10.1093/ biolreprod/7.1.9.

Ganda, E. K., Bisinotto, R. S., Lima, S. F., Kronauer, K., Decter, D. H., Oikonomou, G., Schukken, Y. H., Bicalho, R. C. Longitudinal metagenomic profiling of bovine milk to assess the impact of intramammary treatment using a third-generation cephalosporin. *Sci. Rep.* 2016, 6, 37565.

Galvão, K. N., Bicalho, R. C., Jeon, S. J. Symposium review: The uterine microbiome associated with the development of uterine disease in dairy cows. *J. Dairy Sci.* 2019, 102, 11786–11797.

Greenhalgh, K., Meyer, K. M., Aagaard, K. M., Wilmes, P. The human gut microbiome in health: Establishment and resilience of microbiota over a lifetime. *Environ. Microbiol.* 2016, 18, 2103–2116.

Goto, H.; Qadis, A. Q., Kim, Y. H., Ikuta, K., Ichijo, T., Sato, S. Effects of a bacterial probiotic on ruminal pH and volatile fatty acids during subacute ruminal acidosis (SARA) in cattle. *J. Vet. Med. Sci.* 2016, 16-0211.

Huang, Y., Li, Y., He, B., Hu, J., Ali Mohsin, M., Yu, H., Wang, P., Zhang, P., Du, Y., Huang, L. The Influence of Ketosis on the Rectal Microbiome of Chinese Holstein Cows. *Pak. Vet. J.* 2019, 39, 175–180.

Horn G. W., Growing Cattle on Winter Wheat Pasture: Management and Herd Health Considerations, *Vet. Clin. North Am. - Food Anim. Pract.* 22 (2006) 335–356. doi:10.1016/j.cvfa.2006.03.008.

Hoque, M. N., Istiaq, A., Clement, R. A., Sultana, M., Crandall, K. A., Siddiki, A. Z., Hossain, M. A. Metagenomic deep sequencing reveals association of microbiome signature with functional biases in bovine mastitis. *Sci. Rep.* 2019, 9, 1–14.

Hu, X., Guo, J., Zhao, C., Jiang, P., Maimai, T., Yanyi, L., Cao, Y., Fu, Y., Zhang, N. The gut microbiota contributes to the development of Staphylococcus aureus-induced mastitis in mice. *ISME J.* 2020, 14, 1–15.

Jeon, S. J., Vieira-Neto, A., Gobikrushanth, M., Daetz, R., Mingoti, R. D., Parize, A. C. B., de Freitas, S. L., da Costa, A. N. L., Bicalho, R. C., Lima, S. Uterine microbiota progression from calving until establishment of metritis in dairy cows. *Appl. Environ. Microbiol.* 2015, 81, 6324–6332.

Jeon, S. J., Cunha, F.; Ma, X., Martinez, N., Vieira-Neto, A., Daetz, R., Bicalho, R. C., Lima, S., Santos, J. E., Jeong, K. C. Uterine microbiota and immune parameters associated with fever in dairy cows with metritis. *PLoS ONE* 2016, 11, e0165740.

Jeon, S. J., Cunha, F., Vieira-Neto, A., Bicalho, R. C., Lima, S., Bicalho, M. L., et al. (2017). Blood as a route of transmission of uterine pathogens from the gut to the uterus in cows. *Microbiome* 5 (1), 109. doi: 10.1186/s40168-017-0328-9.

Khalil A, Batool A, Arif S. Healthy cattle microbiome and dysbiosis in diseased phenotypes. *Ruminants.* 2022 Feb 23;2(1):134-56.

Krehbiel C. R., S. R. Rust, G. Zhang, S. E. Gilliland, Bacterial direct-fed microbials in ruminant diets: Performance response and mode of action, *J. Anim. Sci.* 81 (2003) 120– 680 132. doi:/2003.8114_ suppl_2E120x.

Knudsen, L. R., Karstrup, C. C., Pedersen, H. G., Angen, O., Agerholm, J. S., Rasmussen, E. L., Jensen, T. K., Klitgaard, K., 2016. An investigation of the microbiota in uterine flush samples and endometrial biopsies from dairy cows during the first 7 weeks postpartum. *Theriogenology* 86, 642–650.

Khafipour, E., Li, S., Tun, H., Derakhshani, H., Moossavi, S., Plaizier, J. Effects of grain feeding on microbiota in the digestive tract of cattle. *Anim. Front.* 2016, 6, 13–19.

Karstrup, C. C., Klitgaard, K., Jensen, T. K., Agerholm, J. S., Pedersen, H. G. Presence of bacteria in the endometrium and placentomes of pregnant cows. *Theriogenology* 2017, 99, 41–47.

Krause D. O., Denman S. E., Mackie R. I., Morrison M., Rae A. L., Attwood G. T., et al. Opportunities to improve fiber degradation in the rumen: microbiology, ecology, and genomics. *FEMS Microbiol Rev.* (2003) 27:663– 93. doi: 10.1016/S0168-6445(03)00072-X.

Krishnan, D., Al-Harbi, H., Gibson, J., Olchowy, T., Alawneh, J. On the use of probiotics to improve dairy cattle health and productivity. *Microbiol. Aust.* 2020, 41, 86–90.

Laguardia-Nascimento, M., Branco, K. M., Gasparini, M. R., Giannattasio-Ferraz, S., Leite, L. R., Araujo, F. M., et al. (2015). Vaginal microbiome characterization of nellore cattle using metagenomic analysis. *PloS One* 10 (11), e0143294. doi: 10.1371/ journal.pone.0143294.

Mahalingam, S., Dharumadurai, D., Archunan, G. Vaginal microbiome analysis of buffalo (Bubalus bubalis) during estrous cycle using high-throughput amplicon sequence of 16S rRNA gene. *Symbiosis* 2019, 78, 97–106.

Miranda-CasoLuengo, R., Lu, J., Williams, E. J., Miranda-CasoLuengo, A .A., Carrington, S. D., Evans, A. C., Meijer, W. G. Delayed differentiation of vaginal and uterine microbiomes in dairy cows developing postpartum endometritis. *PLoS ONE* 2019, 14, e0200974.

Moreno, C. G., Luque, A. T., Galvão, K. N., Otero, M. C. Bacterial communities from vagina of dairy healthy heifers and cows with impaired reproductive performance. *Res. Vet. Sci.* 2022, 142, 15–23.

Moreno, I., Codoñer, F. M., Vilella, F., Valbuena, D., Martinez-Blanch, J. F., JimenezAlmazán, J., et al. (2016). Evidence that the endometrial microbiota has an effect on implantation success or failure. *Am. J. Obstet. Gynecol.* 215 (6), 684–703. doi: 10.1016/ j.ajog.2016.09.075.

Moore, S., Ericsson, A., Poock, S., Melendez, P., Lucy, M. Hot topic: 16S rRNA gene sequencing reveals the microbiome of the virgin and pregnant bovine uterus. *J. Dairy Sci.* 2017, 100, 4953–4960.

McCann, J. C., Wickersham, T. A., Loor, J. J. High-throughput methods redefine the rumen microbiome and its relationship with nutrition and metabolism. *Bioinform. Biol. Insights* 2014, 8, BBI-S15389.

Mackie RI. Molecular ecology and diversity in gut microbial ecosystems. Ruminant physiology: digestion, metabolism, growth and reproduction. In book: *Ruminant Physiology: Digestion, Metabolism, Growth and Reproduction* (pp. 61-77) Publisher: CAB International 2000, Editors: P. B. Gronje 2000; 2000. p. 61–77.

Newbold, C. J., Ramos-Morales, E. Review: Ruminal Microbiome and Microbial Metabolome: Effects of Diet and Ruminant Host. *Animal* 2020, 14, s78–s86.

Okumura, R., Takeda, K. Maintenance of intestinal homeostasis by mucosal barriers. *Inflamm. Regen.* 2018, 38, 5.

Otero, C., de Ruiz, C. S., Ibañez, R., Wilde, O., Holgado, A. D. R., Nader-Macias, M. Lactobacilli and Enterococci Isolated from the Bovine Vagina during the Estrous cycle. *Anaerobe* 1999, 5, 305–307.

Otero, C., Saavedra, L., Silva de Ruiz, C., Wilde, O., Holgado, A. R., Nader-Macias, M. E., 2000. Vaginal bacterial microflora modifications during the growth of healthy cows. *Lett. Appl. Microbiol.* 31, 251–254.

Ogunade, I., Pech-Cervantes, A., Schweickart, H. Metatranscriptomic analysis of sub-acute ruminal acidosis in beef cattle. *Animals* 2019, 9, 232.

Plaizier, J. C., Li, S., Tun, H. M., Khafipour, E. Nutritional models of experimentally-induced subacute ruminal acidosis (SARA) differ in their impact on rumen and hindgut bacterial communities in dairy cows. *Front. Microbiol.* 2017, 7, 2128.

Palmer, M. A., O'Connell, N. E. Digital dermatitis in dairy cows: A review of risk factors and potential sources of between-animal variation in susceptibility. *Animals* 2015, 5, 512–535.

Payne, M. S., and Bayatibojakhi, S. (2014). Exploring preterm birth as a polymicrobial disease: An overview of the uterine microbiome. *Front. Immunol.* 5. doi: 10.3389/fimmu.2014.00595.

Peixoto, R. S., Harkins, D. M., & Nelson, K. E. (2021). Advances in microbiome research for animal health. *Annual Review of Animal Biosciences*, 9, 289-311.

Perlman D, Martínez-Álvaro M, Moraïs S, Altshuler I, Hagen LH, Jami E, Roehe R, Pope PB, Mizrahi I. Concepts and consequences of a core gut microbiota for animal growth and development. *Annual Review of Animal Biosciences*. 2022 Feb 15;10:177-201.

Pitta D. W., W. E. Pinchak, N. Indugu, B. Vecchiarelli, R. Sinha, J. D. Fulford, Metagenomic analysis of the rumen microbiome of steers with wheat-induced frothy bloat, *Front. Microbiol.* 7 (2016) 1–11. doi:10.3389/fmicb.2016.00689.

Rasool, A., Porteen, K., Sarath, T., & Joshy, K. (2023). Exploring the Potential of Probiotics and Fecal Microbiota in Maintaining a Healthy Gut Microbiota. *Journal of Scientific Research and Reports*, 29(7), 101–106.

Raza, S., Rabbani, M., Ahmad, N., Sheikh, A. A., Muhammad, K., Akhtar, F., Rehman, H. U., 2013. Uterine microbial flora of Nili-Ravi Buffalo during estrus and its relationship with pregnancy rate in Pakistan. *Buffalo Bull.* 32, 564–567.

Rajanarayanan, S., Archunan, G., 2004. Occurrence of flehmen in male buffaloes (Bubalus bubalis) with special reference to estrus. *Theriogenology* 61, 861–866.

Rowland, I., Gibson, G., Heinken, A., Scott, K., Swann, J., Thiele, I., Tuohy, K. Gut Microbiota Functions: Metabolism of Nutrients and Other Food Components. *Eur. J. Nutr.* 2018, 57, 1–24).

Rönnqvist, D. J. P., Forsgren-Brusk, B. U., and Grahn-Håkansson, E. E. (2006). Lactobacilli in the female genital tract in relation to other genital microbes and vaginal pH. *Acta obstetricia et gynecologica Scandinavica* 85, 6, 726–735. doi: 10.1080/ 00016340600578357.

Rodriguez-Jimenez, S., Haerr, K., Trevisi, E., Loor, J., Cardoso, F., Osorio, J. Prepartal standing behavior as a parameter for early detection of postpartal subclinical ketosis associated with inflammation and liver function biomarkers in peripartal dairy cows. *J. Dairy Sci.* 2018, 101, 8224–8235.

Sheldon, I. M., Cronin, J. G., Bromfield, J. J. Tolerance and innate immunity shape the development of postpartum uterine disease and the impact of endometritis in dairy cattle. *Annu. Rev. Anim. Biosci.* 2019, 7, 361–384. (CrossRef).

Sheldon, I. M., Owens, S. E. Postpartum uterine infection and endometritis in dairy cattle. *Anim. Reprod.* 2018, 14, 622–629.

Swartz, J. D., Lachman, M., Westveer, K., O'Neill, T., Geary, T., Kott, R. W., et al. (2014). Characterization of the vaginal microbiota of ewes and cows reveals a unique microbiota with low levels of lactobacilli and near-neutral pH. *Front. Vet. Sci.* 1 (19). doi: 10.3389/fvets.2014.00019.

Swartz, J. D., Lachman, M., Westveer, K., O'Neill, T., Geary, T., Kott, R. W., Berardinelli, J. G., Hatfield, P. G., Thomson, J. M., Roberts, A., Yeoman, C. J., 2014. Characterization of the vaginal microbiota of ewes and cows reveals a unique microbiota with low levels of lactobacilli and near-neutral pH. *Front. Vet. Sci.* 1, 19.

Tang, G., Kitten, T., Munro, C. L., Wellman, G. C., Mintz, K. P., 2008. EmaA, a potential virulence determinant of Aggregatibacter actinomycetemcomitans in infective endocarditis. *Infect. Immun.* 76, 2316–2324.

Tachedjian, G., Aldunate, M., Bradshaw, C. S., Cone, R. A. The role of lactic acid production by probiotic Lactobacillus species in vaginal health. *Res. Microbiol.* 2017, 168, 782–792.

Wrenn, T. R., Bitman, J., and Wood, J. R. (1968). Influence of local estrogen administration on vaginal glycogen. *Endocrinol* 82 (1), 62–68. doi: 10.1210/endo-82- 1-62.

Whitley, N. C., D. Cazac, B. J. Rude, D. Jackson-O'Brien, S. Parveen, Use of a commercial probiotic supplement in meat goats, *J. Anim. Sci.* 87 (2009) 723–728.].

Wang, M. L., Liu, M. C., Xu, J., An, L. G., Wang, J. F., Zhu, Y. H. Uterine microbiota of dairy cows with clinical and subclinical endometritis. *Front. Microbiol.* 2018, 9, 2691.

Wilkins, L. G. E., Leray M., O'Dea A., Yuen B., Peixoto R. S., et al. 2019. Host-associated microbiomes drive structure and function of marine ecosystems. *PLOS Biol.* 17(11):e3000533.

Zeineldin, M., Barakat, R., Elolimy, A., Salem, A. Z., Elghandour, M. M., Monroy, J. C. Synergetic action between the rumen microbiota and bovine health. *Microb. Pathog.* 2018, 124, 106–115.

Zeineldin, M., Lowe, J., Aldridge, B. Contribution of the mucosal microbiota to bovine respiratory health. *Trends Microbiol.* 2019, 27, 753–770.

Chapter 18

Microbiomes of Companion Animal Health, Disease, and Prevention

Deepthi Balam*

Department of Veterinary Microbiology, College of Veterinary Science, Garividi, Vijayanagaram Dist. Affiliated to Sri Venkateswara Veterinary University, Garividi, Andhra Pradesh, India

Abstract

Microbiome plays a crucial role in canine health in several ways. Gut microbiome aids in digestion and production of vitamins and short- chain fatty acids that are beneficial for gut health. A balanced microbiome supports a healthy immune system by distinguishing beneficial microbes from harmful pathogens. Similarly, microbiome also influences metabolism and nutrient absorption which impacts weight management and overall energy levels. Gut- brain axis plays a pivotal role in influencing behaviour, temperament and mood in dogs. Various disease conditions like inflammatory bowel disease, skin allergies, obesity, hormonal imbalances, etc. have been implicated to dysbiosis. Factors such as diet, environment and antibiotic treatment regimens can all impact the microbiome composition. Hence, maintaining a diverse and balanced microbiome is essential for overall canine health.

Keywords: cat, dog, gut microbiota, microbiome, probiotics

Introduction

Pet humanization has been the global trend, where companion animals are being recognised as extended family members and have become an important part of human life. Hence, the health and well- being of these companion animals have become a prime area of research interest.

The collection of all microbes like bacteria, archeae, fungi, eukaryotes and their genetic material is referred to as Microbiome. These microbes play a vital role in canine health – protecting against pathogens, interacting with the immune system and assisting with metabolism. Microbiome differs from the term Microbiota, as it indicates the microbial

* Corresponding Author's Email: drdeepthib@gmail.com

In: Microbiomes in Animal Nutrition, Health, and Disease
Editor: Tanmoy Rana
ISBN: 979-8-89530-425-9

population present in different ecosystems in the body. Any alteration in the composition of microbiome results in dysbiosis. Microbiome plays a very important role in the host nutrition, metabolism, development, immune function, and behaviour. It can also be used as a biomarker to estimate health and disease. Increased awareness on antimicrobial resistance lead to focus on manipulation of microbiome to enhance the livestock health. Leverage of microbiome has been shown to be a promising tool in improving health and combating disease.

Targeting microbiome health in livestock, pets, and wildlife will each have significant economic and societal impacts. As the world population reliant on animals and their products is roaring (estimated to be 10 billion people by the year 2050 according to a 2017 United Nations report), along with increased awareness to reduce use of antibiotics in foods, stimulus to utilize the microbiome as a tool to promote healthier and more productive livestock has a great demand. Indeed, a growing body of research indicates that production can be increased while making it less reliant on antibiotics, and the microbiome has also shown promise as an effective target in the areas of pet health and animal conservation, helping to reduce morbidity and combat infectious diseases. Reading out a microbiome can be a nebulous task, but this can be a crucial task in assessing the proper health issue that in turn helps in developing effective treatment protocol. Imbalance of typical microbiota, often called as, dysbiosis, can be attributed to negative health conditions. This dysbiosis arises due to multiple reasons: overuse of antibiotics may deplete beneficial microbes, dominance of pathogens (or opportunistic pathogens) or an alteration in microbial diversity (Petersen and Round, 2014).

Dysbiosis may not be problematic all the time, because it may be an adaptive technique developed by the body to combat microbial infection, hence, it may be a positive adaptive shift. For example, primates in captivity are more prone to gut related diseases than the free primates, and a study has proven vast differences in the microbiome of these two populations (Amato et al., 2016). Similarly, canines with inflammatory bowel disease (IBD), showed highly dysbiotic microbiome than the healthy canine groups (Va´zquez-Baeza et al., 2016).

Microbiome in Different Parts of the Canine Body

Microbiome varies significantly across different body systems, reflecting the unique environment and functions of each area (Table 1). Each of these microbiomes has a distinct composition of bacteria, viruses, fungi, and other organisms tailored to its specific environment and function.

Skin

The skin is one of the largest organ harboured by a variety of microbes. It also plays a vital role in acting as a barrier in protecting against physical and chemical trauma and in maintenance of homeostasis. Microbial diversity of the skin includes bacteria of various phyla, fungi and viruses like Papilloma viruses. Most frequently identified bacteria includes Actinobacteria, Bacteroides, Fusobacteria, Cyanobacteria, etc. Various researchers characterized the microbes located on haired skin, mucosal surfaces, mucocutaneous junctions, and, even in the deeper skin layers.

Table 1. Microbiota commonly found in different systems of canine body

S.No	System	Microbiota commonly found	Diseases that cause dysbiosis
1.	GIT	family Ruminococcaceae, Faecalibacterium prausnitzii, family Peptostreptococcaceae, family Lachnospiraceae, Blautia spp., Streptococcus, Lactobacillus, Turicibacter, Catenibacterium, and Coprobacillus, Prevotella, Bacteroides, and Megamonas, Fusobacteria, Escherichia coli	Acute uncomplicated diarrhea, Acute hemorrhagic diarrhea syndrome, Chronic enteropathies, Inflammatory bowel disease
2.	Skin	Pseudomonas aeruginosa Anaerococcus prevotii Staphylococcus pseudintermedius Staphylococcus schleiferi Finegoldia magna Malassezia pachydermatis Peptoniphilus grossensis Fusobacterium nucleatum Streptococcus equinus Bacteroides fragilis	Acne vulgaris, Atopic dermatitis/ Eczema, Psoriasis, Hidradenitis suppurativa, Rosacea, Dandruff and Seborrheic dermatitis, Alopecia areata, Skin cancer, Wound healing.
3.	Ear canal	Pseudomonas spp., Staphylococcus spp., Corynebacterium spp., Streptococcus spp., Porphyromonas spp., Acinetobacter spp., Sphingomonas spp.	Otitis
4.	Conjunctiva	Staphylococcus spp. Bacillus spp. Psychrobacter spp. Acinetobacter spp. Moraxella spp.	Keratoconjunctivitis sicca Auto immune uveitis Age- related macular degeneration
5.	Respiratory tract	Filobacterium spp., Acinetobacter spp., B. bronchiseptica, Mycoplasma spp., Moraxella spp., Staphylococcus spp. and Chlamydia spp.	Canine/ Feline chronic rhinitis Nasal neoplasia Canine chronic bronchitis Feline asthma Fibrotic lung disease
6.	Genito urinary tract	*Pseudomonas spp., Mycoplasma spp., Streptococcus sp., Fusobacterium sp, Bacteroides sp., Helicobacter sp., Prevotella copri, Megamonas sp., Porphyromonas sp., Roseburia sp., Campylobacter sp., Phascolarctobacterium sp.*	Urinary tract infections
7.	Oral cavity	Porphyromonas cangingivalis Porphyromonas gulae Capnocytophaga sp, Bergeyella zoohelcum, Neis-seria animaloris, Peptostreptococcus canis and Neisseria shayeganii	Periodontal disease

In a study, microbial diversity of skin in dogs with atopic dermatitis was studied in comparison to healthy dogs. Illumina sequencing of the samples collected from various sites in the selected cohort proved no significant differences in healthy dogs and those with atopic dermatitis. Main phyla inhabiting human skin are Actinobacteria, Firmicutes, Bacteroidetes, and Proteobacteria. Staphylococcus and Porphyromonas are the predominant bacterial species in both. Staphylococcus is a normal skin commensal, while, Porphyromonas is a commensal of oral cavity. Other genera commonly present on healthy dog skin includes Kocuria, Pseudomonas and Corynebacterium spp. Apart from Malassezia, Blumeria and Epicoccum are the other predominant fungal species identified in the skin based on ITS marker gene analyses. (Chermprapai et al., 2019).

An extensive study of canine cutaneous microbiota revealed existence of microhabitats much similar to human cutaneous microbiota. Canine microbiota is characterized by the few taxa predominantly present in specific sites: sites with rich sebaceous glands (occiput, glabella, alar crease, and manubrium) with *Propionibacterium* spp; moist sites (nare, axilla, and inguinal crease) with *Staphylococcus* and *Corynebacterium* spp; and dry sites (palms and butlock) with gram-negative microorganisms. According to the first extensive study reported, dogs harbor mainly the same phyla as human skin (Rodrigues-Hoffmann, 2014). Fusobacteria was also detected as a main phylum, when only considering the paws, forehead and groins (Song et al., 2013; Pierezan et al., 2016).

Ear Canal

Ear canal is one more congenial niche for several microbiota. Conformation of the ear pinna and presence of hair in the ear canal influences colonization of the ear canal by a wide variety of bacteria and fungi. Among bacteria, higher abundance of phylum Proteobacteria, Actinobacteria, Firmicutes, Bacteroidetes, and Fusobacteria (Borriello et al., 2020; Kasai et al., 2020) was reported. Among fungi, Ascomycota was the most abundant phylum, followed by Basidiomycota (Bradley et al., 2020). Acute or chronic otitis externa is a common inflammatory disease of the external ear canal. Although the primary causes include endocrine disorders, hypersensitivity, foreign bodies or neoplasia, it may also be associated with proliferation of normal microbiota present in the ear canal. Dogs with atopic dermatitis are more prone to develop otitis externa. Microbiota that dominate in the ear canal of dogs include *Malassezia pachydermatis, S. pseudointermedius, S. aureus, P. aeruginosa, Klebsiella* sps., etc.

Conjunctiva

Ocular surfaces are usually well protected from microorganisms due to innate immune mechanisms such as tear film formation, blinking mechanisms, etc. Recent sequencing techniques have shown the richness and diversity of the ocular microbiota. Ocular surfaces of the healthy dogs showed presence of 4 dominant phyla: Proteobacteria, Bacteroidetes, Firmicutes and Actinobacteria (Banks et al., 2020).

Respiratory Tract

Respiratory tract includes two main portions- lower respiratory tract and upper respiratory tract. Upper respiratory tract includes nose, larynx and trachea, while lower respiratory tract includes bronchi, bronchioles, alveolar ducts and alveolar sacs. Primary function of the respiratory system is gas exchange, and the secondary function is to protection against pathogenic and non-pathogenic invasion. Similar to the skin mucosae, nasal mucosa also is harboured by several microorganisms. Oropharynx is predominantly occupied by bacteria of families Pasteurellaceae, Moraxellaceae and Porphyromonadaceae. Bronchio alveolar lavage fluid is dominated by members of genera Streptococcus spp, Cutibacterium spp, Pseudomonas spp and Acinetobacter spp etc (Ericsson et al., 2020). Dogs with Kennel cough (Canine infectious respiratory disease complex) usually have a high bacterial (Bordetella bronchiseptica, Mycoplasma cynos, and Streptococcus equi subsp zooepidemicus) and viral loads.

Genito Urinary Tract

The urinary tract of healthy dogs harbors microorganisms. Proteobacteria was the dominant phyla, with *Pseudomonas* spp., *Acinetobacter* spp., and *Sphingobium* spp. relative abundances up to 80%, 5%, and 4.5% respectively (Burton et al., 2017). Microbiota of male and female genitalia had some similarity with urinary microbiota, and dominance of Pseudomonas spp. and Acinetobacter spp. The microbiota of the male reproductive system is less documented. In turn, the microbiota of females has been characterized in vaginal swabs and endometrial biopsy samples (Rota et al., 2020). The three most abundant phyla in the vagina and endometrium are Proteobacteria, Bacteroidetes, and Fusobacteria (Lyman et al., 2019). A study that included females sampled in different stages of their reproductive cycle showed that Hydrotalea, Ralstonia, Mycoplasma, Fusobacterium, and Streptococcus were the most abundant genera in the vagina, whereas in the endometrium, Pseudomonas, Staphylococcus, and Corynebacterium were the most abundant (Lyman et al., 2019). Indeed, the diversity and richness of microbiota of pre-pubertal females were significantly different from mature dogs. Moreover, the reproductive cycle affects the microbiota of the vagina, having estrus, the richest and diverse bacterial community. In turn, despite the effect of gonadal hormones in the vaginal microbiota, the vaginal bacterial community of dogs subjected to ovariohysterectomy was not different from female dogs in anestrus (Rota et al., 2020).

Oral Cavity

Oral cavity is the first part of GIT, where food is chewed, lubricated and then swallowed. Saliva, mostly released by salivary glands, is implicated in buffering, remineralization, inhibition of demineralization, and lubrification of teeth having also anti-microbial properties (Pasha et al., 2018). Hence, the environmental conditions of the mouth, likely modulate the microbial communities of the oral cavity. Rich and diverse microbiota was recorded in the oral cavity of dogs than the other sites. As reported by several authors, the predominant bacterial composition include Proteobacteria, Firmicutes, Bacteroidetes, and Fusobacteria. Common bacteria included genera Porphyromonas, Fusobacterium, Streptococcus, Capnocytophaga, and members of Pasteurellaceae and Neisseriacea families, despite the variability reported (Ruparell et al., 2020).

Dogs with periodontal disease had a higher abundance of Gram-positive and a decrease of Gram negative bacteria compared to healthy individuals (Davis et al., 2013). Periodontitis is characterized by changes in the microbial community i.e., health-associated taxa, shifts in phyla abundance, periodontitis was associated with a predominance of species of Actinomyces, Peptostreptococcaceae, and Porphyromonas. Benefits of canine oral chews was investigated and they are found to enhance the proportion of supragingival plaque bacteria associated with oral health and to reduce harmful ones that results in periodontitis (Ruparell et al., 2020). Contrarily, the oral administration of a commercial probiotic, containing microencapsulated Enterococcus faecium SF68 (5 £108 CFU/g) did not alter the oral microbiota of dogs, which might be due to a short residency in the oral cavity (Bell et al., 2020).

Gastrointestinal Tract

The GIT is composed of compartments that differ in their anatomy, functionality, different enzyme activities, electrolyte concentrations, gastric emptying, intestinal transit times, pH, secretion and composition of bile, and absorption rates. Gastric peristalsis and acidity creates a harsh environment for colonization of microorganisms. Helicobacter is considered as a common commensal bacterium that predominates not only in the stomach of dogs, also in the colon crypts of healthy individuals (Giaretta et al., 2020).

In the recent years, intestinal microbiota has been the target for several investigations. Microbes degrade proteins and carbohydrates that reach the small intestine escaping host enzymatic digestion (Gilbert et al., 2018). The end fermentation products released include short chain fatty acids (mainly propionate, butyrate, and acetate), lactate from carbohydrates, ammonia, branched-chain fatty acids (BCFA; e.g., isovalerate and iso-butyrate), polyamines (e.g., putrescine, cadaverine, spermine, spermidine), indoles, and p-cresol among others, mainly from protein degradation. However, the immunomodulatory effect of SCFA in the gut is well reported (Minamoto et al., 2019). Proper functioning of the gut largely depends on the microbiome residing here which also contributes to intestinal immunity. When the gut microbial composition is balanced, dendritic cells and M cells recognise the commensal bacteria, inducing IgA production and production of Treg cells to secrete anti-inflammatory and immunomodulatory cytokines. Also, SCFA are sources of energy for enterocytes, stimulate goblet cells to produce mucins for the mucus layer, and strengthen tight junctions. As long as the gut microbiome is balanced, gut epithelial cells secrete antimicrobial peptides on stimulation from the subcutaneous lymphoid cells. Dysbiosis in the gut due to proliferation of enteropathogens triggers inflammatory responses with infiltration of mature dendritic cells, activated macrophages, neutrophils, Th1, and Th17 cells.

A decrease of beneficial bacterial numbers might reduce the production of SCFA, which undermines the protective mucus layer and loss of tight junctions, facilitating the translocation of bacteria, and thus infection. In that sense, the promotion of enrichment of beneficial bacteria such as *Bifidobacterium* spp and *Lactobacillus* spp and decrease of potential pathogens such as *Escherichia coli* and *Clostridium perfringens* are a goal of most interventions.

In dogs, the biogeography of microbial community varies throughout the GIT, likely due to chemical gradients (pH), nutrient availability, and intestinal immunity. Intestinal contents collected from the duodenum, jejunum, ileum, colon, and rectum of dogs showed that phylum Firmicutes were the most diverse and abundant in all intestinal sites, particularly, the order Clostridiales. The order Lactobacilalles was highly represented in the duodenum, jejunum, and colon (Suchodolski et al., 2008). Another study showed that Proteobacteria was the most abundant phyla in the small intestine, and substantially lower in the colon, being orders Enterobacteriales and Campylobacteriales the major components in duodenum and Enterobateriales and Clostridiales in the ileum (Honneffer et al., 2017). Phylum Bacteroidetes and Fusobacteria were reported to have low abundances in the proximal small intestine, increasing in the distal parts of the colon. Phylum Actinobacteria, although present, was reported to have relative abundances lower than 1%, in all sites. In faeces, the abundance of phyla reported, from higher to lower, are frequently Fusobacteria or Bacteroidetes, followed by Firmicutes, Proteobacteria, and Actinobacteria (Jha et al., 2020). At the genus level, the gut microbiota was reported to be dominated by Fusobacterium, Bacteroides, and Prevotella, although the variability in abundances found between individuals, might not entirely support the concept of a core microbiota. Genetically related dogs presented low inter-individual faecal microbial composition than unrelated dogs, highlighting the role of genetics to shape the composition of the canine intestinal microbial community (Vilson et al., 2018).

Factors Influencing the Gut Microbial Composition

1. *Age:* Age is among the most important factors affecting the microbiota of dogs. Colonization of the gut microbiota of puppies may occur inside the uterus but rapidly expands after delivery, being facilitated by vaginal birth, although this is still a subject of controversy. Faecal microbiota of new born pups at the age of day 2 was dominated by Firmicutes, while they reduced by day 21 and replaced by Bacteroidetes and little variation of Fusobacteria and Proteobacteria. Progressive weaning and food introduction after the 21st day likely modulated the gut microbiota, which was supported by the increase of genes responsible for amino acid and energy metabolism (Guard et al., 2017). Data suggest that the shifts in gut microbiota continue during growth, not being entirely clear when the core microbiota of dogs is established. Clear differences could be observed in the faecal microbiota composition in samples collected from German Shepherd dogs at 7 weeks of age and in adulthood (15-18 months), despite diversity remained unaffected. Another study reported in Beagles showed low bacterial diversity at 20 weeks of age compared to 28, 36, 44, and 52, as well as differences in the relative abundance of taxa among all time points (Pereira et al., 2020).
2. *Size:* Gut microbiota is also influenced by the size of the dog as evidenced by several studies. Small dogs showed a higher abundances of Bacteroides, Faecalibacterium, and higher of Collinsella and Lactobacillus in large dogs (Middleton et al., 2017). Higher bacterial diversity in large sized dogs might be due to differences in the physiology of the intestine that potentiates bacterial fermentation (Jha et al., 2020).
3. *Sex:* Gut microbiota does not seem to be affected substantially by the sex of the animal. Only few differences in the relative abundances at family and genus levels were

observed. Nevertheless, a study reported that the gut microbial communities of castrated males and spayed females clustered differently than intact dogs, strengthening the hypothesis of bidirectional interaction between microbiota and the endocrine status of the host (Scarsella et al., 2020).

4. *Environment:* Cohabitation of dogs with humans, will have great influence on the microbial composition of domesticated dogs (Alessandri et al., 2019). Similarities in the composition of the microbiota might be explained by an old habit of humans sharing their food with dogs. Few authors indicate that dogs raised in cities have a higher bacterial diversity compared to the countryside dogs (Vilson et al., 2018). Countryside dogs have a higher microbial diversity, as they usually roam around the gardens in the house and have free access to the environment, whereas the city dogs have a limited access to the environmental microbiota, although there is a possibility that the dogs will be taken for a walk along with the owners.
5. *Pregnancy:* During pregnancy in women, gut microbiota plays a role in weight gain, enhancing absorption of nutrients, increasing fasting-induced adipocyte factor secretion, inducing catabolic pathways, and stimulating the immune system (Nuriel-Ohayon et al., 2016), which likely explains major changes that occurrs during pregnancy. Contrary to this, bitches did not exhibit any significant change in the gut microbiota during the last trimester of pregnancy (Vilson et al., 2018). Moreover, mothers' behavior of eating their puppies' stool also helps in increasing the microbial diversity of the dams. Not just the gut microbiota, but, vaginal microbiota also showed significant diversity during the various stages of pregnancy in dogs.
   Furthermore, pronounced changes in the gut microbiota were mainly observed in the third trimester of pregnancy, with an overall increase in the total bacterial load, as well as relative abundance of both Actinobacteria and Proteobacteria, and a decrease in individual microbiota diversity (Nuriel-Ohayon et al., 2016). Those changes favour the gut colonization of newborns, as mothers' vaginal and gut microbiota might be the primary sources of microbes during delivery, which also occurs with puppies. A study by Zakosek Pipan et al., (2020) showed that type of delivery affects the gut microbiota of puppies, and that weight gain of newborns was higher in vaginal birth compared to cesarian section. Moreover, it showed a possible intrauterine bacterial colonization of the canine fetus, as microbiota was found in the placenta and meconium of respectively 57% and 87% of the newborn puppies enrolled in the study.
6. *Diet:* Gut microbiota is greatly influenced by the diet. On one hand, microbes utilize undigested nutrients from diet as substrates to thrive. On the other hand, the physiological conditions of the GIT respond to diet (e.g., fasted vs fed state, food composition, density, viscosity, and volume) and consequently, affect the gut microbiota. Type of diet fed to the dogs largely influences microbial profile of the fecal samples (e.g., commercial moist complete diet, home-made diet, extruded complete diet, feeding bones and raw food, etc.). One study evaluated the jejunal microbiota pre- and post-prandially reporting that feeding caused transient changes in the diversity of the microbial community. The authors suggested that the changes observed after feeding might be explained by the environmental pressure of GIT (e.g., intestinal motility and secretions) and the availability of fermentable substrates. Withholding food for several days had a significant impact on the structure of jejunal intestinal microbiota (Kasiraj et al., 2016). In line with this, one of the studies reported that an

extreme diet transition, from a standard complete dry dog food to a highly purified diet (with limited complex carbohydrates and crystalline amino acids) promoted a distinct structure of the bacterial community and metabolic pathways, after 4 weeks. The study notes that changes were transitory, as the reintroduction of the standard food reverted the effects (Allaway et al., 2020). This suggests that dietary interventions must be maintained for effects to persist and that manipulations at an early age are likely more effective to establish a desired, resilient, and healthy microbial community through life. Several studies have highlighted the role of diet composition on the microbiota of dogs. Feeding a diet with high fat and low starch content instead of a diet with low fat and high starch content increased fecal bile acids, and the relative abundance genus of Prevotella, despite bacterial richness and diversity remained unaffected (Schauf et al., 2018). Another study confirmed that high-fat diets fed to meet energy requirements promoted changes in fecal bacterial composition, and a transient increase of colonic permeability (Moinard et al., 2020). Whereas when fed 50% above the energy requirements for 8 weeks, the negative effects become evident with a decrease of fecal bacterial diversity, lower insulin sensitivity, and disruption in adipokine levels. Dietary protein and carbohydrates ratio also affected dogs' microbiota and metabolome. Switching to feeding a diet with high protein and low carbohydrate content promoted a larger shift in the microbiome of dogs than a diet with low protein and high carbohydrate content, with a decrease of Prevotella: Bacteroides ratio (Coelho et al., 2018). Indeed, Prevotella has been associated with high carbohydrates diets and Bacteroides with a high protein diet in other studies. Also, Bacteroidetes: Firmicutes ratio was related to protein and carbohydrates dietary content, being lower in high-protein with low-carbohydrate than low-protein with high carbohydrates. In addition to effects in the microbiota, studies reported shifts in genes, namely lower abundance of glycol hydrolases and glycosyl transferases in diets with high-protein and low-carbohydrate content. Contrarily, diets with low-protein and high-carbohydrate content had carbohydrate digestion and absorption as well as mineral absorption pathways overrepresented. The supplementation of commercial dry diets with minced beef induced reversible changes in the fecal microbiota, namely decrease of species diversity, decrease in abundance of Faecal bacterium, and an increase of Dorea, Slackia, and members of the Clostridiaceae family, as well as an increase of fecal concentration of iso-valerate. Iso-valerate is a metabolite of L-leucine degradation produced by Clostridioides difficile110 which suggests an increase of undigested protein reaching the colon due to raw meat supplementation. Bones and Raw Food (BARF) composed of fresh uncooked ingredients like meat, offal, and bones typically are rich in proteins and fats and poor in carbohydrates and fibers, thus having an impact on dogs' microbiota. BARF foods in comparison to extruded foods promoted enrichment of bacteria such as Proteobacteria and Fusobacteria at phylum level and Clostridium at the genus level, as well as a higher dysbiosis index driven by an increase in E. coli and Streptococcus and a decrease in Faecalibacterium. 102 Also, the structure of fecal microbiota of dogs fed commercial foods exhibited higher complexity than raw foods, likely explained by their higher content in fibers and carbohydrates. Moreover, a study reported that consumption of high protein diets (460 g/kg dry matter) for 90 days was associated with increased levels of metabolites linked with inflammation (like 9,10-dihydroxyoctadecanoic acid and 12,13-

dihydroxyoctadecanoic acid), and kidney dysfunction (like urea, 5-hydroxyindole sulfate, and 7-hydroxyindole sulfate, p-cresol sulfate).

Weight loss programs often include transition to diets rich in fibers, which according to a study performed on obese Beagles increased biodiversity, increasing the abundance of Allobaculum and decreasing the abundance of Clostridium, Lactobacillus, and Dorea with no effects on blood and serum biochemistry analyses. Another study showed that when obese dogs were fed with a high fiber and protein diet, Firmicutes: Bacteroidetes ratio increased with abundance of Fusobacterium spp, likely due to the increased protein content. None of the studies evaluated the gut microbiota after discontinuing the weight loss diet, which precludes evaluation of the sole effect of the diet transition, although obesity has indeed been shown to affect gut microbiota. Moreover, the end-fermentation products were not evaluated, which likely limits the conclusions on the impact of microbiota changes.

Influence of Microbiota in Disease Conditions

Chronic Kidney Disease (CKD)

CKD is characterized by progressive functional or structural changes in one or both the kidneys. Uremic disease like CKD is greatly influenced by the gut microbiome profile. Cats with CKD were found to have abnormal gut microbiome composition. Abnormal gut microbiota and progression in CKD are related, via increase in the gut permeability and excessive production of microbiota – generated nephrotoxin (Hall et al., 2020). Human CKD patients can be identified with increased proteolytic microbiome over saccharolytic ones (Yang et al., 2018) and this imbalance usually accelerates the progression of CKD. Saccharolytic microbiome resolves the inflammation induced in CKD, thus, preventing the progression of the disease. In patients with CKD, "Enteric dialysis" is an emerging concept that shows the need of intestinal health in devising treatment strategies for CKD. Kim et al., 2023 analyzed the gut microbiome profiles of healthy dogs and dogs with CKD at different stages. overgrowth of the phylum *Proteobacteria, especially Klebsiella sps.* was observed in dogs with CKD compared with that in the healthy dogs. LPS derived from these bacteria results in increased inflammatory response. A gradual increase in *Collinsella intestinalis* was also observed in this study in dogs with CKD. Collinsella sps. is proven to disrupt the tight junction proteins of eneterocytes, thus, leading to translocation of bacterial products from the gut into systemic circulation. It is generally known that patients with CKD experience increased systemic immune inflammatory responses and intestinal susceptibility. Bacteria of families *Enterobacteriaceae*, *Clostridiaceae*, *Pseudomonadaceae*, and *Bacteroidaceae* was higher in dogs with CKD than in the control group. Genera *Klebsiella* and *Clostridium*, showed a gradual increase in abundance as CKD progressed; It was also reported that there was decrease in the abundance of Ruminococcus sps., which is a butyrate- producing bacteria. Butyrate is a short chain fatty acid that is mainly produced by Sacharolytic fermentation of dietary fiber in the gut.

Heart Diseases

Myxomatous mitral valve disease (MMVD) is one of the common heart diseases in small and medium sized geriatric dogs. Circulating levels of a gut microbial metabolite, trimethylamine

N-oxide (TMAO) is linked to increased risk of major adverse cardiovascular events in humans and MMVD in dogs. Concentrations of both BHBA and acetoacetate, which account for >90% of ketone bodies in man, were increased in different stages (B1, B2, and C/D) of MMVD dogs compared to healthy dogs (Li et al., 2021). Impaired gut health or dysbiosis can also be related to congestive heart failure in humans as well as in dogs. Bacterial translocation through oedematous mucosal layers results in immune stimulation via endotoxins contributing to chronic inflammation, malnutrition and cachexia. Faecal gut microbiome was analysed in groups of healthy and CHF dogs (Seo et al., 2020) and found an increase in the abundance of Proteobacteria, especially, the bacteria of family Enterobacteriaceae.

Neuronal Diseases

Meningoencephalitis of unknown origin (MUO) is an umbrella term used to summarize clinically-diagnosed immune-mediated inflammatory disease of the meninges, spinal cord and, especially, the brain of dogs. Efforts in identifying infectious etiology for MUO in dogs proved unsuccessful many a times. Multiple sclerosis, a condition in humans much similar to MOU, has been linked to depleted populations of *Faecalibacterium prausnitzii*, *Prevotella* spp, or both (Miyake et al., 2015). A study conducted in dogs with MUO has shown that high abundance of *Prevotellaceae* is associated with reduced likelihood of developing MUO (Jeffery et al., 2017). Members of *Prevotellaceae* are butyrate fermenting bacteria, which has been identified as a specific inducer of Treg cell differentiation. It is possible that this butyrate-producing metabolic profile may thus constitute the uniting mechanism of resistance to immune-mediated disease thought to be mediated by both *Prevotellaceae* and *F. prausnitzii*. Similarly, hormones produced by specific gut microbes could also directly influence the immune system.

Diabetes Mellitus

Diabetes mellitus in cats shares several pathophysiological features with human type 2 diabetes mellitus. Feline diabetes mellitus depends on several factors, viz., age of the animal, obesity, impaired insulin secretion and peripheral insulin resistance. However, growing evidence suggests that dysbiosis in gut microbiota has direct influence over development of diabetes mellitus. Short chain fatty acids (butyrate, acetate, and propionate) are produced in the large intestine by certain gut microbes through fermentation of complex polysaccharides, and are known to positively influence glucose and energy metabolism as well as local immune function. Butyrate supplies energy for the colonic epithelial cells, and may increase insulin sensitivity and energy expenditure. Acetate and propionate are mostly used as substrates for lipogenesis and gluconeogenesis, respectively, in the liver. Kieler et al. (2021) proved that cats with diabetes mellitus exhibited decreased richness and diversity of the gut microbiota when compared to healthy lean dogs and that dietary intervention with proper diet in DM cats reduced the difference in the gut microbiota between the DM and lean cats.

Dysbiosis in gut microbiota is also noticed in several acute and chronic diseases, such as exocrine pancreatic insufficiency, acute diarrhoea, idiopathic inflammatory bowel disease and obesity. In cats, decreased bacterial diversity has been associated with acute and chronic diarrheoa.

Assessment of Gut Microbiome

Bacterial Culture

Traditional bacterial culture underestimates the real number of intestinal bacteria, as only a limited number of bacteria can be isolated and identified using standard media and protocols.

DNA Shotgun Sequencing

Deep DNA shotgun sequencing, or metagenomics, aims to sequence extracted DNA in a sample without prior amplification by PCR. Using this approach, various genes are identified, allowing assessments of taxonomy and functional gene categories the microbiome of samples (eg, synthesis of amino acids, vitamins, carbohydrate). Metagenomics has better resolution at the species and strain level compared with 16S rRNA gene sequencing, as multiple genes from one organism can be sequenced and overlapping reads can be assembled to a draft genome. This approach also identifies other members of the intestinal community, such as archaea, fungi, and DNA viruses. Metagenomic analysis of fecal samples from dogs and cats revealed that bacteria makeup approx. 98% of all sequencing reads, archaea 1%, and fungi and DNA viruses (mostly bacteriophages) the remaining 1%. A novel approach, called shallow shotgun metagenomics, could be a potential alternative and uses lower sequence coverage, which is less expensive. Shallow shotgun metagenomics has been used to assess the human microbiome and will likely be more commonly used in companion animals in the future.

16S rRNA Gene Sequencing

Sequencing of 16S rRNA genes is currently the standard approach in microbiome studies and provides useful in-depth information about the microbial composition and how individual bacterial groups or the entire community differ between healthy vs diseased animals or respond to dietary or therapeutic interventions. The 16S rRNA gene consists of several variable regions, each flanked by conserved regions. Bacterial primers are used that amplify these conserved regions and, thus, the variable region in-between. Because conserved regions are targeted, in theory, DNA from known and unknown bacteria present in the sample can be amplified, and then variable regions can be sequenced. This process requires specific library preparations for different sequencing platforms (e.g., Illumina MiSeq, Ion Torrent PGM). The obtained raw sequences are then processed through a bioinformatics pipeline, such as QIIME 2 or Mothur, to eliminate sequences of insufficient quality and erroneous reads, remove chimeric sequences, and compare the final sequences against public databases. Statistical analysis is performed according to the study design. There are several easy to use web-based interfaces that allow statistical analysis and visualization of microbiome data, such as Calypso (http://cgeno me.net/wiki/ index.php/Calypso) or MicrobiomeAnalyst (https://github.com/xia-lab/ Micro biome AnalystR).

Quantitative PCR (Dysbiosis index)

qPCR can be used for quantitation of total bacterial load or to understand the abundance of specific taxa. It is a rapid, affordable and highly reproducible method to quantify specific taxa. These reproducible assays can then be used to compare changes in bacterial abundance across studies and assess the magnitude of changes due to an intervention, as the results can be compared to an existing RI.

Insitu Hybridization And Immunochemistry

To understand the role of bacteria in intestinal inflammation, it is useful to determine the localization of bacteria. This allows evaluating whether bacteria are within the mucus layer, attached to the epithelium, or located intracellularly. Bacteria can be visualized on biopsy slides either by using fluorescence in-situ hybridization (FISH) or immunohistochemistry (IHC). FISH is the most commonly used approach, and probes are available that target all bacteria (eg, universal probes such as EUB338) and specific bacterial taxa. The major advantage of FISH is that it enables one to visualize the localization of bacteria. Disadvantages are that specific probes need to be designed for bacteria of interest, and only a few probes can be used per tissue slide. This makes this approach labor-intensive. Furthermore, expensive microscopy equipment is required, limiting FISH to few specialized laboratories.

Stabilizing/Restoring the Microbiome (Bioengineering)

Microbiome can be manipulated by adopting several approaches (Figure 1). Each approach has its applications and considerations depending on the desired outcome and the specific microbiome being targeted.

Supplementation to Enhance Growth/Activity of Specific Beneficial Microbes

Probiotic and prebiotic development is one of the fastest growing areas of animal microbiome research. Probiotics are live microorganisms that confer good health on the host, while, prebiotics are supplements that enhance the growth of probiotics. Probiotics has multiple advantages, i.e., clears infections caused by specific microbes, relieves diarrhoea, decrease gastric pH, decreases risk of acidosis and rapid recovery from gastrointestinal disease (Ziese et al., 2018). Few probiotics have additional properties of providing nutrients to the host resulting in better feed conversion ratio or growth performance. Probiotics are also known to enhance host immune responses by producing antimicrobial molecules and altering biofilms (Kelly and Salinas, 2017). One of the most widely studied and well known probiotics are Lactic acid bacteria, such as Lactobacillus spps. They are the regular commensals in the gastrointestinal tract that produce bacteriocins and prevents colonization of pathogenic organisms in the GIT.

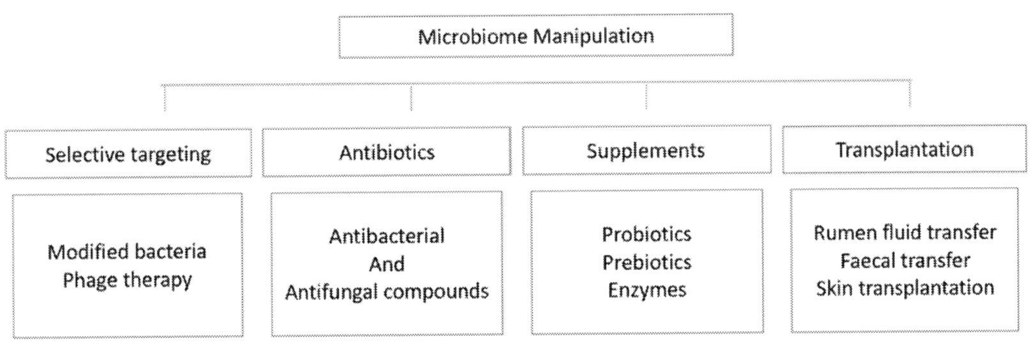

Figure 1. Overview of different modes of Microbiome manipulation.

Community-Level Approach to Reset the Microbiome

Transplantation of a new consortium of microbiota is another key method to correct highly dysbiotic gut environment. Treating digestive issues in cattle by cud transplantation is one of the best example. Similarly, feacal microbiota transplantation (FMT), though an age old practice, was found to have incredible utility in regulating overgrowth of *C.difficile* in human guts (Brandt et al., 2012). However, extensive work has been performed on manipulating the microbiome in chickens, and probiotics and/or fecal transplant approaches have been shown to protect against a number of pathogenic species including *E. coli, Salmonella, Clostridium perfringens, Listeria* spp., and *Campylobacter* spp (Clavijo and Florez, 2018).

Selective Targeting

Instead of altering the whole microbiome, targeting specific member of the microbiome offers several benefits. Phage therapy is one such selective targeting method, where introduction of specific phage targets a specific bacterial strain. Hence, phage therapy offers the advantage of treating specific bacterial infections in livestock or humans with food borne infections such as Salmonellosis, Colibacillosis, Campylobacteriosis, etc (Mills et al., 2017). Editing a specific gene in a specific bacterial strain using CRISPR-Cas technology is an emerging technology with multifarious advantages in animal health (Stout et al., 2017). This technology can be used synonymously with antimicrobials through destruction or deletion of genomic regions that encode for virulence factors (Vercoe et al., 2013). Similarly, bacterial secretion systems can be taken over to deliver therapeutics or agents that induce selective cell death.

Functional Foods

Functional foods, when regularly consumed provides certain beneficial physiological properties and essential nutrients. According to the European Consensus, "food can be regarded as functional if it is satisfactorily demonstrated to beneficially affect one or more target functions in the body, beyond adequate nutritional effects, in a way that is relevant to either an improved state of health and well-being or a reduction in disease risk" (Roberfroid, 2000). Probiotics and prebiotics also are included under functional foods, as they can help prevent or remediate some diseases and improve the overall health and well-being.

Probiotics

Probiotics have become popular supplements in canine feed after they were proven to restore eubiosis in the GI microbiome, especially in animals with compromised microbiota due to disease/antibiotic therapy (Lee et al., 2022). Administration of probiotics involves introducing selected live microorganisms regularly and in sufficient amounts. Strains to be selected as probiotic should be native to the GI tract of target host (Foongsawat et al., 2023). Gut microbiome of canines usually includes *Enterococcus* spp, *Lactobacillus* spp, *Bifidobacterium* spp, etc. Defined probiotic products consists of labelled identified organisms, where as, undefined products include fermented products.

A typical probiotic strain should tolerate fluctuating levels of pH and bile acid concentrations, should adhere to the GI mucosal epithelial cells, produce bactericidal components such as bacteriocins, lactic acid, hydrogen peroxide and maintain competitive inhibition against pathogenic bacteria. *Bacillus subtilis* and *Bacillus licheniformis*, the commonly used probiotic bacteria are known to improve feacal consistency and reduce fetid smell of the faeces (Bastos et al., 2020).

Prebiotics and synbiotics

Indigestible materials that promote proliferation and activity of probiotic bacteria, such as *Lactobacillus* spp. and *Bifidobacterium* spp. in the GI tract. Synbiotics are a combination of prebiotics and probiotics that produce a synergestic effect of one another. Prebiotics include simple or complex carbohydrates such as tagatose, mannanoligosaccharides (MOS), lactulose, fructo-oligosaccharides (FOS), galacto-oligosaccharides, xylo-oligosaccharides, inulin, and polydextrose (Zentek et al., 2002). Few prebiotics such as MOS was proven to be capable of binding with pathogenic *C. perfringens* and have immunostimulatory activity against GI pathogens (Spring et al., 2000). Lactosucrose is known to reduce fecal ammonia and fetid fecal odour. Few prebiotics also promote synthesis of postbiotics.

Postbiotics

Postbiotics confers health benefits to canines when taken regularly and it includes non-viable probiotic or non-probiotic microorganisms (Ren et al. 2011). Postbiotics can directly or indirectly affect body metabolism, secretion of insulin and sensitivity to insulin. Cell-free supernatants of bacterial cultures, that contains secreted metabolites, or they can be para probiotics or probiotics that have inactivated intact or ruptured cells of probiotics strains.

One class of postbiotics include SCFA's (Short chain fatty acids) such as acetate, butyrate and propionate produced from the digestion of food and are important in maintenance of homeostasis in the GI tract, neuroendocrine system and systemic energy availability. Bacterial cell wall components such as peptidoglycan and teichoic acid also acts as postbiotics with immunomodulatory effects. Peptidoglycan components released from *Lactobacillus* spp. prevents the release of inflammatory cytokines. However, the effect of all these are not yet proved in canines. The only study that shows the effect of postbiotics in canines is that supplementation of yeast (*Sacharomyces cerevisiae*) cell wall fraction (YCWF) and YCWF-derived active mannoproteins fraction in the diets of adult and/or elderly dogs (Kroll et al., 2020).

To date, the only cell wall component-related study used on canines is the intestinal health of adult beagle dogs undergoing an abrupt dietary transition (Lin et al., 2020). Similarly, fortification of PUFA's such as omega-3 fatty acids in canine diet promotes decreased production of inflammatory mediators which ultimately decreases inflammation. Phytonutrients derived from plants were found to possess anti-inflammatory, antioxidant, and anti-carcinogenic activities and promotes canine health. Phytosterols, another group of phytonutrients have additional antidiabetic and anti-atherosclerotic activities, apart from the usual antioxidant and anti-inflammatory properties. In hyperlipidemic dogs, phytosterols help in lowering high plasma triglyceride and cholesterol levels (Borin-Crivellenti et al., 2021).

Canines are known to possess unique microbial profile that is greatly affected by their diet and health status (Ziese & Suchodolski, 2021). Hence, identification of "sentinel" bacteria, which are vulnerable to alterations in GI homeostasis, has been used as an indicator of GI functionality. Dysbiosis in GIT may affect several bacterial taxa as revealed by many meta analysis studies.

Conclusion

Understanding and managing the microbiome is increasingly important in veterinary medicine to promote animal health, improve productivity, and enhance overall welfare. Microbiome exerts significant influence over many aspects of health and profoundly affects several key systems of the canine body. The microbiome interacts closely with the immune system, helping to educate it and regulate its responses. There is growing evidence of a gut-brain axis in dogs, where the gut microbiota can influence behaviour, mood, and responses to stress. Changes in the microbiome composition also have been linked to altered behaviour and cognition. Microbiome interacts with and supports the dog's own biological systems, contributing significantly to overall health and well-being.

References

Alessandri, Giulia, Christian Milani, Leonardo Mancabelli, Marta Mangifesta, Gabriele A. Lugli, Alice Viappiani, Sabrina Duranti et al. "Metagenomic dissection of the canine gut microbiota: insights into taxonomic, metabolic and nutritional features." *Environmental Microbiology* 21, no. 4 (2019): 1331-1343.

Allaway, David, Richard Haydock, Zoe N. Lonsdale, Oliver D. Deusch, Ciaran O'Flynn, and Kevin R. Hughes. "Rapid reconstitution of the fecal microbiome after extended diet-induced changes indicates a stable gut microbiome in healthy adult dogs." *Applied and Environmental Microbiology* 86, no. 13 (2020): e00562-20.

Amato, Katherine R., Jessica L. Metcalf, Se Jin Song, Vanessa L. Hale, Jonathan Clayton, Gail Ackermann, Greg Humphrey et al. "Using the gut microbiota as a novel tool for examining colobine primate GI health." *Global Ecology and Conservation* 7 (2016): 225-237.

Banks, Kayla C., Elizabeth A. Giuliano, Susheel B. Busi, Carol R. Reinero, and Aaron C. Ericsson. "Evaluation of healthy canine conjunctival, periocular haired skin, and nasal microbiota compared to conjunctival culture." *Frontiers in Veterinary Science* 7 (2020): 558.

Bastos, Tais Silvino, Daniele Cristina de Lima, Camilla Mariane Menezes Souza, Alex Maiorka, Simone Gisele de Oliveira, Letícia Cardoso Bittencourt, and Ananda Portella Félix. "Bacillus subtilis and Bacillus licheniformis reduce faecal protein catabolites concentration and odour in dogs." *BMC Veterinary Research* 16 (2020): 1-8.

Bell, Sara E., Andrea K. Nash, Brian M. Zanghi, Cynthia M. Otto, and Erin B. Perry. "An assessment of the stability of the canine oral microbiota after probiotic administration in healthy dogs over time." *Frontiers in Veterinary Science* 7 (2020): 616.

Borin-Crivellenti, Sofia, Leandro Z. Crivellenti, Frederico R. de Oliveira, Paula B. Costa, Alef WO Alvarenga, Luara R. Rezende, Fernanda N. Gouvêa, Najla D. Assef, and Luana de O. Branco. "Effect of phytosterols on reducing low-density lipoprotein cholesterol in dogs." *Domestic Animal Endocrinology* 76 (2021): 106610.

Borriello, Giorgia, Rubina Paradiso, Carlotta Catozzi, Roberta Brunetti, Paola Roccabianca, Marita Georgia Riccardi, Bianca Cecere et al. "Cerumen microbial community shifts between healthy and otitis affected dogs." *PLoS One* 15, no. 11 (2020): e0241447.

Bradley, Charles W., Fiona F. Lee, Shelley C. Rankin, Lindsay R. Kalan, Joseph Horwinski, Daniel O. Morris, Elizabeth A. Grice, and Christine L. Cain. "The otic microbiota and mycobiota in a referral population of dogs in eastern USA with otitis externa." *Veterinary Dermatology* 31, no. 3 (2020): 225-e49.

Brandt, Lawrence J., Olga C. Aroniadis, Mark Mellow, Amy Kanatzar, Colleen Kelly, Tina Park, Neil Stollman, Faith Rohlke, and Christina Surawicz. "Long-Term Follow-Up of Colonoscopic Fecal Microbiota Transplant for RecurrentClostridium difficileInfection." *Official journal of the American College of Gastroenterology| ACG* 107, no. 7 (2012): 1079-1087.

Burton, Erin N., Leah A. Cohn, Carol N. Reinero, Hans Rindt, Stephen G. Moore, and Aaron C. Ericsson. "Characterization of the urinary microbiome in healthy dogs." *Plos one* 12, no. 5 (2017): e0177783.

Chermprapai, Suttiwee, Thomas HA Ederveen, Femke Broere, Els M. Broens, Yvette M. Schlotter, Saskia Van Schalkwijk, Jos Boekhorst, Sacha AFT van Hijum, and Victor PMG Rutten. "The bacterial and fungal microbiome of the skin of healthy dogs and dogs with atopic dermatitis and the impact of topical antimicrobial therapy, an exploratory study." *Veterinary Microbiology* 229 (2019): 90-99.

Coelho, Luis Pedro, Jens Roat Kultima, Paul Igor Costea, Coralie Fournier, Yuanlong Pan, Gail Czarnecki-Maulden, Matthew Robert Hayward et al. "Similarity of the dog and human gut microbiomes in gene content and response to diet." *Microbiome* 6 (2018): 1-11.

Davis, Ian J., Corrin Wallis, Oliver Deusch, Alison Colyer, Lisa Milella, Nick Loman, and Stephen Harris. "A cross-sectional survey of bacterial species in plaque from client owned dogs with healthy gingiva, gingivitis or mild periodontitis." *PloS one* 8, no. 12 (2013): e83158.

Ericsson, Aaron C., Alexa R. Personett, Hansjorg Rindt, Megan E. Grobman, and Carol R. Reinero. "Respiratory dysbiosis and population-wide temporal dynamics in canine chronic bronchitis and non-inflammatory respiratory disease." *Plos one* 15, no. 1 (2020): e0228085.

Foongsawat, Ngamlak, Sirinthorn Sunthornthummas, Kwannan Nantavisai, Komwit Surachat, Achariya Rangsiruji, Siriruk Sarawaneeyaruk, Kedvadee Insian, Sirapan Sukontasing, Nuttika Suwannasai, and Onanong Pringsulaka. "Isolation, characterization, and comparative genomics of the novel potential probiotics from canine feces." *Food Science of Animal Resources* 43, no. 4 (2023): 685.

Giaretta, Paula R., Jan S. Suchodolski, Albert E. Jergens, Jörg M. Steiner, Jonathan A. Lidbury, Audrey K. Cook, Mohsen Hanifeh et al. "Bacterial biogeography of the colon in dogs with chronic inflammatory enteropathy." *Veterinary pathology* 57, no. 2 (2020): 258-265.

Gilbert, Myrthe S., Noortje Ijssennagger, Arie K. Kies, and Saskia WC van Mil. "Protein fermentation in the gut; implications for intestinal dysfunction in humans, pigs, and poultry." *American Journal of Physiology-Gastrointestinal and Liver Physiology* (2018).

Guard, Blake C., Hanna Mila, Jörg M. Steiner, Claire Mariani, Jan S. Suchodolski, and Sylvie Chastant-Maillard. "Characterization of the fecal microbiome during neonatal and early pediatric development in puppies." *PloS one* 12, no. 4 (2017): e0175718.

Honneffer, Julia B., Jörg M. Steiner, Jonathan A. Lidbury, and Jan S. Suchodolski. "Variation of the microbiota and metabolome along the canine gastrointestinal tract." *Metabolomics* 13 (2017): 1-20.

Jha, Aashish R., Justin Shmalberg, Jirayu Tanprasertsuk, LeeAnn Perry, Dan Massey, and Ryan W. Honaker. "Characterization of gut microbiomes of household pets in the United States using a direct-to-consumer approach." *PLoS One* 15, no. 2 (2020): e0227289.

Kasai, T., Y. Fukui, K. Aoki, Y. Ishii, and K. Tateda. "Changes in the ear canal microbiota of dogs with otitis externa." *Journal of Applied Microbiology* 130, no. 4 (2021): 1084-1091.

Kelly, Cecelia, and Irene Salinas. "Under pressure: interactions between commensal microbiota and the teleost immune system." *Frontiers in immunology* 8 (2017): 271125.

Kroll, Fernanda Sant Anna, Thaila Cristina Putarov, Leandro Zaine, Katiani Silva Venturini, Caroline Godoy Aoki, João Paulo Fernandes Santos, Vivian Pedrinelli, Thiago Henrique Annibale Vendramini, Márcio Antônio Brunetto, and Aulus Cavalieri Carciofi. "Active fractions of mannoproteins derived from yeast cell wall stimulate innate and acquired immunity of adult and elderly dogs." *Animal feed science and Technology* 261 (2020): 114392.

Lee, Daniel, Tae Wook Goh, Min Geun Kang, Hye Jin Choi, So Young Yeo, Jungwoo Yang, Chul Sung Huh, Yoo Yong Kim, and Younghoon Kim. "Perspectives and advances in probiotics and the gut microbiome in companion animals." *Journal of Animal Science and Technology* 64, no. 2 (2022): 197.

Lin, Ching-Yen, Meredith Q. Carroll, Michael J. Miller, Rodolphe Rabot, and Kelly S. Swanson. "Supplementation of yeast cell wall fraction tends to improve intestinal health in adult dogs undergoing an abrupt diet transition." *Frontiers in Veterinary Science* 7 (2020): 597939.

Lyman, C. C., G. R. Holyoak, K. Meinkoth, X. Wieneke, K. A. Chillemi, and U. DeSilva. "Canine endometrial and vaginal microbiomes reveal distinct and complex ecosystems." *PLoS One* 14, no. 1 (2019): e0210157.

Middleton, Rondo P., Sebastien Lacroix, Marie-Pier Scott-Boyer, Nikola Dordevic, Adam D. Kennedy, Amanda R. Slusky, Jerome Carayol, Christina Petzinger-Germain, Alison Beloshapka, and Jim Kaput. "Metabolic differences between dogs of different body sizes." *Journal of nutrition and metabolism* 2017 (2017).

Mills, Susan, R. Paul Ross, and Colin Hill. "Bacteriocins and bacteriophage; a narrow-minded approach to food and gut microbiology." *FEMS microbiology reviews* 41, no. Supp_1 (2017): S129-S153.

Minamoto, Yasushi, Tomomi Minamoto, Anitha Isaiah, Panpicha Sattasathuchana, Agostino Buono, Venkat R. Rangachari, Isaac H. McNeely, Jonathan Lidbury, Jörg M. Steiner, and Jan S. Suchodolski. "Fecal short-chain fatty acid concentrations and dysbiosis in dogs with chronic enteropathy." *Journal of veterinary internal medicine* 33, no. 4 (2019): 1608-1618.

Moinard, Alex, Cyrielle Payen, Khadija Ouguerram, Agnès André, Juan Hernandez, Amandine Drut, Vincent C. Biourge et al. "Effects of high-fat diet at two energetic levels on fecal microbiota, colonic barrier, and metabolic parameters in dogs." *Frontiers in Veterinary Science* 7 (2020): 566282.

Nuriel-Ohayon, Meital, Hadar Neuman, and Omry Koren. "Microbial changes during pregnancy, birth, and infancy." *Frontiers in microbiology* 7 (2016): 1031.

Pasha, Sabah, Taichi Inui, Iain Chapple, Stephen Harris, Lucy Holcombe, and Melissa M. Grant. "The saliva proteome of dogs: variations within and between breeds and between species." *Proteomics* 18, no. 3-4 (2018): 1700293.

Pereira, Ana Margarida, and Alfonso Clemente. "Dogs' microbiome from tip to toe." *Topics in Companion Animal Medicine* 45 (2021): 100584.

Pereira, Ana Margarida, Carlo Pinna, Giacomo Biagi, Claudio Stefanelli, Margarida RG Maia, Elisabete Matos, Marcela A. Segundo, Antonio JM Fonseca, and Ana Rita J. Cabrita. "Supplemental selenium source on gut health: Insights on fecal microbiome and fermentation products of growing puppies." *FEMS Microbiology Ecology* 96, no. 11 (2020): fiaa212.

Petersen, Charisse, and June L. Round. "Defining dysbiosis and its influence on host immunity and disease." *Cellular microbiology* 16, no. 7 (2014): 1024-1033.

Pierezan, Felipe, Thierry Olivry, Judith S. Paps, Sara D. Lawhon, Jing Wu, Jörg M. Steiner, Jan S. Suchodolski, and Aline Rodrigues Hoffmann. "The skin microbiome in allergen-induced canine atopic dermatitis." *Veterinary Dermatology* 27, no. 5 (2016): 332-e82.

Pipan, Maja Zakošek, Leonida Kajdič, Anja Kalin, Tanja Plavec, and Irena Zdovc. "Do newborn puppies have their own microbiota at birth? Influence of type of birth on newborn puppy microbiota." *Theriogenology* 152 (2020): 18-28.

Ren, Zhihua, Zhiping Zhao, Yangguang Wang, and Kehe Huang. "Preparation of selenium/zinc-enriched probiotics and their effect on blood selenium and zinc concentrations, antioxidant capacities, and intestinal microflora in canine." *Biological Trace Element Research* 141 (2011): 170-183.

Roberfroid, Marcel B. "A European consensus of scientific concepts of functional foods. *Nutrition*. 2000 Jul-Aug;16(7-8):689-91: 689-691.

Rodrigues Hoffmann, Aline, Adam P. Patterson, Alison Diesel, Sara D. Lawhon, Hoai Jaclyn Ly, Christine Elkins Stephenson, Joanne Mansell et al. "The skin microbiome in healthy and allergic dogs." *PloS One* 9, no. 1 (2014): e83197.

Rota, Ada, Michela Corrò, Ilaria Patuzzi, Chiara Milani, Stefania Masia, Eleonora Mastrorilli, Sara Petrin, Alessandra Longo, Angela Del Carro, and Carmen Losasso. "Effect of sterilization on the canine vaginal microbiota: a pilot study." *BMC veterinary research* 16 (2020): 1-10.

Ruparell, Avika, Taichi Inui, Ruth Staunton, Corrin Wallis, Oliver Deusch, and Lucy J. Holcombe. "The canine oral microbiome: variation in bacterial populations across different niches." *BMC microbiology* 20, no. 1 (2020): 42.

Scarsella, Elisa, Bruno Stefanon, Michela Cintio, Danilo Licastro, Sandy Sgorlon, Simeone Dal Monego, and Misa Sandri. "Learning machine approach reveals microbial signatures of diet and sex in dog." *PLoS One* 15, no. 8 (2020): e0237874.

Schauf, Sofia, Gabriel de la Fuente, Charles J. Newbold, Anna Salas-Mani, Celina Torre, Leticia Abecia, and Carlos Castrillo. "Effect of dietary fat to starch content on fecal microbiota composition and activity in dogs." *Journal of Animal Science* 96, no. 9 (2018): 3684-3698.

Song, Se Jin, Christian Lauber, Elizabeth K. Costello, Catherine A. Lozupone, Gregory Humphrey, Donna Berg-Lyons, J. Gregory Caporaso et al. "Cohabiting family members share microbiota with one another and with their dogs." *elife* 2 (2013): e00458.

Spring, P., C. Wenk, K. A. Dawson, and K. E. Newman. "The effects of dietary mannaoligosaccharides on cecal parameters and the concentrations of enteric bacteria in the ceca of salmonella-challenged broiler chicks." *Poultry Science* 79, no. 2 (2000): 205-211.

Stout, Emily, Todd Klaenhammer, and Rodolphe Barrangou. "CRISPR-Cas technologies and applications in food bacteria." *Annual review of food science and technology* 8 (2017): 413-437.

Suchodolski, Jan S., Jennifer Camacho, and Jörg M. Steiner. "Analysis of bacterial diversity in the canine duodenum, jejunum, ileum, and colon by comparative 16S rRNA gene analysis." *FEMS microbiology ecology* 66, no. 3 (2008): 567-578.

Vázquez-Baeza, Yoshiki, Embriette R. Hyde, Jan S. Suchodolski, and Rob Knight. "Dog and human inflammatory bowel disease rely on overlapping yet distinct dysbiosis networks." *Nature microbiology* 1, no. 12 (2016): 1-5.

Vercoe, Reuben B., James T. Chang, Ron L. Dy, Corinda Taylor, Tamzin Gristwood, James S. Clulow, Corinna Richter, Rita Przybilski, Andrew R. Pitman, and Peter C. Fineran. "Cytotoxic chromosomal targeting by CRISPR/Cas systems can reshape bacterial genomes and expel or remodel pathogenicity islands." *PLoS genetics* 9, no. 4 (2013): e1003454.

Vilson, Åsa, Ziad Ramadan, Qinghong Li, Åke Hedhammar, Arleigh Reynolds, Julie Spears, Jeff Labuda et al. "Disentangling factors that shape the gut microbiota in German Shepherd dogs." *PloS one* 13, no. 3 (2018): e0193507.

Zentek, Jürgen, Bettina Marquart, and Tanja Pietrzak. "Intestinal effects of mannanoligosaccharides, transgalactooligosaccharides, lactose and lactulose in dogs." *The Journal of Nutrition* 132, no. 6 (2002): 1682S-1684S.

Ziese, A.L., Suchodolski, J.S., Hartmann, K., Busch, K., Anderson, A., Sarwar, F., Sindern, N. and Unterer, S., 2018. Effect of probiotic treatment on the clinical course, intestinal microbiome, and toxigenic Clostridium perfringens in dogs with acute hemorrhagic diarrhea. *PLoS One*, 13(9), p.e0204691.

Ziese, Anna-Lena, and Jan S. Suchodolski. "Impact of changes in gastrointestinal microbiota in canine and feline digestive diseases." *Veterinary Clinics: Small Animal Practice* 51, no. 1 (2021): 155-169.

Chapter 19

The Impact of Microbiomes on Zoonotic Pathogen Abundance

Jinu Manoj[1,*]
and Manoj Kumar Singh[2]
[1]College Central Laboratory, Lala LajPat Rai University of Veterinary and Animal Sciences, Hisar, Haryana, India
[2]College of Veterinary and Animal Sciences, Sardar Vallabhbhai Patel University of Agriculture and Technology, Meerut, Uttar Pradesh, India

Abstract

Microbiomes are composed of genomes of microbes that live in a particular environment. Each microbiome has its own unique composition and function. The abundance of zoonotic pathogens depends on several factors like geographical location, occupation, reservoir status, vector presence, environmental factors etc. The different microbiomes engage in various forms of interaction with its host organisms. The disruptions in these beneficial interactions can impact disease susceptibility and incidence of zoonotic diseases. Understanding the interactions between zoonotic pathogens and microbiomes is crucial for surveillance, prevention and control of zoonotic diseases. The microbiome has impact on the health of human, animals and the environment, thus acting as a uniting element for One Health. Because of the microbe's high genetic variety, they are able to constantly adapt to evade the host's defense mechanisms and thrive in new ecological niches which results in the emerging zoonoses. Analysis of phylogenomics of zoonotic pathogens is required to determine the evolutionary and transmission processes of zoonoses. Studies are being conducted to investigate the composition and functions of the microbiome with respect to the dynamics of transmission, pathogenicity and host susceptibility to zoonotic diseases. Nevertheless, the development of vaccines using microbiomes helps to enhance animal health.

Keywords: animal, interactions, microbiome, one health, vaccines, zoonoses

---

* Corresponding Author's Email: drjinumanoj@gmail.com

In: Microbiomes in Animal Nutrition, Health, and Disease
Editor: Tanmoy Rana
ISBN: 979-8-89530-425-9

Introduction

The microbiome refers to the community of microorganisms such as fungi, bacteria and viruses and their gene sequences including homologous sequences that exist in a particular habitat and time period. Zoonotic pathogens are infectious agents that can be transmitted between animals and humans. While the microbiome primarily consists of commensal and beneficial organisms that contribute to our health, there are instances where zoonotic pathogens can be part of the microbiome or interact with it. These pathogens may not necessarily reside in the human microbiome but can be introduced through interactions with animals or their environments. Zoonotic pathogens typically reside in animal reservoirs and there is a chance of transmission of these pathogens to humans *via* direct animal contact, consumption of contaminated food and water or through vectors like ticks and mosquitoes and environment.

Abundance of Zoonotic Diseases

Zoonotic pathogens may be bacterial, viral, parasitic or fungal origin or may involve unconventional agents like prions. They pose a serious threat to public health worldwide, due to people's close contact with animals in the natural environment and in agriculture as pets. A significant portion of all newly discovered infectious diseases and many currently recognized diseases falls under the category of zoonotic diseses. Recurrent disease outbreaks like salmonellosis and the Ebola virus can be brought on by some frequent zoonoses. There are still others that have the potential to start pandemics, such the new coronavirus that causes COVID-19. Zoonoses can also cause disruptions in the production and trade of animal products for food and other uses.

Most zoonotic diseases are transmitted to humans from animals. Cattle, buffaloes, goats, sheep, horses, pigs, dogs, cats and other domestic animals act as reservoirs of pathogens of domestic zoonoses and can transmit the diseases to humans. Examples of zoonotic diseases that can be transmitted to humans from domestic animals include rabies, anthrax, tuberculosis, brucellosis, listeriosis, campylobacteriosis, leptospirosis, Japanese encephalitis, rotavirus infection, Crimean-Congo haemorrhagic fever, hendra, nipah and zika virus infectios, toxoplasmosis, hydatidosis, taeniosis, cysticercosis, trichinellosis, toxocariasis, cryptosporidiosis, ancylostomiasis, Q fever, Typhus fever *etc.*

Animals can also get infected from humans and such diseases are known as reverse zoonoses. Reverse zoonoses are those diseases which are transmitted from humans to animals and then back again from animals to humans. Examples of such pathogens include methicillin-resistant *Staphylococcus aureus* (MRSA), *Campylobacter* spp., *Salmonella enterica* Serovar Typhimurium, Influenza A virus, *Cryptosporidium parvum*, *Ascaris lumbricoides* and *Giardia duodenalis*.

In most cases, foodborne pathogens, which cause diarrhoea transmit microbes through food. Zoonotic pathogens cause a large number of foodborne diseases. Both adult and juvenile populations are susceptible to severe infections and mortality due to foodborne diseases. Common foodborne zoonotic pathogens include *Campylobacter* spp., *Salmonella* spp. (*Salmonella enterica* serovar Enteritidis), Shiga toxin-producing *Escherichia coli* and hepatitis E virus. Over 90% of foodborne infections caused by bacteria are caused by *Salmonella* spp.

and *Campylobacter* spp. In addition, *Listeria* spp., *Brucella* spp., *Clostridium* spp., bovine spongiform encephalopathy, calicivirus, norovirus and other hepatitis viruses mainly found in intestines of animals can be transmitted through contaminated food products.

Emerging zoonosis are those that have newly recognized, newly evolved or has occurred previously but shows an increase in incidence or expansion in geographical, host or vector range. Over the past 70 years, at least 250 zoonoses have been identified as emerging and re-emerging zoonotic diseases. With an increasing prevalence and geographic range, many zoonotic diseases have spread quickly around the world. Human are affected as a result of intimate interaction with animals, which may act as reservoirs for emerging and re-emerging zoonotic diseases. Examples of major emerging zoonoses include Ebola virus infection, Nipha virus infection, avian influenza, feline cowpox, norovirus infection, rotavirus infection, hantavirus infection, West Nile fever, canine leptospirosis, MRSA infection, cat scratch disease, severe fever with thrombocytopenia syndrome, Middle East respiratory syndrome, severe acute respiratory syndrome, bovine spongiform encephalopathy and the corona virus pandemic (COVID-19). On the other hand, brucellosis, rabies, Japanese encephalitis, tuberculosis (*M. bovis*) and *Schistosoma japonica* infection are considered to be re-emerging zoonoses in many parts of the world.

Factors for Zoonotic Abundance

Monitoring and understanding the abundance of zoonotic pathogens is critical for public health surveillance and for implementing appropriate prevention and control measures. This includes both monitoring the pathogens themselves and understanding the ecological and behavioral factors that contribute to their transmission. The abundance of zoonotic pathogens in any given environment or population can vary widely depending on several factors which are mentioned below.

- *Geographical Location*: Numerous zoonotic diseases are caused by microorganisms. Most infectious microbes that cause diseases come from wild animals. Infectious diseases arise from the increased interactions between humans, domestic animals (pets and livestock) and wild animals that occur as farms and cities grow and encroach on the areas once inhabited by wild animals. The prevalence of zoonotic pathogens often varies by region due to climate, ecology and human-animal interactions. For example, regions with high livestock density may have higher rates of zoonotic infections like brucellosis or Q fever.
- *Occupation*: Certain occupations and activities can increase the risk of exposure to zoonotic pathogens. Farmers, veterinarians, and those working in animal processing industries are more likely to encounter zoonotic pathogens compared to the general population.
- *Wildlife interactions*: Wildlife can serve as reservoirs for many zoonotic pathogens. Encounters with wild animals or their habitats can increase the risk of exposure to diseases like rabies, hantavirus, or Lyme disease.

- *Livestock and Pets*: Animals kept as livestock or pets can harbor zoonotic pathogens. For example, Salmonella and E. coli can be transmitted through contaminated food products or direct contact with animals.
- *Vector-borne transmission*: Many zoonotic pathogens are transmitted by vectors such as mosquitoes, ticks, and fleas. The abundance of these vectors in an area can significantly influence the prevalence of vector-borne zoonotic diseases like West Nile virus or Lyme disease.
- *Environmental factors*: Environmental conditions such as temperature, humidity, and land use can affect the survival and transmission of zoonotic pathogens. Changes in climate or habitat alteration can impact the distribution and abundance of these pathogens.
- *Public health measures*: The implementation of public health measures such as vaccination programs, vector control efforts, and food safety regulations can influence the prevalence of zoonotic pathogens within human populations.

Host Microbiome Interactions

The microbiome of the diverse community that inhabits the human and animal bodies engages in various forms of interaction with its host organisms. These interactions highlight the dynamic and mutually beneficial relationship between humans and animals and their microbiomes. Disruptions to this relationship such as antibiotic use, diet changes or environmental factors can impact health and disease susceptibility, underscoring the importance of understanding and maintaining a healthy microbiome.

- *Co-evolution*: This refers to the reciprocal evolutionary changes that occur between the host and its microbiome over time. As hosts and microbes adapt to each other's presence and influence, they can undergo genetic changes that benefit both parties. For example, certain gut bacteria may evolve mechanisms to better utilize nutrients from the host's diet, while the host's immune system may adapt to maintain a balance with beneficial microbes.
- *Co-development*: The microbiome and its host often develop together from early stages of life. The establishment and composition of the microbiome in infancy can influence immune system development, nutrient absorption, and overall health throughout the host's life. Factors such as birth mode, milk feeding and early diet can shape the microbiome's composition and function.
- *Co-metabolism*: This involves the metabolic interactions between host cells and microbial cells. Microbes in the gut can produce metabolites such as short-chain fatty acids (SCFAs) from dietary fiber which can benefit the host by providing energy and influencing immune function. In turn, the host provides a habitat and nutrients that support microbial growth and function.
- *Co-regulation*: The host and microbiome also engage in complex regulatory interactions. For example, microbial products and signals can influence host gene expression, immune responses and even behavior. Conversely, the host can regulate

the growth and activity of specific microbial species through mechanisms such as the secretion of antimicrobial peptides or modulation of mucosal surfaces.

Types of Microbiomes

The potential for the spread of pathogens between different species expanded dramatically when humans began to domesticate farm and companion animals and came into greater contact with bushmeat. The predominant body of current scientific literatures suggests that indirect interactions are the primary mechanism for the similarities between the gut microbiota of humans and domesticated animals. Food consumption directly introduces foreign bacteria into the microbiome causing alterations instead of indirect selection to a particular diet. The earliest microbes to colonize the microbiome caused slow modifications that eventually changed the makeup of the microbiome community.

The microbiome has an impact the development of the endocrine and neurological systems. The gut microbiota plays a major role in the formation of the blood brain barrier, neurogenesis, myelination, maturation of microglia and modulation of animal behavior among other neurogenerative processes. Because of this, it is thought that gut bacteria are essential to the development and functionality of the nervous system. The gut microbiome plays a significant role in shaping immune responses and influencing susceptibility to infections, including zoonotic pathogens. While the microbiome generally helps protect against infections, zoonotic pathogens like certain strains of *Salmonella* or *E. coli* can sometimes colonize the gut temporarily, leading to infections if ingested through contaminated food or water.

Microbiomes have a significant impact on the host's nutritional status and way of life by regulating the body's metabolism and functional regulation. The gut microbiota has an active role in the control of several host metabolic pathways and has impact on signal transduction and inflammatory response mechanisms and function as an essential connective tissue between key organs and tissues such the liver, brain, muscle and gut.

The skin microbiome, which consists of diverse communities of microorganisms residing on the skin's surface and in its deeper layers, also interacts with zoonotic pathogens. The skin microbiome acts as a barrier against pathogens, including those with zoonotic potential. Many bacteria residing on the skin produce antimicrobial peptides and other substances that inhibit the growth of pathogens. These antimicrobial compounds contribute to the skin's innate immune defenses. Disruptions in the skin microbiome can facilitate the colonization of pathogens and increase the risk of infection. Personal hygiene practices, such as bathing frequency, use of antimicrobial soaps, and skincare products, can affect the diversity and composition of the skin microbiome. Disruptions to the skin microbiome due to hygiene practices may influence susceptibility to skin infections, including those caused by zoonotic pathogens. Zoonotic pathogens that cause skin infections (like ringworm or certain types of *Staphylococcus*) can potentially interact with the skin microbiome, affecting its composition and potentially causing infections. There is ongoing exploration into how probiotics, prebiotics, and topical treatments can modulate the skin microbiome to enhance its protective functions against pathogens. It highlights the importance of maintaining a healthy skin microbiome as part of overall skin health and immune resilience against infections, including those transmitted from animals to humans. Respiratory and genitourinary microbiomes also play roles in modulating immune responses and can influence susceptibility to respiratory and sexually transmitted zoonotic pathogens, respectively.

- *Vector microbiome interactions*: In cases where zoonotic pathogens are transmitted by vectors such as mosquitoes, ticks, or fleas, the microbiome of the vector can affect the transmission dynamics. Vector microbiota can influence vector competence (the ability of a vector to transmit a pathogen) through mechanisms such as modulation of vector immune responses, production of antimicrobial substances or competition with pathogens for resources. Interspecies transmissions of viruses from bats to humans and other animals can arise from potential contact leading to the spread of dangerous zoonotic illnesses. The exchange of bat microbiomes with those of humans and other animals can happen as a result of adaptable roosting habits, near proximity, partial *eating of fruits etc.*
- *Environmental and ecological factors*: Environmental factors including climate, land use and biodiversity can influence both host and vector microbiomes. Changes in these factors can impact the prevalence, distribution and transmission of zoonotic pathogens. Anthropogenic activities such as agricultural practices, urbanization and wildlife interactions can alter microbiome compositions in both hosts and vectors potentially affecting zoonotic pathogen dynamics. The microbiome plays a critical role in influencing the ecology and epidemiology of zoonotic pathogens. This understanding is essential for developing holistic approaches to prevent and manage zoonotic diseases effectively, taking into account microbiome interactions across host, vector and environmental contexts.
- *Impact of antibiotics and other treatments*: The use of antibiotics and other antimicrobial agents can disrupt microbiome communities in hosts and vectors, potentially altering their susceptibility to zoonotic pathogens. For example, antibiotic treatments can lead to dysbiosis, reducing microbial diversity and altering immune responses. The microbiome can harbor antimicrobial resistance genes that may be shared with zoonotic pathogens, influencing the spread of resistance. Diversity in the microbiome may affect the prevalence and dynamics of resistance genes within microbial communities.

Microbiome and Zoonoses

The impact of the microbiome on zoonotic pathogen abundance is a complex and dynamic relationship. There are several ways in which the microbiome can influence zoonotic infections:

- *Immune system modulation*: The microbiome plays a crucial role in educating and modulating the immune system. It helps maintain immune homeostasis and tolerance to commensal organisms while providing defense against invading pathogens including zoonotic ones. Disruptions in the microbiome such as dysbiosis (microbial imbalance) can weaken immune defenses, potentially increasing susceptibility to zoonotic pathogens. Dysbiosis in the microbiome has been linked to increased systemic inflammation which can impair immune function and exacerbate the severity of infectious diseases including zoonoses.

- *Competitive exclusion*: A healthy microbiome can act as a barrier against colonization by pathogens, including zoonotic ones. This phenomenon is known as colonization resistance and it helps to prevent pathogen establishment and colonization in the host. Commensal microorganisms compete with pathogens for nutrients and colonization sites in various niches such as the gut, skin and mucosal surfaces. This competitive exclusion can limit the abundance of zoonotic pathogens and their ability to establish infection.
- *Production of antimicrobial substances*: Some members of the microbiome produce antimicrobial substances that can inhibit the growth of pathogens. For example, certain strains of bacteria in the gut produce bacteriocins or other antimicrobial peptides that can help defend against invading pathogens.
- *Metabolic interactions*: The metabolic activities of microbiota can influence the host's susceptibility to infections. Microbes in the gut and other body sites contribute to metabolic processes and produce metabolites that can affect host physiology and immunity. These metabolic interactions can indirectly affect susceptibility to zoonotic infections by altering host physiology or immune responses. For example, certain metabolites produced by the microbiome can affect the growth and virulence of pathogens. Conversely, alterations in the microbiome composition due to dietary changes or antibiotic use can impact these metabolic interactions and alter susceptibility to zoonotic pathogens. This highlights the importance of judicious antibiotic use to preserve microbiome health and resilience against infections.
- *Modification of host receptors and adhesion sites*: The microbiome can modify host receptors and adhesion sites on epithelial surfaces, potentially interfering with the attachment and colonization of zoonotic pathogens. This can reduce the likelihood of infection by limiting the ability of pathogens to establish themselves in the host.
- *Environmental interactions*: Environmental factors, such as diet, hygiene practices, and exposure to animals or wildlife can shape the composition and diversity of the human microbiome. These factors in turn, can influence the likelihood of exposure to zoonotic pathogens and the outcome of infection.
- *Vector control*: In cases where zoonotic pathogens are transmitted by vectors (e.g., mosquitoes, ticks) the microbiome of the vector species can also influence pathogen transmission. Research suggests that gut microbiota in vectors can affect their competence as disease vectors, potentially reducing or enhancing their ability to transmit zoonotic pathogens to humans.

Vaccine Development

A prime example of how microbiomes are used to improve animal health is in the production of vaccines. By using a strategy that took animal health into account, Edward Jenner successfully averted and eventually eradicated smallpox in the 18th century through vaccinia vaccine. Following Louis Pasteur's successful development of an efficacious human rabies vaccine, a dog rabies vaccine was developed as a preventive measure. To effectively stop the spread of cholera and haemaorragic septicaemia in animals, *Pasteurella* was creatively incorporated to the vaccination. In order to stop *Plasmodium* from synthesizing proteins, Koch used quinine as a prophylactic strategy. He also introduced bed nets to stop mosquito bites, which stopped the spread of malaria.

Research on Zoonotic Microbiomes

Research on microbiomes is important for tackling zoonotic challenges because it is believed that zoonoses account for 61% of human pathogens roughly. The study of microbiomes provides an enormous amount of information on the diversity of microbial communities, their interactions and impacts on both inside a host and outside environment as a member of an ecological community. The microbiome acts as a uniting element for One Health. The interdependence of human, animal and ecological health highlights the necessity of One Health professionals working together to advance the health and welfare of all species. It is in the background of COVID pandemic that microbiome research, which is based on the principle of One Health, becomes more relevant. The microbiome has impact on the health of human, animals and the environment. The process of changes in species structure of plant, animal and fungal communities all throughout the time period is known as microbiome succession. Emerging zoonotic diseases may be made more likely by the process of mutually beneficial and progressive microbiome succession between host species.

The idea of One Health, which combines environmental, animal and human health has been aided by the microbiome. The fact that microbial genes may now be incorporated into the genomes of humans, animals and plants has encouraged additional research into the connection amongst the microbiome and health from the perspective of natural cycles of information, resources and energy. Recognizing the role of microbiota, involving their dynamic interactions with other microbes and their hosts, can facilitate the development of novel diagnostic methods and interventional strategies that have applications in a wide range of disciplines, including forensics, exobiology, medicine, ecology and agriculture. The possibilities of Big Data technologies like IoT (Internet of Things) and IT data analytics can be utilised to create a spatio-temporal map of the microbiome to learn about and cross-domain interactions and development of innovative products with interdisciplinary research on microbiome. For higher research into the microbiome and to understand the interactions between microbiota (the living microorganisms found in a defined environment), a genomic database can be utilized. One example of this is the Centre of Excellence in Microbiome research at Kerala, which works on six domains, *viz.* the human microbiome, animal microbiome, plant microbiome, aquatic microbiome, environmental microbiome and data labs. It proposes to utilise novel technologies and to encourage and support startups and entrepreneurs to engage in microbiome research. Because of the microbe's high genetic variety, they are able to constantly adapt in order to evade the host immune system and thrive in new ecological niches which constantly results in the emergence of newer zoonotic outbreaks. A thorough analysis of the microbe's phylogenomics is required in order to determine the evolutionary and transmission processes of emerging zoonoses.

Conclusion

Understanding the interactions between zoonotic pathogens and microbiomes is crucial for surveillance, prevention and control of zoonotic diseases effectively. It highlights the interconnectedness of microbiome health with overall immune function and susceptibility to infections, emphasizing the need for integrated approaches in public health and clinical practice. Research into the gut microbiome's role in influencing susceptibility to zoonotic infections is ongoing. Understanding these interactions is crucial for developing strategies to promote a healthy microbiome and mitigate the risks associated with zoonotic bacterial pathogens. Maintaining microbiome diversity through practices such as a balanced diet, probiotic use when appropriate and minimizing unnecessary antibiotic use can support these protective mechanisms. There is growing interest in using probiotics, prebiotics and fecal microbiota transplantation to restore or manipulate the gut microbiome to enhance resistance against zoonotic pathogens. Research continues to explore how microbiome composition and function influence the transmission dynamics, virulence and host susceptibility to zoonotic diseases. Additionally, further research is needed to elucidate specific microbial species and mechanisms that contribute to resilience against zoonotic infections, potentially leading to novel therapeutic approaches or interventions.

References

Apari P and Foldvari G (2023) Domestication and microbiome succession may drive pathogen spillover. *Front. Microbiol.* 14:1102337.doi: 10.3389/fmicb.2023.1102337.

Kozajda A, Miskiewicz E and Jezak K (2024) Zoonotic bacteria in the vicinity of animal farms as a factor disturbing the human microbiome: A review. *Int J Occup Med Environ Health*. 37(2):138-152. doi:10.13075/ijomeh.1896.02003.

Quaresma PF, Martins-Duarte ES and Medeiros LCS (2023) Editorial: One Health Approach in Zoonosis: strategies to control, diagnose and treat neglected diseases. *Front. Cell. Infect. Microbiol.* 13:1227865. doi: 10.3389/fcimb.2023.1227865.

Rajeev R, Prathiviraj R, Kiran GS and Selvin J (2020) Zoonotic evolution and implications of microbiome in viral transmission and infection. *Virus Research* 290: 198175. doi: 10.1016/j.virusres.2020.198175.

Ma L, Zhao H, Wu LB, Cheng Z and Liu C (2023) Impact of the microbiome on human, animal and environmental health from a One Health perspective. *Science in One Health* 2: 100037. doi: 10.1016/j.soh.2023.100037.

Chapter 20

Microbiomes and Antimicrobial Resistance

Chandrasekhar Balasubramaniam[1,*]
Monica Rose Amarlapudi[2]
Akash Sriram[2]
Thulasiraman Parkunan[3]
Diwas Pradhan[2]
Raghu Vishweswaraiah[2]
and Rashmi Hogarehalli Mallappa[2]

[1]Department of Animal Husbandry & Dairying, School of Animal Husbandry, Central University of Odisha, Koraput, India
[2]Molecular Biology Unit, Dairy Microbiology Division, ICAR-National Dairy Research Institute, Karnal, India
[3]Department of Veterinary Physiology and Biochemistry, Faculty of Veterinary and Animal Sciences, Institute of Agricultural Sciences, Rajiv Gandhi South Campus, Barkachha, Banaras Hindu University, Varanasi, U.P., India

Abstract

Antimicrobial Resistance (AMR) has evolved from a natural biological process into a global health crisis, threatening the efficacy of antimicrobial agents and posing significant risks to public health and economies worldwide. Recent estimates suggests that AMR could lead to approximately 2 million premature deaths in India by 2050, while also potentially pushing 24 million people into extreme poverty by 2030. Beyond health impacts, AMR is projected to result in staggering financial costs, including a loss of up to 100 trillion globally. Such alarming statistics have galvanized scientists, policymakers and clinicians to intensify their efforts in addressing this escalating threat. While AMR was initially associated with specific pathogenic microbes, the emerging concept of the "resistome" emphasizes the role of various microbial communities or microbiomes in the spread of resistance. This shift in the perspective underscores the complexity of AMR as a multifaceted issue that extends beyond individual pathogens. Utilizing advanced metagenomics tools has proven to be invaluable in elucidating the dynamics of AMR, highlighting the necessity for a paradigm shift in how we approach this crisis. Instead of

* Corresponding Author's Email: chandrundri93@gmail.com

In: Microbiomes in Animal Nutrition, Health, and Disease
Editor: Tanmoy Rana
ISBN: 979-8-89530-425-9

focussing solely on isolated resistance mechanisms, it is crucial to understand the broader interplay within microbial ecosystems. For effective intervention, global strategies must integrate the concept of the "resistome" into surveillance and monitoring efforts, fostering a comprehensive response to mitigate AMR and safeguard one health for future generations.

Keywords: resistome, antimicrobial resistance, microbiome, AMR

Introduction

Antimicrobial Resistance (AMR) a seemingly harmless natural process has evolved into a dreaded pandemic by ensuring its presence in every nook and corner across the globe. According to a recent projection AMR is expected to take 2 million lives in India alone by 2050. The disease is also anticipated to cause a massive loss at the economic front, as it is expected to cause an estimated loss of 100 trillion dollars to the global exchequer. The economic consequences are further aggravated with the prediction that AMR could force upto 24 million people into extreme poverty by 2030 (U.N. Report). Further AMR is slated to contribute to an average reduction of 2-3.5% in gross domestic product (GDP) (O'Neill, 2016), where the losses due to AMR are estimated to sum up to $85 trillion in GDP and a similar worrying trend is also reflected in international trade where the losses are predicted to the tune of $23 trillion (Syud et al., 2018). These numbers have caused ripples across the scientific community, policy makers and clinicians and hence scientist across the globe are making frantic efforts to find solutions to combat this mushrooming killer. AMR was thought to be the close preserve of few of the clinically relevant pathogenic bugs, however the concept of "resistome" shifts the onus of AMR on microbiomes rather than individual microbes.

Antibiotic Resistance Mechanisms

Initially the term 'Antibiotic' was used to refer to anything that inhibited growth of bacteria in a laboratory experiment. Subsequently, in the current era the term encompasses all naturally occurring or artificially synthesised low molecular weight compounds that inhibit growth of bacteria. Resistance is simply defined as the continued growth of microbes in the presence of cytotoxic concentration of antibiotics. The ancient bacteria carried an inherent defence mechanism (such as antibiotic altering enzymes or efflux pumps) to shield themselves from high concentrations of antibiotics. Hence, it is quite natural that the biosynthetic gene cluster responsible for synthesis of antibiotic must also harbour genes which confer resistance to these antibiotics, and many aspects of the resistome might have developed much before these antibiotics became prevalent in clinical practice (Morrison and Zembower, 2020). Researchers worldwide have made attempts to unravel the mechanisms involved in AMR in pathogens and have successfully illustrated few of them. Conventionally while referring to AMR the focus is on 'acquired resistance' i.e., those that arise from point mutations in chromosome and the 'intrinsic resistance' i.e., those mechanisms that are inherent in the cell that make bacteria resistant to drugs like efflux pumps, modified drug target, modified cell wall proteins and drug inactivating enzymes (Sultan et al., 2018). The various antibiotic resistance mechanisms have been depicted in Figure 1.

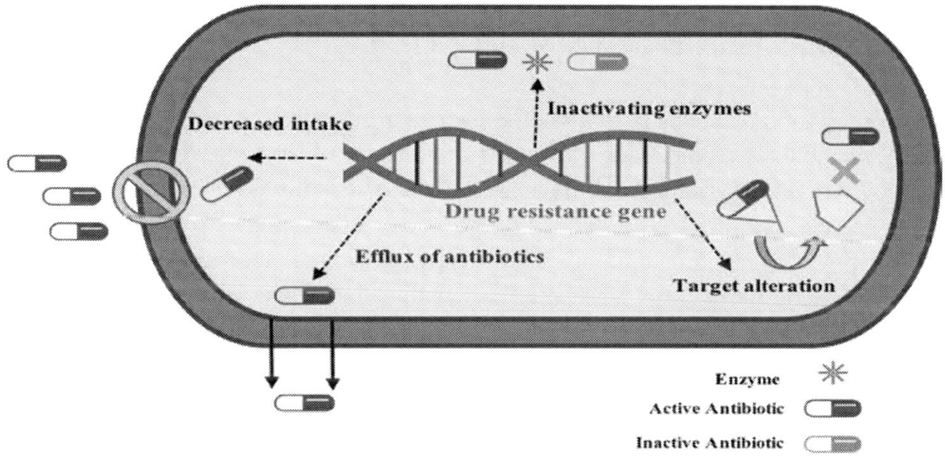

Figure 1. Antibiotic Resistance Mechanisms.

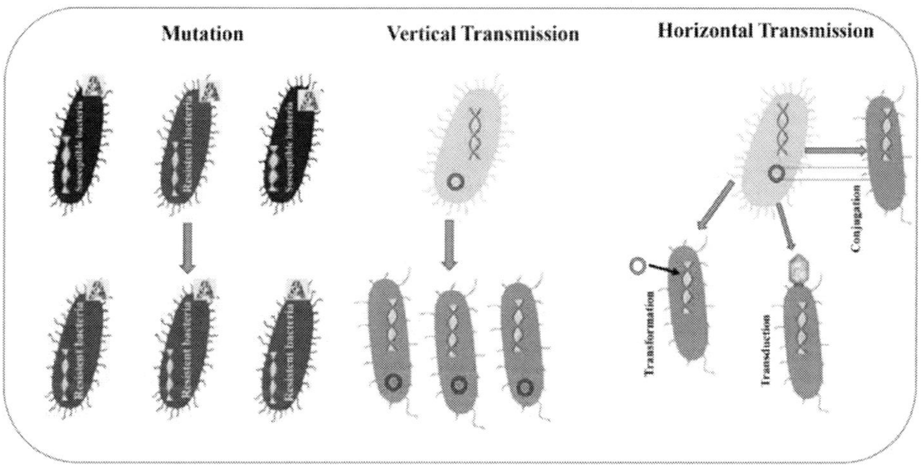

Figure 2. Mechanism of spread of antibiotic resistance (Reproduced from Mancuso et al., 2021).

The intrinsic resistance genes are immobile and have functions other than resistance i.e., they aide in bacterial metabolism and physiology (Cox & Wright., 2013). These genes may become mobile during the course of evolution. The reasons for spread of AMR and the cause of its transformation into a pandemic is traced back to the genetic mechanisms of AMR. Some of the genes conferring resistance are borne on Mobile Genetic Elements (MGE) like plasmids, transposons, integrons and insertion sequences. These are bestowed with the potential for horizontal gene transfer (HGT) and can be transferred from pathogens to other pathogens or their benign counterparts through mechanisms of transduction, transformation and conjugation (Sultan et al., 2018). On the same lines as HGT, vertical gene transfer can also lead to spread of AMR to a considerable extent. These genetic mechanisms have aided in the rapid spread of AMR across various domains of life, plants, animals as well as people and continents. A recent finding, of the presence of AMR gene blaNDM-1 that encodes the protein NDM-1, (first identified in New Delhi) in microbes isolated from the arctic regions i.e., approximately 13,000 Km away in one of the most pristine ecosystems has ratified the claims and reaffirmed the fears of the spread of AMR (McCann et al., 2019). The different modes of spread of resistance are elucidated in Figure 2.

Sources of Antimicrobial Resistant Determinants (ARDs) and Their Origins

AMR predates antibiotic era. It is a well-known fact that most of the antibiotics used are secondary metabolites or their derivatives, of bacteria themselves. Further bacteria evolved over 3.8 billion years ago and antibiotics are at least 100 million years older (Baltz, 2008) Bacteria have been exposed to antibiotics directly or otherwise for an equal time frame. The hypothesis of AMR predating antibiotic era also stems from the fact that bacteria that produce antibiotics must be harbouring resistant genes within the cluster of genes responsible for production of antibiotics as a survival advantage. Lately, the veracity of this hypothesis has been established through confirming the presence of antibiotic resistance (Vancomycin resistance) in 30,000-year-old permafrost soil using PCR assays and high throughput sequencing approaches. Baltz (2008), has illustrated erythromycin, streptomycin and vancomycin biosynthetic pathways emerged over 880, 610 and 240 million years ago respectively. Hall and Barlow (2004) have estimated that, β-lactamases arose over 2 billion years ago. Thanks to the short generation time of bacteria, coupled with the horizontal gene transfer mechanisms and irrational use of antibiotics which have led to an inevitable expansion of resistance.

The environment is the largest source of resistant genes as it harbours a plethora of microbes. The melange of secondary metabolites produced by these microbes i.e., the chemical diversity has a reciprocal effect on the genetic diversity. In other words, the antibiotics and other metabolites induce a selection pressure that leads to expression of resistant genes hence organisms thriving on a cocktail of complex metabolites are the source of diverse antibiotic resistance genes. The environment that are polluted with heavy doses of antibiotics which gain entry through the indiscriminate use of antibiotics in food animal production as they are used as growth promoting antimicrobials (GPA) at sub lethal concentrations binate the problem. The dumping of agriculture waste i.e., waste from poultry, cattle excreta and manure applied on agricultural fields are potent sources of ARD. In a rare quantitative study, tracking changes in antibiotic resistance from soil over time, Graham and colleagues isolated metagenomic DNA from agricultural soil in Netherlands during the period 1940-2008 and used qPCR to quantify the abundance of selected resistant genes. Investigators found that there was invariable increase in the level of all the targeted resistant genes, from the preantibiotic era (1994) till 2008 (Knapp et al., 2009). Hence, suggesting a radical change in the agricultural practices over the time frame favouring the indiscriminate use of antibiotics. Further the authors also found the presence of extended spectrum β - lactamase in these soils even before it achieved clinical significance supporting the hypothesis that resistant genes are harboured by environmental microbes before emerging in clinics (Knapp et al., 2009; Delgado-Baquerizo et al., 2022).

Like soil, water is also polluted with ARD. Water has its own intrinsic resistance conferred by normal microflora. Alternatively, these can also be contributed by run offs from agricultural land and usage of antibiotics in aquaculture (Zhuang et al., 2021). A couple of years back, investigators in India have reported the occurrence of NDM-1 gene in several bacterial species in water source, the most worrying trend being the high prevalence of this gene in chlorinated municipal drinking water samples (Rathinasabapathi et al., 2015).

Animals that are in constant contact with humans are reservoirs of superbugs as a consequence of antibiotic treatment for diseases as well as inter species transfer of resistant bugs from humans. The enteric bacteria in animals gain resistance through frequent exposure to antibiotics either through their use at sub therapeutic concentration as GPA or at lethal levels to cure infectious diseases (Bhatt et al., 2012, Singh et al., 2012, Leclerq et al., 2016, Wang et al., 2017).

Humans are a storehouse of superbugs, and this storehouse is increasingly becoming diverse in the post antibiotic era. The pathogens that inhibit the various anatomical sites of the body have become, and are continuously gaining resistance with the ever-increasing exposure to antibiotics. The gut microbiota which has the maximum diversity of microbes in the human body is the epicentre for evolution and spread of resistance (Theophilus & Taft, 2023). The resistant genes are spread from the pathogens to their commensal counterparts through Horizontal Gene Transfer mechanisms, making humans a prominent reservoir for ARD in the complex vicious nexus of AMR spread involving man, animal and the environment (Singh et al., 2019). Figure 3 depicts the sources of ARDs in the environment.

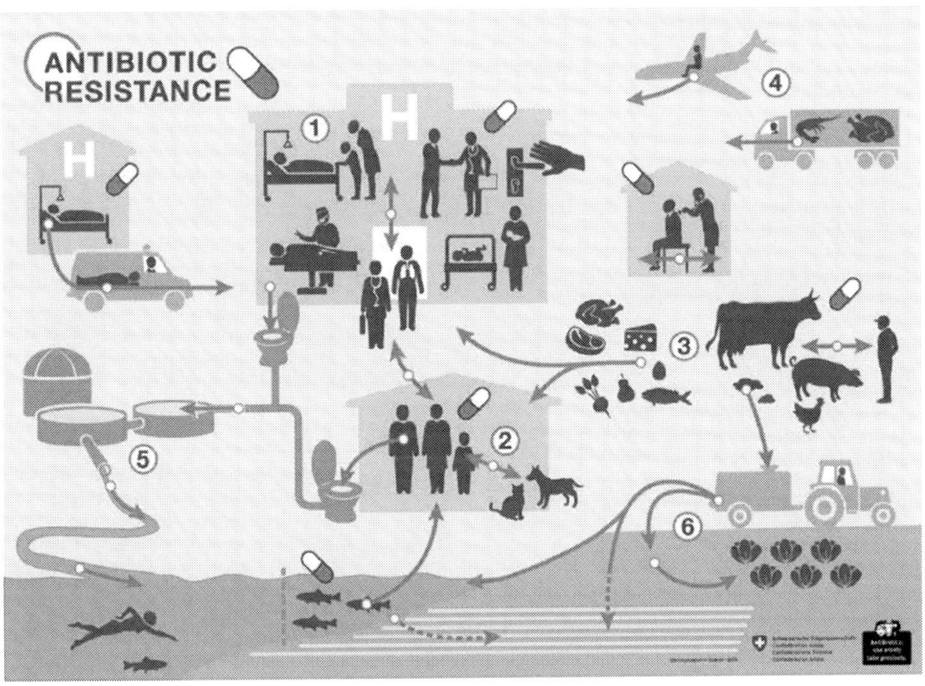

Figure 3. Sources of Antibiotic Resistant Determinants (ARD).

Antimicrobial Resistome

The corroborative data emanating from the researches into the mechanism and spread of AMR have compelled scientists to know the finer details of the mechanisms that bring about resistance. This has led researchers to peep into the evolutionary and genetic mechanism involved in the propagation and spread of AMR. This exploration has led to synthesis of newer concepts and hypothesis signifying a paradigm shift in the conceptual understanding of AMR. A dozen of resistant genes or ARD have been identified to cause AMR and researches hitherto have zeroed on these ones particularly focussing on those that are harboured by clinically significant pathogens. But the fledgling concept of "Resistome" has a different story to tell. It says that resistance is a pan microbial phenomenon that is characteristic of the microbiome

(Wright, 2007). It reiterates that though resistance largely viewed as a clinical problem, has an environmental origin. Given the genetic diversity prevalent in the environment this should not take us by surprise. The Resistome encompasses all the resistant genes that are present in the particular environment at a given time. These genes include known clinical resistant genes, genes closely homologous to known resistance genes which mostly confer same phenotype, genes already confirmed to possess resistance as detected by functional metagenomic screens but that do not share homology to known resistance genes and lastly currently completely unknown genes or in other words "Cryptic" genes, genes for which neither the sequence nor the function is known (Bengtsson-Palme et al., 2015, 2018; Gillings, 2013).

The resistome is a culmination of determinants that confer 'Phenotypic Resistance' and 'Phenotypic Sensitivity' the former includes 'Acquired' and 'Intrinsic' resistance whereas the later involves 'Silent' and 'Proto' resistance. Acquired are those, borne on mobile genetic elements and Intrinsic once are those already embedded in the microbes, these are responsible for resistance in the true sense. 'Silent' ones are represented by ARD that are functional but not expressed and needs mobilization or mutation for activation. Similarly, 'Proto' resistance genes have no activity against antibiotics but can gain it, if favourable mutations occur (Perry et al., 2014). Further, it is predictable that favourable selective pressure induces expression of resistant genes. This obliquely implies that chemical diversity spurs genetic diversity hence indiscriminate use of antibiotics can lead to a diverse resistome. Hence, many factors affecting the environment can drive resistome expansion. These can then be transmitted to the humans and animals keeping the vicious cycle of AMR on the run. Figure 4 diagrammatically illustrates the concept of resistome.

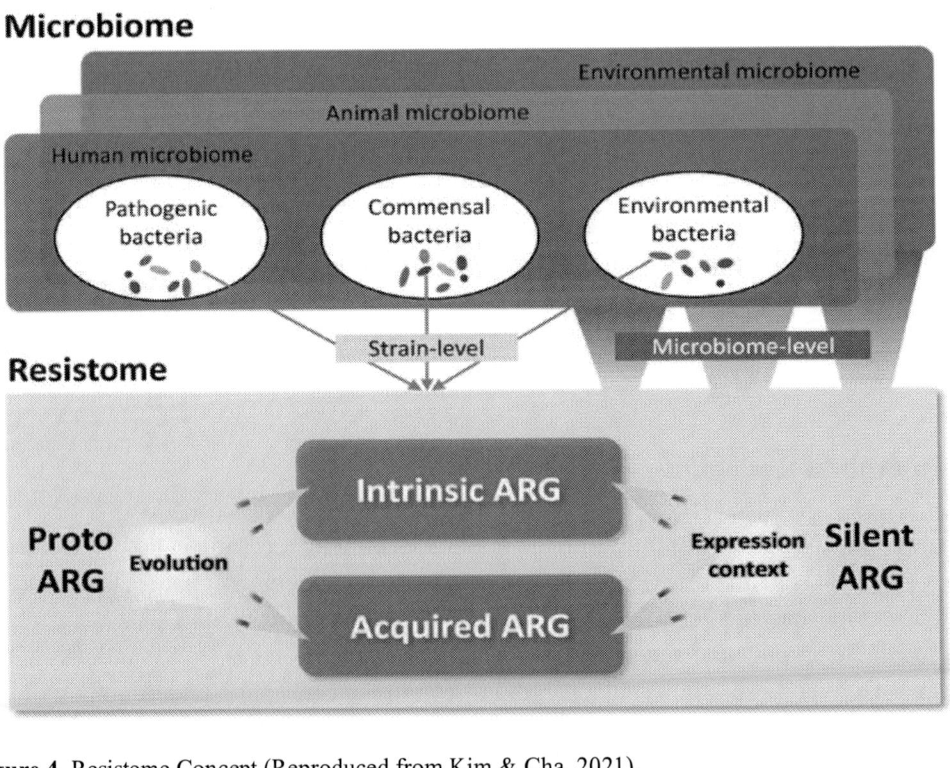

Figure 4. Resistome Concept (Reproduced from Kim & Cha, 2021).

Patterns of Antibiotic Usage

Antibiotics

Bacterial infections have been afflicting mankind since their existence in this planet. They have been implicated in causing several diseases of various scales, ranging from outbreaks to pandemics. These diseases cause morbidity and mortality in humans. Antibiotics are wonder drugs that helped mankind to defend against pathogens and cure diseases by killing or restricting their growth. Antibiotics have been found to be immensely successful in improving health outcomes. They have been instrumental in bettering childhood (under 5) mortality rates, from 216 deaths per 1000 livebirths in 1950 to 39 deaths per 1000 livebirths in 2017 (Burstein et al., 2019). Besides, they have also improved male life expectancy from 48 years to a staggering 71 years within the same time period (GBD, 2017).

Classes of Antibiotics and Mode of Action

Ever since Sir Alexander Fleming discovered "Penicillin", the first antibiotic, the drug discovery marathon kicked off and has been delivering an array of antibiotics encompassing several mechanistic modulations and varied drug targets. The availability of multitude of drugs then called for specific classification systems that could easily guide the clinicians during their treatment protocols to decide upon the most appropriate drug that could cure the disease based on its potency and spectrum of activity. The drugs are classified into various classes, primarily based on the mechanism of action and the spectrum of activity. Table 1 below shows the classification of antibiotics

Table 1. Classification of Antibiotics

Sr.No.	Mechanism of Action	Antibiotic Class	Spectrum of activity	Examples
1.	Inhibit Protein Synthesis (30S)	Aminoglycosides	Gram Negative	Streptomycin, Gentamycin
		Tetracycline	Gram Positive/Negative	Tetracycline, Doxycycline
		Glicylcycline	Gram Positive/Negative	Tigecycline
2.	Inhibit Protein Synthesis (50S)	Macrolides	Gram Positive	Azithromycin, Erythromycin
		Lincosamides	Gram Positive	Clindamycin
		Streptogramins	Gram Positive	Pristinamycin
		Oxazolidinones	Gram Positive	Tedizolid
3.	Inhibit Cell Wall Synthesis	Cephalosporins	Gram Positive/Negative	Ceftriaxone, Cefepime
		Penicillin	Gram Positive/Negative	Ampicillin, Amoxicillin
		Carbapenems	Gram Positive/Negative	Meropenem, Ertapenem
		Glycopeptides	Gram Positive	Vancomycin
		Monobactams	Gram Negative	Aztreonam
		Phosphonic	Gram Positive/Negative	Fosfomycin
4.	Inhibit DNA replication	Fluoroquinolones	Gram Positive/Negative	Ciprofloxacin, Levofloxacin
		Rifamycin	Gram Positive/Negative	Rifampicin
5.	Induces membrane damage in cells	Polymyxin	Gram Negative	Colistin
		Lipopeptides	Gram Positive	Daptomycin
6.	Inhibit Folate synthesis	Sulfonamides	Gram Positive/Negative	Sulfasalazine, Sulfamethoxazole

Pattern of Antibiotic Usage in Animal Health

The global population is constantly on the rise and so will be the demand for food. To meet the nutritional requirements of the citizens, countries across the globe are constantly making policy decisions and pondering on innovative alternatives both in the agriculture and food animal production systems. Added to this conundrum is the challenge of striking a balance between the need for land for agriculture and fodder production and "living space" for human settlements. Practicing "Intensive Farming" has become the order of the day in many developed nations, several developing nations are also viewing this as a promising alternative (Mankhomwa et al., 2022). Maximizing productivity is an essential objective in Intensive farming, hence farmers are banking on finding out cost effective alternatives to attain superior productivity. Antibiotic usage has cropped up as an essential ingredient in such Intensive Farming practices adopted for food animal production (Gilbert et al., 2021; Prack et al., 2023). According to a recent estimate 73% of the total antibiotics sold globally are used for food animal production (Van Boeckel et al., 2017). Such alarming rates of usage of antibiotics will surely have catastrophic repercussions as resistance against these drugs is bound to cross over to humans and environment directly or indirectly. Several studies have hinted at such "cross kingdom" spread of resistance (Tiseo et al., 2020).

Scenario of AMR & AMU Surveillance

Going by the popular adage "Only those that can be measured can be controlled," scientists awakened by the AMR crisis, started modelling surveillance systems to track, trace and quantitate the extent of spread of AMR & AMU. These surveillance systems helped the scientists to effectively measure the immensity of the crisis and also in turn proved helpful to prevent further deterioration of the situation through making informed policy decisions based on concrete scientific data. Denmark as early as 1996, prepared the first official report containing antibiotic sales data under the Danish Integrated Antimicrobial Resistance Monitoring and Research Programme (DANMAP), one of the first such initiatives across the globe. This development was reciprocated in Europe in 2005, by the European Medicine Agencys', European Surveillance of Veterinary Consumption (ESVAC) group wherein antibiotic sales data from 8 European nations namely the Czech Republic, Denmark, Finland, France, the Netherlands, Norway, Sweden, and the UK were compiled. In the North America, surveillance programme was first started in Canada where sales data were published under the Canadian Integrated Program for Antimicrobial Resistance Surveillance (CIPARS). Japan was the first country in Asia where such an effort was initiated under the Japanese Veterinary Antimicrobial Monitoring System (JVARM) in 2000. These early efforts in surveillance programmes were consolidated further, as an increasing number of countries came out with such effective surveillance programmes and officially published their antibiotic sales data. Starting with 5 nations in 2000, growing to 15 in 2015 and as of 2017 there are 41 nations who have established Antibiotic Surveillance programmes (Tiseo et al., 2020).

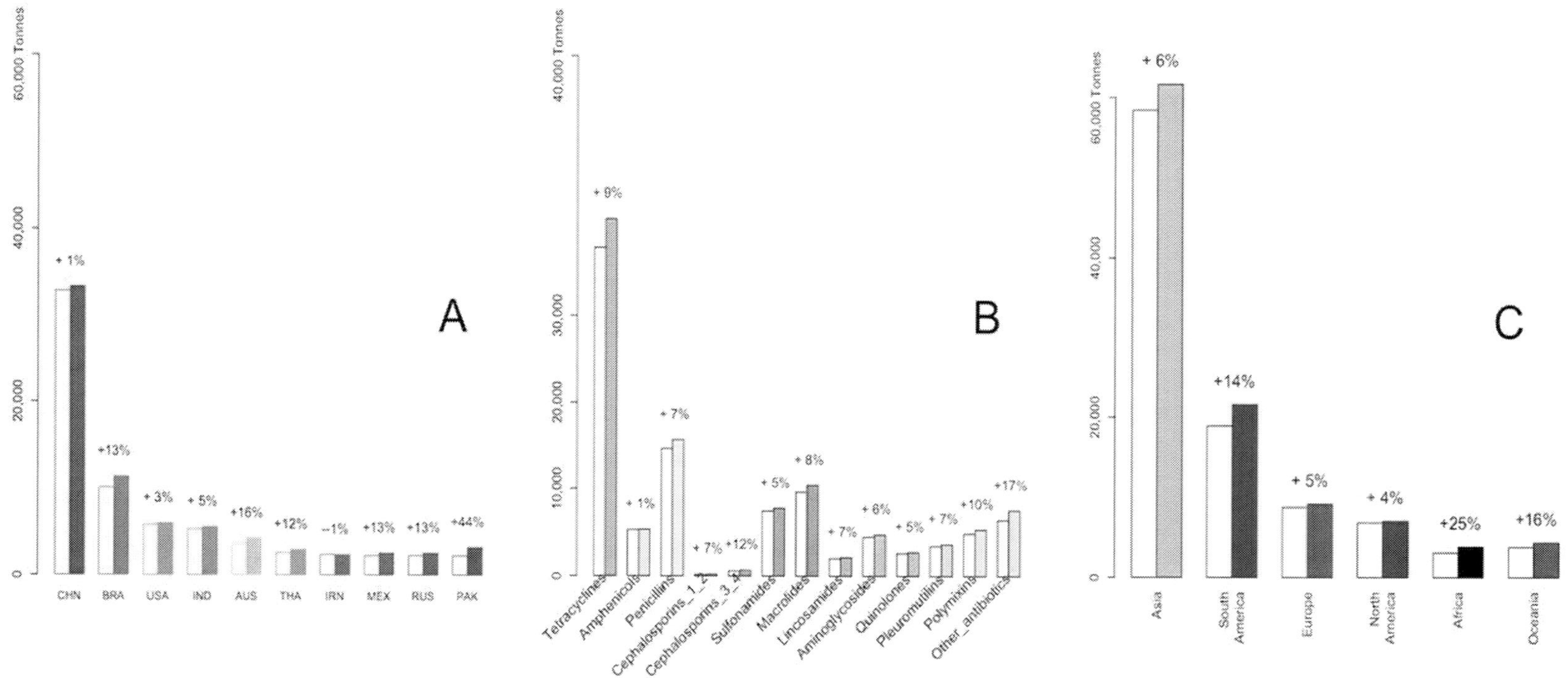

Figure 5. Graph showing trends in antibiotic consumption across A) Country B) Antibiotic Class C) Continent.

A comprehensive analysis of the antibiotic sales data from different countries has shown interesting trends. As of 2020, the top 10 veterinary antimicrobial consumers were: China, Brazil, the USA, India, Australia, Thailand, Iran, Mexico, Russia and Pakistan as illustrated in Figure 5(A). Among these the top 5 countries i.e., China, Brazil, USA, India and Australia accounted for 58% of the global antimicrobial usage (AMU) in food animal production (Mulchandani et al., 2023).

Extrapolating the data, a projection for 2030 shows that these 5 countries will hold their respective positions in global AMU with meagre changes in the levels. Further the estimated global consumption of antimicrobials was pegged at 99,502 tonnes in 2020 and is projected to rise to 1,07,472 tonnes, i.e., an estimated rise of 8.0% by 2030 (Mulchandani *et al.*, 2023). Asia was found to have 67% of the hotspots for antimicrobial use on the contrary Africa had less than 1% of these hotspots as illustrated in Figure 5(C).

The major group of antibiotics that are predominantly used in the animal health sector include Tetracycline, Penicillin, Sulfonamides, Macrolides/Lincosamides, Aminoglycosides, Cephalosporins, Quinolones, Polymyxins and Trimethoprim as depicted in Figure 5 (B). Among these tetracyclines are the class of antibiotics, that have found the maximum usage rate in almost all nations, however, the usage of rest of the antibiotic class varied with countries. The usage of tetracycline was projected to rise by 9% by 2030 (Mulchandani et al., 2023).

Pattern of Antibiotic Usage in Human Health

Bacterial infections have been a major cause of morbidity and mortality in humans. The discovery of antibiotics proved to be a blessing, as these were helpful in curbing infections and could be easily produced from natural sources. It was none other than the inventor of antibiotics himself, Sir Alexander Fleming, who had foreseen the repercussions of antibiotic misuse and had professed in his 1945 Nobel Prize lecture, that,

> "The time may come when penicillin can be bought by anyone in the shops. There is the danger that the ignorant man may easily underdose himself and by exposing his microbes to non-lethal quantities of the drug make them resistant."

With the advancement in processing technologies (that included both Downstream and Upstream processing) these drugs were produced on a massive scale which tremendously reduced the cost. Soon these medicines could be easily accessed at throwaway prices. The ease of accessibility coupled with the reduced cost and promising curative potential, made these wonder drugs extremely popular among the common man. Over the years, the constant use and misuse of these drugs in clinical settings has led to the problem of antibiotic resistance. AMR has gradually snow balled into a health crisis and taken the form of a "Faceless Pandemic."

While antibiotics offer cure for a range of infections i.e., Urinary Tract Infections (UTI), Intraabdominal, Lower Respiratory Tract Infections (LTI), Bloodstream Infections (BSI), Tuberculosis, Skin infections and Diarrhoea, maximum usage of antibiotics is found to be for curing LTI's (Antimicrobial Resistance Collaborators, 2022). According to a recent estimate (2019) more than 1.5 million deaths globally, could be linked to LTI's, arising from treatment failure due to AMR (Antimicrobial Resistance Collaborators, 2022).

Deaths due to AMR can be generally attributed to pathogens that cause disease. It is these superbugs that can resist the antimicrobials and evade death. A careful analysis has shown that deaths due to AMR is mostly caused due to 6 pathogenic superbugs namely *Escherichia coli*, followed by *Staphylococcus aureus*, *Klebsiella pneumoniae*, *Streptococcus pneumoniae*, *Acinetobacter baumannii*, and *Pseudomonas aeruginosa* (De Oliveira et al., 2020).

Antibiotics and Environment

Humans and animals are important components of the ecosystem, any changes to human or animal life is destined to have a direct effect on environmental health. As environment is an inseparable whole from human and animal life, modest changes in the lifestyle of humans and animals have a direct bearing on environment. Humans have been notoriously implicated in degrading the environment. It is an irony that as humans have evolved into more complex organisms, they have caused more harm than good to the surrounding nature. Like climate change which can be solely attributed to human activities, resistance caused due to the use and abuse of antibiotics has also spilled over to the environment.

The major source of antibiotics in the environment are from the waste material originating from humans, animals and pharma industries. Apart from these typical sources, these Resistance Determinants originate due to the application of chemicals falling under three broad chemical classes namely 1) Antimicrobials (antibiotics, antifungals, antiparasitic and antivirals) 2) Heavy Metals 3) Biocides (Disinfectants, Surfactants) as depicted in Figure 6.

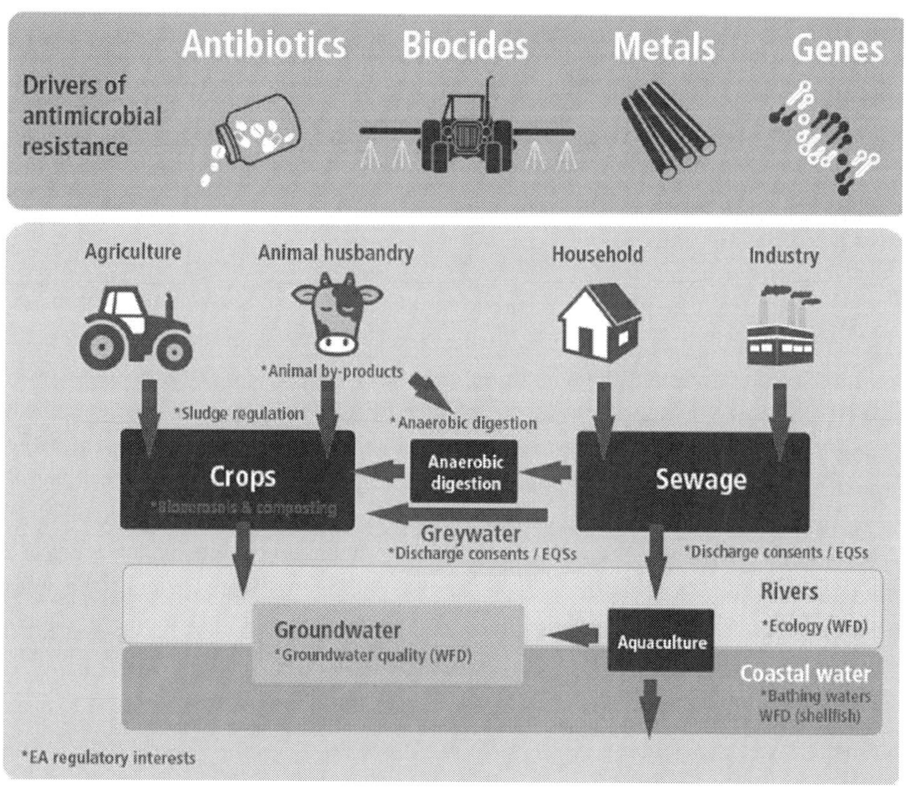

Figure 6. Antibiotic Resistance and Environment.

The three major pathways through which these chemicals gain entry into the environment are 1) Waste water and effluents from municipality and industries 2) Application of manure and sludge on land 3) Use of antimicrobials in aquaculture.

One of the major modes of transmission of antimicrobials is through human and animal waste. Human gut is a reservoir of microbes and these reservoirs upon coming under the influence of antimicrobials can serve as safe havens for evolution of resistant bugs. These superbugs are shed in the human excreta. In addition to ARBs, it has also been found that 70% of the antibiotics consumed pass the human digestive tract unaltered. These wastes ultimately land up in the sewage systems of the cities. The sewage reaches the Waste Water Treatment plants. The mash up of antibiotics present at sub lethal levels coupled with the prevalence of ARBs and ARGs provide the right mix for the evolution of AMR. The nutrient rich environment of the sludge provides conducive atmosphere for growth of microbes and is perfect spot for resistance development and hence forms a "Hotspot" (Kraemer et al., 2019).

Animal waste also undergoes a process similar to human waste and even the manure which serve as biological amendments upon application onto agricultural fields can be a potent source of ARGs (Sanz et al., 2022). These ARGs can then reach the soil microflora and give rise to ARB which in turn can also gain entry into the plant kingdom and contaminate plants. Such type of transmission can trigger "Cross Kingdom" spread of AMR. Cross kingdom spread of AMR can aggravate the crisis as even plant based foods, the major source of foods for vegans can become a source of ARGs and ARB.

Aquatic foods are the fastest growing segment in the food animal sector, across the globe. They contribute about 20% of the total animal protein consumed in the human diet. Among all the continents, countries in Asia contribute (68%), the largest share to global production of aquatic foods (Schar et al., 2021). Aquaculture accounts for 5.7% of the total antibiotics used in food animal sector. These antibiotics from the fish farms can find their way into larger water bodies like rivers and seas to which they are generally flushed out. These sources can then serve as centres for AMR emergence by aiding in the propagation of resistant bacteria.

Antibiotic Resistant Bacteria

Studies on AMR across the globe have predominantly focussed on pathogenic bugs as it was against them the antibiotics were initially designed. Hence, pathogenic bugs are the most concerning groups of resistant bugs as these can lead to treatment failures and affect human health. WHO in its endeavour to curtail the menace of AMR has chalked out several strategies one among this is publishing various reports concerning the status of ARB. In 2017, WHO came up with the list of "Priority Pathogens" those were antibiotic resistant. This was a first of its kind report aimed at guiding drug discoverers on the potential directions on which the drug discovery pipeline should be channelised (http://www.who.int/mediacentre/news/releases/2017/bacteria-antibiotics-needed/en/). The target pathogens were categorized into three categories based on the scale of urgency needed for new antibiotic development, which were

Priority 1: CRITICAL

They are mainly Multi Drug Resistant bugs that can cause life threatening diseases and can evade antibiotic effects through more than one mechanism or are resistant to more than one antibiotic. They are namely:

- Acinetobacter baumannii, carbapenem-resistant
- Pseudomonas aeruginosa, carbapenem-resistant
- Enterobacteriaceae, carbapenem-resistant, ESBL-producing

Priority 2: HIGH

They comprise drug resistant pathogens that cause more common diseases. They are namely:

- *Enterococcus faecium*, vancomycin-resistant
- *Staphylococcus aureus*, methicillin-resistant, vancomycin-intermediate and resistant
- *Helicobacter pylori*, clarithromycin-resistant
- *Campylobacter* spp., fluoroquinolone-resistant
- *Salmonellae*, fluoroquinolone-resistant
- *Neisseria gonorrhoeae*, cephalosporin-resistant, fluoroquinolone-resistant

Priority 3: MEDIUM

They also comprise drug resistant pathogens that cause more common diseases. They are namely:

- *Streptococcus pneumoniae*, penicillin-non-susceptible
- *Haemophilus influenzae*, ampicillin-resistant
- *Shigella* spp., fluoroquinolone-resistant

One Health Approach

It is evident from the omnipresence of AMR that though antibiotics were notionally designed to treat and cure human diseases and later animals, resistance against them have inadvertently spilled over to all three spheres of humans, animals and environment and is severely impacting their respective health by influencing treatment outcomes. Hence, it can be perceived that health should no more be thought off from the human perspective, rather it should involve even animals and environment. Such multidimensional approach to health can only help in mitigating current and future threats form emerging and resurging zoonotic diseases.

The birth of the concept of "One Health" can be traced back to more than two centuries ago where the concepts of "One Medicine" was introduced, which was later expanded to include "One World, One Health" gradually settling to its current coinage of "One Health"

(Atlas, 2013). The term gained prominence in 2003-2004, with the outbreak of Severe Acquired Respiratory Syndrome (SARS), the first readily communicable disease to emerge in 21st century. Subsequently, it was also associated with curtailing the spread of avian influenza (H5N1). These pandemics provided the first instances where it was recognized widely that zoonotic pathogens can jump animal barriers and cause devastating impact to human health and economies.

The US Centre for Disease Control and Prevention and One Health Commission define One Health as: 'One Health is defined as a collaborative, multisectoral, and transdisciplinary approach—working at the local, regional, national, and global levels—with the goal of achieving optimal health outcomes recognizing the interconnection between people, animals, plants, and their shared environment' (Mackenzie and Jeggo, 2019).

Lately in December 2021, the WHO, FAO, OIE and UNEP jointly constituted an Advisory Panel "One Health High Level Expert Panel" (OHHLEP) which defines One Health as:

> "One Health is an integrated, unifying approach that aims to sustainably balance and optimize the health of people, animals and ecosystems.
> It recognizes the health of humans, domestic and wild animals, plants, and the wider environment (including ecosystems) are closely linked and inter-dependent.
> The approach mobilizes multiple sectors, disciplines and communities at varying levels of society to work together to foster well-being and tackle threats to health and ecosystems, while addressing the collective need for clean water, energy and air, safe and nutritious food, taking action on climate change, and contributing to sustainable development."
> The OHHLEP came out with a logo to summarize the definition, as illustrated in Figure 7. (https://www.who.int/news/item/01-12-2021-tripartite-and-unep-support-ohhlep-s-definition-of-one-health)

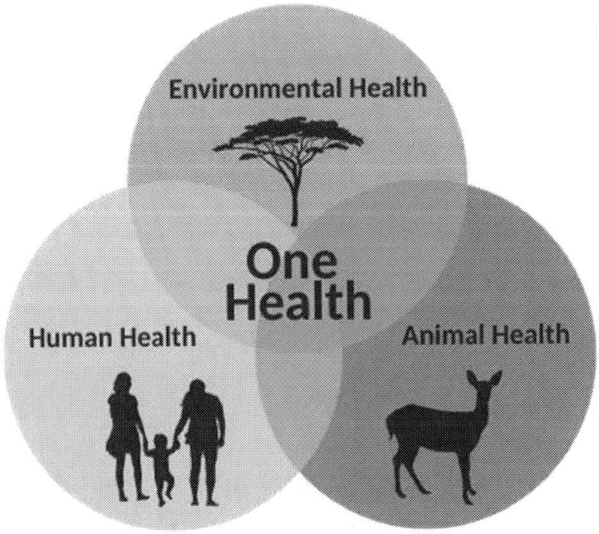

Figure 7. WHO One Health Logo.

Table 2. Resistome studies in humans

Year of Study	Source	Families, Class, Genus and Species of Bacteria found	Resistance Gene found	Sequencing Platform used	Reference
2013	Human gut	Bacteria	Chloramphenicol, aminoglycoside and tetracycline resistance genes. Three novel resistance genes including a 16S rRNA methylase conferring aminoglycoside resistance, and two tetracycline resistance proteins nearly identical to a bifidobacterial MFS transporter	Illumina HiSeq 2000	(Moore et al., 2013)
2015	Human gut	Proteobacteria, Actinobacteria, Bacteroidetes and Firmicutes	CTX-M-15 gene, OXA-1 and TEM (beta-lactamases), aph(3'')-Ib and aph(6)-Id genes and tetracycline resistance genes tet (Q) and tet(X)	Illumina HiSeq 2000	(Bengtsson-Palme et al., 2015a,b)
2015	Gut and oral Microbiome	Bacteria	Erythromycin resistance genes, efflux pumps, chloramphenicol acetyltransferase (CatB1), betalactamases	Illumina MiSeq	(Zaura et al., 2015)
2016	Human gut	Proteobacteria with Escherichia coli and Klebsiella most prevalent	Beta-lactam, multidrug efflux pumps, fluoroquinolone and tetracycline resistance genes	Illumina MiSeq	(Millan et al., 2016)
2016	Bronchial aspirates	Bacteria and fungi	mecA	IlluminaHiseq	(Hilton et al., 2016)
2016	Human gut	Bacteria	blaTEM-124-like (extended spectrum beta lactamase), fosA (fosfomycin), ant(6)-Ia, ermB, lnuB, tetL and tetU conferring resistance to aminoglycosides, macrolides, lincosamides, streptogramin B and tetracycline respectively	454 pyrosequencing	(Jitwasinkul et al., 2016)
2018	Human gut	E. coli	Aminoglycoside-, fluoroquinolone-, beta-lactam- and macrolideresistance genes	Illumina HiSeq	(Richter et al., 2018)
2019	Human gut	Escherichia, Bacteroides & Clostridia	Colistin (mcr-1), Tigecycline (tetX)	Illumina MiSeq	(Zeng et al.,2019)
2022	Human gut	Lactobacillus sp.	dfrA, poxtA, ImrB, BJP-1		(Moradi et al., 2022)
2022	Human gut	Streptococcus thermophilus, Lactococcus lactis	Aminoglycoside resistant gene ant(6)-Ia; erm(B); Tetracycline resistance gene tet(40), tet(Q), tet(W)	Illumina NovaSeq 6000	(Stege et al., 2022)

Table 3. Resistome studies in animals

Year of Study	Source	Families, Class, Genus and Species of Bacteria found	Resistance Gene found	Sequencing Platform used	Reference
2012	Buffalo rumen	Bacteria (Firmicutes predominant)	multidrug resistance efflux pumps, fluoroquinolone and acriflavin resistance genes	454 Life Sciences Technology	(Singh et al., 2012)
2012	Cattle milk	Escherichia coli, Pseudomonas aeruginosa, Pseudomonas mendocina, Shigella flexneri, Bacillus cereus, Staphylococcus aureus, Klebsiella pneumonia, Staphylococcus epidermidis	Fluoroquinolone, copper and cobalt–zinc–cadmium, mercury, arsenic, erythromycin and fosfomycin resistance genes	454 GS-FLX Technology	(Bhatt et al., 2012)
2014	Buffalo rumen	Bacteria, viruses	Fluoroquinolone resistance genes, multidrug resistance efflux pumps, methicillin resistance (In Staphylococci)	Ion Torrent	(Reddy et al., 2014)
2015	Dairy cow faeces	Bacteria	Beta-lactam resistance genes	Illumina HiSeq	(Chambers et al., 2015)
2016	Rattus norvegicus faecal samples	Enterococcus spp.	vanB genes (vancomycin resistance gene)	Illumina HiSeq 2000	(Hansen et al., 2016)
2016	Pig manure	Proteobacteria, Firmicutes	17 distinct tetracycline resistance genes. Two new tet genes: tet(59) (encoding a tetracycline efflux pump) and tet(W/N/W)	Illumina HiSeq 125 bp pair-end sequencing and PacBio SMRT Sequencing	(Leclercq et al., 2016)
2017	Animal manure	Firmicutes, Actinobacteria, Bacteroidetes and Proteobacteria	vanR, tetracycline, fluoroquinolonone resistance genes, APH(3''), msbA, drrA, macB, macA, MFS-1 and emrB	Illumina MiSeq Platform	(Wang et al., 2017)
2019	Dairy Faeces	Enterobacteriaceae, Enterococcaceae, Bacteroidaceae	Tetracycline, Methicillin, Vancomycin and Fosfomycin	Illumina MiSeq	(Liu et al., 2019)
2019	Dairy manure, Faeces	Firmicutes Proteobacteria	Tetracycline, Aminoglycoside, Macrolide	Illumina MiSeq	(Rovira et al., 2019)
2021	Dairy Cattle rumen	NIL	macB, msbA, parY, rpoB2, tetQ and TaeA	MinION, Oxford Nanopore	(López -Catalina et al., 2021)
2022	Cattle, Poultry and Swine	Ruminiclostridium spp., Bacteriodes spp.	blaAIM-1, blaCAM-1, blaGIM-2, and blaHMB-1	Illumina NextSeq 500	(de Carvalho et al., 2022)
2023	Swine	Firmicutes, Bacteroidetes, Proteobacteria and Actinobacteriota	aacA-aphD, ermT, lnuB, strB, sul2, tetA, tetM and blaTEM	Illumina MiSeq	(Tams et al., 2023)

Table 4. Resistome Studies in Environment

Year of Study	Source	Families, Class, Genus and Species of Bacteria found	Resistance Gene found	Sequencing Platform used	Reference
2014	Polluted lake (Indian lake)	Bacteria	The sul2 and qnrD genes	Illumina HiSeq2000	(Bengtsson-Palme et al., 2014)
2015	Paddy soil	Cyanobacteria, Xanthomonadales, Myxococcales and Methylococcales (oxic layer); Clostridia, Actinobacteria, Geobacter, Anaeromyxobacter, Anaerolineae and methanogenic archaea (anoxic zone)	Antibiotic resistance, secondary metabolite production	454 GS Junior system (454 Life Sciences)	(Kim and Liesack, 2015)
2016	Animal faeces, manure and soil samples collected from dairy farms	Bacteroides	Multidrug transporters vancomycin, tetracycline, bacitracin, betalactam and MLS resistance genes	Ion Torrent	(Pitta et al., 2016)
2018	Waste water treatment Systems	Alcaligenes, Paracoccus and Acidovorax	ermF, carA, msrA, sul123, gyrA, grlB	Illumina MiSeq	(Gonzalez-Martinez et al., 2018)
2019	Lettuce	Proteobacteria, Actinobacteria,	sul1, tet(W)	Illumina MiSeq	(Fogler et al., 2019)
2022	River	Limnohabitans, Acidimicrobium, and Candidatus Methylopumilus	uppP, drfE and macB	Illumina HiSeq 4000	(Wang et al., 2022)
2023	Swine Farm	Clostridium, Streptococcus, Acinetobacter and Bacillus	dfrA1, dfrA12, dfrA14, dfrA15, dfrA17, blaTEM1, tetM, mcr1, ermB and sul1	Illumina Novaseq 600C	(Gao et al., 2023)

The Resistome Surveillance and Mapping

As the resistome concept is gaining traction, researchers have started exploring extensively to find out corroborative evidences to give the concept a firm footing. With the advent of the science of metagenomics, several omics tools are being exploited to understand the nitty-gritties of resistome. The bioprospecting of novel resistant genes is made possible through the advancing science of metagenomics and it has given an insight into the resistant gene from the non-culturable microflora apart from the culturable ones. Further, the functional metagenomics has also proved indispensable in resistome surveillance and mapping. The high throughput sequencing techniques have been put to efficient use to decode sequences of resistant genes and have aided in creating resistant gene databases. Studies involving resistome mapping and surveillance have extensively used metagenomics to understand relationship between resistome and environments as well as to track the modes of transmission and the interrelationships between the hosts. Several efforts have been made in this regard. Few of the resistome studies in humans, animals and environment have been enlisted in Tables 2, 3 and 4 respectively.

Conclusion

AMR is a strange phenomenon that is influenced by multifactorial causes. It is clearly evident from the concept of "Resistome" that, cause of AMR should not be attributed to individual pathogenic bugs rather, a larger consortium of microbes i.e., the microbiome has to be held accountable for this man-made menace. The insights gained through advanced metagenomics tools has significantly helped in understanding the phenomenon of AMR from several dimensions and has contributed greatly in unravelling the finer details causing this dreaded pandemic. It is high time that we change our outlook and start viewing AMR through the prism of "Resistome" rather than "Resistance". Further, the global response to mitigate AMR should also realign and repurpose itself by making "Resistome" as a cornerstone, to design surveillance and monitoring mechanisms to tackle AMR.

Acknowledgments

We are grateful to the Director, ICAR-NDRI, Karnal, Haryana and the Vice Chancellor, Central University of Odisha, Koraput, Odisha, for providing the necessary facilities for the preparation of this book chapter.

Conflict of Interest

The authors declare that there is no conflict of interest.

References

Antimicrobial Resistance Collaborators (2022). Global burden of bacterial antimicrobial resistance in 2019: a systematic analysis. *Lancet* (London, England), **399**(10325), 629–655.

Atlas R. M. (2013). One Health: its origins and future. *Current Topics in Microbiology and Immunology*, 365, 1–13.

Baltz R. H. (2008). Renaissance in antibacterial discovery from actinomycetes. *Current opinion in pharmacology*, **8**(5), 557–563.

Bengtsson-Palme[a], J., & Larsson, D. J. (2015). Antibiotic resistance genes in the environment: prioritizing risks. *Nature Reviews Microbiology*, **13**(6), 396.

Bengtsson-Palme[b], J., Angelin, M., Huss, M., Kjellqvist, S., Kristiansson, E., Palmgren, H., Larsson, D. G., & Johansson, A. (2015). The Human Gut Microbiome as a Transporter of Antibiotic Resistance Genes between Continents. *Antimicrobial Agents and Chemotherapy,* **59**(10), 6551–6560.

Bengtsson-Palme, J., Boulund, F., Fick, J., Kristiansson, E., & Larsson, D. G. (2014). Shotgun metagenomics reveals a wide array of antibiotic resistance genes and mobile elements in a polluted lake in India. *Frontiers in Microbiology*, **5**, 648.

Bengtsson-Palme, J., Kristiansson, E., & Larsson, D. G. J. (2018). Environmental factors influencing the development and spread of antibiotic resistance. *FEMS Microbiology Reviews*, **42**(1), fux053.

Bhatt, V. D., Ahir, V. B., Koringa, P. G., Jakhesara, S. J., Rank, D. N., Nauriyal, D. S., Kunjadia P. A & Joshi, C. G. (2012). Milk microbiome signatures of subclinical mastitis-affected cattle analysed by shotgun sequencing. *Journal of Applied Microbiology*, ***112***(4), 639-650.

Burstein, R., Henry, N. J., Collison, M. L., Marczak, L. B., Sligar, A., Watson, S., Marquez, N., Abbasalizad-Farhangi, M., Abbasi, M., Abd-Allah, F., Abdoli, A., Abdollahi, M., Abdollahpour, I., Abdulkader, R. S., Abrigo, M. R. M., Acharya, D., Adebayo, O. M., Adekanmbi, V., Adham, D., Afshari, M., … Hay, S. I. (2019). Mapping 123 million neonatal, infant and child deaths between 2000 and 2017. *Nature*, **574**(7778), 353–358.

Chambers, L., Yang, Y., Littier, H., Ray, P., Zhang, T., Pruden, A., Strickland, M., & Knowlton, K. (2015). Metagenomic Analysis of Antibiotic Resistance Genes in Dairy Cow Feces following Therapeutic Administration of Third Generation Cephalosporin. *PloS one*, **10**(8), e0133764.

Cox, G., & Wright, G. D. (2013). Intrinsic antibiotic resistance: mechanisms, origins, challenges and solutions. *International Journal of Medical Microbiology*, *303*(6-7), 287-292.

de Carvalho, F. M., Valiatti, T. B., Santos, F. F., Silveira, A. C. O., Guimarães, A. P. C., Gerber, A. L., Souza, C. O., Cassu Corsi, D., Brasiliense, D. M., Castelo-Branco, D. S. C. M., Anzai, E. K., Bessa-Neto, F. O., Guedes, G. M. M., de Souza, G. H. A., Lemos, L. N., Ferraz, L. F. C., Bahia, M. N. M., Vaz, M. S. M., da Silva, R. G. B., Veiga, R., … Gales, A. C. (2022). Exploring the Bacteriome and Resistome of Humans and Food-Producing Animals in Brazil. *Microbiology Spectrum*, **10**(5), e0056522.

De Oliveira, D. M. P., Forde, B. M., Kidd, T. J., Harris, P. N. A., Schembri, M. A., Beatson, S. A., Paterson, D. L., & Walker, M. J. (2020). Antimicrobial Resistance in ESKAPE Pathogens. *Clinical Microbiology Reviews*, **33**(3), e00181-19.

Delgado-Baquerizo, M., Hu, H. W., Maestre, F. T., Guerra, C. A., Eisenhauer, N., Eldridge, D. J., Zhu, Y. G., Chen, Q. L., Trivedi, P., Du, S., Makhalanyane, T. P., Verma, J. P., Gozalo, B., Ochoa, V., Asensio, S., Wang, L., Zaady, E., Illán, J. G., Siebe, C., Grebenc, T., … He, J. Z. (2022). The global distribution and environmental drivers of the soil antibiotic resistome. *Microbiome*, **10**(1), 219.

Fogler, K., Guron, G. K., Wind, L. L., Keenum, I. M., Hession, W. C., Krometis, L. A., ... & Pruden, A. (2019). Microbiota and antibiotic resistome of lettuce leaves and radishes grown in soils receiving manure-based amendments derived from antibiotic-treated cows. *Frontiers in Sustainable Food Systems*, *3*, 22.

Gao, F. Z., He, L. Y., Bai, H., He, L. X., Zhang, M., Chen, Z. Y., Liu, Y. S., & Ying, G. G. (2023). Airborne bacterial community and antibiotic resistome in the swine farming environment: Metagenomic insights into livestock relevance, pathogen hosts and public risks. *Environment International*, 172, 107751.

Gilbert, W., Thomas, L. F., Coyne, L., & Rushton, J. (2021). Review: Mitigating the risks posed by intensification in livestock production: the examples of antimicrobial resistance and zoonoses. *Animal : An International Journal of Animal Bioscience*, **15**(2), 100123.

Gillings, M. R. (2013). Evolutionary consequences of antibiotic use for the resistome, mobilome and microbial pangenome. *Frontiers in Microbiology*, *4*, 4.

Gonzalez-Martinez, A., Sihvonen, M., Muñoz-Palazon, B., Rodriguez-Sanchez, A., Mikola, A., & Vahala, R. (2018). Microbial ecology of full-scale wastewater treatment systems in the Polar Arctic Circle: Archaea, Bacteria and Fungi. *Scientific Reports*, **8**(1), 2208.

Hall, B. G., & Barlow, M. (2004). Evolution of the serine β-lactamases: past, present and future. *Drug Resistance Updates*, *7*(2), 111-123.

Hansen, T. A., Joshi, T., Larsen, A. R., Andersen, P. S., Harms, K., Mollerup, S., Willerslev, E., Fuursted, K., Nielsen, L. P., & Hansen, A. J. (2016). Vancomycin gene selection in the microbiome of urban Rattus norvegicus from hospital environment. *Evolution, Medicine, and Public Health*, **2016**(1), 219–226.

Hilton, S. K., Castro-Nallar, E., Pérez-Losada, M., Toma, I., McCaffrey, T. A., Hoffman, E. P., Siegel, M. O., Simon, G. L., Johnson, W. E., & Crandall, K. A. (2016). Metataxonomic and Metagenomic Approaches vs. Culture-Based Techniques for Clinical Pathology. *Frontiers in Microbiology*, **7**, 484.

Jitwasinkul, T., Suriyaphol, P., Tangphatsornruang, S., Hansen, M. A., Hansen, L. H., Sørensen, S. J., Permpikul, C., Rongrungruang, Y., & Tribuddharat, C. (2016). Plasmid metagenomics reveals multiple antibiotic resistance gene classes among the gut microbiomes of hospitalised patients. *Journal of Global Antimicrobial Resistance*, **6**, 57–66.

Kim, Y., & Liesack, W. (2015). Differential assemblage of functional units in paddy soil microbiomes. *PloS One*, **10**(4), e0122221.

Knapp, C. W., Dolfing, J., Ehlert, P. A., & Graham, D. W. (2009). Evidence of increasing antibiotic resistance gene abundances in archived soils since 1940. *Environmental Science & Technology*, **44**(2), 580-587.

Kraemer, S. A., Ramachandran, A., & Perron, G. G. (2019). Antibiotic Pollution in the Environment: From Microbial Ecology to Public Policy. *Microorganisms*, **7**(6), 180.

Leclercq, S. O., Wang, C., Sui, Z., Wu, H., Zhu, B., Deng, Y., & Feng, J. (2016). A multiplayer game: species of Clostridium, Acinetobacter, and Pseudomonas are responsible for the persistence of antibiotic resistance genes in manure-treated soils. *Environmental Microbiology*, **18**(10), 3494-3508.

Liu, J., Taft, D. H., Maldonado-Gomez, M. X., Johnson, D., Treiber, M. L., Lemay, D. G., DePeters, E. J., & Mills, D. A. (2019). The fecal resistome of dairy cattle is associated with diet during nursing. *Nature Communications*, **10**(1), 4406.

López-Catalina, A., Atxaerandio, R., García-Rodríguez, A., Goiri, I., Gutierrez-Rivas, M., Jiménez-Montero, J. A., & González-Recio, O. (2021). Characterisation of the rumen resistome in Spanish dairy cattle. *Animal Microbiome*, **3**(1), 63.

Mackenzie, J. S., & Jeggo, M. (2019). The One Health Approach-Why Is It So Important? *Tropical Medicine and Infectious Disease*, **4**(2), 88.

Mancuso, G., Midiri, A., Gerace, E., & Biondo, C. (2021). Bacterial Antibiotic Resistance: The Most Critical Pathogens. *Pathogens* (Basel, Switzerland), **10**(10), 1310.

Mankhomwa, J., Tolhurst, R., M'biya, E., Chikowe, I., Banda, P., Mussa, J., Mwasikakata, H., Simpson, V., Feasey, N., & MacPherson, E. E. (2022). A Qualitative Study of Antibiotic Use Practices in Intensive Small-Scale Farming in Urban and Peri-Urban Blantyre, Malawi: Implications for Antimicrobial Resistance. *Frontiers in Veterinary Science*, **9**, 876513.

McCann, C. M., Christgen, B., Roberts, J. A., Su, J. Q., Arnold, K. E., Gray, N. D., ... & Graham, D. W. (2019). Understanding drivers of antibiotic resistance genes in High Arctic soil ecosystems. *Environment International*, **125**, 497-504.

Millan, B., Park, H., Hotte, N., Mathieu, O., Burguiere, P., Tompkins, T. A., ... & Madsen, K. L. (2016). Fecal microbial transplants reduce antibiotic-resistant genes in patients with recurrent Clostridium difficile infection. *Clinical Infectious Diseases*, **62**(12), 1479-1486.

Moore, A. M., Patel, S., Forsberg, K. J., Wang, B., Bentley, G., Razia, Y., Qin, X., Tarr, P. I., & Dantas, G. (2013). Pediatric fecal microbiota harbor diverse and novel antibiotic resistance genes. *PloS One*, **8**(11), e78822.

Moradi, J., Fathollahi, M., Halimi, S., Alvandi, A., Abiri, R., Vaziri, S., & Rezaei, A. (2022). Characterization of the resistome in *Lactobacillus* genomic sequences from the human gut. *Journal of Global Antimicrobial Resistance*, 30, 451–458.

Morrison, L., & Zembower, T. R. (2020). Antimicrobial Resistance. *Gastrointestinal Endoscopy Clinics of North America,* **30**(4), 619–635.

Mulchandani, R., Wang, Y., Gilbert, M., & Van Boeckel, T. P. (2023). Global trends in antimicrobial use in food-producing animals: 2020 to 2030. *PLOS Global Public Health*, **3**(2), e0001305.

O'Neill, J. (2016). Review on antimicrobial resistance: tackling drug-resistant infections globally: final report and recommendations.

Perry, J. A., Westman, E. L., & Wright, G. D. (2014). The antibiotic resistome: what's new?. *Current Opinion in Microbiology*, **21**, 45-50.

Pitta, D. W., Dou, Z., Kumar, S., Indugu, N., Toth, J. D., Vecchiarelli, B., & Bhukya, B. (2016). Metagenomic evidence of the prevalence and distribution patterns of antimicrobial resistance genes in dairy agroecosystems. *Foodborne Pathogens and Disease*, **13**(6), 296-302.

Prack McCormick, B., Quiroga, M. P., Álvarez, V. E., Centrón, D., & Tittonell, P. (2023). Antimicrobial resistance dissemination associated with intensive animal production practices in Argentina: A systematic review and meta-analysis. *Revista Argentina de Microbiologia*, **55**(1), 25–42.

Rathinasabapathi, P., Hiremath, D. S., Arunraj, R., & Parani, M. (2015). Molecular Detection of New Delhi Metallo-Beta-Lactamase-1 (NDM-1) Positive Bacteria from Environmental and Drinking Water Samples by Loop Mediated Isothermal Amplification of bla NDM-1. *Indian Journal of Microbiology*, **55**(4), 400–405.

Reddy, B., Singh, K. M., Patel, A. K., Antony, A., Panchasara, H. J., & Joshi, C. G. (2014). Insights into resistome and stress responses genes in Bubalus bubalis rumen through metagenomic analysis. *Molecular Biology Reports*, **41**(10), 6405-6417.

Richter, T. K. S., Michalski, J. M., Zanetti, L., Tennant, S. M., Chen, W. H., & Rasko, D. A. (2018). Responses of the Human Gut Escherichia coli Population to Pathogen and Antibiotic Disturbances. *mSystems*, **3**(4), e00047-18.

Rovira, P., McAllister, T., Lakin, S. M., Cook, S. R., Doster, E., Noyes, N. R., Weinroth, M. D., Yang, X., Parker, J. K., Boucher, C., Booker, C. W., Woerner, D. R., Belk, K. E., & Morley, P. S. (2019). Characterization of the Microbial Resistome in Conventional and "Raised Without Antibiotics" Beef and Dairy Production Systems. *Frontiers in Microbiology*, **10**, 1980.

Sanz, C., Casadoi, M., Tadic, Đ., Pastor-López, E. J., Navarro-Martin, L., Parera, J., Tugues, J., Ortiz, C. A., Bayona, J. M., & Piña, B. (2022). Impact of organic soil amendments in antibiotic levels, antibiotic resistance gene loads, and microbiome composition in corn fields and crops. *Environmental Research*, **214**(Pt 1), 113760.

Schar, D., Zhao, C., Wang, Y., Larsson, D. G. J., Gilbert, M., & Van Boeckel, T. P. (2021). Twenty-year trends in antimicrobial resistance from aquaculture and fisheries in Asia. *Nature Communications*, **12**(1), 5384.

Singh, K. M., Ahir, V. B., Tripathi, A. K., Ramani, U. V., Sajnani, M., Koringa, P. G., Jhakesara, S., Pandya, P. R., Rank, D. N., Murthy, D. S., Kothari, R. K & Joshi, C. G. (2012). Metagenomic analysis of Surti buffalo (Bubalus bubalis) rumen: a preliminary study. *Molecular Biology Reports*, **39**(4), 4841-4848.

Singh, S., Verma, N., & Taneja, N. (2019). The human gut resistome: Current concepts & future prospects. *The Indian Journal of Medical Research*, **150**(4), 345.

Stege, P. B., Hordijk, J., Shetty, S. A., Visser, M., Viveen, M. C., Rogers, M. R. C., Gijsbers, E., Dierikx, C. M., van der Plaats, R. Q. J., van Duijkeren, E., Franz, E., Willems, R. J. L., Fuentes, S., & Paganelli, F. L. (2022). Impact of long-term dietary habits on the human gut resistome in the Dutch population. *Scientific Reports*, **12**(1), 1892.

Sultan, I., Rahman, S., Jan, A. T., Siddiqui, M. T., Mondal, A. H., & Haq, Q. M. R. (2018). Antibiotics, Resistome and Resistance Mechanisms: A Bacterial Perspective. *Frontiers in Microbiology*, **9**, 2066

Syud Amer Ahmed, Enis Barış, Delfin S. Go, Hans Lofgren, Israel Osorio-Rodarte, Karen Thierfelder (2018). Assessing the global poverty effects of antimicrobial resistance. *World Development*, **111**:148-160.

Tams, K. W., Larsen, I., Hansen, J. E., Spiegelhauer, H., Strøm-Hansen, A. D., Rasmussen, S., Ingham, A. C., Kalmar, L., Kean, I. R. L., Angen, Ø., Holmes, M. A., Pedersen, K., Jelsbak, L., Folkesson, A., Larsen, A. R., & Strube, M. L. (2023). The effects of antibiotic use on the dynamics of the microbiome and resistome in pigs. *Animal Microbiome*, **5**(1), 39.

Theophilus, R. J., & Taft, D. H. (2023). Antimicrobial Resistance Genes (ARGs), the Gut Microbiome, and Infant Nutrition. *Nutrients*, **15**(14), 3177.

Tiseo, K., Huber, L., Gilbert, M., Robinson, T. P., & Van Boeckel, T. P. (2020). Global Trends in Antimicrobial Use in Food Animals from 2017 to 2030. *Antibiotics (Basel, Switzerland),* **9**(12), 918.

Van Boeckel, T. P., Glennon, E. E., Chen, D., Gilbert, M., Robinson, T. P., Grenfell, B. T., Levin, S. A., Bonhoeffer, S., & Laxminarayan, R. (2017). Reducing antimicrobial use in food animals. *Science (New York, N.Y.),* **357**(6358), 1350–1352.

Wang, J., Pan, R., Dong, P., Liu, S., Chen, Q., Borthwick, A. G. L., Sun, L., Xu, N., & Ni, J. (2022). Supercarriers of antibiotic resistome in a world's large river. *Microbiome*, **10**(1), 111.

Wang, Y., Zhang, R., Li, J., Wu, Z., Yin, W., Schwarz, S., Tyrrell, J. M., Zheng, Y., Wang, S., Shen, Z., Liu, Z., Liu, J., Lei, L., Li, M., Zhang, Q., Wu, C., Zhang, Q., Wu, Y., Walsh, T. R., & Shen, J. (2017). Comprehensive resistome analysis reveals the prevalence of NDM and MCR-1 in Chinese poultry production. *Nature Microbiology*, **2**, 16260.

Wright G. D. (2007). The antibiotic resistome: the nexus of chemical and genetic diversity. Nature reviews. *Microbiology*, **5**(3), 175–186.

Zaura, E., Brandt, B. W., Teixeira de Mattos, M. J., Buijs, M. J., Caspers, M. P., Rashid, M. U., Weintraub, A., Nord, C. E., Savell, A., Hu, Y., Coates, A. R., Hubank, M., Spratt, D. A., Wilson, M., Keijser, B. J., & Crielaard, W. (2015). Same Exposure but Two Radically Different Responses to Antibiotics: Resilience of the Salivary Microbiome versus Long-Term Microbial Shifts in Feces. *mBio*, **6**(6), e01693-15.

Zeng, J., Pan, Y., Yang, J., Hou, M., Zeng, Z., & Xiong, W. (2019). Metagenomic insights into the distribution of antibiotic resistome between the gut-associated environments and the pristine environments. *Environment International*, **126**, 346-354.

Zhuang, M., Achmon, Y., Cao, Y., Liang, X., Chen, L., Wang, H., Siame, B. A., & Leung, K. Y. (2021). Distribution of antibiotic resistance genes in the environment. *Environmental pollution (Barking) Essex : 1987),* **285**, 117402.

https://www.who.int/news/item/01-12-2021-tripartite-and-unep-support-ohhlep-s-definition-of-one-health

https://www.who.int/news/item/27-02-2017-who-publishes-list-of-bacteria-for-which-new-antibiotics-are-urgently-needed.

Chapter 21

Animal Microbiomes and the One Health Approach

Khumtya Debbarma[1,*]
Mrinmoyee Sarma[2]
Monoshree Sarma[3]
and Das Adwitiya[4]
[1]Department of Veterinary Pharmacology and Toxicology, College of Veterinary Science, Guwahati, Assam, India
[2]Department of Veterinary Public Health and Epidemiology, Institute of Veterinary Science and Animal Husbandry, Siksha o Anusandhan, Bhubaneswar, Odisha, India
[3]Department of Veterinary Pharmacology and Toxicology, Institute of Veterinary Science and Animal Husbandry, Siksha o Anusandhan, Bhubaneswar, Odisha, India
[4]Division of Microbiology, Indian Veterinary Research Institute, Bengaluru, Karnataka, India

Abstract

The intricate relationship between animal microbiomes and their health has gained considerable attention in recent years, particularly within the framework of the One Health approach, which recognizes the interconnectedness of human, animal, and environmental health. Animal microbiomes, comprising diverse communities of microorganisms residing in various body sites, play a crucial role in maintaining host health by influencing immune function, nutrient absorption, and resistance to pathogens. Disruptions in these microbial communities, often caused by environmental changes, antibiotic use, or shifts in diet, can lead to dysbiosis, contributing to the emergence and spread of zoonotic diseases. The One Health perspective emphasizes the need for a holistic understanding of these microbial ecosystems to develop strategies for preventing and controlling diseases that impact both animal and human populations.

Keywords: microbiome, microorganisms, pathogens, resistance, antibiotics

---

* Corresponding Author's Email: khumtyadebbarma999@gmail.com

In: Microbiomes in Animal Nutrition, Health, and Disease
Editor: Tanmoy Rana
ISBN: 979-8-89530-425-9

Introduction

The concept of "One Health" emphasizes the triad relationship of human, animal, and environmental health, acknowledging that the health of each domain is interdependent (Weese, 2015). A critical component of this interrelationship is the microbiome – the vast collection of microorganisms, including bacteria, viruses, fungi, and their genes, that inhabit the bodies of animals and humans, as well as various environments (Chaves et al., 2022). The animal microbiome, in particular, has emerged as a significant area of study due to its profound influence on the health and disease states of both animals and humans (Larsen and Dai, 2015).

Animal microbiomes play a vital role in various physiological processes, such as digestion, immunity, and even behavior. They contribute to the overall well-being of the host and can influence susceptibility to diseases (Weese, 2015). The intricate balance of microbial communities within animals is essential for maintaining health, and disruptions to this balance – whether through disease, diet, environmental changes, or antibiotic use – can have far-reaching consequences (Gilbert and Stephens, 2018). Understanding these microbial communities provides insights into the prevention and treatment of diseases, the enhancement of animal health, and the improvement of food safety and security (Chaves et al., 2022).

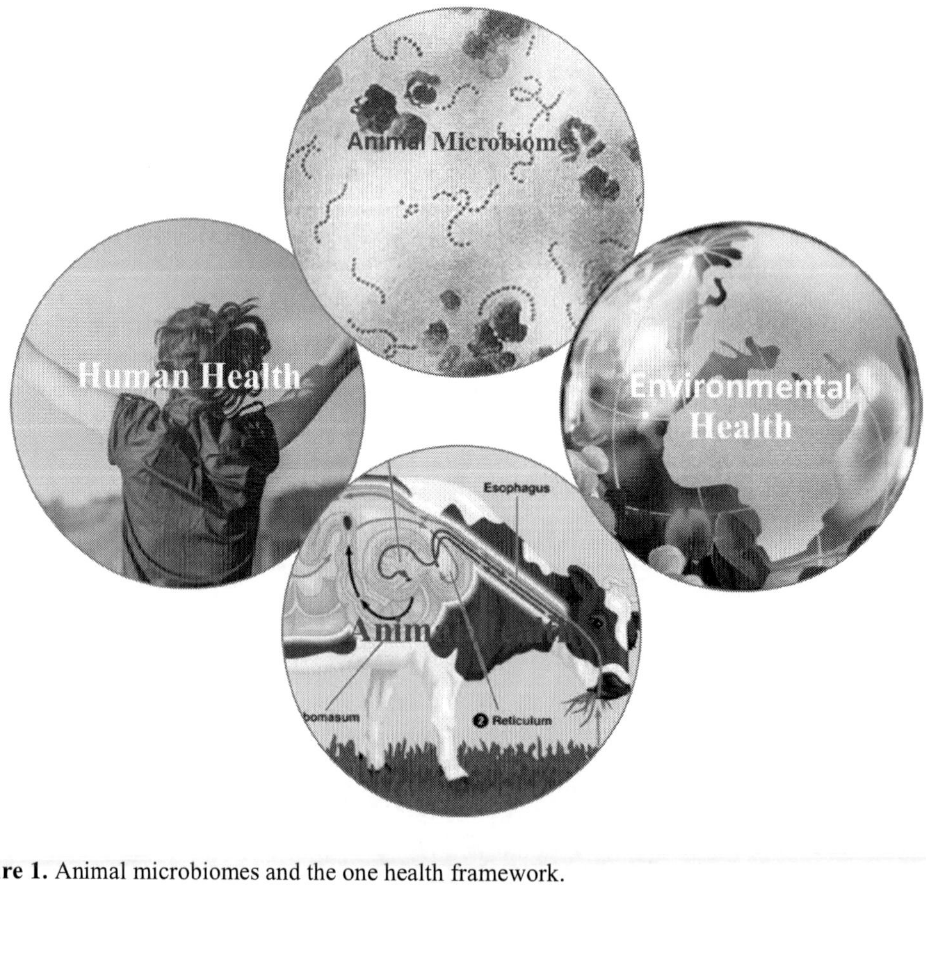

Figure 1. Animal microbiomes and the one health framework.

In the context of One Health, the study of animal microbiomes transcends individual species and ecosystems, highlighting the necessity for a multidisciplinary approach to health (Weese, 2015). For instance, zoonotic diseases, which are diseases transmitted from animals to humans, are often linked to changes or imbalances in the microbiomes of animal reservoirs. Studying these microbiomes can help in identifying potential pathogens before they cause outbreaks, thereby enhancing disease surveillance and prevention strategies (Larsen and Dai, 2015).

Moreover, the environment, including soil, water, and air, serves as both a reservoir and a conduit for microbial communities. Agricultural practices, wildlife interactions, and human activities can all influence the composition and function of these microbiomes (Weese, 2015). For example, the use of antibiotics in livestock not only affects the microbiome of the treated animals but can also lead to the spread of antibiotic-resistant bacteria in the environment, posing a significant public health risk (Gilbert and Stephens, 2018).

This chapter delves into the multifaceted roles of animal microbiomes within the One Health framework (as shown in Figure 1). It explores the composition and functions of these microbiomes, the factors influencing their dynamics, and their implications for health and disease (Young and Wargo, 2018). By integrating knowledge from veterinary science, microbiology, ecology, and public health, this chapter aims to provide a comprehensive understanding of how animal microbiomes can be harnessed to improve health outcomes across species and ecosystems (Chaves et al., 2022).

Understanding Microbiomes

Definition of Microbiomes

Microbiomes refer to the collective genomes of microorganisms that inhabit a particular environment, as well as the community of microbes themselves. In animals, microbiomes are found in various sites, including the gut, skin, respiratory tract, and mucosal surfaces. These microorganisms are involved in numerous physiological processes, including digestion, immunity, and protection against pathogens (Larsen and Dai, 2015).

Relevance to the One Health Framework

Microbiomes play a crucial role in the One Health framework due to their integral impact on health and disease across different domains. In humans, the gut microbiome, for example, is essential for digestion, immunity, and overall health (Larsen and Dai, 2015). Disruptions in the gut microbiome can lead to a range of health issues, including metabolic disorders and inflammatory diseases (Young and Wargo, 2018). Similarly, animal microbiomes influence animal health, productivity, and disease resistance. Environmental microbiomes, including those in soil and water, contribute to ecosystem health and affect the spread of pathogens (Gilbert and Stephens, 2018).

Microbiomes and Animal Diseases

Understanding the impact of microbiomes on both infectious and non-infectious diseases is essential for advancing veterinary medicine and improving animal welfare.

Infectious Diseases

Microbiomes significantly impact the susceptibility and progression of infectious diseases in animals. The dynamic interactions between the host, resident microbiota, and invading pathogens can determine the outcome of infections (Larsen and Dai, 2015).

Pathogen Interactions

Microbiomes provide a first line of defense against pathogenic microbes through mechanisms such as competitive exclusion, production of antimicrobial compounds, and modulation of the host immune response. Disruptions in these microbial communities, often termed dysbiosis, can compromise this protective barrier, allowing pathogens to thrive and cause disease (see Figure 2).

Clostridium difficile Infections

One of the most studied examples of pathogen-microbiome interactions is *Clostridium difficile* infection (CDI) in animals. *C. difficile* is a spore-forming bacterium that can cause severe diarrhea and colitis, particularly following antibiotic treatment which disrupts the gut microbiome (McEwen and Collignon, 2018). Antibiotics reduce the diversity and abundance of beneficial gut bacteria, creating a niche for *C. difficile* to colonize and produce toxins that damage the intestinal lining. Restoring a healthy gut microbiome through fecal microbiota transplantation (FMT) or probiotics has shown promise in treating and preventing CDI in both humans and animals (Chaves et al., 2022).

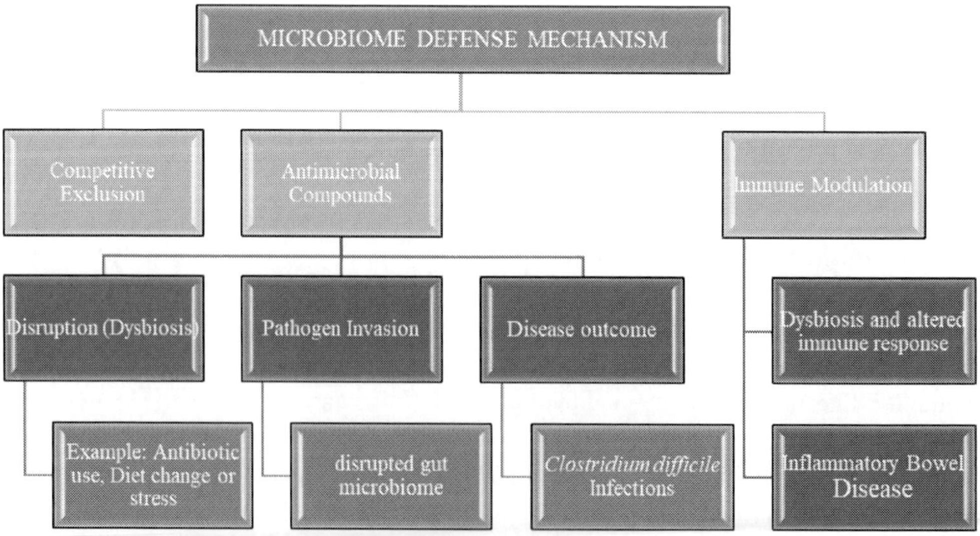

Figure 2. Pathogen interactions and microbiome defense.

Zoonotic Diseases

Zoonotic diseases, which can be transmitted from animals to humans, often involve complex microbiome interactions. The animal microbiome can act as a reservoir for pathogens and antibiotic resistance genes, posing significant public health risks.

Antibiotic Resistant Bacteria

The transmission of antibiotic-resistant bacteria between animals and humans is a growing concern in both veterinary and human medicine. The widespread use of antibiotics in livestock can select for resistant bacteria within the animal microbiome (Gilbert and Stephens, 2018). These resistant bacteria can be transmitted to humans through direct contact, consumption of contaminated food products, or environmental pathways (Young and Wargo, 2018). For instance, methicillin-resistant Staphylococcus aureus (MRSA) has been found in livestock and can cause severe infections in humans. Managing the use of antibiotics and understanding microbiome dynamics are critical for controlling the spread of antibiotic-resistant pathogens (Chaves et al., 2022).

Non-Infectious Diseases

Beyond infectious diseases, the microbiome also influences a range of non-infectious conditions, particularly those involving immune and metabolic functions.

Autoimmune Diseases

The gut microbiome plays a pivotal role in regulating the host immune system. Dysbiosis can disrupt immune homeostasis and contribute to the development of autoimmune diseases, where the immune system mistakenly attacks the body's tissues.

Inflammatory Bowel Disease (IBD)

IBD, which includes Crohn's disease and ulcerative colitis, is characterized by chronic inflammation of the gastrointestinal tract. While the exact cause of IBD is unknown, dysbiosis is thought to trigger abnormal immune responses against intestinal microbes (Larsen and Dai, 2015). Studies have shown that individuals with IBD have reduced microbial diversity and an overrepresentation of pro-inflammatory bacterial species. Similar patterns of dysbiosis have been observed in animals with IBD (shown in Figure 2), such as dogs and cats, suggesting that targeting the microbiome could be a therapeutic strategy.

Rheumatoid Arthritis (RA)

RA is an autoimmune disease that primarily affects joints. Recent research indicates that the gut microbiome may influence the onset and progression of RA. Dysbiosis in the gut can lead to systemic inflammation and the activation of auto-reactive immune cells, which then target joint tissues. Experimental studies in animal models of RA have shown that modulating the gut microbiome with probiotics or dietary interventions can reduce disease severity, highlighting the potential of microbiome-based therapies (Chaves et al., 2022).

Metabolic Disorders

The microbiome's influence on metabolic processes is profound, affecting energy balance, nutrient absorption, and fat storage. Dysbiosis is linked to various metabolic disorders, including obesity, type 2 diabetes, and metabolic syndrome.

Obesity

The gut microbiome contributes to obesity through its effects on energy harvest from the diet and regulation of fat storage. Studies in animals have demonstrated that gut microbiota composition differs between obese and lean individuals. For example, obesity is associated with a higher proportion of Firmicutes and a lower proportion of Bacteroidetes (Larsen and Dai, 2015). Transferring the microbiota from obese animals to germ-free mice can induce weight gain, indicating a causal role of the microbiome in obesity. Dietary interventions, prebiotics, and probiotics that restore a healthy microbiome composition have been shown to promote weight loss and improve metabolic health in animals (Gilbert and Stephens, 2018).

Type 2 Diabetes

Type 2 diabetes is characterized by insulin resistance and impaired glucose metabolism. The gut microbiome influences glucose homeostasis through the production of short-chain fatty acids (SCFAs) and modulation of systemic inflammation. Dysbiosis can reduce SCFA production and increase the permeability of the gut lining, allowing pro-inflammatory substances to enter the bloodstream and exacerbate insulin resistance (Young and Wargo, 2018). Animal studies have shown that microbiome-targeted therapies, such as FMT and SCFA supplementation, can improve glucose tolerance and insulin sensitivity (Chaves et al., 2022).

Types of Animal Microbiomes and Its Impact on Health

Animal microbiomes consist of diverse microbial communities residing in different body sites, each playing a unique role in maintaining health and influencing disease. Here are some key types of animal microbiomes and their impact on health (see Figure 3):

1. *Gut Microbiome:* The gut microbiome is composed of trillions of microorganisms residing in the gastrointestinal tract. The diversity and abundance of these microorganisms can vary significantly between different sections of the gut (e.g., stomach, small intestine, colon).
   Functions:
   - Digestive Health: Gut microbes help in the breakdown of complex carbohydrates, fibres, and other indigestible components of the diet into short-chain fatty acids (SCFAs) and other metabolites that provide energy to the host.
   - Nutrient Synthesis: Certain gut bacteria synthesize essential vitamins (e.g., vitamin K, B vitamins) that the host cannot produce on its own.
   - Immune Regulation: The gut microbiome educates and modulates the host's immune system, helping to prevent excessive inflammatory responses and autoimmune conditions.

- Metabolic Health: It influences host metabolism and energy balance, with implications for conditions such as obesity, diabetes, and metabolic syndrome.

2. *Skin Microbiome:* The skin microbiome consists of bacteria, fungi, viruses, and mites that live on the skin's surface, within hair follicles, and in sweat and sebaceous glands.
   Functions:
   - Barrier Protection: The skin microbiome acts as a first line of defense against pathogenic microorganisms by outcompeting them for resources and producing antimicrobial compounds.
   - Skin Health: It helps maintain the integrity of the skin barrier, preventing conditions like eczema and psoriasis.
   - Immune Interaction: Skin microbes interact with the host's immune system, helping to regulate local immune responses and prevent inflammatory skin disorders.
3. *Respiratory Microbiome:* This microbiome includes microorganisms inhabiting the nasal passages, throat, and lungs. The composition of the respiratory microbiome can change based on factors such as environment, health status, and exposure to pathogens.
   Functions:
   - Infection Defense: A healthy respiratory microbiome helps prevent colonization and infection by pathogenic bacteria and viruses by maintaining a balanced microbial community.
   - Inflammation Regulation: It modulates the host's immune responses to inhaled particles and pathogens, potentially influencing conditions like asthma, chronic obstructive pulmonary disease (COPD), and respiratory infections.
4. *Reproductive Microbiome:* The reproductive microbiome includes microorganisms residing in the reproductive tracts of both male and female animals. This microbiome can vary based on species, hormonal cycles, and environmental factors.
   Functions:
   - Reproductive Health: The reproductive microbiome plays a crucial role in maintaining a healthy reproductive tract by preventing infections and supporting normal reproductive functions.
   - Fertility: It can influence fertility and successful reproduction through its effects on sperm quality, vaginal health, and pregnancy outcomes.
   - Disease Prevention: An imbalance in the reproductive microbiome can lead to conditions such as bacterial vaginosis, sexually transmitted infections (STIs), and infertility.
5. *Oral Microbiome:* The oral microbiome comprises bacteria, viruses, fungi, and protozoa that inhabit the mouth, including the teeth, gums, tongue, and saliva (Chaves et al., 2022).
   Functions:
   - Oral Health: It helps maintain oral hygiene by preventing the colonization of pathogenic microbes that cause dental caries, periodontal disease, and oral infections.
   - Systemic Health: The oral microbiome can influence systemic health by acting as a gateway to the body. For example, oral pathogens can enter the bloodstream and

contribute to systemic conditions like cardiovascular disease, diabetes, and respiratory infections.

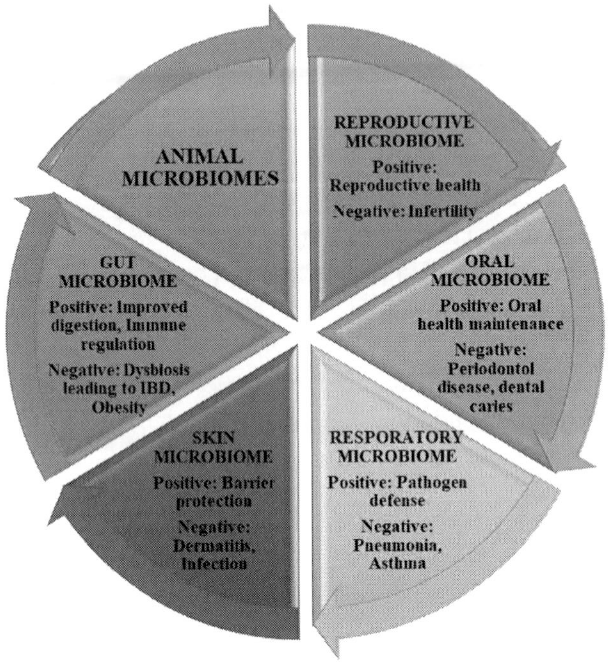

Figure 3. Types of animal microbiomes and its impact on health.

Microbial Diversity

Microbial diversity refers to the variety and abundance of microorganisms present in a given environment. A diverse microbiome is generally considered to be more resilient and capable of maintaining balance, whereas reduced diversity can be associated with various health issues. For example, a diverse gut microbiome is crucial for effective digestion and immune system function (Gilbert and Stephens, 2018).

One Health Approach

Interconnection of Human, Animal, and Environmental Health

The One Health approach emphasizes the interconnectedness of human, animal, and environmental health. Understanding how microbiomes bridge these domains is essential for effective health management (Figure 4).

- *Shared Microbiomes*: Humans and animals can share microbial communities through direct contact or environmental exposures. This sharing can facilitate the transmission of diseases and antibiotic-resistant bacteria.

- *Environmental Impact*: Environmental factors such as pollution, climate change, and habitat destruction can affect microbiomes. For example, changes in the environment can alter the composition of animal microbiomes, influencing their health and the health of humans who come into contact with them.

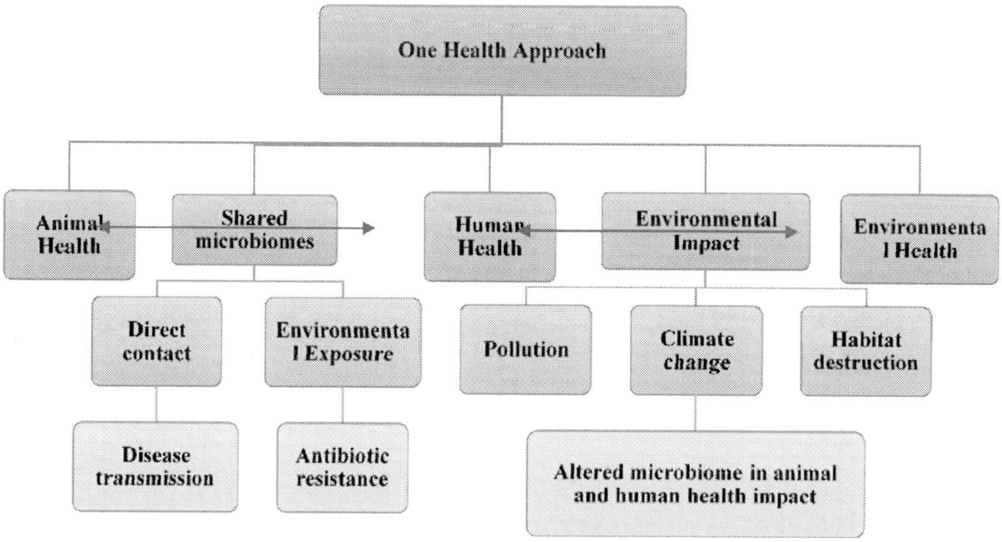

Figure 4. One health approach showing how microbiomes bridges and impact health outcomes

Case Studies of Animal Microbiomes and One Health

Dairy Cow Microbiome and Mastitis Prevention

One notable case study examined the role of the microbiome in dairy cows, particularly in the context of mastitis, a common and costly infection of the udder. Researchers found that specific microbial communities in the udder and on the skin of cows could predict susceptibility to mastitis (Larsen and Dai, 2015). By manipulating these microbial communities through probiotics and improved hygiene practices, the incidence of mastitis significantly decreased. This approach not only improved animal health and welfare but also reduced the need for antibiotics, highlighting the interconnected benefits to animal and human health through the One Health framework (Chaves et al., 2022).

Poultry Gut Microbiome and Food Safety

Another significant case study involves the gut microbiome of poultry and its implications for food safety. Researchers discovered that specific strains of beneficial bacteria in the poultry gut could inhibit the colonization of harmful pathogens such as Salmonella and Campylobacter. By administering these probiotics to poultry, the prevalence of these pathogens in the birds and, consequently, in the meat consumed by humans, was reduced. This study underscores the importance of managing animal microbiomes to enhance food safety and public health, aligning with One Health principles (Gilbert and Stephens, 2018).

Canine Microbiome and Zoonotic Disease Control

In the realm of companion animals, a case study on the canine microbiome provided insights into zoonotic disease control. Dogs can harbor zoonotic pathogens that pose risks to human health (McEwen and Collignon, 2018). Researchers investigated the impact of diet on the canine gut microbiome and its role in the shedding of zoonotic bacteria like Campylobacter. They found that specific dietary interventions could alter the gut microbiome, reducing the shedding of harmful bacteria (Larsen and Dai, 2015). This approach not only improved canine health but also mitigated risks to human health, demonstrating the One Health approach's potential in controlling zoonotic diseases (Chaves et al., 2022).

Wildlife Microbiome and Ecosystem Health

A fascinating case study focused on the microbiome of wildlife species and its impact on ecosystem health. In African elephants, researchers studied the gut microbiome's role in nutrient absorption and overall health. Changes in the microbiome due to habitat loss and altered diets were linked to poor health and reduced reproductive success. Conservation efforts that included restoring natural diets and habitats helped restore the natural microbiome, improving the health and population stability of elephants. This study illustrates the broader environmental and ecological implications of microbiome research and its relevance to One Health by maintaining healthy wildlife populations and ecosystems (McEwen and Collignon, 2018).

Aquatic Animal Microbiome and Aquaculture Sustainability

In aquaculture, a case study explored the microbiome of farmed fish and its implications for sustainable practices. Researchers found that the microbiome of fish could be manipulated to enhance growth, disease resistance, and feed efficiency. By optimizing the microbial communities through probiotics and prebiotics, fish health and productivity improved, reducing the need for antibiotics and minimizing environmental impacts (McEwen and Collignon, 2018). This case highlights how microbiome management in aquaculture can contribute to animal health, human food security, and environmental sustainability, reinforcing the interconnected goals of the One Health approach (Gilbert and Stephens, 2018).

These case studies collectively demonstrate the critical role of microbiomes in animal health and their broader implications for human health and environmental sustainability. They emphasize the necessity of integrated approaches in managing microbiomes to achieve the holistic health goals promoted by the One Health framework (Chaves et al., 2022).

Conclusion

Animal microbiomes are a critical component of the One Health framework, significantly impacting the health of animals, humans, and the environment. Continued research and interdisciplinary collaboration are essential for advancing our understanding of microbiomes and improving health outcomes. Embracing the One Health approach will enable more effective strategies for managing and enhancing health across all domains. By recognizing and addressing the interconnectedness of microbial communities across different species and ecosystems, we can develop comprehensive interventions that promote the well-being of all organisms and the environments they inhabit.

References

Chaves, F. L., Gomes, N. C. M., & Silva, R. M. P. (2022). The interplay between microbiomes and infectious diseases: Perspectives for the One Health approach. *Frontiers in Microbiology, 13*, 841084. https://doi.org/10.3389/fmicb.2022.841084.

Gilbert, J. A., & Stephens, B. (2018). Microbiology of the built environment. *Nature Reviews Microbiology, 16*(11), 661-670. https://doi.org/10.1038/s41579-018-0065-5.

Larsen, P. E., & Dai, Y. (2015). One Health: Interdependencies of human, animal, and ecosystem health. *Environmental Microbiology Reports, 7*(1), 97-104. https://doi.org/10.1111/1758-2229.12241.

McEwen, S. A., & Collignon, P. J. (2018). Antimicrobial resistance: A One Health perspective. *Microbiology Spectrum, 6*(2). https://doi.org/10.1128/microbiolspec.ARBA-0009-2017.

Weese, J. S. (2015). The canine and feline intestinal microbiota and the One Health perspective. *Journal of Veterinary Internal Medicine, 29*(1), 32-42. https://doi.org/10.1111/jvim.12543.

Young, V. B., & Wargo, J. A. (2018). Modulation of the gut microbiota by diet, probiotics, and prebiotics. *Gastroenterology, 154*(2), 337-349. https://doi.org/10.1053/j.gastro.2017.06.056.

Chapter 22

Bioinformatics in Microbiomes

Felix Uchenna Samuel[1,*]
Maureen Chioma Samuel[2]
Clement Akotsa-Mensah[3]
Ama Adadzewa Eshun[4]
Emmanual Okoh[4]
and Masur Dandalla Salisu[5]

[1]Animal Science Program, Alabama Cooperative Extension System, Alabama A&M University, Normal, USA
[2]Crestwood Medical Centre, Huntsville AL, USA
[3]Integrated Pest Managament Program, Alabama Cooperative Extension System, Alabama A&M University, Normal, USA
[4]Department of Food and Animal Science, Alabama A&M University, Normal, USA
[5]Department of Plant and soil Science, Alabama A&M University, Normal, USA
[6]Artificial Insemination Unit, National Animal Production Research Institute Shika-Zaria, Nigeria

Abstract

Bioinformatics plays a crucial role in advancing our understanding of microbiomes in animal health and disease by providing sophisticated tools and methods for analyzing complex microbial communities. The application of bioinformatics in this field involves several key areas, including sequencing technologies, data processing, and functional analysis. High-throughput sequencing methods such as 16S rRNA gene sequencing, metagenomics, and metatranscriptomics generate extensive data on microbial diversity, function, and gene expression. Bioinformatics pipelines process these large datasets, performing tasks such as quality control, sequence alignment, taxonomic classification, and functional annotation. In animal health, bioinformatics helps elucidate the role of the microbiome in maintaining health, identifying pathogenic microbes, and understanding disease mechanisms. For instance, analyzing microbiome profiles can reveal microbial imbalances associated with conditions like gastrointestinal disorders, respiratory diseases, and metabolic disorders. Furthermore, bioinformatics tools enable the exploration of microbial interactions and their impact on host immunity and disease susceptibility.

* Corresponding Author's Email: felix.samuel@aamu.edu

In: Microbiomes in Animal Nutrition, Health, and Disease
Editor: Tanmoy Rana
ISBN: 979-8-89530-425-9

Challenges in microbiome bioinformatics include managing data complexity, ensuring accurate taxonomic and functional annotation, and interpreting dynamic microbial changes. Despite these challenges, advancements in computational methods and sequencing technologies continue to enhance our ability to link microbial profiles with animal health outcomes. Overall, bioinformatics is integral to leveraging microbiome research for improving animal health, developing diagnostic tools, and creating targeted therapeutic strategies.

Keywords: microbiome analysis, bioinformatics tools, sequencing technologies, animal health, disease mechanisms

Introduction

One of the most important areas of biological research at the moment is the study of microbiomes, namely, complex communities of microorganisms inhabiting different environments. Microbiomes can be found in diverse habitats, such as the human/animal gut, soils, oceans, and even extreme environments like deep-sea hydrothermal vents. In this way, microbial communities play a very important role in the functioning of ecosystems and hosts, as well as in health and disease. In fact, knowledge of the microbiomes has grown tremendously in the last couple of years due to next-generation sequencing techniques and bioinformatics tools that have helped analyze and make sense of large volumes of data produced by microbiome studies. This comprehensive review aims at exploring the key bioinformatics tools, sequencing technologies, and data analysis methods that are pivotal in microbiome research (Schneider, 2024). This has really revolutionized the study of microbiomes, particularly in relation to the next-generation sequencing technologies, which make it possible to sequence genetic material directly from a whole microbial community in environmental samples, unlike the traditional ways, which require cultivation of the organisms in the laboratory. With this new discovery, a great diversity of microorganisms has been observed, most of which were unknown before and are not culturable. However, such sequencing efforts produce high volumes of data, which then require sophisticated computational tools and methods to process, analyze, and interpret the data for meaningful answers. This has been the case with the development and use of various bioinformatics tools in microbiome research. Bioinformatics provides core tools in the study of microbiomes, able to deal with large scale data, run complex statistical analyses, and present results in meaningful ways. These tools can help answer basic questions regarding the composition, function, and dynamics of microbial communities, as well as their interactions with the host or environment. Because of the power of bioinformatics, it is now possible for researchers to recognize microbes, identify what they are capable of doing functionally, and learn about the variations of these communities as a consequence of different effects, such as dietary changes, diseases, and environmental variations. This literature focuses on key bioinformatics tools used in microbiome research, including QIIME, Mothur, MEGAN, DADA2, and Phyloseq. All these tools are specially enabled in the processing and analyzing of microbiome data from quality control of sequences to taxonomic classification, diversity analysis, and data visualization. This review will consider, in greater detail, sequencing technologies commonly applied to microbiome studies, such as 16S rRNA gene sequencing, metagenomics, and metatranscriptomics, together with the associated bioinformatics pipelines.

Finally, it will present statistical and computational approaches applied in the analysis of microbiome data, including methods for analyzing diversity, community structure, and network analysis—all critical for deriving actionable insights from microbiome studies (Nikoloudaki et al., 2024).

Important Bioinformatic Tools in Microbiome Studies

The analysis of microbiome data requires a diverse set of bioinformatics tools, each designed to address specific challenges associated with the processing and interpretation of high-throughput sequencing data. These tools have become essential for tasks ranging from quality control and sequence alignment to taxonomic classification, functional annotation, and visualization (Pinto & Bhatt, 2024; Zhao et al., 2024). Below, we discuss some of the most widely used bioinformatics tools in microbiome research.

QIIME (Quantitative Insights into Microbial Ecology)

QIIME is one of the most widely used bioinformatics pipelines for analyzing 16S rRNA gene sequences, a common method for profiling bacterial and archaeal communities. Developed by Caporaso et al. in 2010, QIIME was designed to facilitate the analysis of microbial communities by providing a comprehensive suite of tools for processing raw sequencing data, performing taxonomic classification, and conducting diversity analyses. One of the key features of QIIME is its ability to handle large datasets, making it suitable for studies involving hundreds or thousands of samples. The pipeline supports various steps in the analysis workflow, including quality filtering, OTU (Operational Taxonomic Unit) clustering, phylogenetic tree construction, and the calculation of alpha and beta diversity metrics. QIIME also integrates with several external databases, such as SILVA and Greengenes, for accurate taxonomic assignment. QIIME's flexibility and modular design allow researchers to customize their analysis pipelines according to the specific needs of their studies. For example, users can choose between different OTU clustering algorithms, such as UCLUST, VSEARCH, or DADA2, depending on their preferences for speed and accuracy. Additionally, QIIME provides tools for visualizing the results of microbial community analyses, including bar plots, heatmaps, and ordination plots, which help researchers interpret their findings and communicate them effectively (Figures 1 and 2).

Mothur

Mothur is another versatile and widely used platform for analyzing microbial sequence data, particularly 16S rRNA gene sequences. Developed by Schloss et al. in 2009, Mothur was designed to provide a comprehensive suite of tools for microbial ecology studies, with a focus on reproducibility and transparency. Like QIIME, Mothur supports a wide range of analyses, from quality control and sequence alignment to taxonomic classification and diversity analysis. One of Mothur's strengths is its robust suite of tools for sequence alignment and clustering.

Mothur uses the Needleman-Wunsch algorithm for aligning sequences to a reference database, ensuring high accuracy in the identification of OTUs. Additionally, Mothur offers several clustering algorithms, including hierarchical clustering, nearest-neighbor clustering, and average-neighbor clustering, allowing researchers to choose the method that best suits their data. Mothur also provides tools for conducting phylogenetic analyses, which are essential for understanding the evolutionary relationships between microorganisms. Researchers can use Mothur to construct phylogenetic trees, calculate phylogenetic diversity metrics, and perform comparative analyses between microbial communities. Furthermore, Mothur's ability to process large datasets makes it suitable for studies involving complex environmental samples, such as soil, water, and gut microbiomes. One of the key advantages of Mothur is its focus on reproducibility. The platform encourages users to create standardized workflows that can be easily shared and replicated by other researchers. This emphasis on transparency and reproducibility has made Mothur a popular choice for microbiome studies, particularly those involving large-scale surveys of microbial diversity (Figure 2).

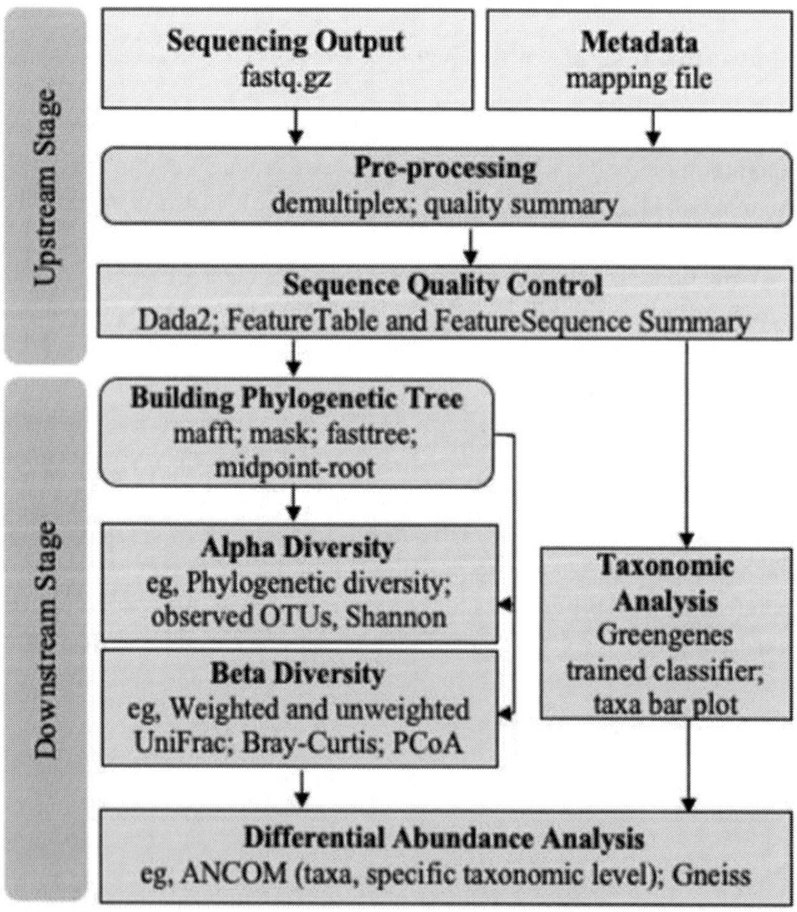

Figure 1. Quantitative Insights Into Microbial Ecology workflow.

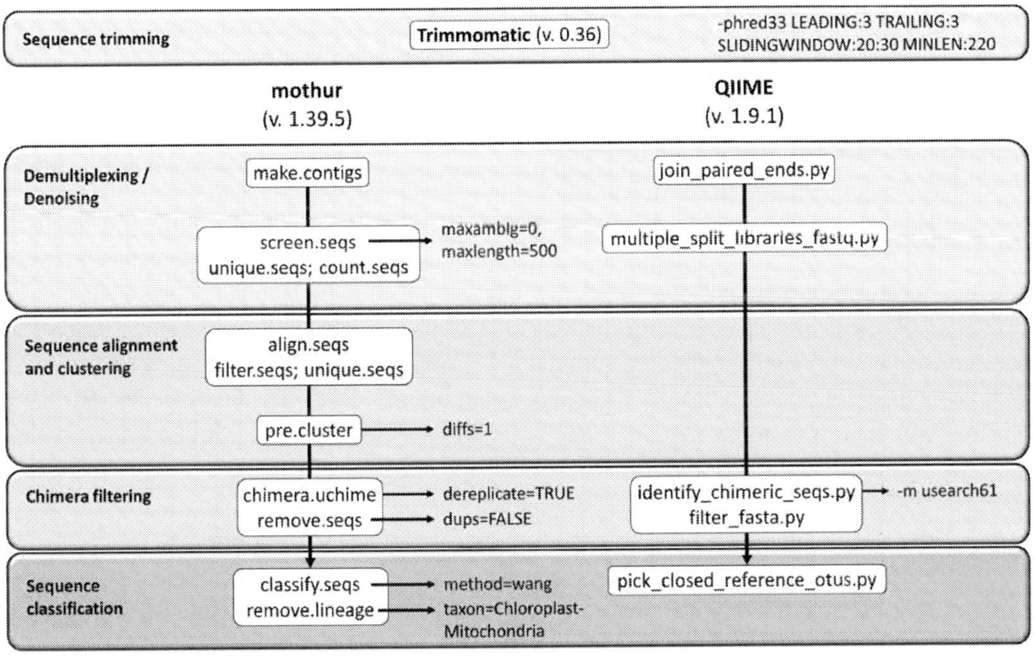

Figure 2. Mothur and Quantitative Insights Into Microbial Ecology comparism.

MEGAN (MEtaGenome ANalyzer)

MEGAN is a bioinformatics tool specifically designed for the analysis of metagenomic data, which involves the sequencing of all genetic material within a sample. Developed by Huson et al. in 2007, MEGAN allows researchers to explore the taxonomic and functional composition of microbial communities by assigning sequences to known taxa and genes. One of the key features of MEGAN is its ability to visualize the results of metagenomic analyses in an interactive and intuitive manner. The software provides several visualization tools, including taxonomic trees, heatmaps, and bubble plots, which allow researchers to explore the composition and functional potential of microbial communities. MEGAN also supports the integration of multiple data types, such as metagenomic, metatranscriptomic, and metaproteomic data, providing a comprehensive view of the microbiome. MEGAN's taxonomic classification algorithm, known as the Lowest Common Ancestor (LCA) algorithm, assigns reads to the lowest common ancestor of a set of reference sequences, ensuring accurate and reliable taxonomic classification. This approach is particularly useful for metagenomic studies, where sequences may originate from a wide range of taxa, including those that are poorly represented in reference databases. In addition to taxonomic classification, MEGAN provides tools for functional annotation, allowing researchers to assign sequences to known genes and pathways. This functionality is essential for understanding the metabolic potential of microbial communities and identifying key functional genes that may be involved in processes such as nutrient cycling, antibiotic resistance, and pathogenesis.

DADA2

DADA2 is a bioinformatics tool specifically designed for denoising 16S rRNA sequence data, allowing for the identification of exact sequence variants (ESVs) rather than OTUs. Developed by Callahan et al. in 2016, DADA2 offers a higher resolution of microbial diversity by distinguishing between closely related sequences that may differ by only a single nucleotide. One of the key advantages of DADA2 is its ability to correct sequencing errors, which are common in high-throughput sequencing data. The algorithm models the error rates associated with each base call and uses this information to infer the true biological sequences present in the sample. This approach reduces the number of spurious sequences and increases the accuracy of taxonomic classification. DADA2's denoising algorithm allows for the detection of ESVs, which are exact representations of the sequences present in the sample. Unlike OTUs, which group sequences based on a predefined similarity threshold (typically 97%), ESVs provide a more precise measure of microbial diversity. This increased resolution is particularly useful for studies involving complex microbial communities, where small differences in sequence composition can have significant ecological implications.

DADA2 is integrated with several other bioinformatics tools, including QIIME and Phyloseq, allowing for seamless integration into existing analysis pipelines. The tool also supports various downstream analyses, including diversity analysis, differential abundance testing, and phylogenetic tree construction, making it a powerful tool for microbiome research.

Phyloseq

Phyloseq is an R-based package designed for the comprehensive statistical analysis and visualization of microbiome data. Developed by McMurdie and Holmes in 2013, Phyloseq provides a unified framework for integrating microbiome data into a single object, facilitating downstream analyses and visualization. One of the key features of Phyloseq is its ability to import data from various bioinformatics pipelines, including QIIME, Mothur, and DADA2. This flexibility allows researchers to use Phyloseq as a central hub for their microbiome analyses, regardless of the tools used for initial sequence processing. Once the data is imported, Phyloseq provides a wide range of functions for analyzing and visualizing microbial communities. Phyloseq supports various diversity analyses, including alpha and beta diversity metrics, which measure the richness and evenness of microbial communities within and between samples. The package also provides tools for ordination and clustering, allowing researchers to explore the relationships between microbial communities across samples. Phyloseq's visualization capabilities include bar plots, heatmaps, ordination plots, and phylogenetic trees, providing a comprehensive view of the microbiome. In addition to diversity analysis, Phyloseq supports differential abundance testing, which allows researchers to identify taxa that differ significantly between conditions or groups. This functionality is essential for understanding the ecological and functional implications of microbial communities and identifying key taxa that may be associated with health or disease. Phyloseq's integration with other R packages, such as ggplot2 and vegan, allows for extensive customization of plots and analyses, making it a versatile tool for microbiome research. The package's modular design and user-friendly interface have made it a popular choice for researchers seeking to perform in-depth statistical analyses of microbiome data (Figure 3).

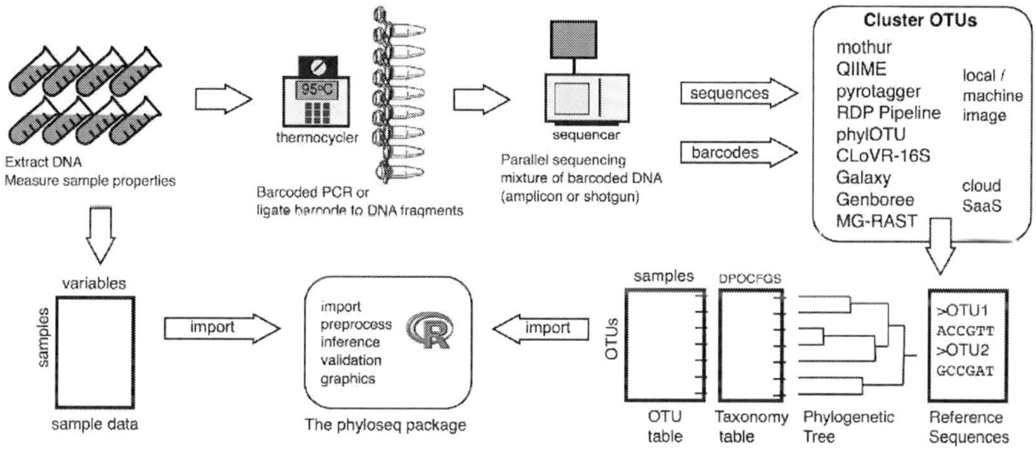

Figure 3. Phyloseq workflow.

Sequencing Techniques and Bioinformatics Processing Frameworks

In the past years, the study of the microbiome has undergone tremendous evolution, driven in part by the rapid development in sequencing technologies and bioinformatics tools. Such technologies currently allow investigators to investigate, with unprecedented resolution, the complexity of microbial communities that underlie animal health and disease. Sequencing technologies like 16S rRNA gene sequencing, metagenomics, and metatranscriptomics all yield information that characterizes different aspects of microbial communities and their functional dynamics (Kumar & Yadav, 2024; Rao & Sunkad, 2024; Tilgam et al., 2024; Mota et al., 2024).

16S rRNA Gene Sequencing

16S rRNA gene sequencing is a cornerstone of microbiome research, particularly in the study of bacterial and archaeal communities. This technique targets the 16S ribosomal RNA gene, which is highly conserved across bacteria and archaea but contains hypervariable regions that provide species-specific signatures. By sequencing these hypervariable regions, researchers can identify and classify microorganisms within a sample (Figure 4).

Sequencing Technology

The most commonly used platforms for 16S rRNA sequencing include Illumina's MiSeq and HiSeq systems, which provide high-throughput sequencing capabilities with high accuracy and relatively low cost. Other platforms, such as the Ion Torrent and PacBio, offer alternative approaches with varying read lengths and error profiles.

Illumina Sequencing

Illumina sequencing, also known as sequencing by synthesis (SBS), is a high-throughput sequencing technology that has transformed genomic research with its ability to generate vast amounts of data quickly and accurately. Since its introduction, Illumina sequencing has become the dominant platform in genomics, widely used for a range of applications including whole-genome sequencing, targeted resequencing, RNA sequencing, and microbiome analysis.

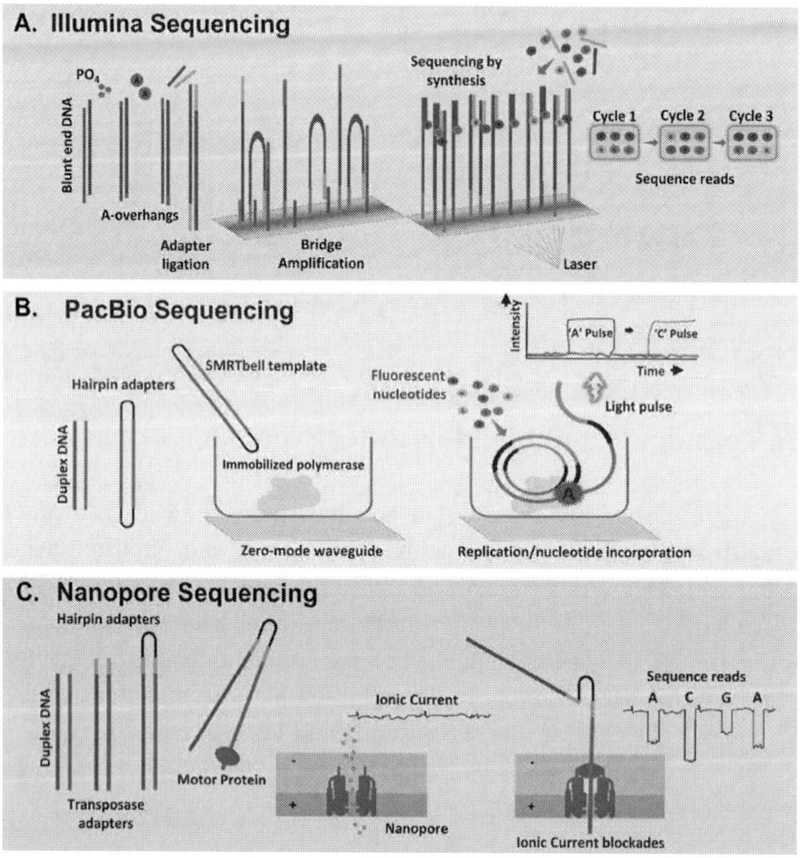

Figure 4. Different kinds of 16S rRNA Gene Sequencing.

Principles of Sequencing by Synthesis (SBS)

The core principle behind Illumina sequencing is sequencing by synthesis (SBS). This method involves the following steps:

- *Library Preparation*: The process begins with the preparation of a DNA or RNA library. The nucleic acids are fragmented into smaller pieces, and specific adapters are added to both ends of each fragment. These adapters are essential for the sequencing process as they bind the fragments to the sequencing flow cell and facilitate the sequencing reactions.
- *Cluster Generation*: The prepared library is then loaded onto an Illumina flow cell. The flow cell is a glass surface coated with oligonucleotides complementary to the adapters. Each DNA fragment on the flow cell undergoes bridge amplification, where it is clonally amplified to form clusters of identical DNA fragments. This amplification process results in a dense array of clusters, each representing a different DNA fragment.

- *Sequencing:* During the sequencing phase, fluorescently labeled nucleotides are added one at a time to the growing DNA strand. Each nucleotide addition is accompanied by the emission of a fluorescent signal, which is captured by a camera. This process is repeated for each nucleotide, generating a sequence of fluorescent signals that correspond to the DNA sequence.
- *Data Acquisition and Analysis*: The emitted fluorescence is detected and recorded, producing digital data that represents the sequence of the DNA fragments. The data is then processed and analyzed using bioinformatics tools to reconstruct the original DNA sequence, identify variants, and perform other analyses.

Advantages of Illumina Sequencing

Illumina sequencing offers several advantages that make it a preferred choice for many applications:

- *High Throughput*: Illumina platforms are capable of generating millions of sequences in a single run. This high throughput enables researchers to perform large-scale studies, such as whole-genome sequencing and comprehensive microbiome analyses, with a high level of detail and statistical power.
- *Accuracy:* Illumina sequencing is known for its high accuracy and low error rates. The technology's ability to detect single nucleotide changes with precision makes it suitable for applications requiring high-resolution data, such as variant discovery and transcriptome analysis.
- *Cost-Effectiveness*: Over time, the cost per base of sequencing has decreased significantly with Illumina technology. This cost reduction has made high-throughput sequencing accessible to a broader range of research projects and applications.
- *Versatility:* Illumina platforms can be used for a wide variety of sequencing applications, including whole-genome sequencing, targeted resequencing, RNA sequencing, and epigenomics. This versatility makes Illumina technology a valuable tool for diverse research needs.
- *Scalability:* Illumina offers a range of sequencing platforms with varying capacities, from benchtop sequencers for smaller studies to high-throughput sequencers for large-scale projects. This scalability allows researchers to choose a platform that best fits their specific requirements and budget.

Challenges of Illumina Sequencing Technology

Despite its many advantages, Illumina sequencing technology also faces some challenges:

- *Short Read Lengths*: Illumina sequencing typically generates short reads (up to 300 bp), which can complicate the assembly of complex genomes and the resolution of structural variants. Long-read technologies, such as PacBio and Oxford Nanopore, can complement Illumina sequencing to address this limitation.
- *Data Management*: The high throughput of Illumina sequencing generates large volumes of data, requiring robust data management and storage solutions. Efficient data processing pipelines and high-performance computing resources are essential for handling and analyzing the data.

- *Error Correction*: While Illumina sequencing is highly accurate, occasional errors can occur, particularly in homopolymeric regions. Error correction tools and algorithms are used to improve the accuracy of the sequencing data.

Ion Torrent Sequencing

Ion Torrent sequencing represents a significant advancement in the field of high-throughput sequencing technologies, offering a unique approach based on semiconductor technology. Unlike traditional sequencing methods that rely on optical detection of fluorescently labeled nucleotides, Ion Torrent sequencing leverages semiconductor chips to detect changes in pH associated with nucleotide incorporation. This technology provides several advantages, including rapid turnaround times, lower costs, and the ability to produce high-throughput sequencing data suitable for various applications, particularly those requiring short-read lengths

Principles of Ion Torrent Sequencing

Ion Torrent sequencing is based on a semiconductor-based detection system that measures changes in pH during nucleotide incorporation. The fundamental principle involves detecting the release of protons (H^+ ions) when nucleotides are added to a growing DNA strand. Each nucleotide incorporation event causes a slight change in the local pH, which is detected by a semiconductor chip.

Semiconductor Sequencing Technology

The core of Ion Torrent sequencing technology is the semiconductor chip, which contains millions of individual wells, each capable of holding a single DNA molecule. The sequencing process involves the following key steps:

- *Library Preparation:* DNA samples are first fragmented into smaller pieces, and adapter sequences are added to the ends of the fragments. These adapters are essential for attaching the DNA fragments to the semiconductor chip and for enabling the sequencing reactions.
- *Template Amplification*: The prepared DNA library is then loaded onto the semiconductor chip. Each DNA fragment is clonally amplified within a well, creating a cluster of identical DNA molecules. This amplification step ensures that there is sufficient signal for detection during sequencing.
- *Sequencing Reactions*: The sequencing process involves cyclically adding nucleotides (A, T, C, G) to the DNA template. As each nucleotide is incorporated into the growing DNA strand, a proton is released, causing a local decrease in pH. The semiconductor chip detects these pH changes and records the nucleotide sequence.
- *Data Analysis:* The detected pH changes are converted into nucleotide sequences using bioinformatics tools. The sequencing data is then analyzed to determine the composition of the DNA sample and to identify any genetic variations.

Advantages of Ion Torrent Sequencing

Ion Torrent sequencing offers several advantages over traditional sequencing technologies:

- *Rapid Turnaround Times*: The semiconductor-based detection system allows for faster sequencing compared to optical-based methods. This rapid turnaround is particularly useful for applications requiring quick results, such as clinical diagnostics and real-time monitoring of microbial communities.
- *Cost-Effectiveness*: The use of semiconductor chips and the absence of fluorescent labeling reduce the overall cost of sequencing. This cost-effectiveness makes Ion Torrent sequencing accessible for a wide range of research and clinical applications.
- *High Throughput*: Ion Torrent technology provides high-throughput sequencing capabilities, enabling the generation of large volumes of data in a single run. This high throughput is beneficial for studies involving large sample sizes or complex genomic analyses.
- *Short-Read Lengths*: Ion Torrent sequencing is well-suited for applications that require short-read lengths, such as targeted sequencing and amplicon sequencing. The technology provides accurate and reliable data for these applications, allowing for detailed analysis of specific genomic regions.

Challenges of Ion Torrent Sequencing in Microbiome Research

Ion Torrent sequencing, while innovative and advantageous in many aspects, also presents several challenges when applied to microbiome research. These challenges can impact the accuracy, reliability, and interpretability of sequencing data. Understanding these limitations is crucial for optimizing the use of Ion Torrent technology in microbiome studies and improving the quality of the results obtained. Below are some key challenges associated with Ion Torrent sequencing in microbiome research:

Short Read Lengths

Limited Read Length

One of the primary challenges of Ion Torrent sequencing is its typically short read lengths. Ion Torrent technology generally produces reads of around 200-400 base pairs. This limitation can pose several issues:

- *Reduced Assembly Accuracy*: Short reads can complicate the assembly of complex microbial genomes, particularly in highly diverse samples where long contigs are necessary to accurately reconstruct genomic sequences. This challenge is especially significant in microbiome studies where the microbial community is often highly diverse and includes species with highly similar genomes.
- *Limited Resolution for Long Reads*: Short reads may not capture long-range genomic features, such as structural variations or large insertions/deletions, which can be important for understanding microbial functions and interactions.
- *Difficulty in Resolving Complex Regions*: Short reads can struggle to resolve complex or repetitive regions within microbial genomes. In microbiome research, this can lead to difficulties in accurately identifying and characterizing microbial species, especially those with highly similar or repetitive sequences.

Insertion/Deletion Errors

Ion Torrent sequencing has been noted for its error profile, particularly in the form of insertion and deletion (indel) errors. These errors arise during nucleotide incorporation and can affect the accuracy of the sequencing data:

- *Impact on Variant Calling*: Indel errors can complicate the identification of genetic variants, such as single nucleotide polymorphisms (SNPs) and structural variations. Accurate variant calling is essential for understanding microbial diversity and function.
- *Challenges in Alignment*: Indel errors can introduce discrepancies in read alignment, leading to misalignment of sequences and affecting the overall accuracy of microbial community profiling.

Lower Accuracy Compared to Other Platforms

While Ion Torrent sequencing is cost-effective and provides rapid results, its accuracy in base calling is generally lower compared to platforms like Illumina. This reduced accuracy can impact the reliability of microbiome data and may necessitate additional quality control steps to mitigate errors.

Data Processing Demands

The complexity of microbiome data generated by Ion Torrent sequencing can be substantial. The challenges include:

- *Data Management:* Managing large volumes of sequencing data requires substantial computational resources and storage capacity. Handling these data sets efficiently is crucial for accurate downstream analysis.
- *Bioinformatics Challenges*: The data generated by Ion Torrent sequencing requires sophisticated bioinformatics tools and pipelines for processing and analysis. This includes quality control, read alignment, taxonomic classification, and diversity analysis. Developing and maintaining these pipelines can be resource-intensive and require expertise.

Difficulty in Accurate Taxonomic Classification

Accurate taxonomic classification of microorganisms in microbiome studies can be challenging with Ion Torrent sequencing due to:

- *Database Limitations*: Taxonomic classification relies on reference databases that may not be comprehensive or up-to-date. For rare or novel microbial species, the available databases may lack sufficient reference sequences, impacting classification accuracy.
- *Short Read Lengths*: The limited length of Ion Torrent reads can make it difficult to confidently classify microorganisms at finer taxonomic levels, such as species or strain identification.

Detection of Low-Abundance Microbes
Ion Torrent sequencing may struggle with detecting low-abundance microorganisms within a complex microbiome sample. This issue is particularly relevant in studies aiming to identify rare or less abundant microbial species:

- *Lower Sensitivity*: The sequencing depth required to detect low-abundance species can be substantial. Inadequate sequencing depth may result in incomplete representation of the microbial community.
- *Biases in Detection*: The technology may introduce biases that affect the detection of certain microbial taxa, leading to skewed representations of microbial diversity.

Cost of Equipment and Consumables
While Ion Torrent sequencing offers cost advantages compared to some other sequencing technologies, the overall cost of equipment, reagents, and consumables can still be significant. The initial investment in Ion Torrent sequencing platforms and associated infrastructure can be high, which may limit accessibility for some research laboratories. Regular maintenance, consumables, and reagents contribute to ongoing operational costs, which can impact the overall budget for microbiome studies.

Instrumentation and Calibration
Maintaining the performance of Ion Torrent sequencing instruments requires careful calibration and regular maintenance. Variations in instrument sensitivity and performance can affect sequencing results. Regular calibration and quality control are necessary to ensure reliable data. Operating and troubleshooting Ion Torrent sequencing platforms requires specialized training and expertise. Ensuring that personnel are adequately trained is essential for obtaining high-quality data.

Environmental and Sample Handling Issues
Handling and preparing samples for Ion Torrent sequencing can introduce variability. Contamination during sample preparation can affect the quality of sequencing data. Rigorous protocols are required to minimize contamination and ensure accurate results. Proper handling and storage of samples are crucial to prevent degradation or loss of quality before sequencing.

PacBio Sequencing
Pacific Biosciences (PacBio) has pioneered a unique sequencing technology known as Single Molecule Real-Time (SMRT) sequencing. This technology offers distinct advantages over other sequencing methods, particularly in terms of read length and accuracy in complex genomic contexts. In this section, we delve into the technical details of PacBio sequencing, its operational principles, and its implications for microbiome research, highlighting its strengths and limitations.

Technical Principles of PacBio SMRT Sequencing

PacBio SMRT sequencing is distinguished by its capability to produce long reads, a feature achieved through its innovative sequencing chemistry and platform design. The core technology involves the following components:

- *Zero-Mode Waveguide (ZMW) Technology*: At the heart of PacBio sequencing is the Zero-Mode Waveguide, a nanophotonic structure that allows the observation of individual DNA polymerase enzymes as they incorporate nucleotides into a growing DNA strand. Each ZMW can accommodate a single DNA molecule, enabling real-time observation of the sequencing process.
- *Single-Molecule Real-Time (SMRT) Sequencing*: The SMRT technology utilizes a DNA polymerase enzyme that adds fluorescently labeled nucleotides to a DNA template strand. As each nucleotide is incorporated, a fluorescent signal is emitted, which is detected by the ZMW. This process provides a continuous sequence read without the need for amplification.
- *Circular Consensus Sequencing (CCS):* To improve accuracy, PacBio sequencing often employs Circular Consensus Sequencing. This method involves circularizing the DNA template and sequencing it multiple times in a circular fashion. The multiple passes through the same DNA molecule enable the generation of highly accurate consensus sequences by correcting errors inherent in single-pass reads.

Read Length and Accuracy

One of the most significant advantages of PacBio sequencing is its ability to produce very long reads, often exceeding 10,000 base pairs (bp) and reaching up to 60,000 bp in some cases. Long reads are particularly beneficial for:

- *Complex Genome Assembly*: Long reads improve the assembly of complex genomes and microbial communities by spanning repetitive regions and structural variants that are difficult to resolve with short-read technologies.
- *Enhanced Resolution*: For microbial communities, long reads can provide better resolution of closely related species and strain-level variation, which is crucial for accurate taxonomic classification and functional annotation.

However, while PacBio sequencing excels in read length, it does have some limitations:

- *Error Rate:* PacBio sequencing traditionally exhibits a higher error rate compared to short-read technologies like Illumina. The errors are often insertion and deletion errors, which can affect the accuracy of base calling. Circular Consensus Sequencing helps mitigate this issue by generating consensus reads with reduced error rates, but it may still be higher compared to the error rates of Illumina sequencing.
- *Cost and Throughput*: PacBio sequencing is more expensive than Illumina sequencing due to the sophisticated technology and reagents required. Additionally, the throughput

is generally lower, which may impact the overall cost-efficiency for large-scale studies.

Bioinformatics Framework

Bioinformatics Framework for 16S rRNA Gene Sequencing

16S rRNA gene sequencing has become a cornerstone in microbiome research, enabling detailed characterization of microbial communities across various environments. The bioinformatics framework for 16S rRNA sequencing involves multiple steps, each crucial for ensuring accurate and meaningful results. This comprehensive overview will delve into the key components of the bioinformatics pipeline for 16S rRNA gene sequencing, including quality filtering, sequence alignment, OTU clustering or ESV detection, taxonomic classification, and diversity analysis (Gao et al., 2021; Zhang, et al.,2023) (Figure 5).

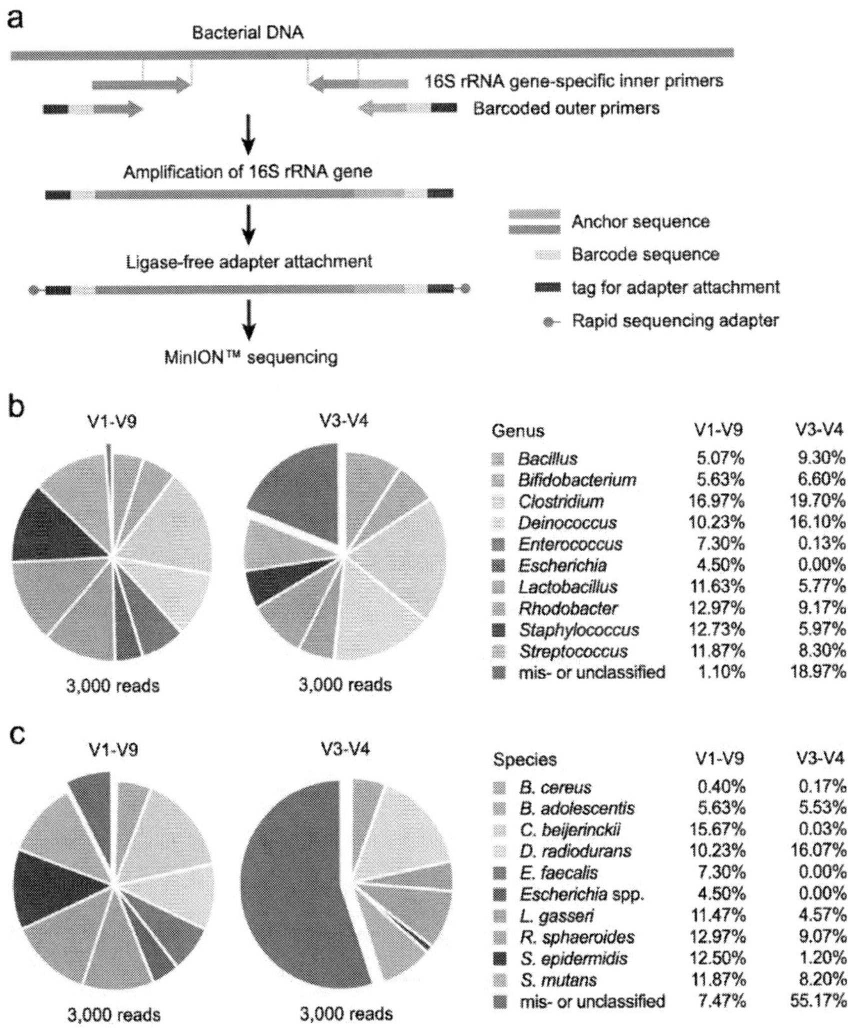

Figure 5. 16S rRNA gene sequencing workflow.

Quality Filtering

Quality filtering is a critical first step in the 16S rRNA gene sequencing pipeline. Raw sequencing reads often contain errors introduced during the sequencing process, such as low-quality bases, adapter contamination, or sequencing artifacts. These errors can compromise the accuracy of downstream analyses, making effective quality filtering essential for reliable results.

Tools for Quality Filtering

Several bioinformatics tools are commonly used for quality filtering, each offering distinct features and capabilities:

- *Trimmomatic:* Trimmomatic is a widely used tool for trimming and filtering sequencing reads. It removes low-quality bases from the ends of reads, trims adapter sequences, and filters out reads that do not meet quality criteria. Trimmomatic employs a sliding window approach, adjusting the quality of bases dynamically to ensure that only high-quality reads are retained.
- *Cutadapt:* Cutadapt is another popular tool for adapter trimming and quality control. It detects and removes adapter sequences from reads, which is crucial for accurate sequence alignment and subsequent analysis. Cutadapt also allows for the removal of low-quality reads based on user-defined thresholds.
- *Fastp:* Fastp is a versatile tool that combines quality filtering with additional features, such as adapter trimming, quality control, and read length filtering. It is designed for high-throughput processing and offers a user-friendly interface for comprehensive quality assessment.

Sequence Alignment

Sequence alignment is the process of comparing sequencing reads to reference databases to identify and classify microorganisms. Accurate alignment is essential for the correct assignment of sequences to microbial taxa and for the reliable interpretation of microbiome data.

Tools for Sequence Alignment

Several tools are commonly used for sequence alignment and clustering in 16S rRNA gene sequencing:

- *USEARCH: USEARCH* is a versatile tool for sequence alignment and clustering. It performs pairwise alignment of sequences and clusters them based on similarity thresholds. USEARCH is known for its efficiency and accuracy in handling large datasets, making it suitable for high-throughput sequencing projects.
- *VSEARCH*: VSEARCH is an open-source alternative to USEARCH with similar functionalities. It performs sequence clustering, alignment, and dereplication, and is widely used in microbiome studies for its ability to handle large-scale data efficiently.
- *BLAST:* BLAST (Basic Local Alignment Search Tool) is a widely used tool for sequence alignment and similarity searching. It compares query sequences against a reference database and provides alignment scores and statistics. While BLAST is not

specifically designed for 16S rRNA sequencing, it can be used for taxonomic classification and functional annotation.

Operational Taxonomic Units (OTUs) Clustering or ESV Detection

Operational Taxonomic Units (OTUs) are clusters of sequencing reads that share a high degree of similarity. OTU clustering is used to group sequences into taxonomic units, allowing researchers to estimate microbial diversity and abundance.

Clustering Methods: OTU clustering involves defining a similarity threshold, typically 97% or 99%, to group sequences into OTUs. Sequences that meet or exceed the similarity threshold are assigned to the same OTU. This method helps reduce the complexity of the data and facilitates downstream analyses.

Exact Sequence Variants (ESVs)

Exact Sequence Variants (ESVs) are individual sequence variants that provide higher resolution compared to OTUs. ESVs represent unique sequences without clustering, allowing for more precise identification and quantification of microbial taxa.

- *DADA2*: DADA2 is a widely used tool for ESV detection. It employs an error-correction algorithm to identify exact sequence variants from high-throughput sequencing data. DADA2 improves the accuracy of sequence variant detection and reduces the impact of sequencing errors.
- *Deblur*: Deblur is another tool for ESV detection that uses a statistical model to distinguish true biological sequences from sequencing errors. It provides high-resolution data and is particularly useful for identifying rare or low-abundance microbial taxa.

Taxonomic Classification

Taxonomic classification involves assigning sequences to specific taxa based on reference databases. Accurate classification is crucial for understanding the composition of microbial communities and for identifying species or genera of interest.

Tools for Taxonomic Classification

Several tools and databases are available for taxonomic classification:

- *SILVA Database:* The SILVA database is a comprehensive resource for ribosomal RNA gene sequences. It provides curated reference sequences for bacteria, archaea, and eukaryotes, making it a valuable tool for taxonomic classification.
- *Greengenes Database*: Greengenes is another widely used reference database for 16S rRNA gene sequences. It includes curated sequences and taxonomic information for a broad range of microbial taxa.
- *RDP Classifier:* The Ribosomal Database Project (RDP) Classifier is a tool for classifying 16S rRNA gene sequences using a reference database. It employs a naive Bayes classifier to assign sequences to taxonomic levels and provides confidence scores for classifications.

- *Kraken:* Kraken is a taxonomic classification tool that uses a k-mer-based approach to assign sequences to taxonomic categories. It is designed for high-speed classification and can handle large datasets efficiently.

Diversity Analysis
Diversity analysis is used to assess the richness, evenness, and structure of microbial communities. It provides insights into the diversity of microorganisms within and between samples and helps identify patterns and relationships in microbiome data.

Alpha Diversity in Microbiome

Alpha diversity is a fundamental concept in microbiome research, providing insights into the diversity of microbial communities within a single sample. It encompasses various metrics that help researchers understand the richness, evenness, and overall structure of microbial communities. This detailed overview explores the key metrics of alpha diversity, including species richness, the Shannon index, and the Simpson index, highlighting their scientific principles, applications, and implications in microbiome research.

Species Richness
Species richness refers to the total number of unique microbial taxa present in a sample. It is a measure of the diversity of species without considering their relative abundances. In microbiome studies, species richness provides a count of the distinct species identified within a sample, which is crucial for understanding microbial diversity.

- *Species Richness Calculation*: Species richness is typically calculated by counting the number of unique operational taxonomic units (OTUs) or exact sequence variants (ESVs) identified in the sample. OTUs are clusters of sequences that share a certain level of similarity, often used as proxies for species in microbial ecology. ESVs represent exact sequences identified without clustering, providing a higher resolution of species identification.
- *Applications*: Species richness is used to compare the microbial diversity across different samples or conditions. It helps in understanding the impact of environmental changes, treatments, or host factors on microbial diversity. For example, comparing species richness between healthy and diseased states can reveal shifts in microbial community composition associated with health conditions.

Limitations

- *Ignores Abundance*: Species richness does not account for the relative abundances of different species. Two communities with the same species richness can differ significantly in their microbial composition if the species are present in different proportions.
- *Sensitive to Sampling Effort*: The accuracy of species richness estimates can be influenced by the depth of sequencing and the completeness of sampling. Inadequate

sampling may lead to underestimation of species richness due to the presence of rare or undetected taxa.

Shannon Index

The Shannon index, also known as the Shannon-Weaver index, is a metric that combines both species richness and evenness. It reflects the uncertainty in predicting the species of a randomly chosen individual from a sample. This index provides a measure of the entropy or disorder within a microbial community.

Formula: The Shannon index (H') is calculated using the formula:

$$H' = -\sum_{i=1}^{S}(P_i \,.\, ln(P_i\,))$$

where S is the total number of species in the sample, and Pi is the proportion of individuals belonging to species I.

Interpretation: A higher Shannon index value indicates greater diversity, with both high species richness and even distribution of species. Conversely, a lower Shannon index suggests lower diversity, with either fewer species or more uneven distribution of species.

Applications

- *Community Comparison*: The Shannon index is widely used to compare microbial diversity between different samples or conditions. For example, comparing the Shannon index of gut microbiota between healthy and diseased individuals can provide insights into the impact of disease on microbial diversity.
- *Ecological Studies*: The Shannon index is employed in ecological studies to assess the impact of environmental factors on microbial communities. It helps in understanding how changes in habitat or environmental conditions influence microbial diversity and community structure.

Limitations

- *Sensitivity to Dominant Species*: While the Shannon index accounts for both richness and evenness, it can be sensitive to dominant species. In cases where a few species dominate the community, the Shannon index may not fully capture the impact of these dominant species on overall diversity.
- *Comparison Across Studies*: Comparing Shannon index values across different studies or environments can be challenging due to variations in sampling methods, sequencing depth, and data processing techniques. Standardization of methods is essential for accurate comparisons.

Simpson Index

The Simpson index measures the probability that two randomly selected individuals from a sample belong to the same species. It provides insights into species dominance and evenness within a community.

Formula: The Simpson index (D) is calculated using the formula:

$$D = \sum_{i=1}^{S} (P^2{}_i)$$

where Pi is the proportion of individuals belonging to species i, and S is the total number of species.

The Simpson index can also be expressed as the reciprocal index (1/D) or the complement (1 - D) to provide different perspectives on diversity.

Interpretation: A lower Simpson index value indicates higher diversity, with a more even distribution of species. Conversely, a higher Simpson index value suggests lower diversity, with more dominance by a few species.

Applications

- *Community Structure Analysis*: The Simpson index is used to analyze the structure of microbial communities, particularly in understanding species dominance and the distribution of microbial taxa. It is useful for identifying communities with high or low dominance and evenness.
- *Impact Assessment*: In environmental and ecological studies, the Simpson index helps assess the impact of various factors on microbial community structure. For example, it can reveal how pollution or habitat changes influence microbial diversity and dominance.

Limitations

- *Sensitivity to Dominant Species*: Similar to the Shannon index, the Simpson index is sensitive to dominant species. It may not fully capture the diversity in communities where a few species are highly dominant, potentially underestimating the impact of less abundant species.
- *Limited Insight into Richness:* The Simpson index focuses more on species dominance and evenness and less on species richness. As a result, it may not provide a complete picture of the overall microbial diversity in a sample.

Comparative Analysis and Integration: Combining Metrics for Comprehensive Insight

Using a combination of alpha diversity metrics provides a more comprehensive understanding of microbial diversity. Each metric offers unique insights into different aspects of microbial communities:

- *Species Richness*: Provides a count of unique taxa but does not account for their relative abundances.

- *Shannon Index*: Combines richness and evenness, offering a measure of uncertainty in predicting species but can be sensitive to dominant taxa.
- *Simpson Index*: Focuses on species dominance and evenness, providing insights into the structure of the community but may overlook the richness of less abundant taxa.

Integrating with Beta Diversity

Alpha diversity metrics are often complemented by beta diversity analyses, which assess the diversity between different samples or environments. Integrating alpha and beta diversity provides a more complete picture of microbial diversity and community dynamics across different conditions or treatments.

Tools for Calculating Alpha Diversity in Microbiome Research

Several bioinformatics tools and software packages are available to calculate and visualize alpha diversity. This section delves into two widely used tools—QIIME and Phyloseq—providing a detailed scientific overview of their functionalities, applications, and methodologies for alpha diversity analysis.

QIIME: Quantitative Insights into Microbial Ecology

QIIME (Quantitative Insights Into Microbial Ecology) is a comprehensive open-source software package designed for the analysis of microbial communities, particularly those characterized by high-throughput sequencing of marker genes such as 16S rRNA. QIIME provides a suite of tools for calculating alpha diversity, including the Shannon and Simpson indices, and offers various visualization and statistical analysis options.

Alpha Diversity Metrics in QIIME

Shannon Index

The Shannon index (H') is a widely used metric that combines both species richness and evenness into a single value. It reflects the uncertainty of predicting the identity of a randomly selected individual from a sample.

Simpson Index

The Simpson index measures the probability that two randomly selected individuals from a sample belong to the same species. It provides a measure of dominance, where a lower value indicates higher diversity.

QIIME's Alpha Diversity Workflow

Data Processing and Input

QIIME requires input data in specific formats, including sequence files and metadata. Raw sequencing data must be processed to produce feature tables, which list the counts of different operational taxonomic units (OTUs) or exact sequence variants (ESVs) across samples.

Quality Control

QIIME includes tools for quality control of sequencing data, including filtering of low-quality sequences, removal of chimeric sequences, and subsampling (rarefaction) to standardize sequencing depth across samples. These steps are crucial for ensuring accurate and comparable alpha diversity metrics.

Calculating Alpha Diversity

Once quality control is complete, QIIME calculates alpha diversity metrics using built-in commands. For example, the alpha_diversity.py script calculates various diversity indices, including Shannon and Simpson, from feature tables. Users can specify parameters for the calculation, such as the metric to be used and the subsampling depth.

Visualization and Statistical Analysis

QIIME offers visualization tools to present alpha diversity results. These include boxplots, bar charts, and rarefaction curves. Statistical tests, such as ANOVA or Kruskal-Wallis tests, can be performed to compare alpha diversity across different groups or conditions.

Phyloseq

Phyloseq is an R-based software package designed for the analysis and visualization of microbiome data. It integrates diverse data types, including taxonomic, phylogenetic, and functional data, into a single object for comprehensive analysis. Phyloseq supports various alpha diversity metrics and offers numerous visualization options, making it a powerful tool for microbiome research.

Alpha Diversity Metrics in Phyloseq

Shannon Index

Phyloseq calculates the Shannon index using the estimate_richness function, which supports a variety of diversity metrics. The Shannon index in Phyloseq is computed similarly to QIIME, incorporating both species richness and evenness:

Simpson Index

Phyloseq also calculates the Simpson index, with options for both the original Simpson index and its reciprocal. The calculation reflects species dominance and is sensitive to the presence of dominant species:

Phyloseq's Alpha Diversity Workflow

Data Integration

Phyloseq integrates multiple types of microbiome data, including OTU tables, taxonomic classifications, and sample metadata. Users import data using functions such as import_biom or import_qiime, which facilitate the conversion of data from other formats into Phyloseq-compatible objects.

Quality Control

Phyloseq supports quality control steps similar to those in QIIME, including filtering of low-abundance taxa and rarefaction. The prune_samples and prune_taxa functions help in removing unwanted taxa or samples, ensuring data quality and consistency.

Calculating Alpha Diversity

Alpha diversity metrics are calculated using the estimate_richness function, which allows users to specify the metrics of interest. The function computes diversity indices from the integrated feature table and provides a data frame with alpha diversity estimates.

Visualization and Statistical Analysis

Phyloseq excels in visualization, offering tools to create rarefaction curves, diversity histograms, and boxplots. The plot_richness function generates visualizations of alpha diversity metrics across samples, while the ggplot2 package can be used for more customized plotting. Statistical analysis, including comparisons between groups, can be performed using R's statistical functions and packages, such as vegan for diversity analysis.

Comparative Analysis of QIIME and Phyloseq

Flexibility and Integration

- *QIIME:* Primarily focused on microbial ecology analysis, QIIME provides a streamlined workflow for alpha diversity calculations and visualizations. It is well-suited for users who prefer a command-line interface and need a comprehensive suite of tools for microbial community analysis.
- *Phyloseq:* Offers greater flexibility and integration within the R environment, allowing for advanced statistical analysis and custom visualizations. Phyloseq's ability to integrate various data types and its compatibility with R-based statistical and graphical tools provide a broader range of analytical options.

Ease of Use

- *QIIME:* May require users to be familiar with command-line interfaces and scripting. Its extensive documentation and tutorials can help users navigate its functionalities.

- *Phyloseq:* Users need a working knowledge of R programming and statistical analysis. However, Phyloseq's integration with R facilitates advanced analyses and custom visualizations, making it a powerful tool for users with R experience.

Visualization Capabilities

- *QIIME:* Provides built-in visualization tools, including boxplots, bar charts, and rarefaction curves. These visualizations are suitable for general analysis but may be less customizable.
- *Phyloseq:* Offers extensive visualization options through integration with R's plotting libraries. Users can create highly customized plots and integrate alpha diversity results with other types of analyses, such as beta diversity or functional annotations.

Statistical Analysis

- *QIIME*: Includes basic statistical tests for comparing alpha diversity across groups, such as ANOVA or Kruskal-Wallis tests. Additional statistical analyses may require external tools or scripts.
- *Phyloseq:* Leverages R's statistical capabilities, providing a wide range of options for hypothesis testing and data exploration. The integration with R's ecosystem allows for advanced statistical analyses and modeling.

Beta Diversity in Microbiome Research

Beta diversity is a fundamental concept in microbial ecology that measures the diversity of microbial communities between different samples or environments. It provides insights into the differences and similarities in community composition, helping researchers understand the variation in microbial communities across different habitats, conditions, or treatments.

Principles of Beta Diversity

Beta diversity quantifies the variation in species composition between different samples, contrasting with alpha diversity, which measures the diversity within a single sample. By assessing beta diversity, researchers can determine how distinct or similar microbial communities are across different environments or conditions. The metrics used to evaluate beta diversity can be broadly categorized into dissimilarity metrics and ordination techniques.

Dissimilarity Metrics

Dissimilarity metrics quantify the differences between microbial communities based on their species composition and abundance. These metrics provide a numerical value representing the extent of dissimilarity between pairs of samples.

Bray-Curtis Dissimilarity

The Bray-Curtis dissimilarity index is one of the most widely used metrics for quantifying community dissimilarity based on species abundance data. It is defined as:

$$D_{BC} = \frac{\Sigma(xi - yi)}{\Sigma(xi + yi)}$$

where xi and yi are the abundances of species i in samples x and y, respectively. The Bray-Curtis dissimilarity ranges from 0 to 1, where 0 indicates identical communities and 1 indicates completely dissimilar communities.

Strengths of Bray-Curtis Dissimilarity

- *Sensitivity to Abundance:* It accounts for the abundance of species, providing a measure that reflects both species presence/absence and relative abundances.
- *Widely Used:* It is a standard metric used in many ecological studies and is well-supported by various analytical tools.

Limitations

Ignores Species Identity: It does not account for the identity of the species, only their relative abundances. This can be a limitation when comparing communities with very different taxonomic compositions.

Unweighted UniFrac

Unweighted UniFrac is a phylogenetic metric that measures dissimilarity based on the evolutionary relationships between taxa. It compares the phylogenetic tree (or tree-like structure) of microbial communities to assess how distinct the communities are in terms of their evolutionary history.

The Unweighted UniFrac distance is calculated as:

$$D_{UW} = \frac{\Sigma_{branches} \min(p_x, p_y)}{\Sigma_{branches} \max(p_x, p_y)}$$

where Px and Py are the relative abundances of the branches in the phylogenetic tree for samples x and y. Unweighted UniFrac measures the proportion of branch lengths in the phylogenetic tree that are unique to each sample.

Strengths

- *Phylogenetic Consideration:* It incorporates phylogenetic relationships, providing insights into the evolutionary differences between microbial communities.
- *Non-Dependence on Abundance:* Unweighted UniFrac focuses on the presence or absence of taxa rather than their abundance, which can be useful when comparing communities with varying abundance distributions.

Limitations

- *Phylogenetic Tree Construction:* Accurate Unweighted UniFrac analysis relies on the availability of a comprehensive and accurate phylogenetic tree, which may not always be available for all microbial taxa.
- *Less Sensitive to Abundance:* It does not account for the relative abundance of taxa, which can be a limitation in certain studies.

Weighted UniFrac

In contrast to Unweighted UniFrac, Weighted UniFrac incorporates both the presence/absence and the relative abundances of taxa. The formula for Weighted UniFrac is:

$$D_W = \frac{\sum_{\text{branches}} (p_x - p_y)}{\sum_{\text{branches}} (p_x + p_y)}$$

where Px and Py are the relative abundances of the branches in the phylogenetic tree for samples x and y. Weighted UniFrac provides a more nuanced view of community differences by accounting for both taxonomic and abundance information.

Strengths of Weighted UniFrac

- *Combination of Presence/Absence and Abundance*: It integrates both phylogenetic relationships and abundance data, offering a more comprehensive assessment of community dissimilarity.
- *Enhanced Sensitivity:* Provides sensitivity to differences in microbial abundances, which can be important for understanding community dynamics.

Limitations Weighted UniFrac

- *Complexity:* The incorporation of both presence/absence and abundance data makes the analysis more complex and computationally intensive.
- *Dependence on Accurate Abundance Data*: Accurate abundance data is essential for reliable results, and errors in abundance estimation can affect the outcome.

Ordination Techniques

Ordination techniques are used to visualize and interpret the relationships between samples based on beta diversity metrics. These techniques reduce the dimensionality of the data and represent complex community structures in a more interpretable form.

Principal Coordinate Analysis (PCA)

Principal Coordinate Analysis (PCA) is a linear ordination technique that transforms beta diversity distance matrices into a lower-dimensional space while retaining the most significant variations. PCA identifies the principal components (axes) that capture the maximum variance in the data.

Process Principal Coordinate Analysis (PCA)

- *Distance Matrix Calculation*: PCA starts with a distance matrix, such as Bray-Curtis or UniFrac distances, which represents the dissimilarities between samples.
- *Eigenvalue Decomposition:* PCA performs eigenvalue decomposition on the distance matrix to identify principal components.
- *Visualization:* The resulting principal components are plotted to visualize the relationships between samples, with each axis representing a principal component.

Strengths

- *Interpretability*: PCA provides a straightforward and interpretable representation of beta diversity, making it easier to visualize patterns and relationships.
- *Dimensionality Reduction:* It reduces the complexity of the data, allowing researchers to focus on the most significant sources of variation.

Limitations Principal Coordinate Analysis (PCA)

- Linear Assumptions: PCA assumes linear relationships between variables, which may not always capture the complexity of microbial community data.
- Overemphasis on Variance: PCA emphasizes the components that explain the most variance, which may not always align with biologically relevant differences.

Non-Metric Multidimensional Scaling (NMDS)

Non-Metric Multidimensional Scaling (NMDS) is a non-linear ordination technique that visualizes the similarity or dissimilarity between samples in a reduced-dimensional space. Unlike PCA, NMDS does not assume linear relationships and can represent complex patterns more accurately.

Process Non-Metric Multidimensional Scaling (NMDS)

- *Distance Matrix Input*: NMDS starts with a distance matrix, such as Bray-Curtis or UniFrac distances.
- *Optimization:* NMDS uses an iterative optimization algorithm to position samples in a lower-dimensional space, aiming to preserve the rank order of distances between samples.
- *Visualization:* The resulting NMDS plot shows the spatial arrangement of samples, with similar samples positioned closer together.

Strengths Non-Metric Multidimensional Scaling (NMDS)

- *Non-Linear Relationships*: NMDS captures non-linear relationships and complex patterns in beta diversity, providing a more nuanced view of community structure.
- *Flexibility:* It can handle a variety of distance metrics and is not constrained by linear assumptions.

Limitations Non-Metric Multidimensional Scaling (NMDS)

- *Interpretability*: NMDS plots can be more challenging to interpret compared to PCA, as the axes do not have straightforward biological meanings.
- *Computational Intensity:* NMDS can be computationally intensive, especially for large datasets, and may require multiple runs to ensure convergence.

Tools for Beta Diversity Analysis

Several bioinformatics tools and software packages are available for calculating and visualizing beta diversity metrics. Two prominent tools used in microbiome research are QIIME and Phyloseq.

QIIME (Quantitative Insights into Microbial Ecology)

QIIME is a widely used software package for analyzing microbiome data, including beta diversity analysis. It provides a comprehensive suite of tools for processing sequencing data, calculating beta diversity metrics, and visualizing results.

- *Diversity Metrics Calculation*: QIIME supports the calculation of various beta diversity metrics, including Bray-Curtis dissimilarity and UniFrac distances. Users can generate distance matrices and perform pairwise comparisons between samples.
- *Ordination and Visualization*: QIIME offers tools for performing PCA and NMDS, allowing users to visualize the relationships between samples based on beta diversity metrics. It provides interactive plots and graphical representations to aid in data interpretation.
- *Statistical Testing*: QIIME includes tools for statistical testing, such as Permutational Multivariate Analysis of Variance (PERMANOVA), to assess the significance of observed differences in beta diversity.

Strengths of QIIME (Quantitative Insights into Microbial Ecology)

- *Comprehensive Toolset*: QIIME offers a wide range of functionalities for microbiome analysis, including beta diversity metrics, ordination, and statistical testing.
- *User-Friendly Interface*: The software provides a user-friendly interface and extensive documentation, making it accessible for researchers with varying levels of expertise.

Limitations

- *Complexity:* QIIME can be complex to set up and configure, particularly for users new to microbiome analysis.
- *Computational Requirements*: The analysis of large datasets may require significant computational resources and time.

Phyloseq
Phyloseq is an R package designed for the analysis and visualization of microbiome data. It integrates microbiome data into a unified object and provides tools for calculating and visualizing beta diversity metrics.

- *Data Integration:* Phyloseq allows users to import and integrate data from various sources, including QIIME, and combine taxonomic, abundance, and sample metadata.
- *Beta Diversity Analysis:* The package supports the calculation of beta diversity metrics, such as Bray-Curtis dissimilarity and UniFrac distances. It provides functions for generating distance matrices and performing ordination analyses.
- *Visualization*: Phyloseq offers tools for creating diverse visualizations, including heatmaps, ordination plots, and distance-based clustering. It enables researchers to explore and interpret beta diversity patterns effectively.

Strengths Phyloseq

- *Flexible and Comprehensive*: Phyloseq provides a flexible and comprehensive environment for microbiome data analysis, with extensive visualization options and integration capabilities.
- *R Integration:* As an R package, Phyloseq leverages the power of R for statistical analysis and customization, allowing researchers to tailor analyses to their specific needs.

Limitations

- *Learning Curve:* Users may face a learning curve when working with R and Phyloseq, particularly if they are not familiar with the R programming language.
- *Computational Requirements*: Similar to QIIME, analyzing large datasets may require significant computational resources.

Challenges and Considerations in Microbiome Sequencing

Handling High-Dimensional Data

Microbiome studies generate high-dimensional data characterized by extensive sequence information and detailed taxonomic classifications. This complexity poses significant challenges for data management, processing, and interpretation.

High-Dimensional Nature of Microbiome Data
Microbiome sequencing projects can produce vast quantities of data, including thousands of sequences from diverse microorganisms. Each sample may contain millions of reads, and the data can span multiple taxonomic levels, from phyla to species. The high-dimensional nature of this data arises from:

- *Sequencing Depth*: High-throughput sequencing technologies produce millions of reads per sample, resulting in a large dataset that includes numerous microbial taxa and their relative abundances.
- *Taxonomic Resolution*: Microbiome data often include detailed taxonomic information at multiple levels (e.g., genus, species, strain), leading to a complex dataset with many dimensions.
- *Diversity Metrics*: Metrics such as alpha diversity (within-sample diversity) and beta diversity (between-sample diversity) further increase the dimensionality of the data, requiring sophisticated analytical methods.

Data Management

Efficient management of high-dimensional microbiome data is critical for ensuring data integrity and usability:

- *Data Storage*: Storing large-scale sequencing data requires substantial computational resources and storage capacity. Efficient data management systems must handle the volume and complexity of the data while maintaining accessibility and security.
- *Data Organization*: Organizing data into structured formats that facilitate analysis is essential. This includes metadata management, data annotation, and ensuring consistency across different datasets.
- *Data Integration:* Combining data from different sources or experiments requires careful consideration of data formats and compatibility. Standardization of data formats and integration protocols can improve data coherence and usability.

Data Processing and Analysis

Processing and analyzing high-dimensional microbiome data involve several key steps:

- *Quality Control*: Initial quality control involves filtering out low-quality sequences, removing contaminants, and correcting errors. Tools such as DADA2 or USEARCH are commonly used for this purpose.
- *Normalization:* Normalization of sequencing data is essential to account for differences in sequencing depth and sample size. Techniques such as rarefaction, relative abundance, and normalization algorithms help standardize data for comparative analysis.
- *Dimensionality Reduction*: Techniques such as principal coordinates analysis (PCA), t-distributed stochastic neighbor embedding (t-SNE), and uniform manifold approximation and projection (UMAP) can reduce data dimensionality and visualize complex relationships among microbial communities.

Addressing Sequencing Errors

Sequencing errors can significantly impact the accuracy and reliability of microbiome data. These errors can manifest as base substitutions, insertions, deletions, or other artifacts, which can affect downstream analyses.

Types of Sequencing Errors

- *Base Substitutions*: Incorrect base calls during sequencing can lead to erroneous nucleotide substitutions, affecting the accuracy of sequence data and taxonomic identification.
- *Insertions and Deletions*: Errors involving insertions or deletions (indels) can disrupt the sequence alignment and affect variant calling. These errors are particularly challenging in regions with repetitive sequences or homopolymers.
- *Chimeras:* Chimeric sequences, resulting from the erroneous merging of two or more sequences during PCR amplification, can create misleading or artificial sequences.

Quality Control and Error Correction

Effective quality control and error correction strategies are essential for minimizing the impact of sequencing errors:

- *Pre-Processing Tools*: Tools like FASTQC, Trimmomatic, and Cutadapt are used to assess and filter raw sequencing data, removing low-quality reads and trimming adapter sequences.
- *Error Correction Algorithms*: Algorithms such as DADA2 and Deblur provide denoising and error correction capabilities. These tools use statistical models to distinguish true biological sequences from sequencing artifacts.
- *Validation and Replication*: Validating sequencing results through replicate experiments or independent methods can help confirm the accuracy of data and mitigate the impact of sequencing errors.

Database Limitations

Reference databases are crucial for taxonomic classification in microbiome studies, but they have inherent limitations that can affect the accuracy and completeness of microbial identification.

Limitations of Reference Databases

- *Coverage and Completeness*: Many reference databases may lack comprehensive coverage of microbial taxa, particularly for less-studied or novel microorganisms. This can result in incomplete or inaccurate taxonomic classification.
- *Database Updates*: Reference databases are regularly updated with new sequences and taxonomic information. However, there may be delays in incorporating the latest data, leading to potential discrepancies in classification.
- *Taxonomic Resolution*: Databases may vary in their taxonomic resolution, with some providing information only at higher taxonomic levels (e.g., phylum or genus) while others offer more detailed species-level classifications.

Addressing Database Limitations

Strategies to address database limitations include:

- *Using Multiple Databases*: Combining information from multiple reference databases can improve coverage and accuracy. Tools like QIIME and Mothur allow for integration with various databases to enhance taxonomic classification.
- *Updating Databases*: Regular updates and curation of reference databases are essential for maintaining accuracy. Researchers and institutions can contribute to database development by submitting new sequences and taxonomic information.
- *In-House Databases*: Creating custom reference databases based on specific research needs or local microbiome data can provide more relevant and accurate taxonomic classifications.

Integration of Multi-Omics Data

Integrating data from various omics approaches, such as 16S rRNA gene sequencing, metagenomics, and metatranscriptomics, can provide a more comprehensive understanding of microbial communities. However, this integration involves several challenges.

Multi-Omics Approaches

The advent of high-throughput sequencing technologies has revolutionized microbiome research, providing unprecedented insights into the diversity and functional potential of microbial communities. Among the various approaches, metagenomics and metatranscriptomics stand out for their ability to explore the genetic material and gene expression profiles of microorganisms in environmental and clinical samples. 16S rRNA Gene Sequencing: Provides information on microbial community composition and diversity based on the 16S rRNA gene. It is valuable for identifying bacterial taxa and understanding community structure.

Metagenomics: Principles of Metagenomics

Metagenomics involves the sequencing of all genetic material present in a sample, providing a comprehensive snapshot of the microbial community's genetic makeup. Unlike traditional microbiological methods that often focus on culturing individual species, metagenomics captures the full spectrum of microbial diversity, including those that are unculturable or present in low abundance.

- *Whole-Genome Sequencing*: Metagenomics captures the entirety of the genetic material, including microbial DNA, RNA, and host DNA, depending on the extraction and sequencing protocols used. This approach allows researchers to obtain a detailed view of the microbial community's genetic repertoire, including genes encoding enzymes, metabolic pathways, and potential virulence factors.
- *Shotgun Sequencing*: In metagenomic shotgun sequencing, DNA is fragmented into smaller pieces, which are then sequenced randomly. This technique enables the assembly of microbial genomes from the resulting sequence data, allowing for the reconstruction of genomic sequences and the identification of microbial species.

Methodologies in Metagenomics

Metagenomics encompasses several key methodologies that facilitate the analysis of microbial communities:

- *Sample Collection and DNA Extraction*: The quality and representativeness of metagenomic data depend on the proper collection of samples and extraction of genetic material. Environmental or clinical samples are collected using sterile techniques, and DNA is extracted using methods that preserve the integrity of microbial DNA. Sample preparation may involve steps to remove host DNA or contaminants, depending on the research goals.
- *Sequencing Platforms*: Various sequencing technologies can be employed in metagenomics, including Illumina, Ion Torrent, and Oxford Nanopore. Each platform offers different advantages in terms of read length, accuracy, and cost. For instance, Illumina sequencing provides high-throughput, short-read sequences, while Oxford Nanopore sequencing offers long reads that can improve genome assembly.
- *Data Processing and Assembly*: The raw sequencing data is processed to remove low-quality reads and contaminants. Sequence reads are then assembled into contiguous sequences (contigs) using bioinformatics tools such as SPAdes or MetaSPAdes. Assembly quality is assessed through metrics such as N50, which reflects the length of the contigs.
- *Functional Annotation*: Once assembled, the contigs are annotated to predict the functional potential of the microbial community. Tools such as PROKKA or MetaGeneAnnotator are used to identify genes and assign functions based on reference databases like KEGG, COG, or Pfam.
- *Taxonomic Classification*: Taxonomic classification involves identifying the microbial species present in the sample. Tools such as Kraken, Kaiju, and MEGAN can assign taxonomic labels to sequence reads based on reference databases. Classification is crucial for understanding the composition and diversity of microbial communities.

Applications of Metagenomics

Metagenomics has broad applications across various fields, including environmental science, medicine, and agriculture:

- *Environmental Microbiomes*: Metagenomics provides insights into microbial communities in diverse environments, such as soil, water, and the atmosphere. Studies have revealed the roles of microorganisms in biogeochemical cycles, pollutant degradation, and ecosystem health.
- *Human Microbiome*: In clinical research, metagenomics is used to explore the human microbiome, including the gut, skin, and oral microbiota. Understanding the microbial diversity and functional potential of these communities is crucial for identifying biomarkers of health and disease.
- *Disease Diagnosis and Treatment*: Metagenomic analysis can identify pathogens in clinical samples, including rare or novel microorganisms that are difficult to detect using traditional methods. This approach has applications in diagnosing infectious diseases, tracking outbreaks, and guiding personalized treatment strategies.
- *Agricultural Microbiomes*: In agriculture, metagenomics is employed to study soil and plant microbiomes. Insights into microbial communities associated with crop health and soil fertility can inform sustainable farming practices and the development of microbial inoculants.

Metatranscriptomics: Principles of Metatranscriptomics

Metatranscriptomics involves the analysis of RNA transcripts within a microbial community, providing insights into the actively expressed genes and functional activities of microorganisms. By focusing on RNA rather than DNA, metatranscriptomics offers a snapshot of the functional state of microbial communities.

- *RNA Extraction*: The first step in metatranscriptomics is the extraction of total RNA from the sample. RNA extraction protocols must effectively isolate and preserve the integrity of RNA, including mRNA, rRNA, and non-coding RNAs.
- *cDNA Library Construction*: RNA is reverse-transcribed into complementary DNA (cDNA) to create a library for sequencing. This step involves converting RNA transcripts into a stable DNA format suitable for sequencing technologies.
- *Sequencing Platforms*: Similar to metagenomics, metatranscriptomics can use various sequencing platforms, including Illumina and Oxford Nanopore. The choice of platform affects the read length, coverage, and sensitivity of transcript detection.

Methodologies in Metatranscriptomics

Metatranscriptomics encompasses several methodologies for analyzing gene expression profiles:

- *Quality Control and Filtering*: The raw sequencing data is processed to remove low-quality reads and contaminants. Quality control steps ensure the reliability of transcriptomic data and minimize potential biases.
- *Read Mapping*: cDNA reads are mapped to reference genomes or metagenomic assemblies to identify expressed genes. Tools such as HISAT2, STAR, or Bowtie2 are commonly used for read mapping. Accurate mapping is essential for quantifying gene expression levels and identifying differentially expressed genes.
- *Expression Quantification*: Gene expression levels are quantified based on read counts mapped to individual genes. Tools such as HTSeq, featureCounts, or Kallisto are used for quantification. The resulting data provides insights into the relative abundance of transcripts and functional activity.
- *Functional Annotation*: Functional annotation involves assigning functions to differentially expressed genes based on reference databases. Tools such as BLAST, InterProScan, or EggNOG can be used to predict gene functions and identify pathways involved in microbial metabolism.
- *Differential Expression Analysis*: Differential expression analysis compares gene expression levels between different conditions or time points. Tools such as DESeq2, edgeR, or limma are used to identify genes with significant changes in expression. This analysis helps in understanding how microbial communities respond to environmental changes or treatments.

Applications of Metatranscriptomics

Metatranscriptomics has diverse applications in understanding microbial functions and interactions:

- *Microbial Function and Metabolism*: By analyzing actively transcribed genes, metatranscriptomics reveals the metabolic pathways and functional activities of microorganisms. This approach helps elucidate the roles of microbial communities in processes such as nutrient cycling, degradation of organic matter, and production of secondary metabolites.
- *Host-Microbe Interactions*: In clinical research, metatranscriptomics provides insights into the interactions between microorganisms and host tissues. Studying gene expression profiles in host-associated microbiomes can reveal mechanisms of host-microbe interactions, including pathogenicity and symbiosis.
- *Environmental Responses*: Metatranscriptomics can be used to study how microbial communities respond to environmental changes, such as shifts in nutrient availability, temperature, or pH. This approach helps understand the adaptability and resilience of microbial communities in response to environmental stressors.
- *Biotechnology and Synthetic Biology*: In biotechnology, metatranscriptomics aids in the discovery of novel enzymes and metabolic pathways with potential applications in industrial processes. Understanding gene expression profiles can inform the development of engineered microorganisms for biotechnological applications.

Challenges in Integration

- *Data Compatibility*: Combining data from different omics approaches requires addressing differences in data formats, resolution, and scale. Standardization and integration protocols are necessary to ensure compatibility.
- *Analysis Methods*: Different omics approaches require distinct analysis methods. Integrating data involves aligning analytical techniques to produce coherent and meaningful results. For example, integrating 16S rRNA sequencing with metagenomics may require combining taxonomic and functional data.
- *Interpretation and Validation*: Interpreting integrated data involves understanding the relationships between microbial composition, function, and activity. Validation of findings through independent experiments or complementary methods is essential for confirming results.

Strategies for Successful Integration

- *Multi-Omics Platforms*: Utilizing integrated multi-omics platforms or frameworks designed for data integration can streamline the process. Tools like MetaPhlAn, HUMAnN, and MG-RAST offer functionalities for integrating taxonomic and functional data.
- *Statistical and Computational Approaches*: Employing advanced statistical and computational methods, such as machine learning algorithms and network analysis, can help integrate and interpret multi-omics data. These approaches can uncover correlations and interactions between different data types.
- *Collaborative Efforts*: Collaborating with experts in different omics fields can enhance the integration process. Multi-disciplinary teams can provide insights into data interpretation and ensure that integration approaches are scientifically robust.

Applications in Animal Health and Disease

16S rRNA gene sequencing has been widely used to investigate the gut microbiomes of various animals, including livestock, companion animals, and wildlife. It has provided insights into how microbial communities influence health, growth, and disease susceptibility. For example, studies have shown that alterations in the gut microbiome can impact immune function, digestion, and resistance to pathogens.

- *Metagenomics:* Metagenomics involves the sequencing of all genetic material within a sample, providing a comprehensive view of the microbial community's composition and functional potential. Unlike 16S rRNA sequencing, which targets a specific gene, metagenomics captures the entire genomic content, including genes from bacteria, archaea, viruses, and fungi. Metagenomics is typically performed using high-throughput sequencing platforms such as Illumina, which provide deep sequencing coverage and high accuracy. Long-read technologies, such as PacBio and Oxford Nanopore, are also increasingly used to improve the assembly of complex metagenomic data.
- *Illumina Sequencing:* Illumina platforms are commonly used for metagenomic sequencing due to their high throughput and accuracy. The technology generates short reads, which are assembled into contigs during the analysis process.
- *PacBio Sequencing*: PacBio's SMRT sequencing provides long reads that can improve the assembly of metagenomic data and reduce the complexity of microbial genomes.
- *Oxford Nanopore Sequencing*: Oxford Nanopore technology offers long-read sequencing with the advantage of real-time data acquisition. It can be particularly useful for assembling complex metagenomic samples and characterizing microbial genomes.
- *Bioinformatics framework*: The bioinformatics pipeline for metagenomic data includes several steps includin Quality Control (Raw sequencing reads are assessed for quality and filtered to remove low-quality sequences. Tools such as FastQC and Trimmomatic are used for quality control), Assembly (Reads are assembled into contigs to reconstruct microbial genomes. Tools like SOAPdenovo, MEGAHIT, and SPAdes are commonly used for assembly), ene Prediction (Genes are predicted within the assembled contigs using tools such as Prodigal and MetaGeneMark. This step is crucial for identifying functional elements within the metagenomic data), Functional Annotation (Genes are annotated with functional information using databases such as KEGG, COG, and eggNOG. This step provides insights into the metabolic capabilities of the microbial community) and Taxonomic Classification (Sequences are classified into taxa using tools like MEGAN, Kraken, and Centrifuge. These tools match sequences to reference databases to determine the taxonomic composition of the sample).

Applications in Animal Health and Disease

Metagenomics has been used to explore the functional potential of microbial communities in various animal systems. For example, metagenomic studies have identified genes associated with antibiotic resistance, pathogenicity, and nutrient metabolism. This information is valuable for understanding the impact of microbial communities on animal health, disease outbreaks, and the efficacy of treatments.

Metatranscriptomics

Metatranscriptomics analyzes the actively transcribed genes within a microbial community, providing insights into the functional activity of the microbiome. By sequencing RNA rather than DNA, metatranscriptomics captures the gene expression profiles of microorganisms, revealing their active metabolic processes and responses to environmental changes. Metatranscriptomics is performed using high-throughput sequencing platforms similar to those used for metagenomics. Illumina sequencing is commonly used due to its high throughput and accuracy. Long-read technologies can also be applied to improve the resolution of transcriptomic data.

- *Illumina Sequencing*: Illumina platforms are widely used for metatranscriptomics due to their ability to provide deep sequencing coverage and high accuracy.
- *Long-Read Sequencing*: Technologies like PacBio and Oxford Nanopore can be used to capture full-length transcripts, which improves the accuracy of transcript quantification and functional annotation. The bioinformatics pipeline for metatranscriptomics involves several key steps including; Quality Control (Raw sequencing reads are filtered and cleaned to remove low-quality sequences and contaminants. Tools like FastQC and Trimmomatic are used for quality control), Read Mapping (Reads are mapped to reference genomes or transcriptomes using tools such as HISAT2, STAR, or Bowtie2. Accurate read mapping is essential for quantifying gene expression), Differential Expression Analysis (Differential expression analysis identifies genes that are differentially expressed between conditions or time points. Tools like DESeq2, edgeR, and Cuffdiff are commonly used for this analysis) and Functional Annotation (Transcripts are annotated with functional information using databases such as GO, KEGG, and PFAM. This step provides insights into the biological functions and pathways represented in the transcriptomic data).

Applications in Animal Health and Disease

Metatranscriptomics has been used to study the functional activity of microbial communities in various animal systems. For example, metatranscriptomic studies have revealed how microbial communities respond to dietary changes, environmental stressors, and infections. This information is valuable for understanding the functional dynamics of the microbiome and its impact on animal health and disease.

References

Gao, Bei, Liang Chi, Yixin Zhu, Xiaochun Shi, Pengcheng Tu, Bing Li, Jun Yin, Nan Gao, Weishou Shen, and Bernd Schnabl. "An introduction to next generation sequencing bioinformatic analysis in gut microbiome studies." *Biomolecules* 11, no. 4 (2021): 530.

Kumar, Ajay, and Anuj Yadav. "Next Generation Sequencing in Metagenomics and Metatranscriptomics." In *Multi-Omics Analysis of the Human Microbiome: From Technology to Clinical Applications*, pp. 49-75. Singapore: Springer Nature Singapore, 2024.

Mota, Vera T., Tiago P. Delforno, Jaqueline C. Ribeiro, Marcelo Zaiat, and Valéria M. de Oliveira. "Understanding microbiome dynamics and functional responses during acidogenic fermentation of sucrose and sugarcane vinasse through metatranscriptomic analysis." *Environmental Research* 246 (2024): 118150.

Nikoloudaki, Olga, Francis Aheto, Raffaella Di Cagno, and Marco Gobbetti. "Synthetic microbial communities: A gateway to understanding resistance, resilience, and functionality in spontaneously fermented food microbiomes." *Food Research International* (2024): 114780.

Pinto, Yishay, and Ami S. Bhatt. "Sequencing-based analysis of microbiomes." *Nature Reviews Genetics* (2024): 1-17.

Rao, M. Kasi, and Gururaj Sunkad. "Metaomics approaches to unravel the functioning of multispecies microbial communities." In *Microbiome Drivers of Ecosystem Function*, pp. 395-416. Academic Press, 2024.

Schneider, Tamar. "The microbiome function in a host organism: A medical puzzle or an essential ecological environment?." *Biological Theory* 19, no. 1 (2024): 44-55.

Tilgam, Jyotsana, Deepanshu Jayaswal, Mushineni Ashajyothi, Jyoti Prakash Singh, Adarsh Kumar, and Hillol Chakdar. "Meta-omics approaches for understanding and exploring soil microbial communities for sustainable agriculture." In *Applications of Metagenomics*, pp. 3-22. Academic Press, 2024.

Zhang, Wenke, Xiaoqian Fan, Haobo Shi, Jian Li, Mingqian Zhang, Jin Zhao, and Xiaoquan Su. "Comprehensive assessment of 16S rRNA gene amplicon sequencing for microbiome profiling across multiple habitats." *Microbiology Spectrum* 11, no. 3 (2023): e00563-23.

Zhao, Yanmei, Zihan Dai, Wanying Zhang, Zelin Lei, Ai-Jie Wang, Lu Fan, and Shu-Hong Gao. "High-throughput sequencing-based bioinformatics identification technologies for emerging biological risk factors." In *Water Security: Big Data-Driven Risk Identification, Assessment and Control of Emerging Contaminants*, pp. 251-272. Elsevier, 2024.

Chapter 23

The Epigenetic Modulation of the Gut Microbiome

Felix Uchenna Samuel[1,*]
Henry Olanrewaju Jegede[2]
John Fred Odhiambo[3]
and Ifetoluwani Oguntoyinbo[4]
[1]Animal Science Program: Alabama Cooperative Extension, Alabama A&M University, USA
[2]Tilad Environmental Consultant, Ritadh, Saudi Arabia
[3]Department of Food and Animal Science, College of Agricultural, Live and Natural Science Alabama A&M University, USA
[4]Animal and Comparative Biomedical Sciences, University of Arizona, Tucson, USA

Abstract

The gut microbiome plays a pivotal role in animal health by influencing various physiological processes, including digestion, immunity, and metabolism. Recent research has highlighted the intricate interplay between epigenetic mechanisms and the composition and function of the gut microbiota. Epigenetic modifications, such as DNA methylation, histone modifications, and non-coding RNA regulation, dynamically regulate gene expression patterns without altering the underlying DNA sequence. These mechanisms are increasingly recognized for their ability to shape microbial diversity and community structure within the gut. In animal models, epigenetic changes induced by diet, environmental factors, and host genetics have been shown to impact the gut microbiome composition. For instance, dietary components rich in methyl donors can alter DNA methylation patterns in both the host and gut microbes, influencing microbial growth and metabolism. Conversely, perturbations in gut microbiota composition can lead to dysregulation of host epigenetic mechanisms, potentially exacerbating disease susceptibility. Understanding the epigenetic modulation of the gut microbiome is crucial for elucidating mechanisms underlying various animal diseases. Studies have implicated epigenetic alterations in gut microbiota dysbiosis associated with conditions such as inflammatory bowel disease, obesity, and metabolic disorders in animals. Furthermore, epigenetic interventions, including dietary modifications and microbial-based therapies, hold promise for manipulating the gut microbiome to promote animal health and mitigate disease risks.

* Corresponding Author's Email: felixsam75@yahoo.com

In: Microbiomes in Animal Nutrition, Health, and Disease
Editor: Tanmoy Rana
ISBN: 979-8-89530-425-9

Keywords: epigenetics, animal health, Disease, DNA methylation, Host-microbe interactions, microbial diversity

Introduction: Unveiling the Dynamic Interplay and the Gut Microbiome

The two major terminologies - Gut Microbiome and Epigenetics have been at the forefront of scientific research in recent years, with growing evidence highlighting their intricate interconnections in shaping various aspects of human health and disease (Alam et al. 2017; Li et al. 2016). The gut microbiome, which encompasses the vast array of microbes residing within the gastrointestinal tract, has been recognized as a critical player in orchestrating diverse physiological and biochemical processes (Alam et al. 2017; Clarke et al. 2019; Gibbons 2019; Li et al. 2016). Concurrently, epigenetic mechanisms, such as the methylation of DNA, histone modifications, and the regulation of non-coding RNA, have emerged as essential modulators of gene expression and cellular function (Alam et al. 2017).

Intriguingly, the gut microbiome and epigenetic alterations appear to be intimately linked, with the former exerting significant influence on the latter through various pathways (Alam et al. 2017). This intricate interplay has been implicated in the pathogenesis of a wide range of disorders, including metabolic, immunological, and neurological conditions (Alam et al. 2017; Li et al. 2016).

- *The Gut Microbiome:* A Hidden World Influencing Human Health: The gut microbiome, composed of an estimated 100 trillion microbial cells, represents a vast and dynamic ecosystem that plays a crucial role in human physiology (Clarke et al. 2019). These microbes engage in a myriad of symbiotic and commensal relationships with their human host, contributing to the regulation of immune function, nutrient absorption, and even the production of various neurotransmitters like serotonin (Alam et al. 2017). Recent studies have further suggested a direct effect of the gut microbiome on mental illness and brain function (Rogers et al. 2016).
- *Epigenetics:* Orchestrators of Gene Expression: Epigenetics refers to the heritable changes in gene expression that occur without alterations in the DNA sequence. This happens by (Begolli et al. 2019; Zheng et al. 2014) the addition of chemical groups, such as methyl groups, to DNA or histone proteins after the translation process. Epigenetic modifications can be influenced by various environmental factors, including nutrition (Begolli et al. 2019), drugs (Jessberger et al. 2007), stress (Franklin et al. 2010; Weiss et al. 2011). These factors can often lead to long-term changes in gene function and metabolism in affected individuals (Zheng et al. 2014). The field of epigenetics has multifaceted interfaces, and in this text, we will further discuss their roles in the modulation of microbiomes. In addition, epigenetic mechanisms can affect the microbiome, leading to changes in gene expression and cellular function within the gut ecosystem (Begolli et al. 2019; Krautkramer et al. 2017;Zheng et al. 2014).
- *The Intricate Gut Microbiome-Epigenetics Interplay*: Mounting evidence suggests a profound impact of the gut microbiome on epigenetic processes which can potentially lead to alterations in gene expression and the metabolic pathways (Li et al. 2016). This dynamic interplay has been implicated in the pathogenesis of a wide range of disorders,

including autoimmune disease (Chen, Sun, and Zhang 2017), metabolic syndromes such as type 2 diabetes (Ling, and Groop 2009), and neurological conditions (Rogers et al. 2016).

- *Navigating the Epigenetic Landscape of the Gut*: An understanding of the mechanisms by which the gut microbiome shapes the epigenetic landscape and influences human health is a crucial step in developing targeted therapeutic interventions and personalized approaches to disease prevention and management. These two would definitely have a vice-versa relationship with each other, and their interplay can have a significant impact on various aspects of human health and disease.

Epigenetic Mechanisms - Deciphering the Language of Gene Regulation

The mammalian genome is a complex and highly regulated system, with epigenetic modifications playing a pivotal role in fine-tuning gene expression patterns. The influence of various environmental factors, including the gut microbiome on these epigenetic mechanisms cannot be overstated (Alam et al. 2017; Choi et al. 2013; Xia et al. 2023).

- *DNA Methylation*: One of the well-studied epigenetic mechanisms, DNA methylation involves the addition of methyl groups to cytosine residues, often resulting in transcriptional silencing or activation of specific genes. Dietary factors, such as folate, vitamin B12, and choline, have been shown to influence DNA methylation patterns, through the donation of the methyl group, potentially contributing to the development of chronic diseases. (Kadayifci, Zheng, and Pan 2018)
- *Histone Modifications*: Present in chromosomes, histones, the protein structures around which DNA is wrapped, undergo a variety of post-translational modifications, such as acetylation, citrullination, methylation, phosphorylation, sumoylation, and crotonylation. These modifications can significantly alter chromatin structure and gene expression. Certain nutrients, including butyrate, a short-chain fatty acid produced by gut microbes, have been found to modulate histone acetylation, thereby influencing cellular processes (Delage and Dashwood 2008)
- *RNA-mediated targeting*: Non-coding RNAs, such as microRNAs and long non-coding RNAs, have emerged as important epigenetic regulators, capable of silencing gene expression through mechanisms like mRNA degradation and translational repression (Paul et al. 2015). These molecules are sensitive to environmental cues, including dietary factors, and can be influenced by the gut microbiome. The gut microbiome has been found to influence the expression and activity of these non-coding RNAs, which can further impact epigenetic processes and disease susceptibility.
- *Chromatin remodelling*: The organization of chromatin, the complex of DNA and histone proteins, is another crucial epigenetic mechanism. This process, mediated by ATP-dependent chromatin remodeling complexes, can dynamically alter the accessibility of DNA and thereby influence gene expression patterns. These chromatin remodeling complexes utilize the energy of ATP hydrolysis to mobilize, reposition, or remove nucleosomes, effectively modulating the compaction and structure of

chromatin. This allows for the selective exposure or shielding of regulatory regions, enabling or repressing transcription as required by cellular and environmental cues (Lezzerini and Riedel 2017).

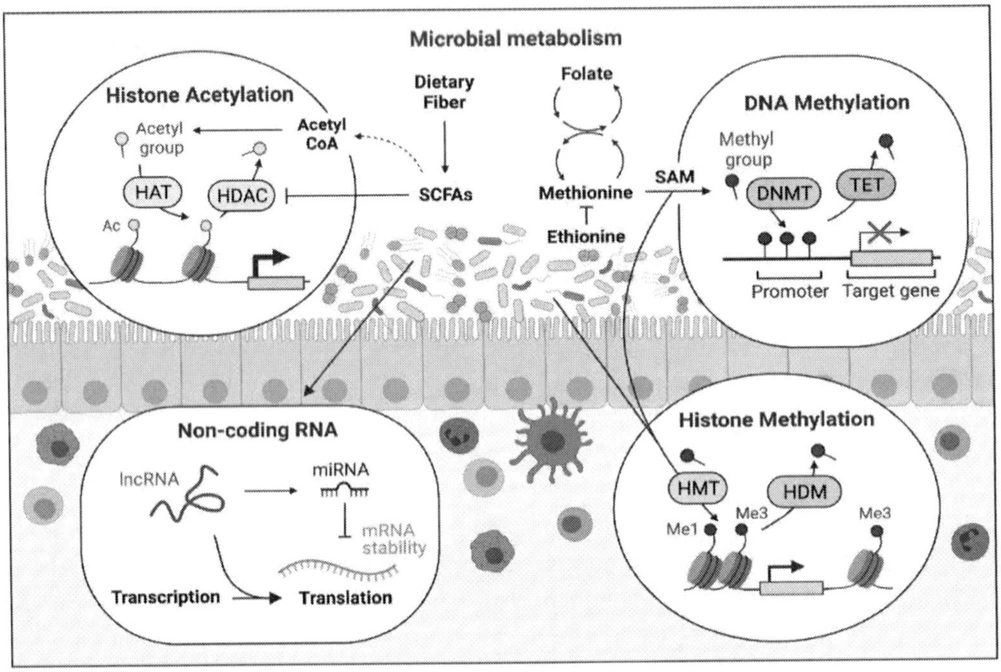

Figure 1. Intestinal microbiota interact with the mammalian epigenome through production of epigenetic substrates or regulators of chromatin-modifying enzymes (Woo and Alenghat, 2022).

Epigenetic Orchestration of the Gut Microbiome-Shaping Microbial Diversity and Function

The bidirectional relationship that exists between the gut microbiome and epigenetic mechanisms has been unveiled in recent studies. These studies show that the microbiome can modulate epigenetic processes, and in turn, epigenetic alterations can shape the composition and function of the gut microbial community (Alam et al. 2017; Delage and Dashwood 2008; Kadayifci et al. 2018; Paul et al. 2015).

- *Host Epigenetics as a Gatekeeper*: The epigenetic landscape of the host can influence the composition and diversity of the gut microbiome. Epigenetic modifications, such as DNA methylation and histone acetylation, can regulate the expression of genes involved in host-microbe interactions, immune responses, and intestinal barrier function, thereby shaping the gut microbial ecosystem. For instance, studies have demonstrated that dietary patterns and specific nutrients can alter the methylation status of genes associated with the immune system, which in turn can impact the gut microbiome and susceptibility to inflammatory conditions (Delage and Dashwood 2008; Paul et al. 2015).

Table 1. Epigenetic Sculptors of Gut Microbiome Dynamics: Mechanisms, Impacts, and Implications

Epigenetic Mechanism	Description	Impact on Gut Microbiome	Examples	Potential Benefits	Potential Risks
DNA Methylation	Addition of methyl groups to DNA, typically at CpG sites.	Alters gene expression by silencing or activating genes involved in microbial interactions and immune responses.	Methylation changes in genes encoding antimicrobial peptides or mucin production can influence microbial colonization.	Can suppress harmful inflammation and support immune tolerance.	May silence beneficial genes, potentially leading to increased susceptibility to infections.
Histone Modification	Chemical modifications to histone proteins (e.g., acetylation, methylation).	Modifies chromatin structure, influencing the accessibility of DNA to transcription machinery and thereby affecting gene expression.	Acetylation of histones in genes related to immune regulation can impact gut microbial composition.	Enhances the expression of beneficial genes involved in gut health and immunity.	Aberrant modifications can lead to dysregulated gene expression and disease.
RNA-mediated targeting	RNA molecules that do not encode proteins but regulate gene expression at the transcriptional and post-transcriptional levels.	Can influence microbial gene expression directly or host genes involved in microbial recognition and immune responses.	miRNAs that target mRNAs of immune-related genes can modulate the host response to gut microbiota.	Can fine-tune immune responses and maintain microbial balance.	Dysregulation may lead to improper immune responses or chronic inflammation.
Chromatin Remodeling	Structural changes to chromatin that affect its compactness and accessibility.	Regulates the expression of genes involved in maintaining gut barrier integrity and microbial interactions.	Changes in chromatin structure in intestinal epithelial cells can affect susceptibility to dysbiosis.	Maintains gut barrier integrity and supports healthy microbial interactions.	Can increase susceptibility to dysbiosis if remodeling leads to barrier dysfunction.
Microbial Metabolites	Metabolites produced by gut microbiota that can act as epigenetic modulators.	Influence host epigenetic mechanisms, impacting gene expression and overall gut health.	Short-chain fatty acids (e.g., butyrate) produced by gut bacteria can inhibit histone deacetylases, promoting anti-inflammatory gene expression.	Supports anti-inflammatory pathways and gut health.	Overproduction or imbalances in metabolites can disrupt epigenetic regulation.
Environmental Influences	Factors such as diet, stress, and toxins that can cause epigenetic changes.	Induce epigenetic modifications that affect gut microbiome composition and function.	High-fat diets can lead to DNA methylation changes that alter gut microbiota and increase inflammation.	Can promote beneficial epigenetic changes through a healthy diet and lifestyle.	Adverse environmental factors can induce harmful epigenetic changes, increasing disease risk.

These processes can create a feedback loop, where alterations in the gut microbiome can further impact the host epigenome thus controlling the microbial colonization and growth. This bidirectional relationship allows the gut microbiome to influence the host's epigenetic landscape, which in turn can shape the composition and diversity of the microbial community within the gut. The intricate interplay between the microbiome and epigenetics can have far-reaching implications on various aspects of human health, from the regulation of metabolism and immune system modulation to neurological processes (Jia, 2020).

- *Epigenetic influence on Microbial metabolism and function*: The gut microbiome can produce a diverse array of metabolites, some of which have the potential to modulate epigenetic mechanisms in the host (Paul et al. 2015). Shaping the production of key metabolites, such as short-chain fatty acids (SCFAs), vitamins, and neurotransmitters, the gut microbiome can indirectly influence epigenetic processes (D'Aquila et al. 2020). Short-chain fatty acids, like propionate and butyrate, have been shown to inhibit histone deacetylases, resulting in increased acetylation of histones, leading to altered gene expression patterns that can influence host health. Furthermore, the gut microbiome's ability to produce vitamins and co-factors, such as folate and vitamin B12, can impact DNA methylation dynamics in the host.
- *Epigenetics and the immune system*: The influence of gut microbiome on the host's immune system is another crucial aspect of this complex interplay. Epigenetic modifications can regulate the expression of genes involved in immune function, including those responsible for the development and differentiation of immune cells, such as T cells and B cells (Héninger et al. 2015). Fine-tuning the host's immune system to the commensal microbes can therefore be controlled through epigenetic mechanisms that promote immune tolerance (Table 1). This process of epigenetically-mediated immune tolerance allows the host to maintain a balanced and regulated immune response to the gut microbiome, preventing excessive inflammation or autoimmunity. Through a rather complex communication between the gut microbiome and the host's epigenome, this inter relationship is able to play a pivotal role in shaping the immune system and maintaining homeostasis within the gut ecosystem (El Aidy et al. 2012; Lee and Mazmanian 2010).

Microbial Feedback - How the Gut Microbiome Shapes Host Epigenetics

A direct influence of the gut microbiome on the host's epigenetic landscape can also occur directly through the production of various metabolites that can act as epigenetic regulators. These include:

1. *Microbial Metabolites as Epigenetic Modulators*: For epigenetic-modifying enzymes to catalyze alterations to the host chromatin, they require appropriate substrates (Woo and Alenghat, 2022). For instance, a typical requirement for DNA and histone methyltransferases (DNMTs and HMTs) and histone acetyltransferases (HATs) to carry out their catalytic functions are methyl donors for DNMTs and HMTs and acetyl donors for HATs. While many of these donor substrates can be generated from host-

intrinsic pathways, the role of the gut microbiota is being increasingly appreciated since it serves as an additional source of these molecules. Indeed, the microbiota can synthesize a plethora of biological compounds including several that serve as epigenetic substrates, cofactors or regulators of epigenetic enzyme activity (Miro-Blanch and Yanes, 2019;Shock et al. 2021). In particular, the intestinal microbiota is known to generate folate and other B vitamins (B2, B12) that donate methyl groups for DNA or histone methylation (Figure 1). Folate, produced by numerous commensal microbes including probiotic *Bifidobacterium* and *Lactobacillus species* that participate in one-carbon metabolism to generate S-adenosylmethionine (SAM), is an essential nutrient and the primary substrate for DNA and histone methylation (Rossi et al. 2011). Commensal microbes can also metabolize dietary methionine into SAM (Chen, 2016, Chen, 2017). Thus, changes in bacterial composition can influence SAM availability and correspondingly shift the DNA or histone methylation status in the host. Short-chain fatty acids (SCFAs) (discussed in more detail below) represent another important group of epigenetically relevant molecules that are exclusively produced by commensal microbes through fermentation of complex non-digestible carbohydrates and fiber. Importantly, the inhibition of deacetylase activity of histone deacetylases (HDACs) by SCFAs, results in chromatin changes usually associated with increased expression of target genes

- Short-chain fatty acids (SCFAs): As stated earlier, they are produced by the fermentation of dietary fiber by gut microbes, SCFAs like acetate, butyrate, and propionate can inhibit histone deacetylases, leading to increased histone acetylation and altered gene expression patterns (D'Aquila et al. 2020;Krautkramer et al. 2016).
- Vitamins and cofactors: The gut microbiome can synthesize essential vitamins, such as folate, vitamin B12, and choline, which serve as cofactors for enzymes involved in DNA methylation and histone modifications.
- Alteration of the gut microbiome and its metabolic output can therefore have widespread epigenetic consequences in the host, impacting a variety of physiological processes and disease susceptibility (Alam et al. 2017; D'Aquila et al. 2020; Krautkramer et al. 2016; Krautkramer, et al. 2017).

2. *Dysbiosis and Epigenetic Disruption:* Imbalances and alterations in the gut microbial community, known as dysbiosis, have been linked to various disease states, including inflammatory bowel disease, metabolic disorders, and neurological conditions (Alam et al. 2017; Krautkramer et al. 2017; Shenderov 2016; Smolińska and O'Mahony 2016). These dysbiotic states can contribute to epigenetic disruption, either through the production of deleterious metabolites or the depletion of beneficial microbial products. The resulting epigenetic changes can lead to persistent alterations in gene expression, contributing to the development and progression of these diseases (Alam et al. 2017; Gargano et al. 2022; Krautkramer et al. 2017; Smolińska and O'Mahony 2016).

The Interplay between Dysbiosis and Epigenetic Disruption

1. *Impact of Dysbiosis on Epigenetics:* The relationship between the gut flora and epigenetics ensures that the impact of dysbiosis can be quite significant on the host's epigenome. Some of the impacts are further discussed below.
    - *Microbial Metabolites*: Gut microbiota produce metabolites (e.g., short-chain fatty acids, SCFAs) that can influence epigenetic marks. SCFAs, such as butyrate, can inhibit histone deacetylases (HDACs), leading to increased histone acetylation and altered gene expression (D'Aquila et al. 2020; Krautkramer et al. 2016).
    - *Inflammation:* Dysbiosis can trigger inflammation, which influences the epigenome. Chronic inflammation can lead to the activation of signaling pathways that modify epigenetic marks (Suzuki et al. 2015), thereby altering the expression of genes involved in immune responses and tissue repair.
    - *Immune System*: Dysbiosis can affect the immune system, leading to changes in immune cell function and epigenetic regulation (Everard and Cani 2013). Immune cells rely on specific epigenetic modifications for proper differentiation and function.
2. *Impact of Epigenetic Disruption on Dysbiosis:* Conversely, epigenetic changes can also affect the overall growth and maintenance of the gut microbiomes in several ways.
    - *Gene Expression*: Epigenetic changes can influence the expression of genes involved in maintaining the gut barrier, immune responses, and microbial recognition. Disruption in these processes can lead to an altered gut environment, favoring dysbiosis (Qin and Wade 2018).
    - *Host-Microbe Interactions*: Epigenetic modifications can affect the host's ability to interact with and regulate its microbiota. For instance, changes in the expression of pattern recognition receptors (PRRs) arising from epigenetic modifications can influence microbial composition and abundance.
    - *Metabolic Pathways*: Epigenetic regulation of metabolic pathways can impact the availability of nutrients and metabolites in the gut, thereby influencing the composition and function of the gut microbiota.

Examples and Evidence

Diet-Induced Dysbiosis

- *High-Fat Diet*: Consumption of a high-fat diet can lead to dysbiosis and is associated with changes in DNA methylation patterns in the host (Kaczmarek, Musaad, and Holscher 2017). These epigenetic changes can alter the expression of genes involved in metabolism and inflammation, exacerbating the effects of dysbiosis.
- *Fiber Deficiency*: The consumption of low-fiber diets can reduce the production of SCFAs by gut microbiota, resulting in decreased histone acetylation in the host and increased inflammation (Sonnenburg et al. 2016). This epigenetic disruption can further contribute to gut barrier dysfunction and dysbiosis.

Disease States

- *Inflammatory Bowel Disease (IBD):* The gut epithelium of patients with IBD exhibit both dysbiosis and widespread epigenetic changes. Inflammation-driven epigenetic modifications can alter gene expression patterns, contributing to disease pathology and the maintenance of dysbiosis.
- *Obesity and Metabolic Syndrome*: Dysbiosis is common in obesity and metabolic syndrome, conditions also associated with significant epigenetic alterations in metabolic tissues (Boulange et al. 2016). Epigenetic changes in genes involved in glucose and lipid metabolism can contribute to the persistence of dysbiosis and metabolic dysfunction. In summary, the gut microbiome and the host's epigenome engage in a complex, bidirectional relationship.

Microbial Signals and Epigenetic Regulation of Host Immunity

The gut microbiome can also influence the epigenome of the host's immune cells, modulating their function and responsiveness. Specific microbial signals, such as bacterial lipopolysaccharides and short-chain fatty acids, can induce epigenetic modifications in immune cells, leading to altered gene expression profiles and immune cell differentiation (Abt and Artis 2009; Gargano et al. 2022). This process allows the gut microbiome to fine-tune the host's immune system, promoting tolerance towards commensal microbes and preventing excessive inflammation (Alam et al. 2017; Yadav 2016).

Regulation of Immune Cells by SCFAs

SCFAs play a huge role in mediating between the mucosal immune cell populations and the gut microbiota. Notably, immune cell regulation by SCFAs extends to innate and adaptive immune compartments locally in the intestine as well as systemically. This occurs through the inhibition of HDAC and the activation of G-protein coupled receptors (GPR) (Macia et al. 2015; Maslowski et al. 2009). Treating intestinal or bone marrow-derived macrophages with butyrate led to the inhibition of HDAC activity and an increase in histone H3 acetylation at secondary response genes and mediators of STAT6 signaling (Chang, 2014; Ji et al. 2016). However, contrary to the predicted inhibitory effect of SCFAs on HDACs, a decreased response to butyrate during the expression of these same genes was observed, reflecting an anti-inflammatory M2 macrophage state that attenuated intestinal inflammation in response to DSS-induced colitis (Chang, 2014; Ji et al. 2016). Additionally, butyrate, and propionate (to a lesser degree), promoted monocyte to macrophage differentiation and activation of an antimicrobial transcriptional program through HDAC3 inhibition thereby improving the defense system against enteric pathogens (Schulthess et al. 2019). The production and acetylation of IL-22 histone by ILCs and CD4+ T cells was also promoted by SCFAs through the activation of GPR41 and inhibition of HDAC resulting in enhanced protection from intestinal inflammation during an infection (Yang et al. 2020). The inhibition of HDACs and the activation of GPR109A by butyrate in colonic macrophages and dendritic cells, enables them to promote anti-inflammatory regulatory T cells (Tregs) (Arpaia, 2013; Kaisar et al. 2017; Singh et al. 2014). This means that the control of epigenetic mechanisms and transcriptional responses in immune cells by SCFAs aids an overall anti-inflammatory phenotype during chemically induced colitis and pathogenic infection. Tregs, defined by their expression of the transcription factor Foxp3 act to suppress anomalous inflammatory and effector responses, thereby playing a crucial role in immune homeostasis and maintenance of microbiota tolerance (Boulangé, 2016).

Treatment with Clostridia-derived butyrate, or colonization with the species was reported by Furusawa et al. (2013) to increase the frequency of circulating Tregs by inducing acetylation on histone H3 within regulatory regions of the Foxp3 locus. The acetylation of Foxp3 proteins post-transcriptionally also occurred in peripheral CD4+ T cells, thereby enhancing expression and protein stability (Arpaia, 2013). In accordance with the above findings, the deletion of HDAC3, HDAC6, or HDAC9 or treatment with SCFA propionate caused an increased H3K9 acetylation at the Foxp3 locus and Foxp3 protein stability enhancement in mice (Beier, 2012; Smith et al. 2013; Wang et al. 2015). Acetate, another microbiota-derived SCFA, also inhibited HDAC activity and increased overall acetylation of histones H3 (K9) and H4 (K5/8/16) in CD3/CD28 activated primary human CD4+ T cells (Bolduc, 2017). HDAC inhibition by acetate is still however somewhat controversial (Waldecker et al. 2008; Arpaia, 2013; ugder, Kamareddine, and Watnick 2021).

Elevations in butyrate levels induced the expression and H3 acetylation of Th1-associated genes in T cells under Treg-inducing conditions. The expression of the transcription factors for Th17 and Th2 (RORγt and Gata3 respectively), were however not induced (Kespohl et al. 2017). Administration of higher concentrations of butyrate to GF mice resulted in elevated IFNγ production. This was reported to exacerbate disease in DSS-induced colitis. Similarly, another study found that butyrate promoted Th1, but suppressed Th17 polarization from naïve T cells in vivo and in culture by inhibiting HDAC activity independently from GPR43 activation to induce expression of T-bet and the anti-inflammatory cytokine IL-10 (Chen, 2019). Wu et al. (2020) reported HDAC3 inhibition in IECs in response to *Faecalibacterium prausnitzii* colonization. Also, pentanoate, an SCFA, also reduced IL-17A production by CD4+ T cells through HDACs. It also induced IL-10 gene histone acetylation and expression in CD4+ T cells and regulatory B cells, thus dampening T-cell mediated autoimmune responses (Luu et al. 2019). SCFAs exert context and concentration dependent effects which promote a low Th17/high Treg phenotype that is protective against intestinal inflammation. CD8+ cytotoxic T lymphocytes (CTLs) possess cancer targeting abilities and defend against intracellular pathogens. Recent findings demonstrate that some SCFAs, can increase effector molecule expression in splenic and mesenteric CD8+ T cells independently from GPR41 or GPR43 activation, subsequently leading to an enhanced anti-tumor activity (Luu et al. 2018). It was also discovered that *Megasphaera massiliensis*- derived pentanoate and butyrate improved the anti-tumor reactivity of antigen-specific CD8+ T cells through an increase in their production of effector cytokines IFNγ and TNF-α through HDAC inhibition (Luu et al. 2021).

Furthermore, studies have shown that increased SCFAs from high-fiber feeding encouraged CD8+ T cell activation and later transition into memory cells during viral infections in mice (Bachem, 2019; Trompette et al. 2018). Microbiota-derived SCFAs also contributed to intestinal B cell differentiation and antibody production by inhibiting HDAC activity invariably leading to increased histone acetylation and gene expression of multiple genes associated with B cell function (Kim et al. 2016). These effects were observed in human and mouse B cells treated specifically with butyrate and propionate with a dose-dependent counteraction of HDAC-mediated H3K9 deacetylation and transcriptional repression of miRNAs that target genes involved in B cell differentiation (Sanchez et al. 2020).

Additionally, normal B cell intrinsic functions such as immunoglobulin class switching and somatic hypermutation were restricted by butyrate and propionate. This shows that, microbiota-derived SCFAs have potent immunomodulatory effects on both innate and adaptive immune cells in the host that actively maintain homeostasis and dampen inflammation in the intestine.

Regulation of Immune Cells by MicroRNAs

Microbiota has been observed to induced lncRNAs and were also found to be highly expressed in the spleen and thymus, suggesting that microbiota-dependent lncRNAs may be involved in host immune regulation (Liang et al. 2015).

Transgenerational Epigenetic Inheritance

Evidence from current studies suggests that the gut microbiome may also play a role in transgenerational epigenetic inheritance, with epigenetic modifications being passed down across generations (Alam et al. 2017; Krautkramer et al. 2016; Otaru et al. 2021; Paul et al. 2015). This is possibly mediated through the transmission of microbial metabolites or other signals that can influence the epigenome of the offspring. This could potentially lead to long-term health consequences.

- *Inheritance Pathways*: The pathways by which these modified epigenomes can be passed down from one generation to another are:
    - *Germline Transmission*: The germline cells (sperm and egg) can serve as means by which epigenetic modifications can be passed down thus, affecting the development and phenotype of the next generation.
    - *Germline Reprogramming*: Although most epigenetic marks are erased during the formation of germ cells to reset the epigenome for the next generation, some marks can escape this reprogramming and be inherited.
    - *Somatic to Germline*: Environmental factors have been known to affect somatic cells. These changes can sometimes influence germline cells, leading to transgenerational inheritance (Feng and Chen 2015).

Examples of Transgenerational Epigenetic Inheritance

- *Agouti Mice:* In mice, the coat color and susceptibility to obesity and diabetes can be influenced by epigenetic changes at the Agouti gene, which can be transmitted to offspring depending on the mother's diet (Jirtle 2014).
- *Human Studies:* Epidemiological studies have suggested that grandchildren of individuals exposed to famine or extreme stress may exhibit health effects due to inherited epigenetic changes. For example, the Dutch Hunger Winter study showed that prenatal exposure to famine was associated with transgenerational effects on health.

Future Horizons - Harnessing the Epigenetic-Microbiome Axis for Animal Health and Disease

1. *Personalized Nutrition and Therapeutics*: Understanding the multifaceted interaction between the host epigenome and the gut microbiome can lead to the development and advancement of tailored nutritional and therapeutic strategies to modulate the microbiome and its epigenetic effects.
2. *Early-Life Interventions*: Targeting the microbiome and epigenome during critical developmental windows, such as pregnancy and early childhood, may have long-lasting impacts on health and disease prevention.
3. *Biomarker Discovery*: The identification of specific epigenetic and microbial signatures associated with different disease states will aid in the development of novel diagnostic and prognostic biomarkers.

Conclusion

In conclusion, the gut microbiome and the host epigenome are deeply intertwined, with each influencing the other in complex and dynamic ways. An understanding of these interactions is crucial for advancing our knowledge of health, disease, and the potential for personalized preventative and therapeutic approaches (Bordoni and Gabbianelli 2019; Choi et al. 2013; Lismer and Kimmins 2023; Shenderov 2016).

References

Abt, Michael C., and David Artis. "The intestinal microbiota in health and disease: the influence of microbial products on immune cell homeostasis." *Current Opinion in Gastroenterology* 25, no. 6 (2009): 496-502. https://doi.org/10.1097/mog.0b013e328331b6b4.

Alam, Reza, Hamid M. Abdolmaleky, and Jin-Rong Zhou. "Microbiome, inflammation, epigenetic alterations, and mental diseases." *American Journal of Medical Genetics Part B: Neuropsychiatric Genetics* 174, no. 6 (2017): 651-660. https://doi.org/10.1002/ajmg.b.32567.

Arpaia, N. "Metabolites produced by commensal bacteria promote peripheral regulatory T-cell generation." *Nature* 504, no. 7480 (2013): 451-455.

Bachem, A. "Microbiota-derived short-chain fatty acids promote the memory potential of antigen-activated CD8+ T cells." *Immunity* 51, no. 2 (2019): 285-297.

Begolli, Rodiola, Nikos Sideris, and Antonis Giakountis. "LncRNAs as chromatin regulators in cancer: from molecular function to clinical potential." *Cancers* 11, no. 10 (2019): 1524.1524-1524. https://doi.org/10.3390/cancers11101524.

Beier, Ulf H. "Histone deacetylases 6 and 9 and sirtuin-1 control Foxp3+ regulatory T cell function through shared and isoform-specific mechanisms." *Science Signaling* 5, no. 229 (2012): ra45-ra45.

Bolduc, J. "Epigenetic metabolite acetate inhibits class I/II histone deacetylases, promotes histone acetylation, and increases HIV-1 integration in CD4+ T cells." Journal of Virology 91, no. 16 (2017): 10-1128.

Bordoni, Laura, and Rosita Gabbianelli. "Primers on nutrigenetics and nutri(epi)genomics: Origins and development of precision nutrition." B*iochimie* 160 (2019): 156-171. https://doi.org/10.1016/j.biochi.2019.03.006.

Boulangé, C. "Impact of the gut microbiota on inflammation, obesity, and metabolic disease." *Genome Medicine* 8 (2016): 1-12.

Chang, PV. "The microbial metabolite butyrate regulates intestinal macrophage function via histone deacetylase inhibition." *Proceedings of the National Academy of Sciences* 111, no. 6 (2014): 2247-2252.

Chen, B. "Integration of microbiome and epigenome to decipher the pathogenesis of autoimmune diseases." *Journal of Autoimmunity* 83 (2017): 31-42.

Chen, H. "Progress in the microbial production of S-adenosyl-L-methionine." *World Journal of Microbiology and Biotechnology* 32 (2016): 1-8.

Chen, L "Microbiota metabolite butyrate differentially regulates Th1 and Th17 cells' differentiation and function in induction of colitis." *Inflammatory Bowel Diseases* 25, no. 9 (2019): 1450-1461.

Choi, Sang-Woon, et al. "Nutritional epigenomics: a portal to disease prevention." *Advances in Nutrition* 4, no. 5 (2013): 530-532. https://doi.org/10.3945/an.113.004168.

Clarke, Gerard, et al. "Gut reactions: breaking down xenobiotic–microbiome interactions." *Pharmacological Reviews* 71, no. 2 (2019): 198-224. https://doi.org/10.1124/pr.118.015768.

D'Aquila, Patrizia, et al. "Gut microbiota as important mediator between diet and DNA methylation and histone modifications in the host." *Nutrients* 12, no. 3 (2020): 597.https://doi.org/10.3390/nu12030597.

Delage, Barbara, and Roderick H. Dashwood. "Dietary manipulation of histone structure and function." *Annual Review of Nutrition* 28, no. 1 (2008): 347-366. https://doi.org/10.1146/annurev.nutr.28.061807.155354.

El Aidy, Sahar, et al. "Temporal and spatial interplay of microbiota and intestinal mucosa drive establishment of immune homeostasis in conventionalized mice." *Mucosal Immunology* 5, no. 5 (2012): 567-579.

Everard, Amandine, and Patrice D. Cani. "Diabetes, obesity and gut microbiota." *Best Practice & Research Clinical Gastroenterology* 27, no. 1 (2013): 73-83.

Feng, Lijuan, and Xin Chen. "Epigenetic regulation of germ cells—remember or forget?." *Current Opinion in Genetics & Development* 31 (2015): 20-27. https://doi.org/10.1016/j.gde.2015.04.003.

Franklin, Tamara B., et al. "Epigenetic transmission of the impact of early stress across generations." *Biological Psychiatry* 68, no. 5 (2010): 408-415.

Furusawa, Yukihiro, et al. "Commensal microbe-derived butyrate induces the differentiation of colonic regulatory T cells." *Nature* 504, no. 7480 (2013): 446-450.

Gargano, Francesca, et al. "Proinflammatory mucosal-associated invariant CD8+ T cells react to gut flora yeasts and infiltrate multiple sclerosis brain." *Frontiers in Immunology* 13 (2022). https://doi.org/10.3389/fimmu.2022.890298.

Gibbons, Sean M. "Defining microbiome health through a host lens." *MSystems* 4, no. 3 (2019): 10-1128. https://doi.org/10.1128/msystems.00155-19.

Héninger, Erika, Timothy EG Krueger, and Joshua M. Lang. "Augmenting antitumor immune responses with epigenetic modifying agents." *Frontiers in Immunology* 6 (2015): 29. https://doi.org/10.3389/fimmu.2015.00029.

Jessberger, Sebastian, et al. "Epigenetic modulation of seizure-induced neurogenesis and cognitive decline." *Journal of Neuroscience* 27, no. 22 (2007): 5967-5975.

Ji, Jian, et al. "Microbial metabolite butyrate facilitates M2 macrophage polarization and function." *Scientific Reports* 6, no. 1 (2016): 24838.

Jia W. Obesity, metabolic syndrome and bariatric surgery: A narrative review. *J Diabetes Investig*. 2020 Mar;11(2):294-296.

Jirtle, Randy L. "The Agouti mouse: a biosensor for environmental epigenomics studies investigating the developmental origins of health and disease." *Epigenomics* 6, no. 5 (2014): 447-450. https://doi.org/10.2217/epi.14.58.

Jugder, Bat-Erdene, Layla Kamareddine, and Paula I. Watnick. "Microbiota-derived acetate activates intestinal innate immunity via the Tip60 histone acetyltransferase complex. "*Immunity* 54, no. 8 (2021): 1683-1697.

Kaczmarek, Jennifer L., et al. "Time of day and eating behaviors are associated with the composition and function of the human gastrointestinal microbiota." *The American Journal of Clinical Nutrition* 106, no. 5 (2017): 1220-1231.

Kadayifci, Fatma Zehra, et al. "Molecular mechanisms underlying the link between diet and DNA methylation." *International Journal of Molecular Sciences* 19, no. 12 (2018): 4055. https://doi.org/10.3390/ijms19124055.

Kaisar, Maria MM, et al. "Butyrate conditions human dendritic cells to prime type 1 regulatory T cells via both histone deacetylase inhibition and G protein-coupled receptor 109A signaling." *Frontiers in Immunology* 8 (2017): 1429.

Kespohl, Meike, et al. "The microbial metabolite butyrate induces expression of Th1-associated factors in CD4+ T cells." *Frontiers in Immunology* 8 (2017): 1036.

Kim, Myunghoo, et al. "Gut microbial metabolites fuel host antibody responses." *Cell Host & Microbe* 28, no. 2 (2020): 202-212.

Krueger, Tim EG, and Joshua M. Lang. "Harnessing the power of epigenetic modifying drugs to enhance immunotherapy approaches for solid tumors." *Frontiers in Cell and Developmental Biology* 7 (2019): 47. https://doi.org/10.3389/fcell.2019.00047.

Levy, Maayan, et al. "Microbiota-modulated metabolites shape the intestinal microenvironment by regulating NLRP6 inflammasome signaling." *Cell* 163, no. 6 (2015): 1428-1443.

Li, Xiao, et al. "Butyrate promotes slow-twitch myofiber formation and mitochondrial biogenesis in finishing pigs via gut microbiota." *Journal of Agricultural and Food Chemistry* 68, no. 37 (2020): 10100-10110.

Lukens, John R., et al. "Dietary modulation of the microbiome affects autoinflammatory disease." *Nature* 516, no. 7530 (2014): 246-249.

Ma, Nan, et al. "Butyrate promotes the recovering of intestinal wound healing through its positive effect on the tight junctions." *Journal of Animal Science and Biotechnology* 12, no. 1 (2021): 1-13.

Macfarlane, George T., et al. "Chemotaxonomic analysis of bacterial populations colonizing the rectal mucosa in patients with ulcerative colitis." *Clinical Infectious Diseases* 38, no. 12 (2004): 1690-1699. https://doi.org/10.1086/421394.

Machiels, Kathleen, et al. "A decrease of the butyrate-producing species Roseburia hominis and Faecalibacterium prausnitzii defines dysbiosis in patients with ulcerative colitis." *Gut* 63, no. 8 (2014): 1275-1283.

Manor, Ohad, and Eran Segal. "Gut microbiota in human metabolic health and disease." *Nature Reviews Microbiology* 15, no. 6 (2017): 328-342.

Maslowski, Kendle M., et al. "Regulation of inflammatory responses by gut microbiota and chemoattractant receptor GPR43." *Nature* 461, no. 7268 (2009): 1282-1286.

Mayer, Emeran A., et al. "Gut microbes and the brain: paradigm shift in neuroscience." *The Journal of Neuroscience* 34, no. 46 (2014): 15490-15496.

Morgan, Xavier C., et al. "Transcriptional reprogramming and redox signaling underlie enhanced antibody-dependent cellular cytotoxicity against cancer targets induced by a metabolite of selenium." *Journal of Immunology* 188, no. 3 (2012): 718-726.

Nagpal, Ravinder, et al. "Gut microbiome and aging: Physiological and mechanistic insights." *Nutrition and Healthy Aging* 4, no. 4 (2018): 267-285.

Ni, Yongqing, et al. "Combinatorial treatment with natural compounds in prostate cancer inhibits prostate tumor growth and leads to key modulations of cancer cell metabolism." *NPJ Precision Oncology* 5, no. 1 (2021): 1-14.

O'Connor, Anne, and Diarmuid O'Shea. "The influence of aging on the gut microbiome and immune system." *The Journal of Nutrition, Health & Aging* 23, no. 6 (2019): 487-494.

Ohnmacht, Caspar, et al. "Intestinal microbiota, evolution of the immune system and the bad reputation of pro-inflammatory immunity." *Cellular Microbiology* 16, no. 10 (2014): 1315-1323.

Pagnini, Cristiano, et al. "Probiotics promote gut health through stimulation of epithelial innate immunity." *Proceedings of the National Academy of Sciences* 107, no. 1 (2010): 454-459.

Perrin-Cocon, Laurence, et al. "HDAC inhibition and Zika virus infection." *Virology* 533 (2019): 115-127.

Pham, Tho X., et al. "Epigenetic mechanisms and microbiota as a toolbox for understanding the development of microbiota-modulated cardiovascular diseases." *Microorganisms* 9, no. 3 (2021): 576.

Qin, Yingrui, et al. "A metagenome-wide association study of gut microbiota in type 2 diabetes." *Nature* 490, no. 7418 (2012): 55-60.

Raposo, Ana, et al. "Histone deacetylase inhibition modulates histone acetylation at gene promoter regions and affects genome-wide gene transcription in Schistosoma mansoni." *PLOS Neglected Tropical Diseases* 13, no. 12 (2019): e0007926.

Rastelli, Eugenio, et al. "Pleiotropic activity of green tea polyphenols in alleviating diseases associated with metabolic syndrome." *Frontiers in Pharmacology* 11 (2020): 1-22.

Rea, Denise, et al. "Microbiota effects on cancer: from risks to therapies." *Oncotarget* 9, no. 30 (2018): 17915-17927.

Reik, Wolf, and Emma Greenberg. "Transposon silencing and imprint establishment in mammalian germ cells." *Cold Spring Harbor Symposia on Quantitative Biology* 69 (2004): 381-387.

Rescigno, Maria. "The intestinal epithelial barrier in the control of homeostasis and immunity." *Trends in Immunology* 32, no. 6 (2011): 256-264.

Riaz Rajoka, Muhammad Shahid, et al. "Lactobacillus rhamnosus GG induced protective effect on allergic airway inflammation is associated with gut microbiota." *Cellular Immunology* 340 (2019): 103915.

Round, June L., and Sarkis K. Mazmanian. "The gut microbiota shapes intestinal immune responses during health and disease." *Nature Reviews Immunology* 9, no. 5 (2009): 313-323.

Rückert, Christine, et al. "Discovery of histone deacetylase inhibitors with potent activity against multiple myeloma." *Leukemia* 23, no. 12 (2009): 2352-2362.

Ryu, Dongryeol, and Myeongjin Kim. "Microbial synthesis of human microbiota-derived L-methionine." *Applied Microbiology and Biotechnology* 99, no. 9 (2015): 3881-3889.

Scheer, Frank AJL, and Steven A. Shea. "Human circadian system causes a morning peak in pro-inflammatory responses." *Nature Immunology* 18, no. 8 (2017): 832-832.

Schulthess, J. "The short chain fatty acid butyrate imprints an antimicrobial program in macrophages." *Immunity* 50, no. 2 (2019): 432-445.

Shao, Yanqin, et al. "Dysbiosis of gut microbiota contributes to chronic stress-induced inflammatory response in rats." *Life Sciences 246* (2020): 117403.

Shi, Liang, et al. "Histone demethylase JMJD2B coordinates H3K4/H3K9 methylation and promotes hormonally responsive breast carcinogenesis." *Proceedings of the National Academy of Sciences* 108, no. 18 (2011): 7541-7546.

Smith, Philip M., et al. "Microbiota and host immune responses: a love–hate relationship." *Immunology* 147, no. 1 (2016): 1-10.

Sokol, Harry, et al. "Faecalibacterium prausnitzii is an anti-inflammatory commensal bacterium identified by gut microbiota analysis of Crohn disease patients." *Proceedings of the National Academy of Sciences* 105, no. 43 (2008): 16731-16736.

Sparrow, Sara, et al. "Diet, microbiota, and microbial metabolites in colon cancer risk in rural Africans and African Americans." *The American Journal of Clinical Nutrition* 98, no. 1 (2013): 111-120.

Taubert, Martin, et al. "Functional role of Faecalibacterium prausnitzii-like bacteria in rats with spontaneous type 1 diabetes." *British Journal of Nutrition* 117, no. 12 (2017): 1741-1749.

Tilg, Herbert, and Arthur Kaser. "Gut microbiome, obesity, and metabolic dysfunction." *The Journal of Clinical Investigation* 121, no. 6 (2011): 2126-2132.

Ting, Kang, et al. "Gut microbiota and metabolism in patients with alcohol use disorder." *Microbial Cell* 5, no. 11 (2018): 482-498.

Tiziani, Stefano, et al. "Metabolomics in cancer drug discovery." *Expert Opinion on Drug Discovery* 5, no. 5 (2010): 433-450.

Turnbaugh, Peter J., et al. "An obesity-associated gut microbiome with increased capacity for energy harvest." *Nature* 444, no. 7122 (2006): 1027-1031.

Vandeputte, Doris, et al. "Stool consistency is strongly associated with gut microbiota richness and composition, enterotypes and bacterial growth rates." *Gut* 65, no. 1 (2016): 57-62.

Villarreal-Calderon, Rodolfo, et al. "Particulate matter, airborne metals and polycyclic aromatic hydrocarbons are associated with respiratory illness in preschool children in Ciudad Juarez, Mexico." *Environmental Research* 135 (2014): 38-45.

Vrieze, Anne, et al. "Transfer of intestinal microbiota from lean donors increases insulin sensitivity in individuals with metabolic syndrome." *Gastroenterology* 143, no. 4 (2012): 913-916.

Wang, Li, et al. "Protein kinase D activation induces mitochondrial fragmentation and dysfunction in cardiomyocytes." *Journal of Biological Chemistry* 292, no. 7 (2017): 2555-2567.

Wang, Qiang, et al. "Berberine induces autophagy in glioblastoma by targeting the AMPK/mTOR/ULK1-pathway." *Oncotarget* 7, no. 41 (2016): 66944-66958.

Weng, Liwei, et al. "Bacterial communities in semen from men of infertile couples: metagenomic sequencing reveals relationships of seminal microbiota to semen quality." *PLOS ONE* 9, no. 10 (2014): e110152.

Wilck, Nicola, et al. "Salt-responsive gut commensal modulates TH17 axis and disease." *Nature* 551, no. 7682 (2017): 585-589.

Willing, Benjamin P., et al. "Tuning of human MAIT cell activation by commensal bacteria species and MR1-dependent T-cell presentation." *Mucosal Immunology* 8, no. 3 (2015): 580-591.

Wu, C. "Induction of intestinal Th17 cells by segmented filamentous bacteria." *Cell* 139, no. 3 (2009): 485-498.

Yadav, Manisha, et al. "Inhibition of TSC2/mTOR activity promotes drug resistance in kidney cancer." *Molecular Cancer Therapeutics* 19, no. 3 (2020): 710-718.

Yamaguchi, Junya, et al. "The impact of microbiota on brain and behavior: mechanisms & therapeutic potential." *Advanced Drug Delivery Reviews* 160 (2020): 327-340.

Yang, T. "Gut dysbiosis is linked to hypertension." *Hypertension* 65, no. 6 (2015): 1331-1340.

Yin, Yanan, et al. "Inducible targeted epigenetic editing using CRISPR/dCas9-based chromatin modifiers." *Cell Stem Cell 22*, no. 6 (2018): 960-973.

Zhang, Yu-Jie, Li S, Gan RY, Zhou T, Xu DP, Li HB. "Impacts of gut bacteria on human health and diseases." *International Journal of Molecular Sciences* 16, no. 4 (2015): 7493-7519.

Zhang, Zhenjiang, et al. "Anti-inflammatory effect of thymol combined with amino acids in lipopolysaccharide-induced RAW264.7 macrophages." *Molecular Immunology* 132 (2021): 92-99.

Zhao, Liping. "The gut microbiota and obesity: from correlation to causality." *Nature Reviews Microbiology* 11, no. 9 (2013): 639-647.

Chapter 24

The Future Perspectives of Microbiomes, Animal Health, and Food Safety

Sina Salajegheh Tazerji[1,*]
Phelipe Magalhães Duarte[2]
Erfan Soroush[1,3]
Soroush Ghasemi[4]
and Fatemeh Shahabinejad[5]

[1]Department of Clinical Sciences, Faculty of Veterinary Medicine, Science and Research Branch, Islamic Azad University, Tehran, Iran
[2]Postgraduate Program in Animal Bioscience, Federal Rural University of Pernambuco - UFRPE, Recife, PE, Brazil
[3]Microbiology Research Center (MRC), Pasteur Institute of Iran, Tehran, Iran
[4]Department of Clinical Sciences, Faculty of Veterinary Medicine, Shahrekord Branch, Islamic Azad University, Shahrekord, Iran
[5]Department of Clinical Sciences, Faculty of Medicine, Kerman Medicine University, Kerman, Iran

Abstract

Livestock breeders use antibiotics to maintain animal health and have a better product. So, nearly 50% of antibiotics produced in the world are used for livestock. Improper and excessive use of antibiotics in livestock causes the creation of antibiotic resistant genes. It is predicted that the death rate due to antibiotic resistance will be more than that of cancer by 2050. Humans are heavily exposed to antibiotic-resistant bacteria through food consumption, which is mainly due to the use of antibiotics in animal husbandry. Also, not observing the period of abstinence from consumption after the injection of antibiotics in livestock and the lack of food safety principles can cause the transfer of antibiotic residues to humans through food.

Considering the constant priority of prevention over treatment, it is possible to reduce the incidence of infectious diseases by modifying and strengthening the gut microbiota in order to minimize the use of antibiotics. Gut microbiota is often considered as reservoirs of resistance due to its widespread presence and long-term exposure to antibiotics. Probiotics play an important role in modifying the gut microbiota and creating a strong barrier to fight

* Corresponding Author's Email: sina.salajegheh@gmail.com; sina.salajegheh@srbiau.ac.ir

In: Microbiomes in Animal Nutrition, Health, and Disease
Editor: Tanmoy Rana
ISBN: 979-8-89530-425-9

pathogens. Many studies have shown the positive results of using probiotics in livestock feed to reduce infectious diseases.
By using probiotics in livestock feed, which plays a significant role in modifying the gut microbiota, it is possible to increase the health of animals, which leads to the production of high-quality food with high food safety, and it will follow human health.

Keywords: animal health, human health, food safety, gut microbiota, antibiotic resistance

Introduction

The human population is expected to reach 9.7 billion people in the world by 2050. This growing world population presents us with serious challenges to ensure that people have access to nutritious, healthy and safe food (FAO 2017). For this reason, people in developed countries have become inclined to consume organic, local and non-GMO foods (O'Connor, E. L., Sims, L., and White 2017). Nowadays, paying attention to the production and distribution of healthy and safe food in societies has become an important and serious matter. In most of the developed countries, the average amount of protein consumed is above 1.4 g/kg on a daily basis, with animal proteins constituting 65-70% of the total protein intake. To respond to the growing demand of animal protein consumption, livestock industries have resorted to various techniques to increase their production. These new techniques are intended to accelerate livestock growth, protect them from disease, improve feed conversion efficiency, and optimize reproduction by limiting birth mortality (Mottet et al. 2017). Foods of animal origin play an important role in the human diet and should be safe for human consumption. Healthy food of animal origin must be free of any animal pathogens and drug residues, especially antibiotics (Makita 2021).

The first goal of livestock breeding is to deliver healthy food for human consumption, taking into account animal welfare and respect for the environment, which is one of the most important researches in the field of food, improving the quality and safety of meat. It has long been well known that pathogens such as *Campylobacter* and *Salmonella* can be transmitted along the food chain and be the source of human disease. In the past, antibiotics were added to animal feed at sub-therapeutic levels to act as growth promoters and on the other hand, they were used to treat bacterial diseases (Dibner and Richards 2005). Antibiotics have undoubtedly played a great role in helping humans to fight disease-causing agents, but their excessive use leads to the resistance of microorganisms to antibiotics, which can endanger food safety (Zinsstag et al. 2005). A large number of microorganisms and their products are frequently shared between humans and animals, which may increase the possibility of interspecies transmission of pathogens, antibiotic resistance genes, and virulence genes (Reperant, Cornaglia, and Osterhaus 2013). However, the global concern about the development of antimicrobial resistance in the case of the transfer of antibiotic resistance genes from animals to human microbiota has led to the prohibition of the use of antibiotics in animal feed in many countries of the world (Gaggìa, Mattarelli, and Biavati 2010). Considering the ban on the use of antibiotics in animal feed as a growth stimulant and the harmful effects of its improper use on human health, there is a need to search for suitable alternatives that can strengthen the natural defense mechanisms of animals and the widespread use of reduce antibiotics and force growers to look for more environmentally efficient methods. One of the ways is the use of feed raw materials or special feed additives, which can have a favorable effect on the performance and health of animals, especially through the modifying of the gut microbiota in maintaining the health of the host. A diverse and balanced gut microbiota creates an efficient barrier against colonization by pathogens, produces efficient metabolites (such as vitamins and short-chain fatty acids) and stimulates the immune system in a non-inflammatory manner. The gut microbiota plays an important role in the health and well-being of the host, which helps absorb energy by facilitating the digestion of complex food and maintaining and improving the immune system. In addition, the gut microbiota protects the host from the development of potential pathogens. This symbiotic relationship has evolved over many generations, and the composition of the microbiota is influenced by factors related to the physiology and ecology of the host, as well as other biotic and abiotic environmental factors (Moeller and Sanders 2020). The main effects of using feed additives to modify gut microbiota instead of using antibiotics are improving the resistance against the colonization of pathogenic bacteria and increasing the immunity of the host's mucosa. Therefore, it leads to the reduction of the pathogen load, the improvement of the health status of animals and the reduction of the risk of pathogens transmitted from food to humans (Choct 2009).

In this chapter, we take a special look at antibiotic resistance and residues, the role of the gut microbiota in animal and human health, ways to improve food safety and, as a result, we discuss the issue of One health.

Harmful Effects of Improper Use of Antibiotics in Human and Animal Health

Antibiotic Resistance

Livestock breeders use antibiotics to maintain animal health and have a better product. Antibiotics are prescribed in two ways orally or intramuscularly in livestock. A wide range of food additives in animal feed are antibiotics, which are used in animal nutrition to increase production efficiency and control various diseases("Disseminating Innovative Solutions for Antibiotic Resistance Management" 2022). According to recent studies, a significant percentage of livestock farmers add antibiotics to the food of healthy animals in order to prevent disease rather than cure it (Landers et al. 2012; Mayorova et al. 2020; Maslennikova and Erofeeva 2020). It is interesting to note that nearly 50% of antibiotics produced in the world are used for livestock. During the breeding period, the use of antibiotics significantly increases the weight of animals, which ultimately shortens the feeding period and reduces feed costs (Erofeeva et al. 2021). Prevention of disease is more effective than treatment. The use of antibiotics should not in any way replace the appropriate methods of animal husbandry, biological security and management, otherwise it may cause antibiotic resistance in animals and humans. The ability and resistance of bacteria against the growth-inhibiting or lethal activity of antibiotics that they were not resistant to in the past is called antibiotic resistance. This antibiotic resistance is due to excessive and unprincipled use of antibiotics in livestock breeding and mechanisms of transfer of antibiotic resistance between bacteria. This resistance transfer phenomenon between humans and animals is possible because they share the same ecosystem. In fact, resistant bacteria and antibiotic resistance genes can circulate between different parts of the ecosystem. Therefore, bacteria can be transmitted between animals and humans through direct or indirect contact such as water, food, and the environment. This applies both to the gut microbiota, which are often considered reservoirs of resistance due to their widespread presence and long-term exposure to antibiotics in the intestines of food animals, and to pathogenic bacteria (Larsson and Flach 2022). Recent studies have shown that multi-resistant bacteria caused by animal husbandry can be found in daily consumed food products. If the treatment of bacterial infections with antibiotics becomes ineffective due to antibiotic resistance, the production of meat, milk and other animal products can suffer serious problems. Similarly, bacteria that have developed resistance to multiple antibiotics can be many times more dangerous if they become pathogenic and are transmitted to humans, as the resulting disease will be more difficult to treat (Mayorova et al. 2020). Antimicrobial resistance in foods is considered a food safety issue. In addition, awareness of the prevalence of antibiotic-resistant food pathogenic bacterial strains is of great importance (Collineau et al. 2019). Bacteria can become resistant to antibiotics using several mechanisms: modifying the antibiotic target, enzymatic degradation of antibiotics, changing the permeability of the bacterial cell wall and alternative pathways to escape from the activity. One of the most common mechanisms of bacterial resistance is the destruction or enzymatic modification of antibiotics. For example, beta-lactamase enzymes that hydrolyze the beta-lactam ring in beta-lactam antibiotics such as cephalosporins, which are of most concern in gram-negative bacteria (Livermore and Woodford 2006). Resistance to aminoglycoside antibiotics is mainly caused by enzymatic degradation caused by acetyltransferases, nucleotidyltransferases and phosphotransferases. Antibiotic resistance by modifying the antibiotic target means modifying the target molecule of the antibiotic, so that the antibiotic loses its binding capacity and thus its activity (Drlica and Zhao 1997). Finally, the antimicrobial resistance of a bacterium can be intrinsic or acquired. Intrinsic antimicrobial resistance is an inherent characteristic of a bacterial species or genus to a specific antibiotic. As a result, antibiotic treatment will not be successful (Holzbauer and Chiller 2006).

Food may be contaminated with antibiotic-resistant bacteria or genes in various ways, and these components may be found in soil, water, human or animal feces, and animal products (Bergogne-Bérézin 1997). The presence of antibiotic-resistant bacteria in foods of animal origin has increased significantly in recent years. Therefore, humans are heavily exposed to antibiotic-resistant bacteria through food consumption, which is mainly due to the use of antibiotics in livestock breeding (Li et al. 2022). For example, the transfer of plasmid-dependent ampicillin resistance genes from *Salmonella typhimurium* to *E. coli* K12 in sterilized milk and ground beef and the transfer of antimicrobial resistance from lactic acid bacteria (LAB) to pathogenic strains of *Lactococcus lactis*, *Enterococcus faecalis, Listeria*, *Staphylococcus aureus*, *Salmonella* spp. and *E. coli* reported in milk (Toomey et al. 2009). Therefore, the food chain has a great influence on the transmission of antibiotic resistant bacteria and their genes, because food is not sterile and usually its food safety is compromised through recontamination or cross contamination during production. Antibiotic-resistant bacteria are transferred from food products to humans, followed by horizontal transfer of resistance genes in the human intestine. Recently, several studies have studied the human gut microbiota along the food production chain to evaluate antibiotic resistance genes (Li et al. 2022; Ferrocino et al. 2023). These microbiome studies, which we will fully address in the following, may help produce safer meat and meat products for animal and human health. Currently, several policies have been considered in the world to reduce the use of antibiotics in food, such as a 50% reduction in the sale of antibiotics for animal breeding and aquaculture by 2030, which may not be enough to effectively control antibiotics (Pennone et al. 2021).

One of the strategic foods in terms of food safety and widely consumed in the world is meat. Meat and other meat products are an important source of protein minerals, vitamins, amino acid compounds along with iron, zinc and B12 and D vitamins in the human diet and play an important role in food culture in many countries (de Araújo et al. 2022). Simultaneously with the increase in meat consumption, the demand for meat products has also increased so that the consumption of meat is expected to reach between 460 and 570 million tons per year by 2050 (Chaudhary and Kumar 2022). In the livestock industry, pork consumption has increased due to high protein quality and cheap price. However, mainly due to the incorrect and excessive use of antibiotics and the fact that antibiotics are allowed in feed in some countries, pigs are considered one of the largest reservoirs of antibiotic-resistant bacteria (Kanokudom et al. 2021). Each stage of the food chain of animal breeding, especially pigs, including feeding, slaughtering and processing, has the potential to affect human and animal health. Contamination reduction measures along the food production chain include advanced disinfection methods in the aforementioned contaminations to increase food safety and reduce the risk of antibiotics and pathogens in final products and the environment (Campos Calero et al. 2018). Disturbance in food safety is the main cause of foodborne diseases in developed and developing countries. In addition, the transfer of antibiotic-resistant bacteria and their genes from farm to home is an additional concern for food safety. It is estimated that antibiotic resistance will cause nearly 10 million deaths per year and cost $100 trillion by 2050 (Das et al. 2019). The microbiota of meat and meat products includes foodborne pathogens and spoilage and technological microorganisms, all of which are responsible for the transfer of antibiotic-resistant bacteria from farm to home. For example, technological microbiota such as coagulase-negative *staphylococci* can acquire and incorporate antimicrobial resistance genes from other bacterial genera through horizontal gene transfer (Conceição, Queiroga, and Laranjo 2023; Chajęcka-Wierzchowska et al. 2015). In addition, there are studies that confirm the transfer of antibiotic resistance genes from food bacteria to the human gut microbiota (Bouchami et al. 2020; Cao et al. 2022; Lawal et al. 2021; Henriques, Cristino, and Fraqueza 2017). A study in 2022 by examining human samples and pig and poultry carcasses identified the presence of antibiotic resistance genes in both humans and carcasses that were resistant to vancomycin, tetracycline and macrolides (Rawat et al. 2024).

The above-mentioned facts highlight the attention to human health and animal health so that food safety is given more attention in order to prevent the creation of bacteria and antibiotic-resistant genes in animals and their subsequent transmission to humans.

Antibiotic Residues

Antibiotics can be produced by living organisms or synthetically in a laboratory. The main role of antibiotics is to inhibit the growth or kill microorganisms (Darwish et al. 2013). One of the global concerns of the use of antibiotics in agriculture and animal husbandry is the presence of residues of these substances in food and the environment. Antibiotic residues are defined as all antibiotic active substances or degradation products, as well as metabolites present in the environment or remaining in food for animals to which the antibiotic in question has been administered (Okaiyeto et al. 2024). Antibiotic residues may cause risks to public health, especially antibiotic resistance, disruption of the natural gut microbiota, carcinogenesis, mutagenesis, and allergic reactions (Rana et al. 2019). Antibiotic residues can be transmitted to humans through milk, honey, fish, meat and eggs (Donoghue 2003). From the perspective of environmental risks, after antibiotic treatment, animals excrete a fraction of the prescribed dose in their living environment. There are significant differences in the half-life of antibiotics: for example, tylosin degrades much faster than oxytetracycline. In addition, antibiotic residues can form during the production of some fermented foods (cheese, yogurt, fermented meats) that inhibit the starter culture. The desire of consumers around the world to consume quality food has increased and they have turned to organic products, which provide them with more safety than conventional products (Gundala and Singh 2021). The use of antibiotics that may accumulate in milk, meat, eggs and honey is not allowed to be consumed by humans. If the use of antibiotics in animals is necessary, a withdrawal period should be observed as long as antibiotic residues remain (A R 2008). The presence of antibiotic residues in meat is considered a significant risk to public health. Cooling and freezing and thermal processing of meat can be mentioned as ways to reduce antibiotic residues in meat. A study in 2020 in Iraq showed that the sheep meat available in supermarkets contain antibiotic residues more than the allowed limit. In this study, chilling the meat reduced a small amount of antibiotic residues and cooking the meat for 45 minutes inactivated the antibiotic residues against bacteria (Al Mashhadany 2020). According to studies, many countries such as Nigeria, Indonesia and Iran have also reported high amounts of antibiotic residues in meat, but the lowest amount has been reported in Spain (Babapour, Azami, and Fartashmehr 2012; Olasoju et al. 2021; Nchima et al. 2017; Ramatla et al. 2017). Milk containing antibiotic residues significantly affects the production process of dairy products, which can cause problems in fermentation. Such milk is considered as a fake food in the market and by consuming it, bacteria resistant to antibiotics can be transferred to the consumers (Brown et al. 2020). The study of antibiotic residues in dairy products shows the presence of gentamicin/neomycin group in pasteurized milk, whipped milk, cheese, yogurt and especially in sour cream. Macrolides have sometimes been detected in cheese and tetracyclines in sour cream. The reported cases were all isolated from confirmed milk for human consumption (Moghadam et al. 2016).

In most countries, eggs are one of the most popular foods in people's diets, and its food safety should be important. A study was conducted in Chile in 2020 and eggs were analyzed for antibiotic residues of tetracyclines, aminoglycosides, beta-lactams and macrolides. The results showed that all the eggs were contaminated with at least one of the mentioned antibiotics (Cornejo et al. 2020). Another study in China showed that 30% of eggs containing antibiotic residues were contaminated with quinolones, sulfonamides, and tetracyclines, which should be monitored more closely on egg food safety (Yang et al. 2020).

Fortunately, today, people have enough information about the consequences of improper use of antibiotics in animal breeding, antibiotic resistance, antibiotic residues, and the transfer of antibiotic resistant genes to the gut microbiota, which leads to increased mortality in humans (Papatsiros et al. 2014; Ganguly et al. 2011). Foods of animal origin are considered as one of the most important reservoirs of antibiotic residues, so they play a very important role in inducing antibiotic resistance worldwide, which endangers food safety and human health. In terms of the prevalence of antibiotic-resistant bacteria of animal origin, antibiotic-resistant genes and antibiotic residues at the global level, great differences have been reported between geographical regions (Founou, Founou, and Essack 2016).

For this reason, paying attention to the correct and optimal use of antibiotics in livestock breeding and careful monitoring of food safety requires everyone's efforts to ensure the health of animals and prevent the creation of antibiotic-resistant bacteria in them. Food safety must be closely monitored to prevent the transfer of antibiotic-resistant genes and bacteria to human gut microbiota.

Gut Microbiota and Human Health

Livestock and livestock products that reach human consumption have an inseparable relationship with the human microbiome in the food safety chain. At each stage of this chain, microorganisms are exchanged between animals and humans to the extent that if food safety principles are not observed, interspecies transmission of antibiotic-resistant microorganisms may occur. In order to reach a healthy and dynamic ecosystem, medicine and veterinary medicine play a prominent role in maintaining human and animal health (Manlove et al. 2016).

Microbiota in the human body consists of a wide spectrum of bacteria, viruses and fungi. Most of these microorganisms are present in the human body as commensals. Some of them are mutualistic and some are opportunistic pathogens. The term microbiota refers to the population of microorganisms that live in different parts of the body of healthy people. More than 10 to 100 trillion bacterial cells live in the human body. This population has various functions in the body, such as defense against pathogenic microorganisms, helping food digestion, destroying toxins, and maturing the immune system. So that human life is inseparable from communication and interaction with these microorganisms (Abdolghanizadeh et al. 2024). The term microbiome was first introduced by Joshua Lederberg in 2001 as the genetic content of microbiota populations in the human body. Among the most important microbiota population of the human body is the gut microbiota (Thursby and Juge 2017). The gut microbiota plays a very important role in the functioning of the immune system and defense against pathogens and diseases. *Bacteroidetes* and *Firmicutes* are among the most important groups of the gut microbiota, and *Clostridiales* including *Lacnopiraceae*, *Rominococcaceae* and *Clostridia* are considered as the most important taxonomic groups that consist of many types of species in digestive sites (Zafar and Saier 2021). The gut microbiota consists of commensal microorganisms, which, despite the many advantages they bring to their hosts, can acquire antibiotic-resistant genes and even cause severe infections at times. Identification of antibiotic resistance genes is one of the most important characteristics studied in the gut microbiota. It is predicted that the death rate due to antibiotic resistance will be more than that of cancer by 2050 (Figure 1, Forecast chart of common deaths in the future) (de Kraker, Stewardson, and Harbarth 2016). Bacteria have always been trying to acquire resistance genes in a competitive environment in order to maintain their survival. These genes are often transferred by bacteriophages, plasmids or transposons. The ecosystem in which the microbiota is present is very complex because there are many organisms and cells in it that are constantly competing with each other and may have created an arms race in the real sense. Antibiotics are one of the most important drugs that are prescribed for various reasons in humans and animals. An increase in the frequency of antibiotic resistant genes following their widespread use has been reported in the microbiota of different parts of the human body, especially in the gut microbiota population. Considering the importance of gut microbiota in regulating various body functions, it is very important to know the effective factors in the formation and spread of resistance genes in this population of human body gut microbiota (Patangia et al. 2022; Fishbein, Mahmud, and Dantas 2023). There are two different patterns of antibiotic resistance in the microbiota population of the human body, especially in the gut microbiota. Resident resistome, which is in the commensal bacterial population, and transitory resistome, which is caused by the presence of transitory microbiota in the gut and their ability to transfer resistance to commensal bacteria or even to maintain the transitory microbiota. This shows that microbiota can be a good source for maintaining antibiotic resistance as well as the horizontal transfer of resistance genes from commensal bacteria to pathogens (Juhász et al. 2021). There are a variety of resistance genes in the gut microbiota, but most of the genes that have been reported include those that cause resistance to tetracycline, beta-lactams, aminoglycosides, glycopeptides, chloramphenicol, and macrolides. Considering the widespread use of tetracycline in livestock, it is possible that the tetracycline resistance gene was transferred from animals to humans (Grossman 2016).

In terms of diet, foods of animal origin influence the human gut microbiome. For example, levels of Firmicutes bacteria, which metabolize plant polysaccharides, were lower in people who ate animal-source foods. In addition, high consumption of animal-source foods rapidly alters the composition of the gut microbial community, increasing levels of deoxycholic acid concentrations in the gut and increasing the likelihood of inflammatory bowel disease (IBD) (David et al. 2014). The effects of animal foods on human health are very different according to its type and quantity. For example, tryptophan is an essential amino acid that can only be obtained from red meat, fish, and eggs. Commensal bacteria express tryptophanase, and tryptophan is degraded to indole, which activates aromatic hydrocarbon receptors and regulates the production of the cytokine IL-22, both essential in the mucosal immune response to pathogens (Wu et al. 2019). In addition, dairy products contain lactic acid bacteria and galacto-oligosaccharides that support the growth of beneficial gut microbiota (such as *Bifidobacterium*) (Petrova et al. 2021).

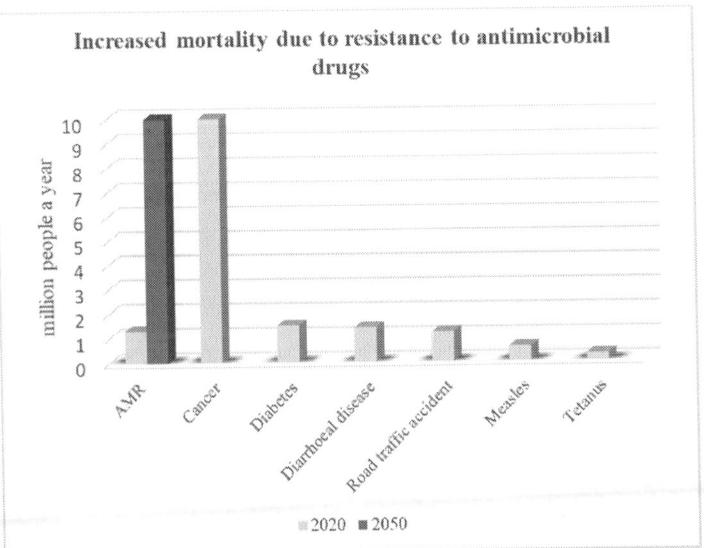

Figure 1. Prediction of mortality caused by the use of antimicrobial drugs (AMR) versus today's common causes of death.

In studies, a close relationship between human and animal microbiomes has been observed. Interestingly, people who work in animal husbandry and are exposed to antibiotic resistance genes have a higher frequency of antibiotic resistance genes in their stool microbiome than other people (Guarner and Malagelada 2003). Researchers have collected antibiotic resistant genes in the suspended particles in the air of livestock farms. For this reason, there is a relationship between human and animal microbiome based on their environment (Luiken et al. 2020).

The mentioned materials are a good reminder that we need healthy food to have a healthy society, and considering the high consumption of animal foods in the world and the high importance of the gut microbiota in human health, more attention should be paid to food safety.

Gut Microbiota and Animal Health

Antibiotic resistance genes have existed everywhere since ancient times, and the natural environment, animals and humans are considered to be the most important sources. These resistance genes are often identified in areas that have been exposed to antibiotics, such as the digestive system. In countries where strict laws are applied for the prescription and consumption of antibiotics in humans and animals, the frequency of antibiotic resistant genes is much less. Adding antibiotics to livestock feed also affects the spread of resistance genes (Jian et al. 2021). In the United States, approximately 80% of antibiotics are used for therapeutic purposes in livestock. As a result, following the implementation of regular treatment programs and prescribing antibiotics, the microbiota of livestock become enriched with antibiotic resistance genes. A study has reported a 28.000-fold increase in antibiotic resistant genes after adding antibiotics to the feed of Chinese pigs. However, new genes that induce resistance to tetracyclines have been reported in the pigs' gut microbiota for which antibiotics have not been prescribed, and their transfer is related to mobile genetic elements (Tian et al. 2021). The identification of antibiotic resistance genes in human feces samples is significantly the result of antibiotics used by livestock. In addition, migratory birds that fly long distances may also be important in the spread of antibiotic resistance genes (Yuan et al. 2021).

The mammalian gut microbiota can be considered a metabolically active organ with a wide biodiversity that can reach 10^{14} cells. ntestinal commensal bacteria are an essential asset for health that have a protective effect on intestinal structure and homeostasis. Although the gut microbiota is complex and the role of most bacteria in providing benefits to the host is not known, bacterial species from the genera *Lactobacillus* and *Bifidobacterium* play an effective role in protecting the gut against infections (O'Hara and Shanahan 2007; Jandhyala et al. 2015). Considering that the proximity of humans to animals in livestock and poultry farms facilitates the transfer of microbiota to each other, maintaining a healthy microbiota as a barrier against pathogens is important for both human and animal health. One of the most important issues in today's science is to identify and recognize the species present in the gut microbiota of different animals. A study of the gut microbiota of 60 species of mammals based on 16S rRNA showed that diet, host phylogeny and gut morphology influence the microbial ecology of the digestive tract (Ley, Hamady, et al. 2008; Ley, Lozupone, et al. 2008). The main microbiota in monogastric animals (such as humans, rabbits, chickens and pigs) are: *Bacteroides*, *Clostridium*, *Bifidobacterium*, *Eubacterium*, *Lactobacillus*, *Enterobacteriaeae*, *Streptococcus*,

Fusobacterium, *Peptostreptococcus* and *Ptopionibacterium,* and in polygastric animals (such as cattle and sheep), the rumen is the most important place for the presence of microbiota, which includes fiber-degrading groups such as: *Fibrobacter*, *Ruminococcus*, *Butyrivibrio* and *Bacteroides* (Mackie et al. 1999).

In the digestive system of humans and animals, the mucous barrier separates the internal environment from the external environment. The mucous layer consists of various mucous secretions, including trefoil peptides, mucin glycoproteins, and surfactant phospholipids. The gut microbiota with its nutritional and protective functions can positively influence the integrity of the intestinal barrier (Van Gompel et al. 2020). Many evidences show that with the increase of pathogenic bacteria and the decrease of beneficial gut microbiota, intestinal inflammation also increases (Andoh and Fujiyama 2006). Physiological and psychological stress that leads to the dysfunction of the intestinal barrier and the increase of intestinal permeability causes pathogens to enter the bloodstream more easily and make animals sick (Gareau, Wine, and Sherman 2009).

Farm animals are subjected to environmental stress, such as incorrect management methods and unprincipled diet, which can cause an imbalance in the gut microbiota and is a major risk factor for pathogen infections. On the other hand, when the animal is under environmental stress, it does not eat, which in the past was covered by adding antibiotics to the feed. In addition, these stresses cause the imbalance of the gut microbiota and increase its permeability, in which case the animal gets infected and needs antibiotic treatment. Since prevention always has priority over treatment, one of the effective preventive ways to increase the immune system of animals and prevent the prescription of antibiotics that have irreparable consequences on animal health and human health is to modify the gut microbiota using probiotics (Chen, Luo, and Yan 2021).

So far, many definitions have been provided for the word probiotic. One of the most widely accepted is 'live microorganisms that, when administered in sufficient quantities, are beneficial to the health of the host'(Suresh, Krishna, and Govender 2012). Among the properties of probiotics, we can mention: regulation of intestinal microbial homeostasis, stabilization of gastric and intestinal barrier function, expression of bacteriocins, immunomodulatory effects, and interference with the colonization of pathogens and pathogenic agents (Mazmanian, Round, and Kasper 2008). *Lactobacillus*, *Enterococcus*, *Bacillus*, *Bifidobacterium* and *Saccharomyces* are actually the most widely used probiotics in the livestock and poultry industry (Wannaprasat et al. 2009). For example, administration of *Enterococcus faecalis* to piglets and sows modified the primary colonization of bacteria in the intestine and reduced enteropathogens (Scharek et al. 2005). In another study, supplementing the diet of newborn pigs with a strain of *Lactobacillus plantarum* led to an increase in the total population of *lactobacilli* in the intestines of weaned pigs (Takahashi et al. 2007). In the case of birds, a variety of probiotic strains have been selected to modulate the gut microbiota of birds and protect against a variety of pathogens. In particular, *Lactobacillus* species have increased recently (Patterson and Burkholder 2003). A study showed that *Lactobacillus*-based probiotics significantly reduce the recovery of *Salmonella Enteritidis* in chick broilers (Higgins et al. 2008). Also, chickens that received *lactobacillus* probiotics in the form of food on the first day of life, mortality due to necrotizing enteritis decreased from 60% to 30% (Hofacre et al. 2003). Many studies have been done on the effects of probiotics in ruminants. Neonatal diarrhea of calves, which is one of the important causes of morbidity and mortality in young ruminants, is mainly caused by enterotoxigenic *E. coli.* A study reported that diarrhea was reduced in calves fed milk containing lactic acid bacteria (von Buenau et al. 2005). In another study, probiotics containing six *Lactobacillus* species of bovine and human origin reduced the number of faecal coliforms, mortality and diarrhea in calves (Timmerman et al. 2005).

As mentioned in the material, the use of probiotics can improve the immune system of animals by modifying and increasing the diversity of the animals' gut microbiota. If the gut microbiota is rich and diverse, it increases the animal's health and, as a result, it is less susceptible to infectious diseases, which reduces the use of antibiotics and ultimately increases food safety and human health.

Conclusion

Indiscriminate use of antibiotics in animal feed causes pathogens to become resistant to antibiotics and antibiotic accumulation in animal body tissues. Antibiotic residues and antibiotic-resistant genes of bacteria are transferred to humans through food and show their irreparable consequences.

The gut microbiota plays a very important and vital role in animal and human health. Strengthening and modifying the gut microbiota of animals will increase the immune system, strengthen the mucosal defense barrier against pathogens and reduce the incidence of infectious diseases. The gut microbiota in human can receive genes of resistance to treatment through foods lacking proper nutritional immunity, which can cause death due to infections.

Considering the priority of prevention over treatment, by strengthening the gut microbiota of animals by using probiotics in their food, the use of antibiotics will be reduced, and first of all, the health of the animal will be improved, and then by producing quality food with high food safety, which does not contain any antibiotic residues and resistance genes, will guarantee human health.

References

Abdolghanizadeh, Sepideh, Elaheh Salmeh, Fatemeh Mirzakhani, Erfan Soroush, Seyed Davar Siadat, and Samira Tarashi. 2024. "Microbiota Insights into Pet Ownership and Human Health." *Research in Veterinary Science* 171 (May): 105220. https://doi.org/10.1016/j.rvsc.2024.105220.

Andoh, Akira, and Yoshihide Fujiyama. 2006. "Therapeutic Approaches Targeting Intestinal Microflora in Inflammatory Bowel Disease." *World Journal of Gastroenterology* 12 (28): 4452–60. https://doi.org/10.3748/wjg.v12.i28.4452.

Araújo, Priscila Dinah de, Wilma Maria Coelho Araújo, Luís Patarata, and Maria João Fraqueza. 2022. "Understanding the Main Factors That Influence Consumer Quality Perception and Attitude towards Meat and Processed Meat Products." *Meat Science* 193 (November): 108952. https://doi.org/10.1016/j.meatsci.2022.108952.

Babapour, A, L Azami, and J Fartashmehr. 2012. "Overview of Antibiotic Residues in Beef and Mutton in Ardebil, North West of Iran." *World Applied Sciences Journal* 19 (January): 1495–1500. https://doi.org/10.5829/idosi.wasj.2012.19.10.1802.

Bergogne-Bérézin, E. 1997. "Who or What Is the Source of Antibiotic Resistance?" *Journal of Medical Microbiology* 46 (6): 461–64.

Bouchami, Ons, Maria João Fraqueza, Nuno Alexandre Faria, Valquíria Alves, Opeyemi Uwangbaoje Lawal, Herminia de Lencastre, and Maria Miragaia. 2020. "Evidence for the Dissemination to Humans of

Methicillin-Resistant Staphylococcus Aureus ST398 through the Pork Production Chain: A Study in a Portuguese Slaughterhouse." *Microorganisms* 8 (12). https://doi.org/10.3390/microorganisms8121892.

Brown, Kelsey, Maina Mugoh, Douglas R Call, and Sylvia Omulo. 2020. "Antibiotic Residues and Antibiotic-Resistant Bacteria Detected in Milk Marketed for Human Consumption in Kibera, Nairobi." *PloS One* 15 (5): e0233413. https://doi.org/10.1371/journal.pone.0233413.

Buenau, R von, L Jaekel, E Schubotz, S Schwarz, T Stroff, and M Krueger. 2005. "Escherichia Coli Strain Nissle 1917: Significant Reduction of Neonatal Calf Diarrhea." *Journal of Dairy Science* 88 (1): 317–23. https://doi.org/https://doi.org/10.3168/jds.S0022-0302(05)72690-4.

Campos Calero, Guillermo, Natacha Caballero Gómez, Nabil Benomar, Beatriz Pérez Montoro, Charles W Knapp, Antonio Gálvez, and Hikmate Abriouel. 2018. "Deciphering Resistome and Virulome Diversity in a Porcine Slaughterhouse and Pork Products Through Its Production Chain." *Frontiers in Microbiology* 9: 2099. https://doi.org/10.3389/fmicb.2018.02099.

Cao, Huiluo, Salim Bougouffa, Tae-Jin Park, Andes Lau, Man-Ki Tong, Kin-Hung Chow, and Pak-Leung Ho. 2022. "Sharing of Antimicrobial Resistance Genes between Humans and Food Animals." *MSystems* 7 (6): e0077522. https://doi.org/10.1128/msystems.00775-22.

Chajęcka-Wierzchowska, Wioleta, Anna Zadernowska, Beata Nalepa, Magda Sierpińska, and Łucja Łaniewska-Trokenheim. 2015. "Coagulase-Negative Staphylococci (CoNS) Isolated from Ready-to-Eat Food of Animal Origin--Phenotypic and Genotypic Antibiotic Resistance." *Food Microbiology* 46 (April): 222–26. https://doi.org/10.1016/j.fm.2014.08.001.

Chaudhary, Poonam, and Yogesh Kumar. 2022. "Recent Advances in Multiplex Molecular Techniques for Meat Species Identification." *Journal of Food Composition and Analysis* 110: 104581. https://doi.org/https://doi.org/10.1016/j.jfca.2022.104581.

Chen, Siyu, Shuyan Luo, and Chao Yan. 2021. "Gut Microbiota Implications for Health and Welfare in Farm Animals: A Review." *Animals : An Open Access Journal from MDPI* 12 (1). https://doi.org/10.3390/ani12010093.

Choct, M. 2009. "Managing Gut Health through Nutrition." *British Poultry Science* 50 (1): 9–15. https://doi.org/10.1080/00071660802538632.

Collineau, Lucie, Patrick Boerlin, Carolee A Carson, Brennan Chapman, Aamir Fazil, Benjamin Hetman, Scott A McEwen, et al. 2019. "Integrating Whole-Genome Sequencing Data Into Quantitative Risk Assessment of Foodborne Antimicrobial Resistance: A Review of Opportunities and Challenges." *Frontiers in Microbiology* 10: 1107. https://doi.org/10.3389/fmicb.2019.01107.

Conceição, Sara, Maria Cristina Queiroga, and Marta Laranjo. 2023. "Antimicrobial Resistance in Bacteria from Meat and Meat Products: A One Health Perspective." *Microorganisms* 11 (10). https://doi.org/10.3390/microorganisms11102581.

Cornejo, Javiera, Ekaterina Pokrant, Francisco Figueroa, Ricardo Riquelme, Pablo Galdames, Francisca Di Pillo, Pedro Jimenez-Bluhm, and Christopher Hamilton-West. 2020. "Assessing Antibiotic Residues in Poultry Eggs from Backyard Production Systems in Chile, First Approach to a Non-Addressed Issue in Farm Animals." *Animals : An Open Access Journal from MDPI* 10 (6). https://doi.org/10.3390/ani10061056.

Darwish, Wageh Sobhy, Elsaid A Eldaly, Mohamed Tharwat El-Abbasy, Yoshinori Ikenaka, Shouta Nakayama, and Mayumi Ishizuka. 2013. "Antibiotic Residues in Food: The African Scenario." *The Japanese Journal of Veterinary Research* 61 Suppl (February): S13-22.

Das, Theerthankar, Denis Paino, Arthika Manoharan, Jessica Farrell, Greg Whiteley, Frederik H Kriel, Trevor Glasbey, and Jim Manos. 2019. "Conditions Under Which Glutathione Disrupts the Biofilms and Improves Antibiotic Efficacy of Both ESKAPE and Non-ESKAPE Species." *Frontiers in Microbiology* 10: 2000. https://doi.org/10.3389/fmicb.2019.02000.

David, Lawrence A, Corinne F Maurice, Rachel N Carmody, David B Gootenberg, Julie E Button, Benjamin E Wolfe, Alisha V Ling, et al. 2014. "Diet Rapidly and Reproducibly Alters the Human Gut Microbiome." *Nature* 505 (7484): 559–63. https://doi.org/10.1038/nature12820.

Dibner, J J, and J D Richards. 2005. "Antibiotic Growth Promoters in Agriculture: History and Mode of Action." *Poultry Science* 84 (4): 634–43. https://doi.org/10.1093/ps/84.4.634.

Donoghue, Dan J. 2003. "Antibiotic Residues in Poultry Tissues and Eggs: Human Health Concerns?" *Poultry Science* 82 (4): 618–21. https://doi.org/10.1093/ps/82.4.618.

Drlica, K, and X Zhao. 1997. "DNA Gyrase, Topoisomerase IV, and the 4-Quinolones." *Microbiology and Molecular Biology Reviews : MMBR* 61 (3): 377–92. https://doi.org/10.1128/mmbr.61.3.377-392.1997.

"Disseminating Innovative Solutions for Antibiotic Resistance Management." 2022. 2022. https://disarmproject.eu/.

Erofeeva, Viktoriya, Yulia Zakirova, Sergey Yablochnikov, Evgeny Prys, and Irina Prys. 2021. "The Use of Antibiotics in Food Technology: The Case Study of Products from Moscow Stores." *E3S Web Conf.* 311. https://doi.org/10.1051/e3sconf/202131110005.

FAO. 2017. "The Future of Food and Agriculture – Trends and Challenges." Rome: Food and Agriculture Organization of the United Nations. 2017.

Ferrocino, Ilario, Kalliopi Rantsiou, Ryan McClure, Tanja Kostic, Rafael Soares Correa de Souza, Lene Lange, Jamie FitzGerald, et al. 2023. "The Need for an Integrated Multi-OMICs Approach in Microbiome Science in the Food System." *Comprehensive Reviews in Food Science and Food Safety* 22 (2): 1082–1103. https://doi.org/10.1111/1541-4337.13103.

Fishbein, Skye RS, Bejan Mahmud, and Gautam Dantas. 2023. "Antibiotic Perturbations to the Gut Microbiome." *Nature Reviews. Microbiology* 21 (12): 772–88. https://doi.org/10.1038/s41579-023-00933-y.

Founou, Luria Leslie, Raspail Carrel Founou, and Sabiha Yusuf Essack. 2016. "Antibiotic Resistance in the Food Chain: A Developing Country-Perspective." *Frontiers in Microbiology* 7: 1881. https://doi.org/10.3389/fmicb.2016.01881.

Gaggìa, Francesca, Paola Mattarelli, and Bruno Biavati. 2010. "Probiotics and Prebiotics in Animal Feeding for Safe Food Production." *International Journal of Food Microbiology* 141 Suppl 1 (July): S15-28. https://doi.org/10.1016/j.ijfoodmicro.2010.02.031.

Ganguly, Nirmal K, NK Arora, Sujith J Chandy, Mohamed Nadeem Fairoze, J P S Gill, Usha Gupta, Shah Hossain, et al. 2011. "Rationalizing Antibiotic Use to Limit Antibiotic Resistance in India." *The Indian Journal of Medical Research* 134 (3): 281–94.

Gareau, Mélanie, Eytan Wine, and Philip Sherman. 2009. "Early Life Stress Induces Both Acute and Chronic Colonic Barrier Dysfunction." *Neoreviews* 10 (April). https://doi.org/10.1542/neo.10-4-e191.

Gompel, Liese Van, Roosmarijn EC Luiken, Rasmus B Hansen, Patrick Munk, Martijn Bouwknegt, Lourens Heres, Gerdit D Greve, et al. 2020. "Description and Determinants of the Faecal Resistome and Microbiome of Farmers and Slaughterhouse Workers: A Metagenome-Wide Cross-Sectional Study." *Environment International* 143 (October): 105939. https://doi.org/10.1016/j.envint.2020.105939.

Grossman, Trudy H. 2016. "Tetracycline Antibiotics and Resistance." *Cold Spring Harbor Perspectives in Medicine* 6 (4): a025387. https://doi.org/10.1101/cshperspect.a025387.

Guarner, Francisco, and Juan-R Malagelada. 2003. "Gut Flora in Health and Disease." *Lancet (London, England)* 361 (9356): 512–19. https://doi.org/10.1016/S0140-6736(03)12489-0.

Gundala, Raghava R, and Anupam Singh. 2021. "What Motivates Consumers to Buy Organic Foods? Results of an Empirical Study in the United States." *PloS One* 16 (9): e0257288. https://doi.org/10.1371/journal.pone.0257288.

Henriques, AR, J Melo Cristino, and MJ Fraqueza. 2017. "Genetic Characterization of Listeria Monocytogenes Isolates from Industrial and Retail Ready-to-Eat Meat-Based Foods and Their Relationship with Clinical Strains from Human Listeriosis in Portugal." *Journal of Food Protection* 80 (4): 551–60. https://doi.org/10.4315/0362-028X.JFP-16-310.

Higgins, SE, JP Higgins, AD Wolfenden, SN Henderson, A Torres-Rodriguez, G Tellez, and B Hargis. 2008. "Evaluation of a Lactobacillus-Based Probiotic Culture for the Reduction of Salmonella Enteritidis in Neonatal Broiler Chicks." *Poultry Science* 87 (1): 27–31. https://doi.org/10.3382/ps.2007-00210.

Hofacre, CL, T Beacorn, S Collett, and G Mathis. 2003. "Using Competitive Exclusion, Mannan-Oligosaccharide and Other Intestinal Products to Control Necrotic Enteritis." *Journal of Applied Poultry Research* 12 (1): 60–64. https://doi.org/https://doi.org/10.1093/japr/12.1.60.

Holzbauer, Stacy, and Tom Chiller. 2006. "Antimicrobial Resistance in Bacteria of Animal Origin." *Emerging Infectious Diseases*. https://doi.org/10.3201/eid1207.060503.

Jandhyala, Sai Manasa, Rupjyoti Talukdar, Chivkula Subramanyam, Harish Vuyyuru, Mitnala Sasikala, and D Nageshwar Reddy. 2015. "Role of the Normal Gut Microbiota." *World Journal of Gastroenterology* 21 (29): 8787–8803. https://doi.org/10.3748/wjg.v21.i29.8787.

Jian, Zonghui, Li Zeng, Taojie Xu, Shuai Sun, Shixiong Yan, Lan Yang, Ying Huang, Junjing Jia, and Tengfei Dou. 2021. "Antibiotic Resistance Genes in Bacteria: Occurrence, Spread, and Control." *Journal of Basic Microbiology* 61 (12): 1049–70. https://doi.org/10.1002/jobm.202100201.

Juhász, János, Balázs Ligeti, Márió Gajdács, Nóra Makra, Eszter Ostorházi, Ferenc Balázs Farkas, Balázs Stercz, et al. 2021. "Colonization Dynamics of Multidrug-Resistant Klebsiella Pneumoniae Are Dictated by Microbiota-Cluster Group Behavior over Individual Antibiotic Susceptibility: A Metataxonomic Analysis." *Antibiotics (Basel, Switzerland)* 10 (3). https://doi.org/10.3390/antibiotics10030268.

Kanokudom, Sitthichai, Thachaporn Assawakongkarat, Yukihiro Akeda, Panan Ratthawongjirakul, Rungtip Chuanchuen, and Nuntaree Chaichanawongsaroj. 2021. "Rapid Detection of Extended Spectrum β-Lactamase Producing Escherichia Coli Isolated from Fresh Pork Meat and Pig Cecum Samples Using Multiplex Recombinase Polymerase Amplification and Lateral Flow Strip Analysis." *PloS One* 16 (3): e0248536. https://doi.org/10.1371/journal.pone.0248536.

Kraker, Marlieke E A de, Andrew J Stewardson, and Stephan Harbarth. 2016. "Will 10 Million People Die a Year Due to Antimicrobial Resistance by 2050?" *PLoS Medicine* 13 (11): e1002184. https://doi.org/10.1371/journal.pmed.1002184.

Landers, Timothy F, Bevin Cohen, Thomas E Wittum, and Elaine L Larson. 2012. "A Review of Antibiotic Use in Food Animals: Perspective, Policy, and Potential." *Public Health Reports (Washington, D.C. : 1974)* 127 (1): 4–22. https://doi.org/10.1177/003335491212700103.

Larsson, DG Joakim, and Carl-Fredrik Flach. 2022. "Antibiotic Resistance in the Environment." *Nature Reviews. Microbiology* 20 (5): 257–69. https://doi.org/10.1038/s41579-021-00649-x.

Lawal, Opeyemi U, Maria J Fraqueza, Ons Bouchami, Peder Worning, Mette D Bartels, Maria L Gonçalves, Paulo Paixão, et al. 2021. "Foodborne Origin and Local and Global Spread of Staphylococcus Saprophyticus Causing Human Urinary Tract Infections." *Emerging Infectious Diseases* 27 (3): 880–93. https://doi.org/10.3201/eid2703.200852.

Ley, Ruth E, Catherine A Lozupone, Micah Hamady, Rob Knight, and Jeffrey I Gordon. 2008. "Worlds within Worlds: Evolution of the Vertebrate Gut Microbiota." *Nature Reviews. Microbiology*. England. https://doi.org/10.1038/nrmicro1978.

Ley, Ruth E, Micah Hamady, Catherine Lozupone, Peter J Turnbaugh, Rob Roy Ramey, J Stephen Bircher, Michael L Schlegel, et al. 2008. "Evolution of Mammals and Their Gut Microbes." *Science (New York, N.Y.)* 320 (5883): 1647–51. https://doi.org/10.1126/science.1155725.

Li, Lili, Yawen Xiao, Chong Wang, Rikke Heidemann Olsen, Hecheng Meng, and Lei Shi. 2022. "Exploring the Resistome, Virulome, Mobilome and Microbiome along Pork Production Chain Using Metagenomics." *International Journal of Food Microbiology* 371 (June): 109674. https://doi.org/10.1016/j.ijfoodmicro.2022.109674.

Livermore, David M, and Neil Woodford. 2006. "The Beta-Lactamase Threat in Enterobacteriaceae, Pseudomonas and Acinetobacter." *Trends in Microbiology* 14 (9): 413–20. https://doi.org/10.1016/j.tim.2006.07.008.

Luiken, Roosmarijn EC, Liese Van Gompel, Alex Bossers, Patrick Munk, Philip Joosten, Rasmus Borup Hansen, Berith E Knudsen, et al. 2020. "Farm Dust Resistomes and Bacterial Microbiomes in European Poultry and Pig Farms." *Environment International* 143 (October): 105971. https://doi.org/10.1016/j.envint.2020.105971.

Mackie, Roderick, Rustam Aminov, Rex Gaskins, and B White. 1999. *Molecular Microbial Ecology in Gut Ecosystems*.

Makita, K. 2021. "Animal Health and Food Safety Risk Assessments." *Revue Scientifique et Technique (International Office of Epizootics)* 40 (2): 533–44. https://doi.org/10.20506/rst.40.2.3243.

Manlove, Kezia R, Josephine G Walker, Meggan E Craft, Kathryn P Huyvaert, Maxwell B Joseph, Ryan S Miller, Pauline Nol, et al. 2016. "'One Health' or Three? Publication Silos Among the One Health Disciplines." *PLOS Biology* 14 (4): e1002448. https://doi.org/10.1371/journal.pbio.1002448.

Mashhadany, Dhary Al. 2020. "Monitoring of Antibiotic Residues among Sheep Meat in Erbil City and Thermal Processing Effect on Their Remnants," April, 217–22. https://doi.org/10.33899/ijvs.2019.125814.1161.

Maslennikova, Olga, and Viktoriya Erofeeva. 2020. "Biological Contamination of Soils in Urbanized Ecosystems by Toxocara Sp. Eggs." *E3S Web Conf.* 169. https://doi.org/10.1051/e3sconf/202016904002.

Mayorova, Yana, Viktor Glebov, Viktoriya Erofeeva, Sergey Yablochnikov, and Bogdan Laver. 2020. "Physiological Evaluation of the Cardiology System of Nonresident Students at Different Training Periods." *E3S Web Conf.* 169. https://doi.org/10.1051/e3sconf/202016904004.

Mazmanian, Sarkis K, June L Round, and Dennis L Kasper. 2008. "A Microbial Symbiosis Factor Prevents Intestinal Inflammatory Disease." *Nature* 453 (7195): 620–25. https://doi.org/10.1038/nature07008.

Moeller, Andrew H, and Jon G Sanders. 2020. "Roles of the Gut Microbiota in the Adaptive Evolution of Mammalian Species." *Philosophical Transactions of the Royal Society of London. Series B, Biological Sciences* 375 (1808): 20190597. https://doi.org/10.1098/rstb.2019.0597.

Moghadam, Mortez Mohammadzadeh, Mostafa Amiri, Hamed Ramezani Awal Riabi, and Hamid Ramezani Awal Riabi. 2016. "Evaluation of Antibiotic Residues in Pasteurized and Raw Milk Distributed in the South of Khorasan-e Razavi Province, Iran." *Journal of Clinical and Diagnostic Research : JCDR* 10 (12): FC31–35. https://doi.org/10.7860/JCDR/2016/21034.9034.

Mottet, Anne, Cees de Haan, Alessandra Falcucci, Giuseppe Tempio, Carolyn Opio, and Pierre Gerber. 2017. "Livestock: On Our Plates or Eating at Our Table? A New Analysis of the Feed/Food Debate." *Global Food Security* 14: 1–8. https://doi.org/https://doi.org/10.1016/j.gfs.2017.01.001.

Nchima, Gilbert, Kennedy Choongo, Bumbangi Flavien, Kaampwe Muzandu, King Nalubamba, Gerald Monga, Henry Kangwa, and John Muma. 2017. "Determination of Oxytetracycline and Sulphamethazine Residues in Marketed Beef from Selected Parts of Zambia to Assess Compliance with Maximum Residual Limits," September.

Nisha, AR. 2008. "Antibiotic Residues - A Global Health Hazard." *Veterinary World* 1 (January). https://doi.org/10.5455/vetworld.2008.375-377.

O'Connor, E. L., Sims, L., and White, K.M. 2017. "Ethical Food Choices: Examining People's Fair Trade Purchasing Decisions." *Food Qual. Prefer* 60: 105–12. https://doi.org/. doi: 10.1016/j.foodqual.2017.04.001.

O'Hara, Ann M, and Fergus Shanahan. 2007. "Gut Microbiota: Mining for Therapeutic Potential." *Clinical Gastroenterology and Hepatology: The Official Clinical Practice Journal of the American Gastroenterological Association* 5 (3): 274–84. https://doi.org/10.1016/j.cgh.2006.12.009.

Okaiyeto, Samuel Ariyo, Parag Prakash Sutar, Chang Chen, Jia-Bao Ni, Jun Wang, Arun S Mujumdar, Jing-Shou Zhang, et al. 2024. "Antibiotic Resistant Bacteria in Food Systems: Current Status, Resistance Mechanisms, and Mitigation Strategies." *Agriculture Communications* 2 (1): 100027. https://doi.org/https://doi.org/10.1016/j.agrcom.2024.100027.

Olasoju, Mary Idowu, Taiwo Israel Olasoju, Oluwawemimo Oluseun Adebowale, and Victoria Olusola Adetunji. 2021. "Knowledge and Practice of Cattle Handlers on Antibiotic Residues in Meat and Milk in Kwara State, Northcentral Nigeria." *PloS One* 16 (10): e0257249. https://doi.org/10.1371/journal.pone.0257249.

Papatsiros, Vassilis, Panagiotis Katsoulos, Konstantinos Koutoulis, Maria Karatzia, A Dedousi, and Georgios Christodoulopoulos. 2014. "Alternatives to Antibiotics for Farm Animals." *CAB Reviews Perspectives in Agriculture Veterinary Science Nutrition and Natural Resources* 8 (January). https://doi.org/10.1079/PAVSNNR20138032.

Patangia, Dhrati V, Cornelius Anthony Ryan, Eugene Dempsey, Reynolds Paul Ross, and Catherine Stanton. 2022. "Impact of Antibiotics on the Human Microbiome and Consequences for Host Health." *MicrobiologyOpen* 11 (1): e1260. https://doi.org/10.1002/mbo3.1260.

Patterson, JA, and KM Burkholder. 2003. "Application of Prebiotics and Probiotics in Poultry Production." *Poultry Science* 82 (4): 627–31. https://doi.org/10.1093/ps/82.4.627.

Pennone, Vincenzo, José Cobo Díaz, M Prieto, and Avelino Alvarez-Ordóñez. 2021. "Application of Genomics and Metagenomics to Improve Food Safety Based on an Enhanced Characterisation of Antimicrobial Resistance." *Current Opinion in Food Science* 43 (December). https://doi.org/10.1016/j.cofs.2021.12.002.

Petrova, Penka, Ivan Ivanov, Lidia Tsigoriyna, Nadezhda Valcheva, Evgenia Vasileva, Tsvetomila Parvanova-Mancheva, Alexander Arsov, and Kaloyan Petrov. 2021. "Traditional Bulgarian Dairy Products: Ethnic Foods with Health Benefits." *Microorganisms* 9 (3). https://doi.org/10.3390/microorganisms9030480.

Ramatla, Tsepo, Lubanza Ngoma, Modupeade Adetunji, and Mulunda Mwanza. 2017. "Evaluation of Antibiotic Residues in Raw Meat Using Different Analytical Methods." *Antibiotics (Basel, Switzerland)* 6 (4). https://doi.org/10.3390/antibiotics6040034.

Rana, Md Shohel, Seung Yun Lee, Hae Jin Kang, and Sun Jin Hur. 2019. "Reducing Veterinary Drug Residues in Animal Products: A Review." *Food Science of Animal Resources* 39 (5): 687–703. https://doi.org/10.5851/kosfa.2019.e65.

Rawat, Nitish, Anjali, Shreyata, Benoy Sabu, Anannya Bandyopadhyay, and Raman Rajagopal. 2024. "Assessment of Antibiotic Resistance in Chicken Meat Labelled as Antibiotic-Free: A Focus on Escherichia Coli and Horizontally Transmissible Antibiotic Resistance Genes." *LWT* 194: 115751. https://doi.org/https://doi.org/10.1016/j.lwt.2024.115751.

Reperant, Leslie A, Giuseppe Cornaglia, and Albert D M E Osterhaus. 2013. "The Importance of Understanding the Human-Animal Interface : From Early Hominins to Global Citizens." *Current Topics in Microbiology and Immunology* 365: 49–81. https://doi.org/10.1007/82_2012_269.

Scharek, L, J Guth, K Reiter, KD Weyrauch, D Taras, P Schwerk, P Schierack, MFG Schmidt, LH Wieler, and K Tedin. 2005. "Influence of a Probiotic Enterococcus Faecium Strain on Development of the Immune System of Sows and Piglets." *Veterinary Immunology and Immunopathology* 105 (1–2): 151–61. https://doi.org/10.1016/j.vetimm.2004.12.022.

Suresh, K, Suresh Babu Naidu Krishna, and Patrick Govender. 2012. "The Use of Probiotics and Safety Concerns: A Review." *African Journal of Microbiology Research* 6 (October): 6871–77. https://doi.org/10.5897/AJMR12.1281.

Takahashi, Shunsuke, Yayoi Egawa, Nobuyuki Simojo, Takamitsu Tsukahara, and Kazunari Ushida. 2007. "Oral Administration of Lactobacillus Plantarum Strain Lq80 to Weaning Piglets Stimulates the Growth of Indigenous Lactobacilli to Modify the Lactobacillal Population." *The Journal of General and Applied Microbiology* 53 (6): 325–32. https://doi.org/10.2323/jgam.53.325.

Thursby, Elizabeth, and Nathalie Juge. 2017. "Introduction to the Human Gut Microbiota." *The Biochemical Journal* 474 (11): 1823–36. https://doi.org/10.1042/BCJ20160510.

Tian, Ming, Xinmiao He, Yanzhong Feng, Wentao Wang, Heshu Chen, Ming Gong, Di Liu, Jihong Liu Clarke, and André van Eerde. 2021. "Pollution by Antibiotics and Antimicrobial Resistance in LiveStock and Poultry Manure in China, and Countermeasures." *Antibiotics (Basel, Switzerland)* 10 (5). https://doi.org/10.3390/antibiotics10050539.

Timmerman, HM, L Mulder, H Everts, DC van Espen, E van der Wal, G Klaassen, SMG Rouwers, R Hartemink, FM Rombouts, and AC Beynen. 2005. "Health and Growth of Veal Calves Fed Milk Replacers with or without Probiotics." *Journal of Dairy Science* 88 (6): 2154–65. https://doi.org/10.3168/jds.S0022-0302(05)72891-5.

Toomey, Niamh, Aine Monaghan, Séamus Fanning, and Declan J Bolton. 2009. "Assessment of Antimicrobial Resistance Transfer between Lactic Acid Bacteria and Potential Foodborne Pathogens Using in Vitro Methods and Mating in a Food Matrix." *Foodborne Pathogens and Disease* 6 (8): 925–33. https://doi.org/10.1089/fpd.2009.0278.

Wannaprasat, Wechsiri, Chailai Koowatananukul, Chanon Ekkapobyotin, and Rungtip Chuanchuen. 2009. "Quality Analysis of Commercial Probiotic Products for Food Animals." *The Southeast Asian Journal of Tropical Medicine and Public Health* 40 (5): 1103–12.

Wu, Yanbei, Jiawei Wan, Uyory Choe, Quynhchi Pham, Norberta W Schoene, Qiang He, Bin Li, Liangli Yu, and Thomas TY Wang. 2019. "Interactions Between Food and Gut Microbiota: Impact on Human Health." *Annual Review of Food Science and Technology* 10 (March): 389–408. https://doi.org/10.1146/annurev-food-032818-121303.

Yang, Yan, Wenqian Qiu, Yuxiang Li, and Lijing Liu. 2020. "Antibiotic Residues in Poultry Food in Fujian Province of China." *Food Additives & Contaminants. Part B, Surveillance* 13 (3): 177–84. https://doi.org/10.1080/19393210.2020.1751309.

Yuan, Yue, Bing Liang, Bo-Wen Jiang, Ling-Wei Zhu, Tie-Cheng Wang, Yuan-Guo Li, Jun Liu, Xue-Jun Guo, Xue Ji, and Yang Sun. 2021. "Migratory Wild Birds Carrying Multidrug-Resistant Escherichia Coli as Potential Transmitters of Antimicrobial Resistance in China." *PloS One* 16 (12): e0261444. https://doi.org/10.1371/journal.pone.0261444.

Zafar, Hassan, and Milton H Jr Saier. 2021. "Gut Bacteroides Species in Health and Disease." *Gut Microbes* 13 (1): 1–20. https://doi.org/10.1080/19490976.2020.1848158.

Zinsstag, Jakob, Esther Schelling, Kaspar Wyss, and Mahamat Bechir Mahamat. 2005. "Potential of Cooperation between Human and Animal Health to Strengthen Health Systems." *Lancet (London, England)* 366 (9503): 2142–45. https://doi.org/10.1016/S0140-6736(05)67731-8.

Index

D

E

F

G

H

I

L

M

N

O

P

R

S

T

V

Z